SECOND EDITION

INDUSTRIAL CONTROL ELECTRONICS

John Webb
Northcentral Technical College
Wausau, Wisconsin

Kevin Greshock
DeVry Institute of Technology
Lombard, Illinois

Merrill, an imprint of
Macmillan Publishing Company
New York

Maxwell Macmillan Canada
Toronto

Maxwell Macmillan International
New York   Oxford   Singapore   Sydney

NOTICE TO THE READER

The publisher and the authors do not warrant or guarantee any of the products and/or equipment described herein nor has the publisher or the authors made any independent analysis in connection with any of the products, equipment, or information used herein. The reader is directed to the manufacturer for any warranty or guarantee for any claim, loss, damages, costs, or expense, arising out of or incurred by the reader in connection with the use or operation of the products and/or equipment.

The reader is expressly advised to adopt all safety precautions that might be indicated by the activities and experiments described herein. The reader assumes all risks in connection with such instructions.

Cover photo: H. Mark Weidman
Editor: Dave Garza
Developmental Editor: Carol Hinklin Robison
Production Editor: Mary Harlan
Art Coordinator: Lorraine Woost
Cover Designer: Robert Vega
Production Buyer: Patricia A. Tonneman

This book was set in Times Roman by Bi-Comp, Inc. and was printed and bound by R. R. Donnelley & Sons Company. The cover was printed by Lehigh Press, Inc.

Macmillan Publishing Company
866 Third Avenue
New York, NY 10022

Macmillan Publishing Company is part of the
Maxwell Communication Group of Companies.

Maxwell Macmillan Canada, Inc.
1200 Eglinton Avenue East, Suite 200
Don Mills, Ontario M3C 3N1

Library of Congress Cataloging-in-Publication Data
Webb, John W.
    Industrial control electronics / John Webb, Kevin Greshock. -- 2nd
  ed.
      p.   cm.
    Includes bibliographical references and index.
    ISBN 0-02-424864-9
    1. Electronic control. I. Greshock, Kevin. II. Title.
  TK7881.2.W43   1993
  629.8′9--dc20   92-16734
                                                                       CIP

Printing: 1 2 3 4 5 6 7 8 9 Year: 3 4 5 6 7
Photo credit: Jo Hall/Macmillan, p. 87.

To our wives and children:
Thelma and Janet
Kathy, Sara, David, Mark, and Paula
Ryan, Colin, and Kaitlyn

MERRILL'S INTERNATIONAL SERIES IN ENGINEERING TECHNOLOGY

INTRODUCTION TO ENGINEERING TECHNOLOGY

Pond, *Introduction to Engineering Technology, 2nd Edition*, 0-02-396031-0

ELECTRONICS TECHNOLOGY

Electronics Reference

Adamson, *The Electronics Dictionary for Technicians*, 0-02-300820-2
Berlin, *The Illustrated Electronics Dictionary*, 0-675-20451-8
Reis, *Becoming an Electronics Technician: Securing Your High-Tech Future*, 0-02-399231-X

DC/AC Circuits

Boylestad, *DC/AC: The Basics*, 0-675-20918-8
Boylestad, *Introductory Circuit Analysis, 6th Edition*, 0-675-21181-6
Ciccarelli, *Circuit Modeling: Exercises and Software, 2nd Edition*, 0-02-322455-X
Floyd, *Electric Circuits Fundamentals, 2nd Edition*, 0-675-21408-4
Floyd, *Electronics Fundamentals: Circuits, Devices, and Applications, 2nd Edition*, 0-675-21310-X
Floyd, *Principles of Electric Circuits, 4th Edition*, 0-02-338501-4
Floyd, *Principles of Electric Circuits: Electron Flow Version, 3rd Edition*, 0-02-338531-6
Keown, *PSpice and Circuit Analysis*, 0-675-22135-8
Monssen, *PSpice with Circuit Analysis* 0-675-21376-2
Tocci, *Introduction to Electric Circuit Analysis, 2nd Edition*, 0-675-20002-4

Devices and Linear Circuits

Berlin & Getz, *Fundamentals of Operational Amplifiers and Linear Integrated Circuits*, 0-675-21002-X
Berube, *Electronic Devices and Circuits Using MICRO-CAP II*, 0-02-309160-6
Berube, *Electronic Devices and Circuits Using MICRO-CAP III*, 0-02-309151-7
Bogart, *Electronic Devices and Circuits, 3rd Edition*, 0-02-311701-X
Tocci, *Electronic Devices: Conventional Flow Version, 3rd Edition*, 0-675-21150-6
Floyd, *Electronic Devices, 3rd Edition*, 0-675-22170-6
Floyd, *Electronic Devices: Electron Flow Version*, 0-02-338540-5
Floyd, *Fundamentals of Linear Circuits*, 0-02-338481-6
Schwartz, *Survey of Electronics, 3rd Edition*, 0-675-20162-4
Stanley, *Operational Amplifiers with Linear Integrated Circuits, 2nd Edition*, 0-675-20660-X
Tocci & Oliver, *Fundamentals of Electronic Devices, 4th Edition*, 0-675-21259-6

Digital Electronics

Floyd, *Digital Fundamentals, 4th Edition*, 0-675-21217-0
McCalla, *Digital Logic and Computer Design*, 0-675-21170-0
Reis, *Digital Electronics through Project Analysis* 0-675-21141-7
Tocci, *Fundamentals of Pulse and Digital Circuits, 3rd Edition*, 0-675-20033-4

Microprocessor Technology

Antonakos, *The 68000 Microprocessor: Hardware and Software Principles and Applications, 2nd Edition*, 0-02-303603-6
Antonakos, *The 8088 Microprocessor*, 0-675-22173-0
Brey, *The Advanced Intel Microprocessors*, 0-02-314245-6
Brey, *The Intel Microprocessors: 8086/8088, 80186, 80286, 80386, and 80486: Architecture, Programming, and Interfacing, 2nd Edition*, 0-675-21309-6
Brey, *Microprocessors and Peripherals: Hardware, Software, Interfacing, and Applications, 2nd Edition*, 0-675-20884-X
Gaonkar, *Microprocessor Architecture, Programming, and Applications with the 8085/8080A, 2nd Edition*, 0-675-20675-6
Gaonkar, *The Z80 Microprocessor: Architecture, Interfacing, Programming, and Design, 2nd Edition*, 0-02-340484-1
Goody, *Programming and Interfacing the 8086/8088 Microprocessor: A Product- Development Laboratory Process*, 0-675-21312-6
MacKenzie, *The 8051 Microcontroller*, 0-02-373650-X
Miller, *The 68000 Family of Microprocessors: Architecture, Programming, and Applications, 2nd Edition*, 0-02-381560-4
Quinn, *The 6800 Microprocessor*, 0-675-20515-8
Subbarao, *16/32 Bit Microprocessors: 68000/68010/68020 Software, Hardware, and Design Applications*, 0-675-21119-0

Electronic Communications

Monaco, *Introduction to Microwave Technology*, 0-675-21030-5
Monaco, *Preparing for the FCC Radio-Telephone Operator's License Examination*, 0-675-21313-4
Schoenbeck, *Electronic Communications: Modulation and Transmission, 2nd Edition*, 0-675-21311-8
Young, *Electronic Communication Techniques, 2nd Edition*, 0-675-21045-3
Zanger & Zanger, *Fiber Optics: Communication and Other Applications*, 0-675-20944-7

Microcomputer Servicing

Adamson, *Microcomputer Repair*, 0-02-300825-3
Asser, Stigliano, & Bahrenburg, *Microcomputer Servicing: Practical Systems and Troubleshooting, 2nd Edition*, 0-02-304241-9
Asser, Stigliano, & Bahrenburg, *Microcomputer Theory and Servicing, 2nd Edition*, 0-02-304231-1

Programming

Adamson, *Applied Pascal for Technology*, 0-675-20771-1
Adamson, *Structured BASIC Applied to Technology, 2nd Edition*, 0-02-300827-X
Adamson, *Structured C for Technology*, 0-675-20993-5
Adamson, *Structured C for Technology (with disk)*, 0-675-21289-8
Nashelsky & Boylestad, *BASIC Applied to Circuit Analysis*, 0-675-20161-6

Instrumentation and Measurement

Berlin & Getz, *Principles of Electronic Instrumentation and Measurement*, 0-675-20449-6
Buchla & McLachlan, *Applied Electronic Instrumentation and Measurement*, 0-675-21162-X
Gillies, *Instrumentation and Measurements for Electronic Technicians, 2nd Edition*, 0-02-343051-6

Transform Analysis

Kulathinal, *Transform Analysis and Electronic Networks with Applications*, 0-675-20765-7

Biomedical Equipment Technology

Aston, *Principles of Biomedical Instrumentation and Measurement*, 0-675-20943-9

Mathematics

Monaco, *Essential Mathematics for Electronics Technicians*, 0-675-21172-7
Davis, *Technical Mathematics*, 0-675-20338-4
Davis, *Technical Mathematics with Calculus*, 0-675-20965-X

INDUSTRIAL ELECTRONICS/ INDUSTRIAL TECHNOLOGY

Bateson, *Introduction to Control System Technology, 4th Edition*, 0-02-306463-3
Fuller, *Robotics: Introduction, Programming, and Projects*, 0-675-21078-X
Goetsch, *Industrial Safety: In the Age of High Technology*, 0-02-344207-7
Goetsch, *Industrial Supervision: In the Age of High Technology*, 0-675-22137-4
Horath, *Computer Numerical Control Programming of Machines*, 0-02-357201-9
Hubert, *Electric Machines: Theory, Operation, Applications, Adjustment, and Control*, 0-675-20765-7
Humphries, *Motors and Controls*, 0-675-20235-3
Hutchins, *Introduction to Quality: Management, Assurance, and Control*, 0-675-20896-3
Laviana, *Basic Computer Numerical Control Programming*, 0-675-21298-7

Reis, *Electronic Project Design and Fabrication, 2nd Edition*, 0-02-399230-1
Rosenblatt & Friedman, *Direct and Alternating Current Machinery, 2nd Edition*, 0-675-20160-8
Smith, *Statistical Process Control and Quality Improvement*, 0-675-21160-3
Webb, *Programmable Logic Controllers: Principles and Applications, 2nd Edition*, 0-02-424970-X
Webb & Greshock, *Industrial Control Electronics, 2nd Edition*, 0-02-424864-9

MECHANICAL/CIVIL TECHNOLOGY

Keyser, *Materials Science in Engineering, 4th Edition*, 0-675-20401-1
Kraut, *Fluid Mechanics for Technicians*, 0-675-21330-4
Mott, *Applied Fluid Mechanics, 3rd Edition*, 0-675-21026-7
Mott, *Machine Elements in Mechanical Design, 2nd Edition*, 0-675-22289-3
Rolle, *Thermodynamics and Heat Power, 3rd Edition*, 0-675-21016-X
Spiegel & Limbrunner, *Applied Statics and Strength of Materials*, 0-675-21123-9
Wolansky & Akers, *Modern Hydraulics: The Basics at Work*, 0-675-20987-0
Wolf, *Statics and Strength of Materials: A Parallel Approach to Understanding Structures*, 0-675-20622-7

DRAFTING TECHNOLOGY

Cooper, *Introduction to VersaCAD*, 0-675-21164-6
Goetsch & Rickman, *Computer-Aided Drafting with AutoCAD*, 0-675-20915-3
Kirkpatrick & Kirkpatrick, *AutoCAD for Interior Design and Space Planning*, 0-02-364455-9
Kirkpatrick, *The AutoCAD Book: Drawing, Modeling, and Applications, 2nd Edition*, 0-675-22288-5
Lamit and Lloyd, *Drafting for Electronics, 2nd Edition*, 0-02-367342-7
Lamit and Paige, *Computer-Aided Design and Drafting*, 0-675-20475-5
Maruggi, *Technical Graphics: Electronics Worktext, 2nd Edition*, 0-675-21378-9
Maruggi, *The Technology of Drafting*, 0-675-20762-2
Sell, *Basic Technical Drawing*, 0-675-21001-1

TECHNICAL WRITING

Croft, *Getting a Job: Resume Writing, Job Application Letters, and Interview Strategies*, 0-675-20917-X
Panares, *A Handbook of English for Technical Students*, 0-675-20650-2
Pfeiffer, *Proposal Writing: The Art of Friendly Persuasion*, 0-675-20988-9
Pfeiffer, *Technical Writing: A Practical Approach*, 0-675-21221-9
Roze, *Technical Communications: The Practical Craft*, 0-675-20641-3
Weisman, *Basic Technical Writing, 6th Edition*, 0-675-21256-1

Preface

GENERAL DESCRIPTION

The electronic control field is constantly expanding and changing, not only in the United States, but worldwide. This textbook has been created to help train students in electronic control. It has been designed to be as state-of-the-art as possible. For example, a chapter on programmable logic controllers, a relatively new subject, has been included. The text concentrates on the application of components and systems. It does not cover such areas as the design of individual components or a mathematical analysis of a complicated servo control system. The application area is the field in which most students will be employed and will need knowledge. Consequently, it is the main thrust of this book.

The subject matter of this text has been selected to match the knowledge needed by technicians and engineers to deal with electronic control systems. The text covers all areas of electronic control systems. Some other texts cover only a portion of the parts that make up a complete system. The complete subject coverage of this book is illustrated in Figure P–1. The figure shows a fundamental process that pumps fluid at different rates. The chapters of this text that apply to each part of this industrial process are shown in circles adjacent to each system section. Note that all aspects of the control system, including peripheral subjects, are covered by this text.

ADDED FEATURES OF THE SECOND EDITION

Technically updated material has been added to numerous chapters in the second edition. Also, some sections have been rewritten for greater clarity. Some chapter exercises have been revised and new exercises added where applicable. The chapter order has been revised for a more logical sequencing. In the first edition chapter 4 covered solid-state devices. In this edition the material has been divided

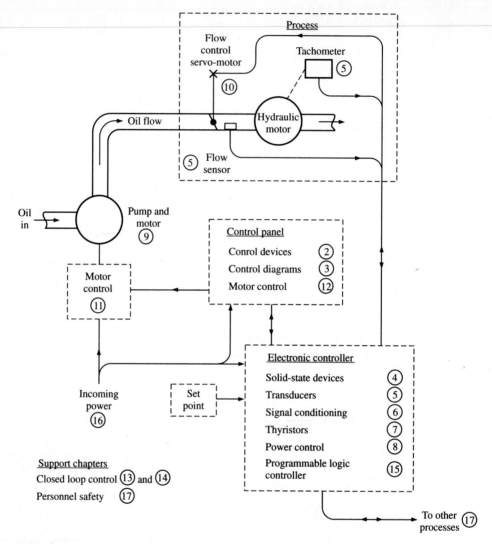

FIGURE P-1
Guide to Chapter Coverage of a Central System.

into two chapters: Chapter 4, which covers diodes through ICs, and Chapter 7, which covers thyristors, SCRs, triacs and other devices, and also includes up-dated material. The chapters on op amps, signal conditioning, and industrial power supplies have been revised and updated.

AUDIENCE

The text is suitable for associate's degree or diploma programs in electrical, electronic, and electromechanical technology. It is appropriate for courses in control systems in these associate's degree areas. The text may also be suitable for sophomore courses in four-year technology and engineering programs. It can also be used for courses in electrical and electronic control in trade and industrial schools. Another possible use is as a general reference book for persons in the electronic control field. For optimum text effectiveness, the student needs some knowledge of basic AC and DC electricity and electronics.

Both authors have industrial experience as well as teaching experience. Their teaching experience includes numerous associate degree courses, adult re-training programs, electrical trades training, and various short electrical and electronic courses. One author has written a textbook on programmable logic controllers. The authors have used their industrial and educational experiences to select material for the text that best serves the needs of the various audiences just listed.

OVERALL ORGANIZATION

The book, which is divided into six parts, starts with basic concepts and components, and progresses to operational systems. The systems covered use the basic concepts and components, along with controlling computers, for process control. The six parts are

   I. Control System Components
  II. Signal Conditioning and Power Control
 III. Motors and Controls
  IV. Closed-Loop Control
   V. Programmable Logic Controllers and Power Distribution Effects
  VI. Safety and Automation

As with any text, it is possible to work with those parts that are most useful for one's purpose. The chapters to be used depend on the students' previous knowledge and on the required content of the course being taught. Some chapters can be omitted or covered only in part. Furthermore, the chapters do not have to be covered in sequence. For example, Chapter 17, on safety, could be covered first, with no loss of effectiveness.

Two particular chapters included in this text are usually not found in similar texts. These are the chapters on power distribution (Chapter 16) and safety (Chapter 17). The authors have found that many electronic control problems are the result of improper functioning of the building power system. The power chapter covers distribution in sufficient detail to convey understanding of the possible effects of incoming power on control system performance. The chapter on safety covers an area not often included in students' training. The authors believe it is an important facet of training.

CHAPTER ORGANIZATION

Each chapter has been organized for convenience and effectiveness of instruction. The organizational features found in each chapter are as follows:

- □ An outline of chapter sections.
- □ A listing of learning objectives.
- □ One student application problem at the beginning of the chapter, with an answer at the end of the chapter. The answer to the application problem relates to the material covered in the chapter. These real-life problems have been chosen to help stimulate student interest.
- □ An introduction.
- □ Chapter material, with actual industrial examples.
- □ Numerous illustrations—photographs, drawings, and diagrams—for each subject.
- □ A chapter summary for review.
- □ Chapter exercises. The exercise questions are not just a rehash of the chapter material. They require the student to apply chapter principles to various situations and problems.

FEATURES OF THE *INSTRUCTOR'S RESOURCE MANUAL*

A 100-page *Instructor's Resource Manual* containing many instructional aids has been prepared for this text and is available to qualified instructors, professors, and course leaders. The manual contains the following:

- □ Section A   General Information
  1. General instructional approach for presenting the material
  2. Suggested course outlines for courses of various lengths
- □ Section B   Chapter Information
  1. General approach to teaching each chapter's material
  2. Suggested audiovisual materials and their sources
  3. Suggested lab equipment and lab training materials and their sources
- □ Section C   Solutions to All Exercises
  Completely worked-out solutions for each chapter exercise—not just the answers alone
- □ Section D   Exam Questions
  Suggested exam formats and types of questions for each chapter
- □ Transparency Masters
  Fifty key illustrations reproduced for classroom use

ACKNOWLEDGMENTS

The authors wish to thank the following organizations for their assistance in preparing this text and its illustrations: ABB Robotics; Allen-Bradley; Amatrol; Banner Engineering Corp.; Best Power Technology; Beswick Communications; Bryant Division/Westinghouse Electric; Bussmann Division/Cooper Industries; Eaton Corp./Cutler-Hammer; Electro Corp.; Entran Devices, Inc.; ETCO Electric Supply; The Foxboro Co.; Glendale Protective Technologies; Greenheck Fan; Hamamatsu Corp.; Intel Corp.; Kavlico Corp.; Klein Tools; L and S Electric; Littlefuse Tracor/Division of Westmark; Lucas Ledex, Inc.; M and I Bank/ First American; Magnetek Drives and Systems; Marathon Electric; Microswitch, A Honeywell Division; Motorola, Inc.; National Fire Protection Association; Newport Electronics, Inc.; Northcentral Technical College; Omega Engineering Co.; Omron Electronics, Inc.; Powermation; Rapistan Corp.; Reed Optical; Rexnord Corp./Stearns Division; Sprague Electric/Semiconductor Group; Square D Company; Superior Electric Co.; Thermometrics, Inc.; Truck Multiprox, Inc.; Wardco Safety, Inc.; Wausau Insurance Co.; Weyerhaeuser Corp.; and Wisconsin Public Service.

The following companies and organizations provided Application Assignment photographs: The Foxboro Corporation (Chapters 5 and 6); Greenheck Fan (Chapter 8); L and S Electric (Chapters 1, 4, 16); Northcentral Technical College (Chapters 2, 3, 9, and 17); Marathon Electric (Chapters 10, 14, and 15); Rapistan Corp. (Chapters 7 and 18); and Weyerhaeuser Corp. (Chapter 11).

The authors acknowledge the use of illustrations from other Merrill texts: Boylestad, *Introductory Circuit Analysis,* 6th ed.; Floyd, *Electronics Fundamentals: Circuits, Devices and Applications* and *Electric Circuit Fundamentals;* Humphries, *Motors and Controls;* and Webb, *Programmable Logic Controllers: Principles and Applications.*

We wish to thank the following reviewers for their helpful comments: Gene Adair, IVY Tech.; Robert Allen, Columbus Tech. Inst., GA; Jim Brice, DeVry Inst. of Tech.—Phoenix; Walter Buchanan, Indiana Univ., Purdue Univ. Ext.; James Davis, Muskingum Area Tech. Coll., OH; Jim DeLoach, DeVry Inst. of Tech.—Atlanta; Dave Dottl, Wisconsin School of Electronics; Ken Ferguson, Midlands Tech. Coll., SC; Eric Liimata, Anne Arundel Comm. Coll., MD; Jill Harlamert, DeVry Inst. of Tech.—Columbus, OH; John Kettula, Milwaukee School of Engineering; Jim Marks, Indiana Vocational Tech. Inst.; George Mason, Indiana Vocational Tech. Inst.; Donald J. Montgomery, ITT Tech. Inst.— LaMesa; Frank Naujokas, DeVry Inst. of Tech.—Lombard; Roy Powell, Chattanooga State Tech. Comm. Coll.; Bill Rice, Catonsville Comm. College—MD; Marvin Rogers, Vermont Tech. Coll.; and Mike Williams, DeVry Inst. of Tech.— Phoenix.

Preparation of the glossary and index was by Thelma Webb.

We thank Jill Harlamert for her contributions to the concepts and organization of this book during the early stages of its development.

Finally, we thank the following at Merrill/Macmillan Publishing for their help: Steve Helba, Dave Garza, Carol Robison, Lorry Woost, and Mary Harlan.

John Webb
Kevin Greshock

Contents

4 Solid-State Devices 87

5 Transducers and Sensors 107

PART II
SIGNAL CONDITIONING AND POWER CONTROL 161

6 Signal Conditioning 163

7 SCRs, Triacs, and Other Thyristors 205

8 Solid-State Power Control 237

PART III
MOTORS AND CONTROLS 271

9 Industrial-Use Motors 273

10   Special-Purpose Motors   315

11   Electrical Power-Control Devices   345

12   Control of Motors   369

PART IV
CLOSED-LOOP CONTROL   399

13   Analog Controllers   401

14   Closed-Loop Systems   429

1

Introduction to Control Electronics

CHAPTER OUTLINE

1–1 Introduction □ **1–2** Reasons for Using Industrial Automatic Control □ **1–3** Levels of Industrial Control: The Control System Triangle □ **1–4** Sections, Components, and Devices of a Control System □ **1–5** Example of the Evolution of a Control System

LEARNING OBJECTIVES

- □ List the reasons for using automatic control, and show how the reasons can vary with the application.
- □ List the various levels of control, and give an application example of each.
- □ List the essential sections of an automated control system, and show how they are interrelated.
- □ Define *open loop, closed loop,* and *feedback* in relation to automated systems.
- □ Construct a block diagram of a control system.

APPLICATION ASSIGNMENT

At the beginning of each chapter, starting with Chapter 2, there will be an application assignment in which a question will be raised concerning the information to be covered in the ensuing chapter material. A technician or engineer will be shown with electrical or electronic process equipment involved in the problem. As you go through the chapter material, you will find the information you need to answer the question. At the end of the chapter, a general answer to the application assignment question will be given.

1–1   INTRODUCTION

This chapter is a general introduction to control electronics and automatic control. In subsequent chapters the concepts of Chapter 1 will be expanded in detail. Specifically, this chapter will present a general outline of the various levels of automatic control. Some general terms applicable to automatic control will be defined and explained. The role of each general section of a control system and its relation to other sections of the machine control system will be outlined.

The reasons that control systems, particularly electronic ones, are used will be listed and discussed. It will be shown that the level of control chosen for a system depends on the process application and the complexity of the required control system. A generalized control system layout will show the various sections of a control system.

Finally, a home heating system will be used as an example of a control system. We will show the operation of a basic heating system and a more complex heating system. Automatic control of both heating systems will be explained. Block control diagrams will illustrate the operation of both heating systems.

1–2   REASONS FOR USING INDUSTRIAL AUTOMATIC CONTROL

In recent years there has been an increasing need to improve the efficiency and effectiveness of industrial machines and other machines. The interface between human operator and machine is slowly but steadily being eliminated. This change is being accomplished by the addition of automated control to industrial processes. In the early 1960s, computers were used effectively for control in the industrial environment. However, those computer systems had little success, because of their expense, their unreliability, and the skepticism of management and operating personnel.

Today computers are readily used in industry. The development of the microprocessor has been a major factor in this change because the microprocessor provides more computer power at lower cost and in less space. These newer, computer-controlled systems use complex programming to regulate, monitor, and control the operation of industrial processes.

The control of automated systems may be electronic, mechanical, hydraulic, or pneumatic. In many instances, combinations of these types of control are used. We will be almost exclusively concerned with electronic control in this text. Electronic control circuits are by far the most common type of automated control in use. All electronics-based automatic control systems have common characteristics. We will discuss a general electronic control system with these common characteristics in this chapter.

Automatic control is used primarily in manufacturing—in machining, welding, assembly, testing, robotics, and other processing areas. Automatic control is also used for heating, ventilating, and air conditioning systems. Additionally, home appliances and office machines have their own internal automated control systems. A recent trend is to use a master automatic control system to coordinate the operation of various individually controlled subparts of a large system. An example of a coordinated automated control system is the manufacturing work cell, discussed in Chapter 18.

Automated process control has many advantages over human control. The advantages may be grouped in seven classifications:

1. reduced overall manufacturing costs
2. computational capabilities
3. fast response time
4. reduced equipment size and cost
5. environmental safety for operating personnel
6. prompt emergency recognition and reaction
7. general convenience

In manufacturing, overall costs can be reduced in many ways, direct and indirect. Space required for manufacturing can be reduced. Product cycle time from order date to shipping date can be shortened, reducing inventory and its cost. Reworking and scrap are reduced when the process is properly set up. Quality can be improved through consistency. Automation lowers labor hours per product unit, lowering unit cost. On the negative side, an automated machine does need maintenance and can break down. However, it does not take breaks or vacations or come in late on Mondays. A robot, a special automatic machine, works without heat or light and does not need insurance or worker's compensation. A robot cannot sue its employer either (but it can hurt people).

An automated device can make computations at a phenomenal rate compared with a human operator. It can quickly determine optimum paths of cutting, for example. Computer control makes fast corrective changes in process levels where a variable control is involved—for example, holding a 4.375 in. diameter. Gear cutting calculations can be made by the machine's computer and automatically followed in the manufacturing process. Computer-aided design and computer-aided manufacturing (CAD/CAM) systems can be hooked up from a design terminal to a machine. A part can be made automatically and directly from a design previously drawn on the CAD/CAM computer.

Automated control systems can respond to a change or error, when detected, in nanoseconds. Human response can take seconds and will not normally be as accurate as that of an automated system. Today's military and commercial aircraft are controlled in part by computers, which work faster and more accurately than a pilot on many flight control maneuvers.

Equipment size and cost improve every year. Control equipment has become more compact as well as more reliable. Large relay panels have been replaced by small electronic boxes with a small keyboard and monitor. Changes in the control system can be made from the keyboard instead of by rewiring a panel. All of these factors tend to decrease the overall cost of automatic control systems. Lower overall costs make automatic control economically feasible when compared with the use of a human operator.

Some of the earliest uses of automation were those that reduced the environmental hazard to the workers involved in a process. Remote, automatic control in hazardous atmospheres helps save lives and reduces the number of injuries and long-term human disabilities. For example, in the nuclear industry, automated systems reduce or eliminate exposure of humans to radiation. Some types of submarines for oceanography are controlled remotely from the water's surface. A common example of environmental personnel protection automation is in the use of a robotic paint spray booth, with the operator at a distance from the paint booth.

A properly designed automated emergency-shutdown system will act faster than a human. An automated system operates in microseconds. A human takes longer. Furthermore, a human operator may be looking at something else when the emergency occurs. The automated device continuously reviews the critical situation that may require shutdown. The automatic control system can also review many critical factors, practically simultaneously. A human can reliably watch only a few.

Finally, automation is convenient. Home appliances with automated control systems are convenient. Automated systems have been expanded to include the computation of production rates, the determination of process inventory, and the monitoring of quality control trends. Computer-organized data from these computations make operating and planning decisions much more accurate and effective.

Note that there is a cost involved in automating a process or device. The cost involves many factors, such as equipment cost, downtime for conversion, and retraining. An analysis has to be made in every case to see if it might be better to automate partially or, perhaps, not to automate at all. In some cases, it might be better to continue with a human-run process. In other cases, it is possible that a human with some automatic help would be more effective.

1–3 LEVELS OF INDUSTRIAL CONTROL: THE CONTROL SYSTEM TRIANGLE

Figure 1–1 is a triangle showing the general levels of control of an industrial factory. At the bottom, only human control is involved. At the top, very involved computer analysis is used. Most smaller industrial operations go up through level 3. Larger factories are increasingly at level 4. Very large, multiplant operations are generally at level 5. This text is concerned primarily with level 2 of the figure and somewhat with level 3. A brief description of each level of control follows.

Level 1 is the machine level. An example is a lathe with manual controls for moving the cutter through its path. Control is manual with cranks. The lathe may have power assists for more cutting power, but control is manual.

Level 2 is reached when electrical or other controls are added. Level 2 can be divided into three sublevels as shown in Figure 1–1. Suppose an electronic, computer-based control operated the lathe. Such a control for automatic feed rate for cutting the metal would be level 2A. This level 2A rate would be automatically

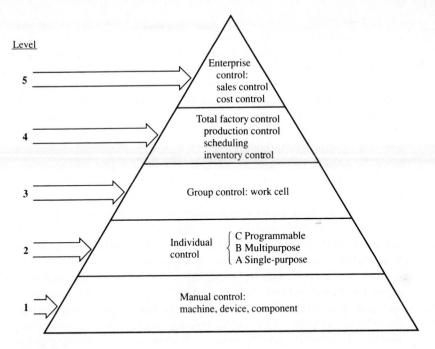

FIGURE 1–1
Control System Level Triangle

set for each part machined. Level 2B is reached when controls are programmed for machining more than one part. The control pattern for each part is called up from a master control as required. The master control has been preprogrammed for each part. The third sublevel is 2C, programmable control. At level 2C the machining pattern for each part is programmed in by an operator. When a part's machining program is first used, each step and motion are recorded as the process progresses. The steps and motions are stored in memory. The next time the same part is to be machined, the part's machining program is recalled from memory. This recall procedure saves redoing the setup each time the part is to be made. Many different parts' machining patterns can be stored and recalled as needed. Note that as you go from A to C of level 2, the cost of installation goes up accordingly. The cost of more control versus the possible benefits must be considered in determining economic feasibility.

Level 3 involves connecting the control of two or more individual machines or devices to work together. An example of this level is controlling a robot to load a lathe as well as controlling the lathe's operation. A master controller is needed to coordinate the two individual controllers of the robot and the lathe. In many manufacturing operations more than two devices are coordinated. An example might be on an automobile assembly line. Conveyors, positioners, robotic welders, and inspection devices are coordinated with a master computer. Such groups of machines and devices, called work cells, are discussed in Chapter 18.

Level 4 of automatic control involves a number of work cells hooked up to a master, coordinating automatic computer control. The entire factory is under the control of a large master computer. The master computer takes an order from sales input, checks for raw material availability in inventory, and prepares a production plan. It then causes the required parts to be made, by running them through the appropriate operations in the plant. The master computer does such other chores as reorder an appropriate amount of raw material as it anticipates the need. The master computer also carries out such tasks as scheduling part manufacturing in a given work cell for maximum machine utilization. One result of proper control at this level is reduced amounts of raw material and in-process inventory. Inventory takes up factory floor space as well adding to costs.

Level 5 is one more step into sophisticated manufacturing. At this level, a computer looks at past demand for each product and predicts the number of items that will be needed at each time of the month or year. Sales forecasts are also factored in. The product to be manufactured is then scheduled and made accordingly. This level gets into an area called *artificial intelligence,* at a high cost. An analysis has to be made to determine whether people can make the same predictions at lower cost and with comparable accuracy. One result of proper use of level 5 would be reduced finished material inventory, if sales and production are matched properly. Another result would be timely availability of a product for prompt delivery. Other considerations, such as the time to buy raw material at lowest cost, would be included in the computer program at level 5.

1–4 SECTIONS, COMPONENTS, AND DEVICES OF A CONTROL SYSTEM

What does a typical control system layout look like? Figure 1–2 is a layout diagram for a simple hydraulic motor speed-control system. This type of layout diagram could be applicable to one axis of control for a large hydraulically controlled robot. The purpose of the control system is to precisely control the oil-flow rate. The flow rate will tend to be affected by variations in back pressure from the device to which the oil is fed. However, the flow rate is to remain unchanged. Changing the control valve setting to increase or decrease oil pressure will compensate for the change in flow and will change the flow back to its original rate. Another purpose of the control system is to change the flow rate promptly and accurately from one value to another when the set-point is changed.

Oil is pumped through the system by a hydraulic pump driven by an electric motor. The oil flow to the hydraulic motor in the process is controlled by a servomotor-driven vane valve. The output oil flow rate is sensed by a flowmeter. The hydraulic motor speed is monitored by a tachometer. The sensed information is sent in electrical form to the electronic controller. The electronic controller checks to see how closely the actual flow conditions match the flow set-point. If

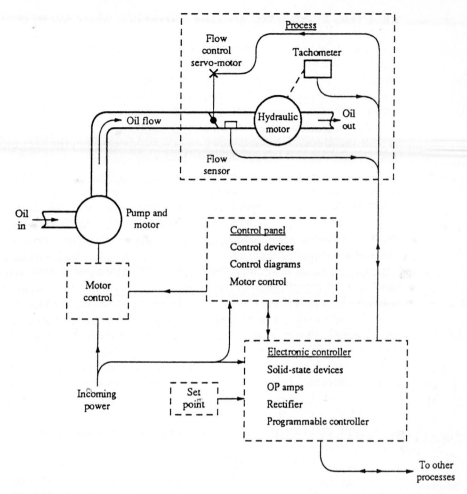

FIGURE 1-2
General Control System Diagram

needed, the oil flow is then adjusted up or down by a signal from the electronic controller to the servovalve. This signal controls the opening or closing of the servovalve. The servovalve aperture controls the flow-control vane, which in turn varies the oil flow.

Electric power comes from the utility company as shown in Figure 1-2. A control panel is used to turn the system ON and OFF. The control panel also controls the electric motor. Appropriate process status indicators, such as lights and meters, are mounted on the control panel. The process may be interfaced with other processes of a master control computer, as shown in Figure 1-2 by a control signal line from the electronic controller.

1–5   EXAMPLE OF THE EVOLUTION OF A CONTROL SYSTEM

Two home heating examples will help illustrate the evolution of increasingly complex controls—from no control to present-day complex automatic control. A fundamental no-control heating system is one in which a person throws logs in a fireplace periodically, without any true control of the heat level. The only real control is that one throws on fewer logs when it gets too hot and more when it gets too cold. This is an *open-loop* heating system.

If the fireplace has a damper and doors, a person in the area can adjust the damper and doors for a comfortable heating level, up or down. This adjustment capability creates what is called a *closed-loop* system. The person doing the adjusting acts as a sensor and also as a correcting device. When too hot, the person closes the damper or doors for less heat. When too cold, he or she does the opposite. This correcting action by the person is called *feedback*. The feedback in this illustration is designated *negative feedback* because as the temperature goes up too high, an adjustment (damper) is made to cause it to go down, the opposite direction. If the person increased the heat (an unlikely situation) by opening the damper further as the temperature went up, the feedback would be called *positive feedback*. These feedback concepts will be expanded on in chapters 13 and 14.

We will now look at two home heating systems that are more complex. They are automatic closed-loop systems with negative feedback. The first is a simple nonblower, gas-fired furnace, shown in Figure 1–3A. A wall thermostat in a room turns the burner ON and OFF. For example, the thermostat is set at 68°. When the room temperature falls below 68°, the furnace goes ON, and the room heats up. Assume that the thermostat has a built-in differential of 1.5°. When the temperature reaches 69.5°, the thermostat shuts off the furnace. The room then cools down, and the process is repeated when the temperature cools to 68°.

Figure 1–3A includes a control diagram below the system to show how the control system operates. Control diagrams will be explained more completely in later chapters.

Figure 1–3B shows a more complicated heating system. The thermostat, TST1, controls the furnace burner as in Figure 1–3A. When the temperature setting is reached on TST1, the furnace burner is turned ON as before. Additionally, another thermostat, TST2, in the furnace heating chamber, controls a blower system. Thermostat TST2, which has a wide differential, normally has two settings, high and low, each individually adjustable. The blower goes on only after the temperature in the heating chamber has reached the high setting of TST2. Blower operation, therefore, begins sometime after the furnace burner goes on. When room temperature reaches the high (69.5°) value, the furnace burner is turned OFF. The blower continues to run until the heating chamber heat supply is reduced to the lower setting of TST2. A control diagram is shown in Figure 1–3B below the system layout. Note that the control diagram is a double version of the one in A, with some added control connections.

The Figure 1–3 heating systems could be expanded further. All furnaces have a third thermostat in their heating chambers, an overtemperature-shutoff

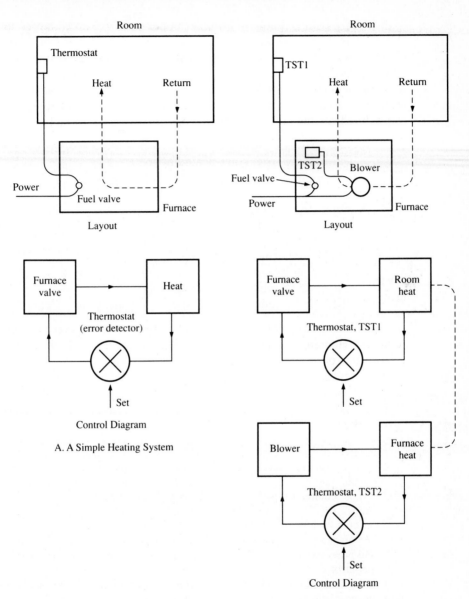

FIGURE 1–3
Home Heating Control Systems

safety thermostat set above the TST2 upper setting. Furthermore, some systems have a fourth thermostat on the outside of the house to be used as an anticipator. Adding both of these thermostats would increase the complexity of control and the circuit diagram.

How does automatic control apply to heating and cooling systems? The two examples already given are very fundamental. For the heating and cooling of a large building, the control system would be expanded to include the following:

- □ zone heating by area or room
- □ heat transfer from solar-heated areas
- □ time-of-day control to reduce heating or cooling at night and thus save fuel costs
- □ anticipation of the time required to heat the building to a comfortable temperature by the time personnel arrive in the morning
- □ air exchange rates that conform to applicable public building codes
- □ humidity control as required
- □ sequencing of motor and fan control to keep peak power demand, and cost, to the lowest level possible
- □ other special requirements, such as air purity for "clean rooms"

Automatic control is the only way to meet the advanced demands of building comfort. Building comfort, of course, is only one example of the need for automatic control in today's complex world.

SUMMARY

- □ Automatic control is used in many areas in today's complex world.
- □ The microprocessor has made automatic control possible at reasonable cost.
- □ Automated industrial processes are the most common use of automatic control.
- □ The most common type of automatic control is electronic.
- □ There are various advantages to automatic control. Major advantages are labor savings, other cost savings, time savings, and personnel safety.
- □ Automatic factory control levels can be divided into five major categories. The automated processes described in this textbook are at level 2, machine control, and level 3, cell control.
- □ The levels of automated control go from manual control to the machine level to the artificial-intelligence, total-factory-control level.
- □ The general sections of an automatically controlled process are (1) the process itself, (2) a sensor to determine the process level, or position, and (3) a controller to carry out appropriate control and correction of the process level and actions.
- □ The home heating system is a good example of the evolution of automated control systems.
- □ A heating and cooling system becomes a very high-level automated system in a large building.
- □ Other types of systems are becoming more automated in the same manner as heating and cooling systems, as their complexity increases.

EXERCISES

1. Why is automatic control so important in today's world?

2. You have invested in a wood pallet plant. Wood is cut by hand on table and band saws. The wood slats are stapled or nailed together by hand with pneumatic staplers and nailers. For shipping, pallets are stacked and banded by hand. What are some of the opportunities and reasons for automating this plant?

3. In your own words, describe the levels of control. You may want to include in your explanation more than the five levels listed in the chapter.

4. What would be some of the problems encountered if you started an industrial automation project at this book's level 4 without first going through levels 2 and 3?

5. What are the essential sections and devices of an automatic control system? How are they interrelated?

6. What is left out of an open-loop system that is included in a closed-loop system?

7. Define *open loop*. Give an example.

8. Define *closed loop*. Given an example.

9. Define *feedback*. Give an example.

10. Explain the difference between positive and negative feedback.

11. Redraw both parts of Figure 1–3*B* to include two more thermostats. Thermostat TST3 is the overtemperature safety device in the furnace heat chamber. Thermostat TST4, outside the house, is an indicating thermostat for anticipating big swings in outside temperature.

12. Draw the layout and control diagram for an automobile speed (cruise) control. Refer to Figure 1–3 for the format.

13. You have a system for control of air conditioning and humidity control. It includes separate temperature and humidity sensors. Draw a layout and a control diagram for the system.

14. Trace the evolution from hand to automatic control of a home appliance of your choice. Examples might be the dishwasher or the automatic washer. Include what might be possible in the future—for example, voice control.

I

CONTROL SYSTEM COMPONENTS

2

Electrical Control Devices

CHAPTER OUTLINE

2–1 Introduction □ **2–2** Basic Principles of Electrical Switching □ **2–3** Toggle, Push-Button, and Selector Switches □ **2–4** Solenoids □ **2–5** Relays— Electromechanical and Solid-State □ **2–6** Timers and Counters □ **2–7** Control Transformers □ **2–8** Potentiometers

LEARNING OBJECTIVES

□ Describe and apply various electrical switching arrangements.
□ List the characteristics of various types of switches.
□ Describe the electrical solenoid and its important application characteristics.
□ Describe the construction and operation of electromechanical and solid-state relays.
□ Differentiate among types of relays.
□ Describe the operation of timers and counters.
□ Draw the electrical symbols for various types of transformers.
□ Design control transformers electrically based on application requirements.
□ List and describe the important application principles of potentiometers.

APPLICATION ASSIGNMENT

You have been given a piece of equipment to repair. You have been told that three of the switches in the device are defective or intermittent. Also, one potentiometer is bad and one control transformer has no output. What considerations should you apply to each component as you proceed with replacement?

2–1   INTRODUCTION

Chapter 2 covers basic inputs and control devices for control electronics. Later chapters will show how these devices are used in control systems. This chapter will begin with electrical switching principles. Switching can involve one circuit or multiple circuits. Switching arrangements are of many types, depending on the switching application involved. Examples of switching devices will be shown and described.

Next, electrical solenoids will be discussed. Then various major types of electromechanical relays will be described and illustrated. Solid-state relays will also be covered. Differences among various types of relays will be discussed and listed.

Other topics covered are nonelectronic timers and counters, the various types of control transformers, and the various basic types of potentiometers (variable resistors).

2–2   BASIC PRINCIPLES OF ELECTRICAL SWITCHING

There are many electrical arrangements for electrical switching. Figure 2–1 illustrates the basic types of switching. The switches shown may be appropriately connected in electrical circuits as shown by the arrows.

The basic electrical switch is the single-pole, single-throw switch, which is shown in Figure 2–1A. This switching arrangement is usually abbreviated *SPST*.

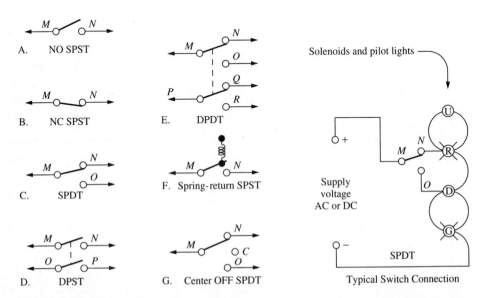

FIGURE 2–1
Electrical Switch Configurations and Typical Connection

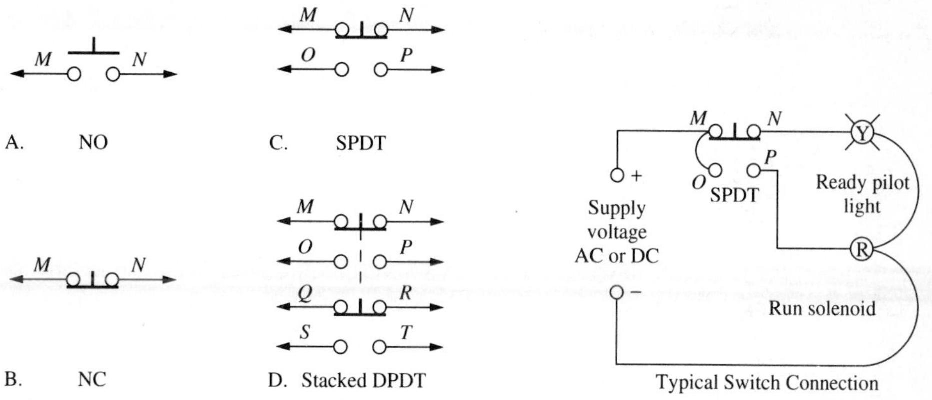

FIGURE 2–2
Push-Button Switch Configurations and Typical Connection

When the switch is in the OFF position, the circuit is electrically open between *M* and *N*. As the switch is moved to the ON position, a connecting circuit is created between points *M* and *N*. This switch, in 2–1*A*, is designated as having a *normally open (NO)* contact. In Figure 2–1*B* another form of the SPST switch is shown: *normally closed (NC)*. The NC SPST switch has a circuit *M* to *N* when OFF and opens *M* to *N* when turned ON. It is important to keep the NO and NC contact designations straight when choosing a switch for an application.

Another type of switch, the single-pole, double-throw switch, is shown in Figure 2–1*C*. This is designated *SPDT*. The circuit from *M* is switched back and forth between *N* and *O* as the switch is turned ON and OFF.

Next, suppose we need to turn two separate circuits ON and OFF simultaneously. We could use two SPST switches. Actually, we use the *DPST* switch, shown in *D*. It consists of two SPST switches in one package. If we need two simultaneously acting double-pole switches, we use the DPDT switch, shown in *E*. This switching arrangement can be expanded to three poles or more as needed. The other types of switching arrangements can also be similarly expanded.

Two other possible variations of switching configurations are *spring-return* and *center-OFF*. A spring-return SPST switch is shown in *F*. Actuating the switch in *F* will close *M* and *N*. However, when the switch is released, it springs back open. It does not stay in the closed position as a regular switch does. The center-OFF type is shown in *G*. It is a three-position switch and can also spring-return to the center, which is the OFF position. Figure 2–1 also shows a typical connection for one switch. Connections will be discussed in detail in Chapter 3.

Push buttons act similarly to the switching arrangements previously shown. The difference is that the push buttons are all spring-return (unless specifically designated otherwise). A few push buttons are NO or NC only, as shown in Figure 2–2*A* and *B*. A more common type of push button is the SPDT shown in *C*. When

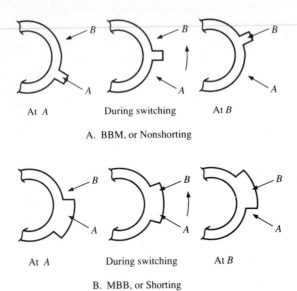

A. BBM, or Nonshorting

B. MBB, or Shorting

FIGURE 2–3
Nonshorting and Shorting Switch Operation

the push button is depressed, the path between M and N is open, and O and P have a circuit. When it is released, M and N have a circuit, and the path between O and P is open electrically. Stacking more switches on a push button can double it up to the DPDT arrangement shown in D. Some push buttons have three, four, or more stacks, increasing their switching capabilities. Figure 2–2 also shows a typical switch connection.

All multiple switching arrangements discussed so far can be classified as *break-before-make* (*BBM*) or *make-before-break* (*MBB*). These two types of switching arrangements are usually called *nonshorting* and *shorting*. If unspecified, a switch is normally of the BBM (nonshorting) type.

Figure 2–3 illustrates the principles of operation of these two general classes of switches. This illustration uses a wafer switch. A metal ring is mounted on a phenolic, nonconducting wafer. Electrical connections are brought in on pads that ride on the metal ring as it rotates. The wafer is rotated to specific positions to accomplish the switching.

The BBM, nonshorting switch is shown in Figure 2–3A. In going from A to B, the circuit is completely broken in the intermediate position, as shown. For the MBB, shorting switch in Figure 2–3B, the rotating metal ring has a larger tang. The tang length exceeds the span between A and B. The circuit is therefore connected to both A and B in the intermediate position.

An illustration will help show where each type of switch is used. Figure 2–4 illustrates a voltmeter and an ammeter. The voltmeter must have a nonshorting

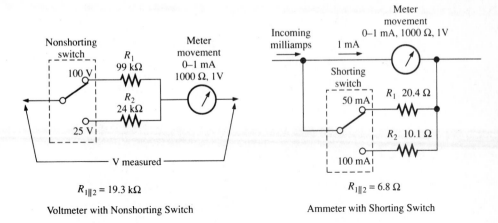

$R_{1\|2} = 19.3 \text{ k}\Omega$

Voltmeter with Nonshorting Switch

$R_{1\|2} = 6.8 \ \Omega$

Ammeter with Shorting Switch

FIGURE 2–4
Meter Applications for Switches

range selector switch. Between ranges, the switch will turn the voltage OFF to the meter movement. If a shorting switch were used instead, resistors would be in parallel. The temporary lower resistance, 19.3 kΩ, would allow excess current to flow into the meter movement. In that case, the meter could have up to five times rated current. Repeated overloads of this magnitude could damage the meter movement.

Conversely, the ammeter must have a shorting switch. If a nonshorting switch were used, the meter movement could have up to 100 times rated current through it. For each amp range, a shunt resistor carries the appropriate current, as shown. If there is no shunt resistor connected in the circuit between range positions, all the line current must go through the meter movement. For line current of 100 mA, this is 100 times rated current. During switching, with the proper shorting switch, we have a lowered resistance, 6.8 Ω. This lower resistance results in a temporary dip in current to the meter movement, causing it no harm.

2–3 TOGGLE, PUSH-BUTTON, AND SELECTOR SWITCHES

The electrical switching devices that perform the tasks discussed in the previous section are available in many forms. One simple type is the single-button ON/OFF switch. Such switches are typically used on small devices with limited panel space. Push once for ON, then once for OFF.

The toggle switch is very common and is typically used in single- and multiple-pole applications. The toggle switch has a small toggle shaft protruding from it. The toggle shaft is pushed up or down to accomplish switching. Small, multiple toggle-type switches are shown in Figure 2–5. The slide switch is used as a low-cost alternative to the toggle switch. A slide switch has a small protruding rectan-

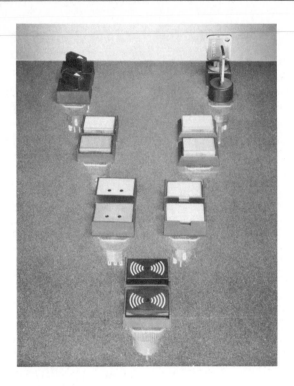

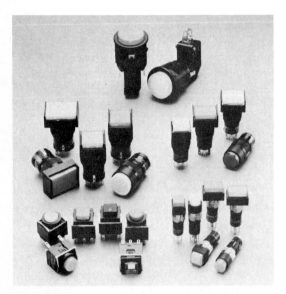

FIGURE 2–5
Major Switch Types (Courtesy of OMRON Electronics, Inc.)

gular button, which is pushed up or down for operation. Multiple stacked slide switches are included in Figure 2–5.

Push-button switches are used in many industrial applications. An assortment of push-button switches is shown in Figure 2–5. Push-button switches come in different colors and with many different identification labels as an option. Some push buttons have "mushroom" heads. These mushroom-head buttons provide easy accessibility for such applications as emergency stop.

Wafer switches are used for multiple-pole applications. A typical wafer switch was shown in Figure 2–3. The switch shown in Figure 2–3 was only a 2-pole switch. A 12-pole, single-throw wafer switch is a more typical configuration. For operation, the rotor of a 12-pole switch is rotated 360°/12, or 30°, between output tabs. A common-lead tab feeds current into the metal insert on the phenolic base. As the switch is rotated 30°, each output terminal is energized in turn through the tang on the metal insert to each tab. Wafer switches may be stacked on a common shaft for multiple switching of various types.

Figure 2–5 includes some typical selector switches. These selector switches can have two, three, or ten or more positions, as illustrated in the figure.

The type of switch chosen for a specific task depends on many factors. Some of these factors are as follows:

1. the configuration as discussed in the previous section (poles and throws)
2. the voltage to be switched and the type of current (AC or DC)
3. the current value to be switched and the current to be carried after switching
4. the required life cycle in number of actuations
5. environmental considerations (moisture, vibration, etc.)
6. the physical size needed
7. options such as built-in pilot light and key locks

First, the configuration for switching must be determined before the switch is chosen. Note that it is well to consider including extra contacts on the switch, if possible, for future needs.

Second, you must consider what voltage you will be switching. Will it be 24 V, 120 V, 220 V, or another rating? Higher voltages require more substantial switches. Most switches handle AC or DC of a given voltage value. However, in higher ratings there could be a design difference between AC or DC, which must be considered. Switches are rated for current and voltage according to NEMA standards. The abbreviation *NEMA* stands for National Electrical Manufacturers Association, an industrial trade organization. The NEMA standards are issued to ensure interchangeability of electrical parts made by different member companies.

Third, current switching rating must be considered. The considerations are similar to those for voltage. Higher currents require larger contact areas and larger switches. In some applications, switching current has one value, and the current-carrying rating after switching is higher. For example, a switch may have a 0.5 A switching rating and a 3.5 A carrying capacity. As with voltages, current switching may be different for AC and DC, especially at higher ratings. Very high current and voltage switching is discussed in Chapter 11.

Factors 4 and 5 must also be considered. How reliable does the switch need to be? How many times will it be operated each hour? A more substantial switch is required if it is to be operated often. Another major consideration is the environment in which the switch operates. For example, most push buttons are rated oiltight for industrial applications. An oiltight switch keeps moisture or oil from entering the contact area. As another environmental example, a slide switch is not appropriate for an application on a panel subject to vibration. Heat and cold extremes must also be considered in some switch applications.

The physical size of a switch will depend on its rating and the number of poles. Another consideration is the space available and the visibility required.

Finally, in considering factor 7, there are a number of options available in the choice of a switch. Push-button switches, particularly, have a large number of options. One option is a built-in pilot light under the button to indicate when the circuit is ON. Another common option is a key lock. The switch may be locked in the ON or OFF position and the key removed. Another option for some push buttons is a built-in time delay. When the button is pushed, the contact closes a preset time later. There are many other possible options for switches. The best sources of information on these options are switch manufacturers' catalogs and a local distributor or sales representative.

2–4   SOLENOIDS

The solenoid is a commonly used electrical device that translates ON/OFF electrical signals to ON/OFF mechanical movements. A common use of solenoids in homes and commercial buildings is in heating systems. A solenoid valve in a furnace turns the gas flow ON and OFF. Other solenoids are used in commercial water lines to control hot water flow. Note that the true solenoid valve does not open to an intermediate position for limited flow. A true solenoid valve is either ON or OFF. Valves that open partially are the servovalves, which are more complicated. The variable servovalve will be discussed in later chapters that describe control system components.

In industrial settings solenoids are used for a large variety of process-control ON/OFF applications. These applications include control of hydraulic and pneumatic flow and the opening and closing of various types of actuators.

Figure 2–6 shows the basic operation of a solenoid in a cutaway view. The plunger is held in the upper position by a spring when the coil is electrically deenergized (unactuated). When appropriate voltage is applied to the electrical coil, a magnetic field is produced in the solenoid frame. The frame is made up of permeable magnetic metal laminations. The resulting magnetic field acts on the plunger, pulling it to the down position against spring pressure. The solenoid plunger has an external protrusion that is fastened to the mechanical device to be operated.

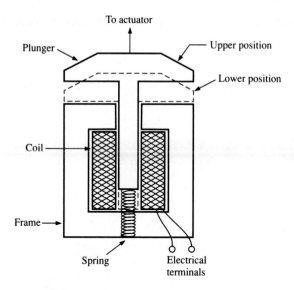

To actuator

Plunger

Upper position

Lower position

Coil

Frame

Spring

Electrical
terminals

FIGURE 2–6
Solenoid Actuator Cutaway View

An illustration of an actual solenoid is in Figure 2–7.

In the choice of a solenoid for an application, the following factors must be considered:

1. the size of the mechanical load to be moved (pounds pull)
2. the stroke distance
3. the environment of operation
4. the type of electrical connection
5. the voltage and the type of current (AC or DC)
6. current values (holding and inrush)
7. the duty cycle in actuations per minute

Factor 1 involves available solenoid pull from a fraction of an ounce to 100 pounds. The pull must be greater than the load by at least 25%. As the required pulling power of the solenoid goes up, the size and the cost of the unit become accordingly higher.

Factor 2 is an obvious application consideration. The stroke of the solenoid must be equal to or greater than the external linear motion required.

Factor 3, the environment, is important, as with all components. The major consideration in industrial situations is whether the solenoid must be oiltight. Another key consideration is vibration. Solenoids can be caused to wear or overheat under excessive vibration.

Factor 4 concerns the way in which electrical connections are made. Connections to solenoids can be by pigtails, plugs, or screw terminals.

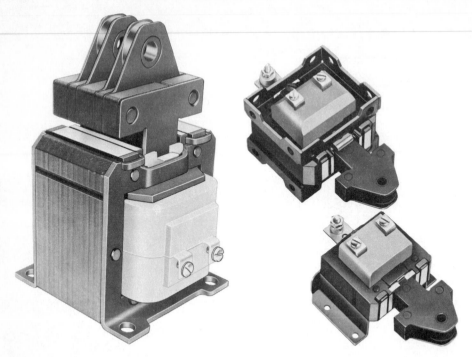

FIGURE 2–7
Industrial Solenoid (Courtesy of Rexnord, Stearns Div.)

Voltage, factor 5, is usually 110 V AC. However, lower-voltage AC, such as 24 V, is often used. You must match the control voltage available. Another important voltage consideration is whether AC or DC voltage is to be used. Solenoids are built specifically for AC operation or for DC operation. For example, a 110 V AC solenoid cannot be used for 110 V DC.

The current of the solenoid, factor 6, is also important. When energized, the solenoid draws rated current. The electrical line and fuse must be sized accordingly. Another factor in considering electrical current requirements is that a solenoid draws from 5 to 15 times its rated current initially, depending on its size. A current surge on a 5 A solenoid could be 45 A. This short-term current must be considered in the fusing of the circuit.

Factor 7, the duty cycle, concerns the frequency of operation of the solenoid. Some solenoids go ON once in a while for a short time. Others are ON for long periods of time and OFF occasionally. Still others are operated a number of times each second. Solenoids that operate at high rates are subject to greater heating and more mechanical stress. It is important that the proper solenoid be chosen for the duty cycle required.

Solenoid failure is normally due either to coil burnout or mechanical failure or both. If the solenoid stroke is incomplete, the electrical coil draws excessive

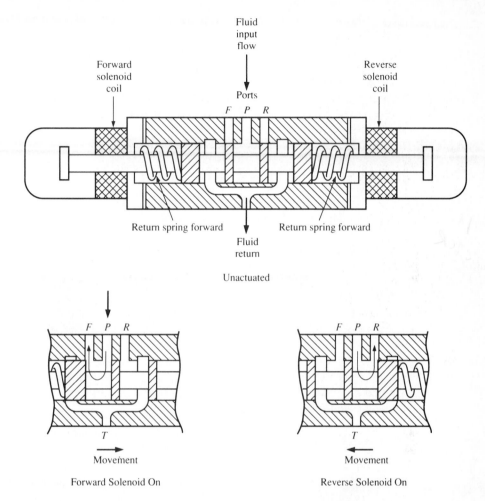

FIGURE 2–8
Double-acting Solenoid

current continuously. The coil then overheats, develops shorted turns, and burns up, if not properly fused. An incomplete stroke can be caused by the wearing out of the mechanical parts of the solenoid. More often, an incomplete stroke occurs when the mechanism to which the solenoid is attached jams in the open or partially open position. Proper electrical fusing, to be described in Chapter 16, is the best prevention for electrical burnout.

Solenoids can be either single-acting or double-acting. The solenoid shown in Figure 2–6 was single-acting. In many hydraulic control valves, double-acting solenoids are used. A layout of a double solenoid is shown in Figure 2–8. Two solenoid coils are connected to a common actuator rod. With neither solenoid coil energized, the rod would be centered, as shown in the "Unactuated" illustration. The figure also shows what happens when the forward solenoid coil is energized. The rod moves to the right. With the rod on the right, fluid is allowed to flow from input port P to output port F. If the reverse solenoid coil is then energized, the rod moves to the left. The fluid flow is then from P to output port R. Flow to port F produces forward motion in the external mechanical system. Fluid flow to port R results in reverse operation. In the center position, fluid flow is from P to T (tank return), and neither F nor R will have fluid flow.

2–5   RELAYS—ELECTROMECHANICAL AND SOLID-STATE

There are a few hundred configurations of relays. This section will cover a representative cross-section of those many types. Relays are used for many control functions. Among their important characteristics for use in control circuits are remote operation, logic operation, control of high voltage from low voltage, and isolation of the control circuit from the switching circuits. The special high-current control relay, the contactor, will be covered in Chapter 11.

The most common type of relay is the electromechanical relay (EMR). The EMR combines the switching principles of Section 2–2 and the solenoid actuation principles of Section 2–3. As the actuator (solenoid) is energized and deenergized, it moves a plunger in, and the plunger is moved out by return spring pressure. The plunger, in turn, opens and closes contacts. Contacts are specified in the unactuated state, as NO or NC. The contacts of a relay can be in multiples as needed, up to a limit determined by relay design. A photograph of an EMR, its symbol, and its connections are shown in Figure 2–9.

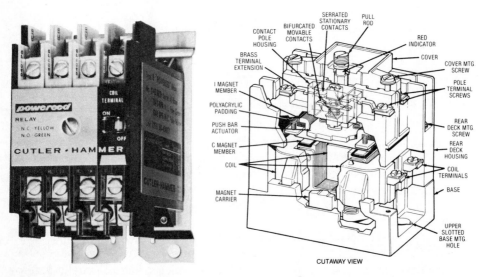

FIGURE 2–9
Electromechanical Relay (Courtesy of Eaton Corp./Cutler-Hammer® Products)

Some EMRs are of the latching relay type (not illustrated). These latch/unlatch relays close when power is applied to the two latching terminals of a coil. When the power is removed from this coil, the relay stays in the latched position. To unlatch, or turn OFF the relay, power is applied to different relay terminals, designated "unlatch." Energizing the unlatching terminals applies power to another coil, which releases the plunger. The plunger then returns to the original position. Latching relays are used in applications where a relay is ON for a long time. Power may be removed during normal ON operation. Note that the latching relay must not be used for a fail-safe circuit configuration. If power has to be applied to turn a circuit or system OFF, the circuit is not fail-safe.

For multiple-pole operation there are a number of rotary relays available. A typical 8-wafer, 12-throw relay is shown in Figure 2–10. The actuating mechanism consists of a solenoid and a detent (hold-in-place) system. The solenoid, when energized, produces rotary motion of 30° through a ratchet mechanism. Note that other numbers of poles require a different rotary angle. The retaining ball bearing in the detent mechanism advances to the next hole position, 30° away. When the solenoid is deenergized, the relay stays in the new position. A ratchet is used to ensure motion in one direction only.

Another common relay used in control circuits is the reed relay. The reed relay is fast-acting, has a long life, uses low power, and is compact. Most reed relays are the plug-in type, illustrated in Figure 2–11. The plug-in feature allows easy replacement in the event of a relay malfunction. The reed relay uses magnetic action for operation, not solenoid action. The magnetic action is also shown in the figure. When the relay coil is energized. it produces a magnetic field. The contact arm material is magnetic. Therefore, the two contact strips become magnetized and attract each other. When the contacts meet, an electrical circuit is completed. When the coil power is removed, the contacts reopen by contact-arm spring action. Some reed relay contacts are "mercury wetted" for longer contact life (at greater expense). There are also latching reed relays available for certain special applications.

With the advent of solid-state devices, a solid-state relay (SSR) has been developed. The SSR has replaced the EMR in many applications. The SSR has no moving parts. The ON and OFF switching is accomplished electronically. A typical SSR is shown in Figure 2–12.

Figure 2–13 is a comparison of the SSR and the EMR. Some of the major advantages of the SSR are long life expectancy, fast switching (response time), and quiet operation. Some of the major advantages of the EMR are multiple contacts, high current capability, no leakage when OFF, and isolation of the coil circuit from the contact circuit.

2–6   TIMERS AND COUNTERS

Timing devices used in industrial processes are of many different configurations. They may be mechanical, pneumatic, motor-driven, or electronic. The electronic types will be discussed in Chapter 15.

A. Graymil Wafer Deck

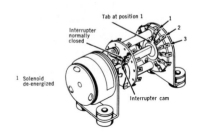

Tab at position 1

Interrupter
normally
closed

1 Solenoid
de-energized

Interrupter cam

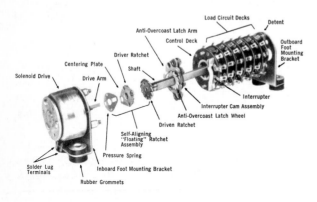

B. Shallcross Wafer Deck

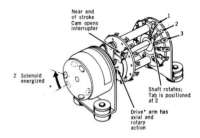

Near end
of stroke
Cam opens
interrupter

2 Solenoid
energized

Shaft rotates;
Tab is positioned
at 2

Drive· arm has
axial and
rotary
action

Load Circuit Decks — Detent

Anti-Overcoast Latch Arm

Control Deck

Outboard
Foot
Mounting
Bracket

Driver Ratchet

Centering Plate

Shaft

Solenoid Drive

Drive Arm

Interrupter

Interrupter Cam Assembly

Anti-Overcoast Latch Wheel

Driven Ratchet

Self-Aligning
"Floating" Ratchet
Assembly

Pressure Spring

Solder Lug
Terminals

Inboard Foot Mounting Bracket

Rubber Grommets

C. Selector Switch
Exploded View

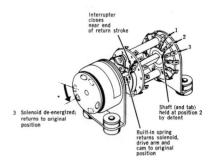

Interrupter
closes
near end
of return stroke

3 Solenoid de-energized;
returns to original
position

Shaft (and tab)
held at position 2
by detent

Built-in spring
returns solenoid,
drive arm and
cam to original
position

D. Switch Operation

FIGURE 2–10
The Rotary Relay (Courtesy of Lucas Ledex, Inc.)

Figure 2–14 shows a motor-driven timer. Mechanical and pneumatic timers have a similar appearance, with a dial for setting the time delay desired. Figure 2–15 illustrates the electrical symbols for the motor-driven timer of Figure 2–14. The same symbols would be applicable to other types of timers. The circle symbol in the figure is for the timer's coil or motor. This element of the timer is used to run the timing interval. When the coil is energized, the timer runs through its preset

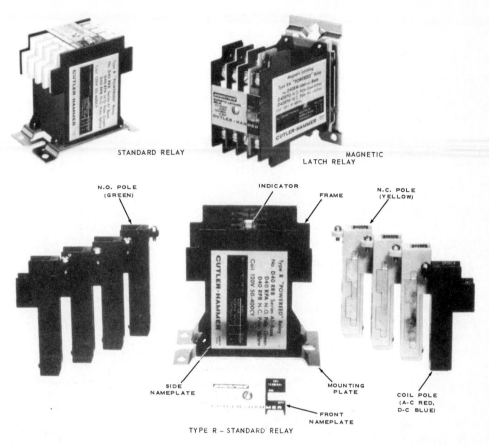

FIGURE 2–11
The Reed Relay (Courtesy of Eaton Corp./Cutler-Hammer® Products)

time interval. At the end of the preset interval, electrical contacts change status from open to closed or from closed to open. The four symbols for the timed contacts are shown in the figure. The contacts are designated NO, timed-closing (NOTC) and NC, timed-opening (NCTO).

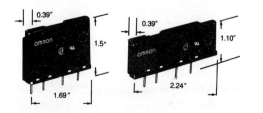
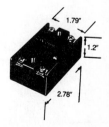

FIGURE 2–12
Solid-State Relay (Courtesy of OMRON Electronics, Inc.)

**Advantages and Disadvantages
of Electromechanical
and Solid State Relays**

Note: Pluses indicate advantages; minuses indicate disadvantages.

	GENERAL CHARACTERISTICS	EMR	SSR
1.	Arcless switching of the load	−	+
2.	Electronic (IC, etc.) compatibility for interfacing	−	+
3.	Effects of temperature	+	−
4.	Shock and vibration resistance	−	+
5.	Immunity to improper functioning because of transients	+	−
6.	Radio frequency switching	+	−
7.	Zero voltage turn-on	−	+
8.	Acoustic noise	−	+
9.	Selection of multipole, multithrow switching capability	+	−
10.	General cost (It should be noted that the cost of SSRs in comparison with EMRs has been decreasing.)	+	−
11.	Ability to stand surge currents	+	−
12.	Response time	−	+
13.	Voltage drop in load circuit	+	−
14.	AC & DC switching with same contacts	+	−
15.	Zero current turn-off	−	+
16.	Leakage current	+	−
17.	Minimum current turn-on	+	−
18.	Life expectancy	−	+

FIGURE 2–13
Comparison of EMRs and SSRs

The most common type of timer is the time-delay ON. After the preset time interval is completed, the contacts change status, from open to closed or from closed to open. When the coil is deenergized, the contacts return to their original open or closed states. Some types of timers are reset by a separate circuit and do not reset when the coil is deenergized. The timed contacts utilized for control can

FIGURE 2–14
Motor-driven Timer (Courtesy of Eagle
Signal Controls Division, Gulf and West-
ern Mfg. Co.)

be either normally open or normally closed. Many timers have both NO and NC
timed contacts in their configurations.

Most timers carry out the timing patterns when the coil is energized. Some
timers, however, carry out the time delay after the coil is deenergized. These
contacts are designated NO, OFF timed-closing (NO/OFF/TC) and NC, OFF
timed-opening (NC/OFF/TO). Their symbols are different, as shown in Figure
2–15. One example of delay after deenergizing is the "night sentinel" in an automobile. The night sentinel keeps the headlights ON for a few minutes after they
have been switched OFF. It provides light for the driver until he or she is safely
inside.

Many timers also contain contacts that act instantly, which are illustrated in
Figure 2–15. These contacts, designated instantaneous, change status immedi-

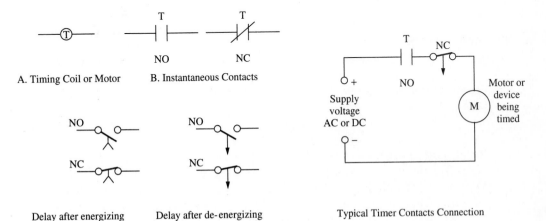

FIGURE 2–15
Timer Symbols and Typical Connection

Combination	NUMBER	Contact used (√)				Status x = ON, o = OFF		
		NCTO	NOTC	INSTNO	INSTNC	Before Timing	During Timing	After Timing
Individual	1	√				x	x	o
	2		√			o	o	x
	3			√		o	x	x
	4				√	x	o	o
Series	5	√		√		o	x	o
	6	√			√	x	o	o
	7		√	√		o	o	x
	8		√		√	o	o	o
Parallel	9	√		√		x	x	x
	10	√			√	x	x	o
	11		√	√		o	x	x
	12		√		√	x	o	x

FIGURE 2–16
Timer Patterns for Various Contact Combinations

ately when the coil is energized. They revert to their original status when the coil is deenergized. These contacts are designated INSTNO and INSTNC. It is possible to obtain a variety of timing patterns by using the instantaneous and timed contacts in combination. The various timing patterns possible are shown in Figure 2–16.

In timing situations, you are concerned with output status at three times: (1) before starting, when the circuit is in the preset state, (2) during timing, and (3) after timing, when the circuit is timed out. You can obtain different timing patterns for these three time periods by using various timer contact combinations. You may use individual contacts only, or contacts in series or parallel. The 12 possible combinations are shown in Figure 2–16. For combination 5, with the NCTO and INSTNO in series, the series circuit is initially open, as INSTNO is open. During the timing interval, both contacts are closed, so the circuit is closed. At the end of the timing, NCTO opens, and the circuit is open. Note the way this pattern differs from both 1 and 3, which use individual contacts only.

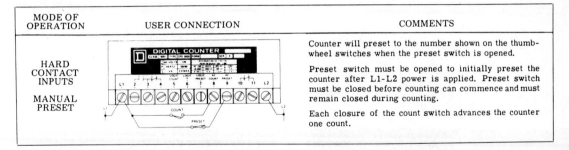

MODE OF OPERATION	USER CONNECTION	COMMENTS
HARD CONTACT INPUTS MANUAL PRESET		Counter will preset to the number shown on the thumb-wheel switches when the preset switch is opened. Preset switch must be opened to initially preset the counter after L1-L2 power is applied. Preset switch must be closed before counting can commence and must remain closed during counting. Each closure of the count switch advances the counter one count.

FIGURE 2–17
Digital Counter (Courtesy of Square D Company)

Figure 2–17 shows a digital counter. Counters can be mechanical, electrical, or digital electronic. Electronic counters will be illustrated in Chapter 15. A counter does not automatically reset to the initial state on the deenergization of its coil, as the timer does. Otherwise, we would count to 1 and reset continually. The counter needs a separate device or connection to reset it. The use of instantaneous and after-count contacts with counters is similar to the use of different timer contacts as illustrated in Figure 2–16. Counters can count up or down, depending on the configuration chosen.

2–7   CONTROL TRANSFORMERS

Transformers are often used to produce control-circuit voltages, such as 120 V AC and 24 V AC. Most industrial plants have line voltages of 440 V AC and some-times 220 V AC. The 440 V or 220 V is applied to the primary winding of the control transformer. The control transformer secondary winding then produces the required control AC voltage. Control voltages of 120 V or 24 V AC are commonly in use. These lower voltages are much safer to use than 440 V or 220 V. Furthermore, the transformer electrically isolates the safer control voltage from the higher line voltages.

Note that in this section only control transformers are covered. Higher-power line transformers will be covered in Chapter 16. Note also that transform-ers function only for AC power and will not work for DC power. Also, a given type of transformer may not function well for nonsinusoidal AC voltages.

Some considerations in the selection or replacement of a control transformer are:

1. the control voltage from the secondary
2. the line voltage of the primary
3. the power rating of the transformer in VA or kVA
4. electrical grounding of the secondary
5. secondary fusing

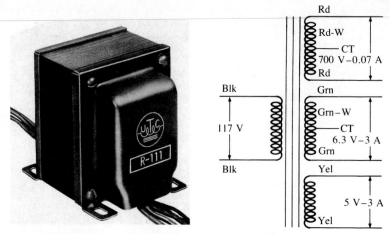

A. Shell-type Power Transformer   B. Schematic Representation

FIGURE 2–18
Control Circuit Transformer (Courtesy of United Transformer Company)

6. loading effects due to relay-coil inrush currents
7. environmental considerations
8. secondary configuration

A typical power transformer is shown in Figure 2–18. This transformer has eight connections, two for the primary and three groups of two for the secondary. Other control transformers have different connections, as will be illustrated.

The schematic diagram for a control transformer is shown in Figure 2–19A. In 2–19B, we have added some typical parameters to the schematic symbol to illustrate factors 1 through 5. The input voltage is 440 V from single-phase line power or from any one of three phases of the line power. The secondary produces

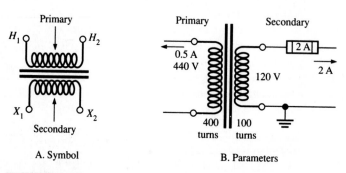

A. Symbol   B. Parameters

FIGURE 2–19
Transformer Symbol and Parameters

120 V. The load requirement is 2 A. A 2 A fuse is needed. The secondary is electrically grounded for safety reasons.

Transformer voltages are produced in the primary and secondary windings in proportion to the number of turns of each. In the example in Figure 2–19B, there are 400 turns in the primary coil. For the secondary to produce 1/4 the voltage, the secondary should have 100 turns. In actual operation the proportions are not quite exact, but direct proportion can be assumed for the purposes of this example. How much current will flow in the primary for a 2 A flow in the secondary? The current relationship is inverse to the number of turns. The primary will have a flow of 0.5, or 1/4, the current in the secondary. The current ratios are not strictly exact, but they can be assumed to be for this discussion.

We may determine the rating of the transformer in VA, volt-amps, or kVA, kilovolt-amps. Volt-amps are determined by simply multiplying the secondary voltage by the maximum current in the secondary. In the example of Figure 2–19B, the rating in volt-amps would be 120 × 2, or 240 VA. To determine kilovolt-amps, divide 240 VA by 1000 to obtain 0.240 kVA. Smaller transformers are rated in volt-amps, and larger transformers are rated in kilovolt-amps for convenience.

Other transformer rating considerations may be found in the NEMA standards. One consideration is oversizing the volt-amps rating for safety of operation, often by 15%. Another consideration is the inrush currents of coils in the control circuit (factor 6). Coils can initially draw 5 to 10 times their rated steady-state current. If coils are energized often, the inrush currents must be considered. The extra inrush current flow, if not considered, will cause the transformer to overheat beyond its rating and ultimately fail. The formula for determining the extra rating value is

$$\text{Total VA} = \frac{(\text{total continuous VA} \times 2) + (\text{inrush VA} \times 2)}{2}$$

Environmental factors, factor 7, include vibration and high temperature in the area where the transformer is used. To reduce vibration, shock mounts may be necessary. In warm environments a higher-rated transformer or a cooling fan is required, depending on the application requirement.

FIGURE 2–20
Tapped Control Transformer

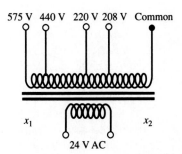

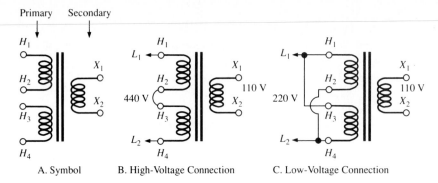

FIGURE 2–21
Dual-Voltage Transformer

Three possible secondary configurations of control transformers are (1) multiple-line voltage (usually dual), (2) multiple-voltage-secondary transformers, and (3) ground-isolation transformers.

Multiple-primary transformers are supplied on equipment to be used in a plant where the line voltage may be 575 V, 440 V, or 220 V. Without such transformers, the equipment supplier would have to supply a different control transformer for each application instead of one standard transformer.

Multiple-line-voltage control transformers can be tapped or multiple-winding. The tapped type is shown in Figure 2–20. The primary has one common connection and multiple connections for the other line connections. The primary winding is tapped at various appropriate points for different line voltages. For the correct 24 V output, the line must be connected to the proper primary connection. This type of transformer is rather expensive. Furthermore, it must be derated for the lower voltages, as only part of the primary is used.

A more commonly used multiple-line-voltage transformer is the dual-primary transformer. The primary has two identical windings, as shown in Figure 2–21. When the primary windings are connected in series, 440 V line produces 110 V for control. When the primary windings are connected in parallel, 220 V line produces the required 110 V for control. The figure shows both possible connections.

Some applications might require 120 V for some devices and 24 V for others. Two control transformers could be used. More often, a transformer with multiple secondaries is used, as shown in Figure 2–22. Use of such a transformer saves space, weight, and cost. Figure 2–22 is a control transformer with both 120 V and 24 V outputs. The line voltage is 480 V with 860 turns in the primary. The two outputs are 120 V, 2 A, and 24 V, 5 A. The total volt-amps in the secondary is 240 + 120 = 360 VA. The primary amps are therefore 360 VA/480 V = 0.75 A. Again, the actual figures will be a little different with an in-depth analysis. This illustration does show the principles involved. The number of turns in each sec-

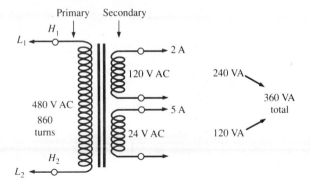

FIGURE 2–22
Dual-Control-Voltage Transformer

ondary can be determined by voltage proportionalities. There are 215 turns for 120 V and 43 for 24 V.

A ground-isolation transformer is a transformer for which neither side of the secondary is electrically grounded. You have the situation of Figure 2–19*B*, but the ground is not used. Isolation transformers are used for personnel safety and for some instrumentation measurement applications. Often such transformers are 1 : 1 voltage-ratio transformers. The primary and secondary windings are identical—for example, 120 V/120 V. The incoming line voltage involves a ground, since it is connected to the building distribution system. However, the secondary does not have an electrical ground connection.

2–8   POTENTIOMETERS

Potentiometers are three-terminal variable resistors. The potentiometer stem is rotated to vary the resistance from the center terminal to either end terminal. The resistance from one end terminal to the movable center terminal varies continuously, from 0 Ω to the potentiometer value, as the stem is rotated. There are two major classifications of potentiometers: carbon-element and wire-wound. Typical examples of these potentiometers are shown in Figure 2–23. There are many physical forms of potentiometers besides those shown. It is also possible to procure tapped resistors instead of potentiometers. However, these tapped resistors have only a few fixed resistance values, depending on the number of taps. Also, tapped resistors are more costly than "pots."

Potentiometers are used extensively in electrical and electronic control circuits. They are used to set input values, adjust zero points, set calibration, and meet other requirements for variables.

The choice of carbon or wire-wound potentiometer depends on the application. The following table gives the major characteristics of each:

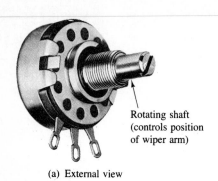

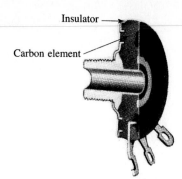

(a) External view   (b) Internal view   (c) Carbon element

A. Molded-Composition Potentiometer

B. Screw-Drive Rheostat

FIGURE 2–23
Potentiometers (A. Courtesy of Allen-Bradley Co.   B. Courtesy of James G. Biddle
Co.)

Power

Carbon	**Wire-Wound**
Values to 100 mΩ	Limited values, usually to 1 kΩ
Low amps and power	Can have high amps and power
Perfectly continuous	Output in small steps
Can be nonlinear	Linear
Low cost	Higher cost
Carbon ages and changes	No aging
Subject to burned spots	Less likelihood of burned spots

The foregoing comparisons are self-explanatory except for continuous/steps characteristic and linearity. As for the continuous/steps concept, suppose a 5 kΩ wire-wound pot is made up of 250 turns of wire. Each turn represents 5000/250, or 20 Ω.

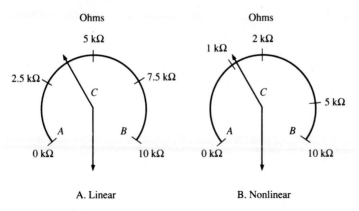

A. Linear   B. Nonlinear

FIGURE 2–24
Linear and Nonlinear Potentiometers

As the variable wiper moves through its rotation, it changes resistance in 20 Ω steps. By comparison, the carbon pot variable resistance value is perfectly continuous, not in steps.

Linearity is another consideration. Some potentiometers are required to be nonlinear in construction. This nonlinearity of a typical carbon pot is shown in Figure 2–24. A typical example of a nonlinear pot requirement is in an audio amplifier. Human hearing is nonlinear. Therefore, the amplifier output is normally nonlinear, usually logarithmic, to match our hearing ability. The audio matching is accomplished by using a logarithmic nonlinear volume-control pot.

To control large amounts of power, a *rheostat,* a high-power wire-wound pot, is often used. The rheostat is a highly inefficient device electrically. It uses a lot of power itself, and it heats up. For AC variable-power control, an autotransformer is often used. In some field applications, the autotransformer and the rheostat may be confused with each other. The confusion usually happens when component replacements are made. Figure 2–25 compares the physical appearance and the symbols of the autotransformer and the rheostat.

This section will conclude with a discussion of a common cause of potentiometer burnout and failure. If potentiometers are not properly protected electrically, they can be destroyed by being adjusted to a value that is too low. To protect the pot, a protective resistor is needed in the circuit. This principle is illustrated in Figure 2–26. Figure 2–26A shows a 10 kΩ, 10 W pot connected in series with a 100 Ω load resistor. The pot setting is varied to vary the current through the 100 Ω load resistor, which in turn gives a variable voltage output. The supply voltage is a fixed 200 V. In Figure 2–26A, the pot is electrically unprotected. Note in the table that the pot dissipates 4 W at its full, 10 kΩ setting. This dissipation is below its 10 W rating, so no pot overloading is present. At the midway, 5 kΩ setting, the pot uses 7.6 W, still below its rating. However, note its watts usage at the 1 kΩ setting—32 W. It will burn out and fail very quickly at that setting. Below 1 kΩ it will burn out even faster.

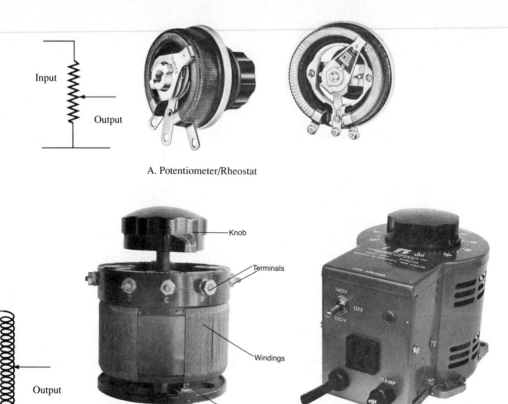

A. Potentiometer/Rheostat

B. Autotransformer

FIGURE 2–25
Comparison of Rheostat and Autotransformer (A. Courtesy of Ohmite Manufacturing
Co.   B. From Humphries, J. T. *Motors and Controls.* Merrill/Macmillan: 1988.
Fig. 4.25, p. 127)

    In Figure 2–26*B* we include a 10 kΩ protective series resistor. Now fewer
watts are used. Note that there is a trade-off. The range of V_{out} will be about half,
which may not be acceptable and may require another compensating circuit revision. However, the pot is protected from burnout. In the table, at the 10 kΩ
setting, the power dissipated in the potentiometer is only 1 W. At 5 kΩ the watts
dissipated are 0.85, and at 1 kΩ they are a low 0.32. These watt values are well
below the 10 W rating of the pot. With protection, the pot will not be electrically overloaded at the 1 kΩ setting, as it was in Figure 2–26*A*. It is also protected below the 1 kΩ setting, as the watts continue to decrease with lower settings.

	Potentiometer Setting		
Formula	10 kΩ	5 kΩ	1 kΩ
$I = \dfrac{E}{R_T}$	$\dfrac{200}{10,100}$ = 20 mA	$\dfrac{200}{5100}$ = 39 mA	$\dfrac{200}{1100}$ = 180 mA
Watts = I^2R	$(20\text{ mA})^2$ × 10 kΩ = 4	$(39\text{ mA})^2$ × 5 kΩ = 7.6	$(180\text{ mA})^2$ × 1 kΩ = 32

	Potentiometer Setting		
Formula	10 kΩ	5 kΩ	1 kΩ
$I = \dfrac{E}{R_T}$	$\dfrac{200}{20,100}$ = 10 mA	$\dfrac{200}{15,100}$ = 13 mA	$\dfrac{200}{11,100}$ = 18 mA
Watts = I^2R	$(10\text{ mA})^2$ × 10 kΩ = 1	$(13\text{ mA})^2$ × 5 kΩ = 0.85	$(18\text{ mA})^2$ × 1 kΩ = 0.32

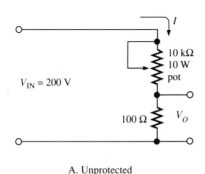

A. Unprotected   B. Protected

FIGURE 2–26
Potentiometer Electrical Protection

SUMMARY

☐ Various electrical switching arrangements are available for control applications—numbers of poles, numbers of throws, and NO or NC status.

☐ Important switching choices include spring-return, center-OFF, and shorting or nonshorting.

☐ There are many types of switches available, such as toggle and push-button. The choice depends on application, environment, cost, reliability, personal preference, and other factors.

☐ Solenoids convert electrically induced magnetic force to mechanical movement.

☐ Solenoids are chosen according to such application factors as the pull force required, the electrical supply available, and the environment. They may be single- or double-acting.

☐ Relays are used in many control systems for logic, isolation, control of high current, and control of multiple circuits.

☐ There are many configurations of relays available. They include electromechanical, reed, rotary, and solid-state.

☐ Many types of nonelectronic timers are available, including mechanical, pneumatic, and motor-operated.

☐ Timers that have both NO and NC timed contacts and both NO and NC instantaneous contacts can be used for a wide variety of timing patterns.

☐ Counters can be mechanical, electrical, or electronic. Counters require both count capability and a separate reset capability.

☐ Control transformers are used to convert a high line voltage to a safer control voltage, usually 120 V AC.

☐ Control transformers may be basic four-wire, dual-voltage, or multiple-voltage. The volt-amps rating required is determined by the power requirements of the control-circuit components.

☐ Potentiometers have many physical forms but are electrically of two basic types: carbon-element and wire-wound.

☐ Potentiometers are chosen according to value, power rating, and linearity requirements. Potentiometers must be electrically protected from overload and burnout.

EXERCISES

1. Draw the electrical symbol for the following switching arrangements:
   a. triple pole, single throw   c. DPST—spring-return
   b. triple pole, double throw   d. DPDT—center-OFF

2. Draw the electrical symbol for a three-pole, double-throw push-button switch.

3. Repeat the analysis for the voltmeter in Figure 2–4. The ranges are 200 V and 50 V. Resistor R_1 is 199 kΩ; R_2 is 49 kΩ. Show that using a shorting switch is a bad practice and using a nonshorting switch is a good practice.

4. Repeat the analysis for the ammeter in Figure 2–4. The ranges are 25 mA and 200 mA. Resistor R_1 is 41.7 Ω; R_2 is 5.03 Ω. Show that using a nonshorting switch is a bad practice and using a shorting switch is a good practice.

5. List two applications for each of the following switches:
   a. toggle
   b. slide
   c. push-button

6. List one application for each of the following switches:
   a. 12-pole wafer switch
   b. three-position selector switch
   c. four-position selector switch

7. You are selecting a solenoid for an application. List the considerations in order of importance for the selection.

8. List two applications for a double-acting solenoid.

9. Draw the electrical symbol for a relay with four NO and four NC contacts.

10. List two applications for each of the following:
    a. an electromechanical relay
    b. a solid-state relay

11. List one application for each of the following:
    a. a latching relay
    b. a reed relay

12. What type of relay would you use for each of the following situations, and why?
    a. 12 circuits, fast-acting, 10 V DC
    b. transistor, 6 V, to control one NC, 120 V AC circuit
    c. 24 V AC to control eight 440 V AC circuits
    d. one million actuations per month for 20 V AC

13. Draw the symbol for each of the following timers:
    a. two NO and three NC time-delay ON contacts, and two NC instantaneous contacts
    b. one NC and two NO time-delay OFF contacts, and one NO and three NC instantaneous contacts

For Exercises 14–18, refer to Figure 2–16 and the accompanying text material.

14. Explain how combinations 1, 6, and 12 function.

15. Explain why combinations 8, 9, and 11 are not usable. Show how they would function.

16. How would the entire figure be different for time-delay OFF?

17. Would a counter functional pattern be like Figure 2–16? Why or why not?

18. What timer combinations would you use to produce the following ON/OFF sequence:
    a. Initially OFF   c. Initially OFF
       Switch ON   Switch ON
       OFF 3 more seconds   OFF 1 second
       ON 2 seconds   ON 3 seconds
       OFF   OFF 2 seconds
    b. Initially ON   ON 4 seconds
       Switch ON   OFF
       OFF 4 seconds
       ON 3 seconds
       OFF

19. Refer to Figure 2–19 on control transformers. Assume the primary is 220 V AC with 325 turns. The secondary is 24 V AC and 4.5 A. Find the primary current, the volt-amps rating, and the number of turns in the secondary.

20. For the transformer in Exercise 19, what if the primary voltage were 220 V DC?

21. How would you construct a control transformer capable of handling three, not two, line voltages? The possible primary AC voltages are 440 V, 220 V, and 110 V. The secondary voltage is 24 V AC. *Hint:* Refer to Figure 2–21.

22. Refer to Figure 2–21. What would happen if the primaries were misconnected for high voltage, H_2 to H_4 and H_1 to H_3?

23. Refer to Figure 2–21. What would happen if the primaries were misconnected for low voltage, H_1 to H_4 and H_2 to H_3?

24. Refer to Figure 2–22. You have a similar control transformer with three secondaries. The primary is 440 V AC with 950 turns instead of 860 turns. The secondaries are 120 V AC, 2.75 A; 24 V AC, 3.5 A; and 12 V AC, 10.5 A. Find primary current, volt-amps rating, and the number of turns for each of the secondaries.

25. List two applications for each of the following:
    a. carbon-element potentiometer
    b. wire-wound potentiometer

26. Refer to Figure 2–24B. What would be the effect in the application of connecting to *B* and *C*, in error, instead of to the proper terminals, *A* and *C*?

27. You want 8 Ω accuracy for a 2250 Ω wire-wound pot. What is the minimum number of turns you must have? What is the minimum number of turns for an accuracy of 2.5%?

28. Refer to Figure 2–26 and the accompanying text explanation. Repeat the calculation for 75 V, a 1 Ω load, and a 1750 Ω, ½ W pot. Would the pot be subject to failure for connection *A*? Explain why or why not.

APPLICATION ASSIGNMENT ANSWER

First, verify that each component is indeed defective. This may involve removing each device electrically from the circuit before checking its operation. If the switches are defective, replace with switches of the same configuration. The replacement switches must have equivalent current, voltage, and breaking characteristics. If the potentiometer must be replaced, use one of the same type (carbon or wire-wound), value, and taper. The important potentiometer considerations are value, wattage, linearity, and electrical protection of the circuit and the potentiometer. Replace the transformer, if defective, with one that has the same voltage ratings. Transformer application considerations are power (VA rating) and the number of different voltages at which the transformer can be used.

3

Control Diagrams

CHAPTER OUTLINE

3–1 Introduction □ 3–2 Commonly Used Symbols □ 3–3 Types of Diagrams and Their Uses and Applications □ 3–4 Schematic Diagrams □ 3–5 Ladder Diagrams and Sequence Listings □ 3–6 Wiring, Interconnecting, and Harness Diagrams □ 3–7 Printed-Circuit Diagrams □ 3–8 Block and Pictorial Diagrams □ 3–9 Troubleshooting Diagrams □ 3–10 Standards Applicable to Control Diagrams □ 3–11 Diagram Revision Systems

LEARNING OBJECTIVES

□ Draw the symbols for commonly used control-diagram components.
□ Recognize those symbols, and name what each represents.
□ List the major types of control diagrams.
□ Define the characteristics of each type of control diagram, and explain its use.
□ Create basic ladder diagrams from a process description.
□ Make up a sequence listing for a given ladder diagram.
□ Create wiring diagrams from ladder-diagram information.
□ Draw interconnection and harness diagrams based on wiring diagrams.
□ Describe block and pictorial diagrams and their uses.
□ Describe the elements of a troubleshooting diagram, and create an effective troubleshooting diagram for a simple device.
□ List major standards systems that apply to control diagrams.
□ Discuss the importance of a diagram revision system, and illustrate its composition.

APPLICATION ASSIGNMENT

You are a technician assigned to a service call. The site of the service problem is 550 miles away. The service problem is intermittent operation of a large printing press. The different colors periodically get out of alignment. What type of control diagrams for the machine would you take on the service call, and why?

3–1   INTRODUCTION

In Chapter 2 many of the devices used in control circuits were discussed. This chapter will cover the electrical connection of those devices to control industrial processes.

First the commonly used symbols for the components will be covered. A general discussion of the various types of diagrams and their use will be next. Methods of creating one type of diagram based on another will be illustrated. The relationship between ladder diagrams and wiring diagrams is especially important for control systems. Ladder diagrams are for logic analysis and troubleshooting. Wiring diagrams are for installation and for physical checks of system wiring. Printed-circuit, block, and pictorial diagrams and their uses will be covered. The use and construction of troubleshooting chart diagrams will also be explained. Standards for format, symbols, and other elements of diagrams will be discussed. Finally, the importance of change documentation for diagrams will be illustrated. The makeup of an effective change system for diagrams and drawings will be explained.

3–2   COMMONLY USED SYMBOLS

Control circuit devices generally fall into five categories. The symbols for the devices will be discussed by category in this section. The five categories are discrete input, analog input, internal controller, discrete output, and analog output. Discrete devices have two states, ON and OFF. Analog devices have or respond to varying electrical signal values between two end limits.

Figure 3–1 shows the symbols for some commonly used discrete-input devices. The most commonly used discrete inputs are switches, push buttons, and status-indicating devices such as limit switches. These and others are illustrated. As explained in Chapter 2, *NO* stands for normally open—that is, open in the unactuated position. When actuated, the NO device will close. The abbreviation *NC* stands for normally closed. When actuated, an NC device goes from the closed position to the open position. Note that for limit switches there are four symbols. Two are straight NO and NC limit switches. The other two are held closed and held open. The designation *held closed* or *held open* means that at the start of a process, the limit switch is mechanically held in the actuated position, ON or OFF.

Analog-input devices are shown in Figure 3–2. They include potentiometers and any other device that gives a varying voltage or current signal to an input of a process. Note that the input voltages are normally assumed to be linear with respect to the input parameter variation. In a linear situation the voltage is directly proportional to the position of an analog device throughout its range. For nonlinear devices proportionality is not constant. Nonlinear devices in a control circuit require special considerations.

Limit Switch

Normally open contact	
Normally open contact held closed	
Normally closed contact	
Normally closed held open contact	

Toggle Switch

NO	NC

Toggle switch spring return	

Motor Centrifugal Switch

Speed (plugging)	Anti-plug

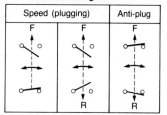

General Switches

Float	NO	
	NC	
Thermal	NO	
	NC	
Pressure	NO	
	NC	
Foot	NO	
	NC	
Liquid Level	NO	
	NC	
Flow	NO	
	NC	
Proximity	NO	
	NC	

Push Buttons

Momentary Contact					Maintained Contact		Illuminated
Single circuit		Double circuit	Mushroom head	Wobble stick	Two single ckt.	One double ckt.	
NO	NC	NO & NC					

Selector

2 Position	3 Position	2 Pos. Sel. Push Button

2 Position:

J   K
A1
A2

	J	K
A1	1	
A2		1

1−contact closed

3 Position:

J K L
A1
A2

	J	K	L
A1	1		
A2			1

1−contact closed

2 Pos. Sel. Push Button:

A   B
1   2
3   4

	Selector Position			
Contacts	A		B	
	Button		Button	
	Free	Depres'd	Free	Depres'd
1−2	1			
3−4		1	1	1

1−contact closed

General Multiple Switch Convention

SPST NO		SPST NC		SPDT		Terms
single break	double break	single break	double break	single break	double break	SPST − single pole single throw
						SPDT − single pole double throw
DPST, 2 NO		DPST, 2 NC		DPDT		DPST − double pole single throw
single break	double break	single break	double break	single break	double break	DPDT − double pole double throw

KEY:
NO = Normally open
NC = Normally closed

FIGURE 3–1
Discrete-Input Device Symbols

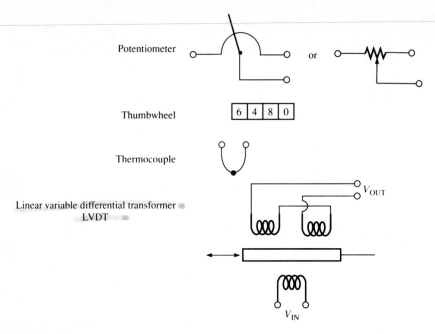

FIGURE 3–2
Analog-Input Device Symbols

Symbols for major controller devices are listed in Figure 3–3. Many of these control devices will be covered in this text, and the symbols shown in Figure 3–3 will be used.

Typical discrete-output devices are shown in Figure 3–4. They include relays, indicating lights, contactors, and motors. Such devices are ON when power is applied and OFF, not actuated, when power to them is OFF.

Analog-output devices include positioning motors, meters, positioners, and stepper motors. Even though the stepper motor is not strictly continuous (it moves in small steps), it is considered an analog output device. Figure 3–5 shows the symbols for some major analog-output devices.

3–3   TYPES OF DIAGRAMS AND THEIR USES AND APPLICATIONS

There are many types of electrical control diagrams, each with a specific purpose. This section is an overview of the types of diagrams to be discussed in detail in the ensuing sections of this chapter. It is important to have the correct type of diagram for each task in control design or analysis. Two examples illustrate the importance of having and using the correct diagram.

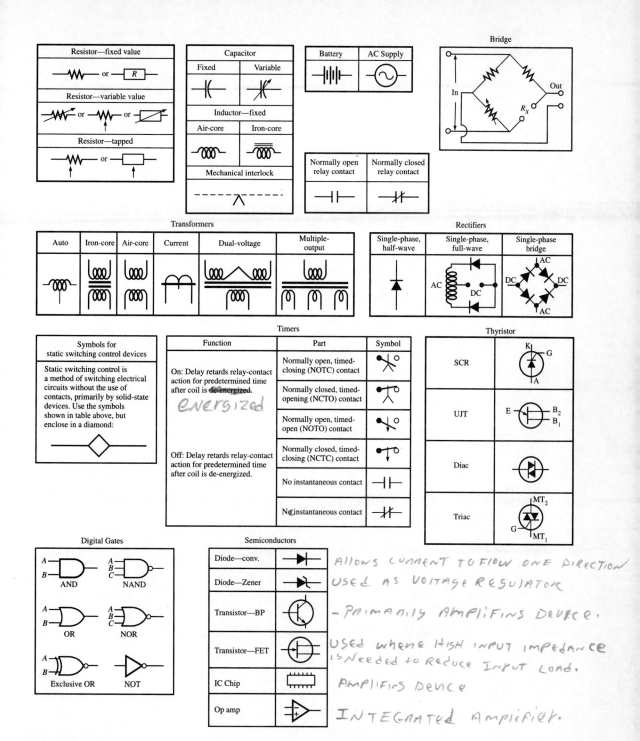

FIGURE 3–3
Controller Device Symbols

51

Timer	Contactor	Solenoid	Relay Coil
TR	M		CR

Annunciator	Bell	Buzzer	Horn, Siren, etc.

Pilot Lights	
Indicate color by letter	
Non push-to-test	Push-to-test
A	R

FIGURE 3–4
Discrete-Output Device Symbols

Valve position
V

Readout	
BCD	1 6 3 7

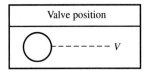

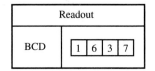

Meters	AC Motors				DC Motors			
Letters indicate type	Single-phase	3-phase squirrel-cage	2-phase 4-wire	Wound-rotor	Armature	Shunt field	Series field	Comm. or compens. field
VM AM						(show 4 loops)	(show 3 loops)	(show 2 loops)

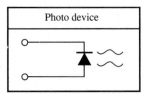

Photo device

FIGURE 3–5
Analog-Output Device Symbols

EXAMPLE 1 A technician was called to a plant where an SCR was periodically burning out after a certain relay was energized. Plant personnel were asked to furnish electrical diagrams for the machine in question. They brought out six drawings, each 3 ft × 4 ft, with many wires and components. At the right and left of each drawing were 30–40 wire ends, which connected to another drawing. The SCR and the relay were four drawings apart. The six drawings were wiring diagrams. Needless to say, analysis was next to impossible using the wiring drawings. What was needed for analysis was a schematic or ladder diagram along with a sequence description.

EXAMPLE 2 A technician was given the task of wiring 35 panels of a given model number. The task would take about three weeks. The technician was given a ladder diagram with 65 lines from which to do the wiring. The technician wired the first panel one way, the next another way, and the others many different ways. All panels should have been wired identically. What was needed was a complete panel wiring diagram to show exactly how the panels were to be wired.

These two examples concern only two of the many types of diagrams available. A description of each type follows:

☐ *Schematic diagram.* Schematic diagrams show the electrical connections to the components in an orderly fashion. The components are shown in symbol form. The diagram has no relation to the physical location of the wires or components. Schematics are used for operational analysis of electronic circuits.

☐ *Ladder diagram.* There are two vertical lines, one on the right and one on the left. Connections are made in straight horizontal lines, which run between the vertical lines. Contacts are on the left, and coils or outputs are on the right. The first horizontal line is for the first occurrence in the sequence of operation. The next operational occurrence follows on the next line. The following lines downward continue to follow the sequence, as nearly as possible. Ladder diagrams are used for circuit operational analysis. They are normally used for electrical, not electronic, circuitry. Examples of ladder diagrams will be shown in the next section.

☐ *Sequence listing.* A sequence listing is not strictly a drawing. It is a written sequence of events and is auxiliary to a ladder or schematic diagram. The listed events (actions) are the actual operating steps for the associated equipment.

☐ *Wiring diagram.* The wiring diagram shows how a group of electrical components, units, and panels are wired. The diagram components are placed in the relative physical positions the actual components will have. The wiring diagram is used for construction and installation purposes.

☐ *Interconnecting diagram.* Only actual interconnecting wiring is shown on interconnecting diagrams. Such diagrams are used in large installations with many interconnecting wires. There are so many wires going to so many different places that including them on a single large wiring diagram would be confusing.

☐ *Harness diagram.* Similar to an interconnecting diagram, a harness diagram shows a specific group of wires in a harness or ribbon cable.

☐ *Printed-circuit diagram.* Essentially a picture of the printed circuit (PC) board, with interconnecting paths and components shown, a printed-circuit diagram is used for PC board construction and analysis.

☐ *Block diagram.* In a block diagram each major functional controller section is shown as a block. Each block is then given a suitable name or description, which is written inside the block. The flow of information by means of electrical signals is then shown by single or multiple connecting lines between blocks, with directional arrows. A block diagram is used for a general description of the operation of a single piece of equipment or of a system.

☐ *Pictorial diagram.* Pictorial diagrams are similar to wiring diagrams. Instead of equivalent symbols, however, a pictorial diagram uses pictorial symbols, which look like the devices. Furthermore, components are shown in physical position as they will actually appear, taking into account actual dimensional perspective. Pictorial diagrams are used in construction and installation.

☐ *Troubleshooting diagram.* For troubleshooting manuals, diagrams of sections of the controller can be shown. A word description of each possible problem and its solution is also included, for ease of troubleshooting. Flowcharts can show the sequence of analysis.

3–4 SCHEMATIC DIAGRAMS

A schematic diagram is an orderly representation of the components and connections of an electrical circuit. An example is shown in Figure 3–6. Such a circuit diagram gives no indication of the actual physical placement of components in the electrical package. The flow of electrical signal is generally from left to right. A radio frequency signal coming from the mixer on the left in Figure 3–6 is modified through the circuit to become an audio output to a speaker on the right. In more complex schematics, parallel sections and their interconnections are shown.

The schematic is the first step in designing an electronic controller. After a rough sketch, the diagram is put into tentative form. After the circuit has proven out, the diagram is put into the form shown in the figure.

The schematic is the master drawing for part numbers, values, and ratings of the components used. The components are identified by letter-number combinations. A master parts list (not shown) would identify the component values and the power, current, and voltage ratings. The ground symbols indicate a common connection to chassis ground.

For analysis of a circuit function, or malfunction, the schematic drawing is an invaluable aid.

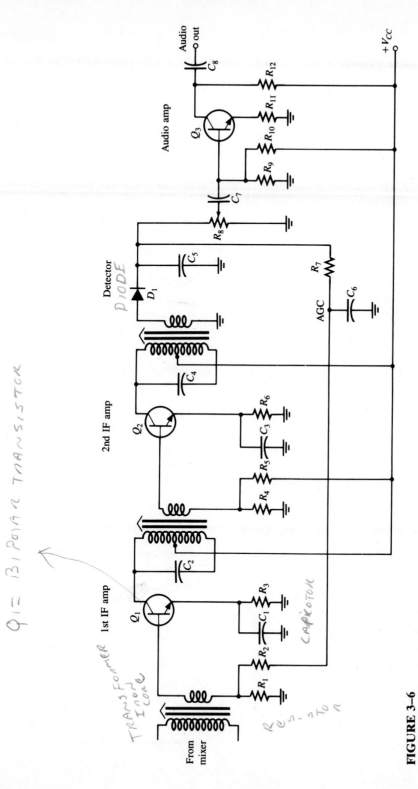

FIGURE 3–6
Typical Schematic Diagram

3–5   LADDER DIAGRAMS AND SEQUENCE LISTINGS

Ladder diagrams are the most commonly used diagrams for nonelectronic control circuits. They are sometimes called *elementary diagrams* or *line diagrams*. Sometimes they are considered a subtype of schematic diagrams. The term *ladder diagrams* will be used in this text. Why are these diagrams called ladder diagrams? They look like a ladder in a way. You start at the top of the ladder and generally work your way down.

Two types of ladder diagrams are used in control systems: the *control ladder diagram* and the *power ladder diagram*. This section will concentrate on control ladder diagrams, with only a fundamental explanation of the power ladder diagram.

Figure 3–7 shows two basic control ladder diagrams. The first one, *A*, is for a single switch that turns a relay output, CR_5, ON and OFF. The second, *B*, is a single-function diagram with parallel lines for control and parallel lines for output. Either or both of two switches turn the output and a pilot light on.

The control ladder diagram of Figure 3–8 has two active functional lines. Some of the common practices for the format of control ladder diagrams are illustrated by this figure. Those practices are as follows:

- □ All coils, pilot lights, and outputs are on the right.
- □ An input line can feed more than one output. If it does, the outputs are connected in parallel.
- □ Switches, contacts, and so on are inserted in the ladder line starting on the left.
- □ Switches, contacts, and so on may be multiple contacts in series, parallel, or series parallel.

FIGURE 3–7
Basic Control Ladder Diagrams

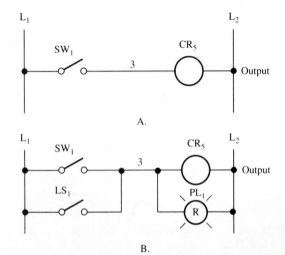

FIGURE 3–8
Two-Function Control Ladder Diagram

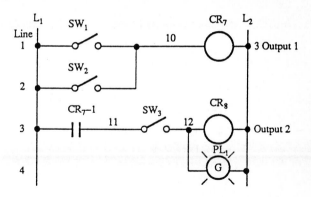

□ Lines are numbered consecutively downward on the left.
□ Every connection node is given a unique identification number.
□ Outputs can be identified by function on the right, in notes.
□ A cross-identification system may be included on the right. The contacts associated with the line's coil or output are identified by line location. In Figure 3–8 the *3* to the right of line 1 indicates that a normally open contact of relay CR_7 (the coil on line 1) is located on line 3. For a normally closed contact, the number would have an asterisk (*) next to it or a bar over it. Figure 3–11 uses the same system on two different lines.
□ Relay contacts are identified by the relay coil number plus a consecutive sequence number. For example, we have included contact CR_7-1. If other relay CR_7 contacts were used, the next would be CR_7-2, and so on.

The control ladder diagram in Figure 3–8 has an operating sequence as follows:

Straight-Through Sequence:
   All switches are open to start; both coils are OFF.
   Close SW_1, SW_2, or both; CR_7 is energized.
   On line 3, CR_7-1 closes, enabling line 3 (CR_8 is still OFF).
   Closing SW_3 energizes CR_8 and pilot light PL_1.
   Opening both SW_1 and SW_2 turns everything OFF.

Alternative Possible Sequence:
   Initially turning on SW_3 causes nothing to energize.
   Opening SW_3 when everything is ON would turn OFF CR_8 and PL_1 only.
(Other sequence possibilities exist.)

Figure 3–9 is an incorrect ladder diagram that contains the same components as are used in Figure 3–8. Will this circuit work? No. First of all, if power could get to point 13, the outputs would not work. Each would have 1/3 control voltage across it. Relays would not pull in, and the light would glow dimly or not at all.

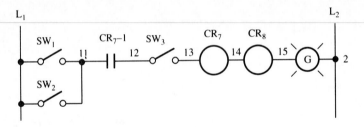

FIGURE 3–9
Incorrect Control Ladder Diagram for Figure 3–8

But the outputs will never go ON anyway. If all switches are closed, no power gets through contact CR_7-1. It cannot close until CR_7 is energized, which is impossible.

The operation of the power ladder diagram in Figure 3–10 is straightforward. When the power contactor coil is energized, the power contacts close, and power is applied to the motor or the load device. Note that the power ladder diagram wiring is shown by thicker lines, to differentiate its wiring from control circuit lines.

Additional sequence requirements may call for the construction of additional control ladder lines. The following functional modifications can be added to the ladder diagram of Figure 3–8:

SW_4 must be ON for CR_7 to go ON.
CR_7 must be OFF for CR_8 to go ON.
CR_9 is turned ON by CR_7, CR_8, and SW_3.

The extended ladder diagram is shown in Figure 3–11. Note that there is a dotted line between the two SW_3 contacts. The dotted line indicates a common single switch with two contacts. (If SW_3 were on the left, only one contact would be needed to run lines 3, 4, and 5.)

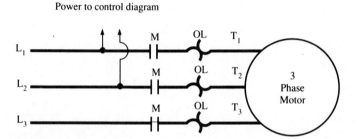

FIGURE 3–10
Power Ladder Diagram

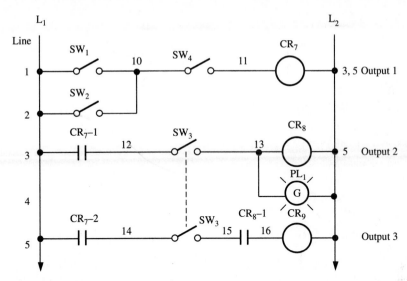

FIGURE 3–11
Extended Control Ladder Diagram for Figure 3–8

An added sequence of operation can be determined from an added ladder line. Such an added ladder line is shown in Figure 3–12. The added sequence based on this additional line would be as follows: CR_7 or CR_8 or both, plus LS_{12} and CR_9, turn on relay output CR_{10}.

3–6 WIRING, INTERCONNECTING, AND HARNESS DIAGRAMS

Wiring diagrams show how equipment is actually wired. Wiring diagrams correspond to ladder diagrams but are in a different form. Figure 3–13 shows three

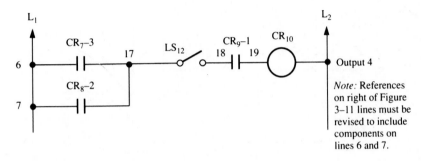

Note: References on right of Figure 3–11 lines must be revised to include components on lines 6 and 7.

FIGURE 3–12
Added Line for Control Ladder Diagram of Figure 3–11

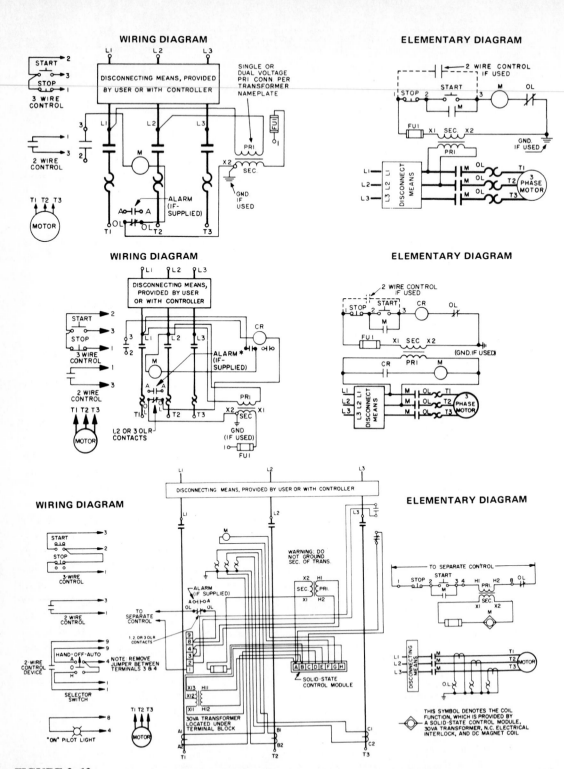

FIGURE 3–13

Typical Ladder (Elementary) and Wiring Diagrams (Courtesy of Square D Company)

ladder diagrams and their corresponding wiring diagrams. These examples happen to be for motor starting and control, but they are representative of all types of wiring diagrams.

The three examples in Figure 3–13 illustrate some of the important characteristics of wiring diagrams:

- The power portion is usually in heavier lines.
- The corresponding node identification numbers from the ladder diagrams are used in the wiring diagrams for ease of construction and comprehension.
- The wiring diagrams follow the physical layout of the electrical equipment. Although they are not to exact dimensions, a technician can follow them easily when looking at the actual devices and panels.
- In some wiring diagrams not all wires are actually shown connected. To show all the connections could result in a cluttered drawing. Instead, arrows with identification numbers are often used. The arrow system is used in the wiring diagrams of Figure 3–13.
- Electrically, a wiring diagram exactly matches the corresponding ladder diagram. There are no extra wires on the wiring diagram that are not on the ladder diagram. Conversely, all ladder wiring is carried out in the wiring diagram. (Some unused spare wires—for future use—may be shown in the wiring diagram.)

How is a wiring diagram made up from a ladder diagram? Figure 3–14 illustrates the three steps in the construction of a wiring diagram. The wiring diagram constructed here is for the ladder diagram of Figure 3–8. First, the panels are outlined by a solid or dotted line as shown in Figure 3–14A. There can be any number of different panels or locations. The panel boundaries must allow enough space to fit in all the components the panel contains.

Second, the components are drawn in the panels in symbolic form, as shown in Figure 3–14B. Next, the appropriate node wire numbers are placed on the symbol terminals. The node numbers are based on the ladder diagram numbers (or letters, if used). This step, number identification, is sometimes omitted. Confusion can occur during construction when terminals are not properly identified, especially in larger circuits.

Finally, as in Figure 3–14C, the wires are drawn to connect the terminals that have the same node-identifying numbers. The diagram shown is for a rather simple circuit. Note that it is cluttered and somewhat hard to follow. A clearer system for indicating connections will be shown in Figure 3–15.

Figures 3–11 and 3–12 in the previous section were more complex ladder diagrams than Figure 3–8. Figure 3–15 shows the wiring diagram for the ladder diagrams of Figures 3–11 and 3–12. The diagram illustrates the following important characteristics of a good, definitive wiring diagram:

- The diagram was started by drawing the four rectangles. There could be many more locations with individual components, requiring additional rectangular enclosures.

A. Panels

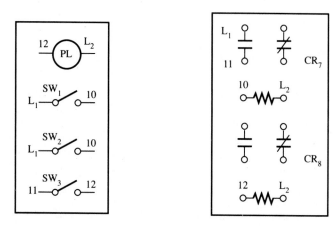

B. Components and Wire (Node) Numbers

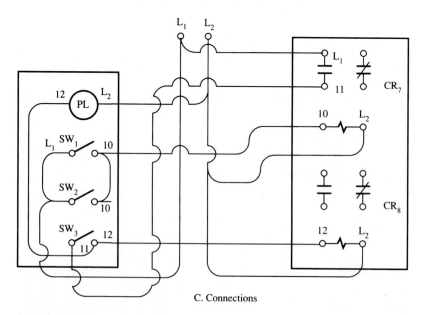

C. Connections

FIGURE 3–14
Wiring Diagram for Figure 3–8

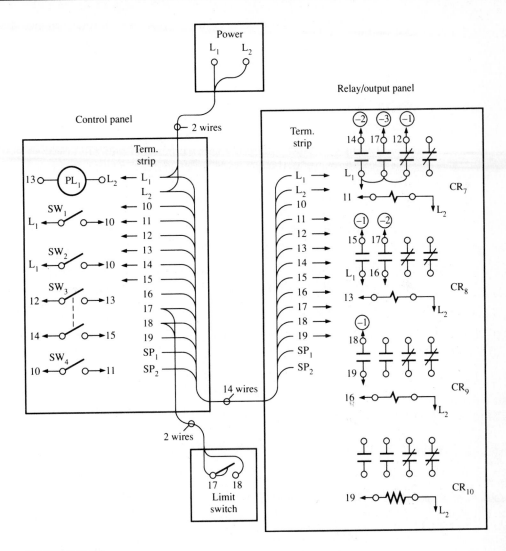

FIGURE 3–15
Wiring Diagram for Figures 3–11 and 3–12

□ Components were then added. The components correspond to actual devices used. For example, the relays shown have two NO and two NC
contacts. The diagram must match the actual relay configuration. If the
relays being used had different contact arrangements, different configurations would be shown in the diagram.

□ Terminal strips were added in the two largest panels. The interconnecting
wires are connected to them. The corresponding opposite-side terminals
of each terminal strip connect to the wires from the various components.

Construction with direct connected wires, without terminal strips, would be extremely messy. Disassembly would be even messier.

☐ The number of wires in each interconnecting group is specified.

☐ Spare wires (SP_1 and SP_2) are included. These are useful for future expansion. If connections between new units are needed someday, new wiring will not be required. Engineering changes that require more wiring occur frequently. Also, if one interconnecting wire were damaged, a spare could easily be used for replacement without adding more wires.

☐ The corresponding relay suffix number is indicated on the relay contacts. For example, the -2 in the circle above the first CR_7 NO contact indicates that the contact is contact CR_7-2. This is the identification number corresponding to the ladder diagram. For any troubleshooting, the technician would know exactly which contact he or she was tracing.

☐ Although color-coded wires are not used in Figure 3–15, they are often used for ease of wire identification.

Figure 3–16 illustrates an interconnection diagram. The connections are for the ladder diagram of Figure 3–15. An interconnection diagram includes only the

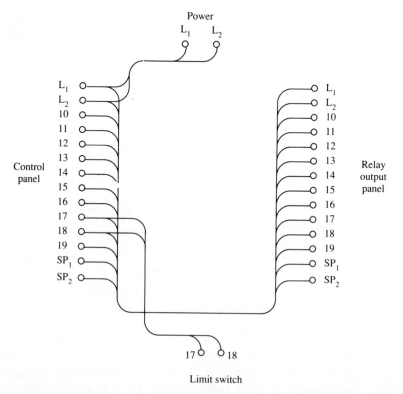

FIGURE 3–16
Interconnection Diagram for Figure 3–15

wiring that interconnects modules of a system—nothing more. For a system the size of that used for illustration, an interconnection diagram is not unnecessary. The interconnection diagram is used more as the number of modules increases. If, for example, 23 modules had to be interconnected, specific interconnection diagrams would be necessary for construction.

A harness assembly diagram is a drawing of the group of wires that interconnect two or more locations. It would be similar to Figure 3–16 without the terminal dots. If the individual harness wires were color-coded, the colors would be designated on the harness drawing and would be used for harness fabrication and installation.

3–7   PRINTED-CIRCUIT DIAGRAMS

Printed-circuit diagrams are based on corresponding electronic schematic diagrams. Design and layout of the printed circuit board is carried out from a schematic diagram. The design procedure is similar to the making of a wiring diagram based on a ladder diagram. The physical layout of the PC board is determined from component size, the number of connections, and the optimum electrical path configuration. The PC board for a large circuit is often designed by a computer program. Figure 3–17 is a diagram of a PC board for a rather involved circuit. The circuit paths, which are on the bottom side of the circuit board, are shown in gray. Component locations, on the top of the board, are shown by dark lines and part identification numbers.

Many PC boards are single-sided boards and have conducting paths on one side only, like the one shown. Some PC boards have conducting paths on both sides. They are called double-sided PC boards. The diagrams of double-sided boards are more complicated and are not shown. Still other PC boards are multilayered. Multilayered boards are used in large computers, among other applications. Multilayered boards can be from two to eight layers thick. The diagrams for such boards, not shown, are even more complicated than those for double-sided boards.

In Figures 3–18 and 3–19 the conversion from a schematic to a printed-circuit-board diagram is shown for a relatively small electronic circuit. Figure 3–18 is a schematic for one type of rectifier circuit and filter. The resulting layout of the corresponding PC board is shown in Figure 3–19. The diagram does not consider component dimensional accuracy, but is adequate for the purpose of illustration. The PC board layout shown is only one of many possible layouts.

Note that there is a problem in the PC layout with one connection, where the jumper is used. One more path could be included, looped around other paths. In other, more complicated circuits, there may be an impossible crossover. An impossible crossover occurs when there is no way to get from one point to another without crossing at least one other path. One solution to the crossover problem, when it occurs, is to include one wire jumper on the board as shown in Figure 3–19. Another solution would be to use a double-sided board, with the jumper connection on the opposite side. For only one crossover it is probably better to

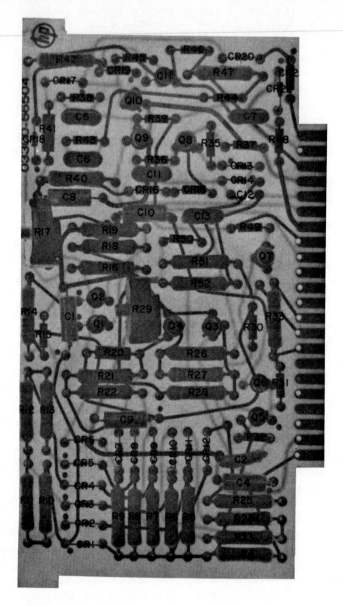

FIGURE 3–17
Printed-Circuit-Board Diagram

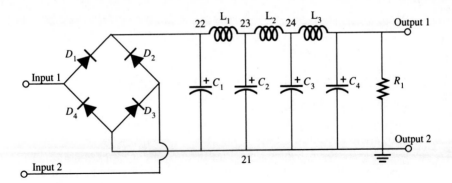

FIGURE 3–18
Rectifier-and-Filter Schematic

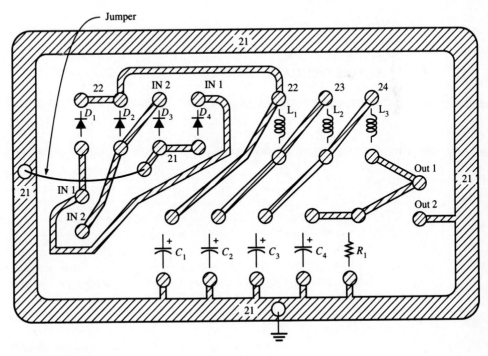

FIGURE 3–19
Printed-Circuit Diagram for Figure 3–18 Schematic

use a jumper and not go to the expense of a two-sided board. However, if there were numerous crossovers, a double-sided board would be practical.

3–8   BLOCK AND PICTORIAL DIAGRAMS

Block diagrams are used for ease of understanding of equipment and system operation. Pictorial diagrams are used for ease of understanding of component and equipment layout.

Figure 3–20 shows one form of equipment block diagram. Each functional group of components is represented by a block. Inside the block there is a description of the function represented. The blocks are connected by lines with directional arrows. The lines and arrows indicate the flow of electrical signals or information. Each electrical or electronic block diagram is based on a corresponding schematic or ladder diagram. The schematic or ladder diagram is broken into groups according to function. For example, in Figure 3–20 the power-supply block represents one function. It represents only the power-supply section of the schematic.

Block diagrams give an overall functional picture of equipment and devices. They are a big help in understanding and explaining how a piece of equipment functions. Block diagrams are also helpful in troubleshooting by helping to isolate a malfunction to a specific block. Once the malfunctioning block is determined, the technician can work with the part of the schematic corresponding to that block's portion of the equipment. Note that the use of block diagrams is not restricted to electrical or electronic systems. Block diagrams can be used for hydraulic, pneumatic, mechanical, electromechanical, and other equipment and systems.

Figure 3–21 is a block diagram for the speedometer and cruise (speed) control of an automobile. To construct a block diagram, first draw the blocks as shown in Figure 3–21. Next, fill in the blocks with a description of their function. Finally, draw lines with arrows between blocks. The lines show the flow of signals and information. More than one line may leave or enter a block. The true block diagram will vary from automobile to automobile. This diagram is only one example of how a block diagram can be constructed.

Pictorial diagrams are similar to wiring diagrams. Instead of using the wiring diagram's electrical or electronic symbol system, the pictorial diagram uses symbols that look like the actual devices. Often the pictorial diagram uses photographs of the components. An example of such a diagram is Figure 3–22.

Often, pictorial symbol drawings are used instead of photographs. Some typical symbols are shown in Figure 3–23. Many other symbols are used. They may be found in other textbooks on the subject. Note that the pictorial symbol for a given component can vary depending on the drawing system used.

Figure 3–24 is a pictorial diagram made from the wiring diagram of Figures 3–8 and 3–14. Changes made include the addition of fuses in the power-supply lines. Also, two terminal blocks of eight terminals each (standard) have been

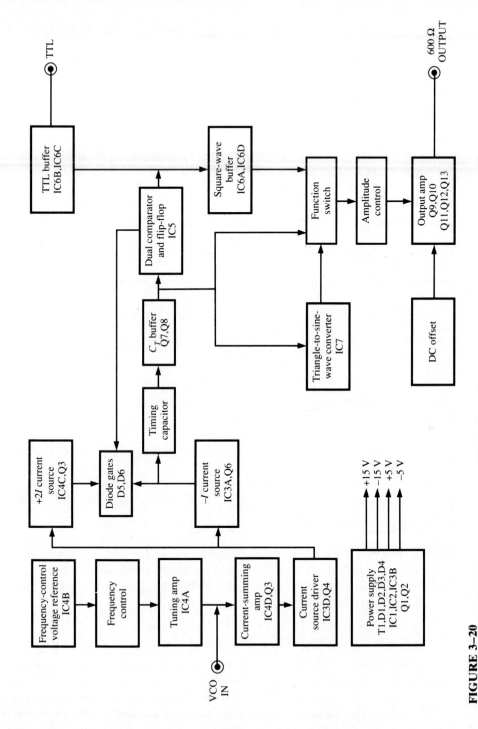

FIGURE 3–20
Typical Equipment Block Diagram

69

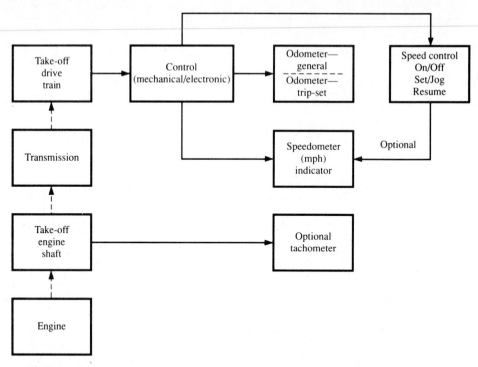

FIGURE 3–21
Automobile Speedometer and Cruise Control Block Diagram

included, even though only five terminals are used. More detail of the actual wiring of the unit is included by showing individual wires in the figure.

3–9 TROUBLESHOOTING DIAGRAMS

When a system or piece of equipment does not work properly, it is helpful to have written material for aid in prompt and accurate analysis. The written troubleshooting material is often in paragraph or tabulated form, as in this example:

    A. If the gas valve does not open, check that the pilot valve is on.
    B. If the pilot valve is on, check the thermostat.
    C. If the thermostat is OK, check the electronic spark control.

Such material is helpful. However, it does not give an overall picture. Furthermore, it gives no opportunity for "branching."

A good way to give troubleshooting information is by a flowchart. Figure 3–25 is such a troubleshooting guide chart. The diagram shown is for analyzing the cause of a relay's failure to operate. Start at the upper left and follow the arrows.

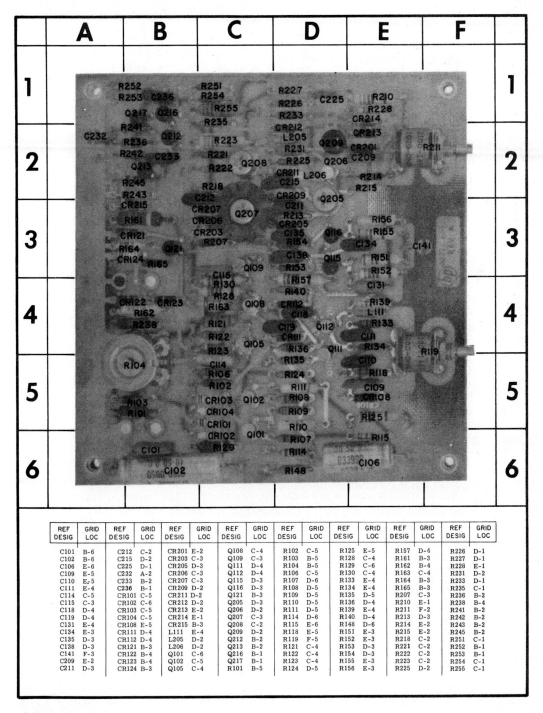

REF DESIG	GRID LOC	REF DESIG	GRID LOC	REF DESIG	GRID LOC	REF DESIG	GRID LOC	REF DESIG	GRID LOC	REF DESIG	GRID LOC	REF DESIG	GRID LOC	REF DESIG	GRID LOC
C101	B-6	C212	C-2	CR201	E-2	Q108	C-4	R102	C-5	R125	E-5	R157	D-4	R226	D-1
C102	B-6	C215	D-2	CR203	C-3	Q109	C-3	R103	B-5	R128	C-4	R161	B-3	R227	D-1
C106	E-6	C225	D-1	CR205	D-3	Q111	D-4	R104	B-5	R129	C-6	R162	B-4	R228	E-1
C109	E-5	C232	A-2	CR206	C-3	Q112	D-4	R106	C-5	R130	C-4	R163	C-4	R231	D-2
C110	E-5	C233	B-2	CR207	C-3	Q115	D-3	R107	D-6	R133	E-4	R164	B-3	R233	D-1
C111	E-4	C236	B-1	CR209	D-2	Q116	D-3	R108	D-5	R134	E-4	R165	B-3	R235	C-1
C114	C-5	CR101	C-5	CR211	D-2	Q121	B-3	R109	D-5	R135	D-5	R207	C-3	R236	B-2
C115	C-3	CR102	C-6	CR212	D-2	Q205	D-2	R110	D-5	R136	D-4	R210	E-1	R238	B-4
C118	D-4	CR103	C-5	CR213	E-2	Q206	D-2	R111	D-5	R139	E-4	R211	F-2	R241	B-2
C119	D-4	CR104	C-5	CR214	E-1	Q208	C-2	R114	D-6	R140	C-4	R213	D-3	R242	B-2
C131	E-4	CR108	E-5	CR215	B-3	Q209	D-2	R115	E-6	R148	D-6	R214	E-2	R243	B-2
C134	E-3	CR111	D-4	L111	E-4	Q212	B-2	R118	E-5	R151	E-3	R215	E-2	R245	B-2
C135	D-3	CR112	D-4	L205	D-2	Q213	B-2	R119	F-5	R152	E-3	R218	C-2	R251	C-1
C138	D-3	CR121	B-3	L206	D-2	Q216	B-1	R121	C-4	R153	D-3	R221	C-2	R252	B-1
C141	F-3	CR122	B-4	Q101	C-6	Q217	B-1	R122	C-4	R154	D-3	R222	C-2	R253	B-1
C209	E-2	CR123	B-4	Q102	C-5	R101	B-5	R123	C-4	R155	E-3	R223	C-2	R254	C-1
C211	D-3	CR124	B-3	Q105	C-4			R124	D-5	R156	E-3	R225	D-2	R255	C-1

FIGURE 3–22
Typical Pictorial Diagram

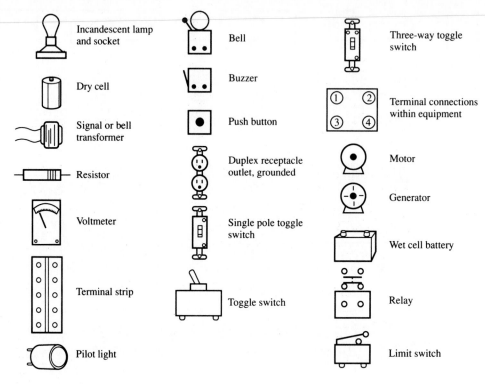

FIGURE 3–23
Typical Pictorial Symbols

After each action, follow the branch corresponding to "OK" or "not OK," depending on the outcome of the action. Troubleshooting charts are arranged so that the most probable causes are near the beginning of the chart. This arrangement prevents working on an unlikely cause first.

Troubleshooting charts are often furnished by the machine manufacturer, but not always. If their inclusion in the machine drawings was not made part of the original machine specifications by the purchaser, they could be omitted.

You can make up a chart of your own, if necessary. Suppose you are to make up a troubleshooting diagram for a blower motor. Such a diagram is shown in Figure 3–26. The blower motor is on an electronic assembly but is powered separately. Place the most likely causes of malfunction at the start, at the upper left. The less likely and more time-consuming procedures are included later in the chart. Note that this diagram follows the OK/not OK format. Following the diagram will speed up troubleshooting, especially for a technician unfamiliar with the product.

Failure data should be gathered for every piece of equipment that has a troubleshooting chart. The data should include the type of failure, the repair

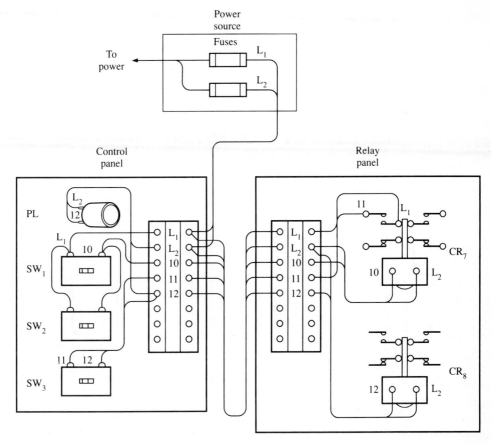

FIGURE 3–24
Pictorial Diagram for Figures 3–8 and 3–14

made, the date, the shift, and other information. From these data the chart can be rearranged if necessary. The failures that are found to occur most often can be moved to the upper left in the diagram if they are not already there.

3–10   STANDARDS APPLICABLE TO CONTROL DIAGRAMS

There are *standards,* sometimes called *codes*, for electrical and electronic diagram construction. Those standards are written instructions for format, symbols, size, and other factors. If diagrams are for company internal use only, the company may set up its own standards. Then everyone in the operation will be using the same symbols and will understand the drawings of everyone else in the company.

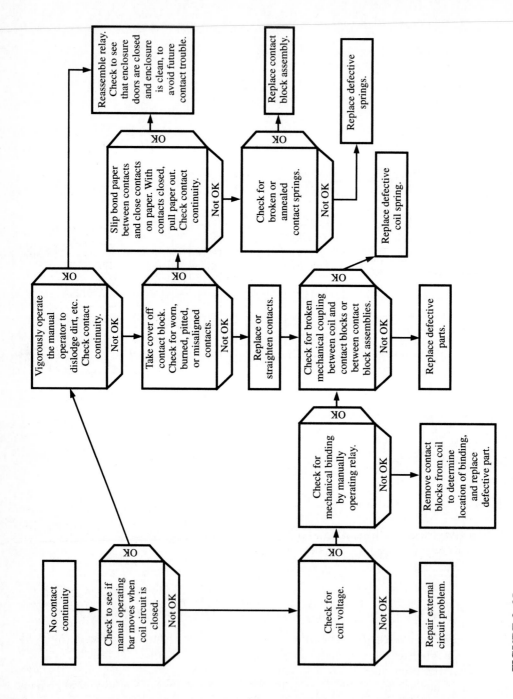

FIGURE 3–25
Relay Troubleshooting Chart

The flowchart contains the following decision and action blocks:

- No contact continuity
- Check to see if manual operating bar moves when coil circuit is closed. — OK / Not OK
- Check for coil voltage. — OK / Not OK
- Repair external circuit problem.
- Check for mechanical binding by manually operating relay. — OK / Not OK
- Remove contact blocks from coil to determine location of binding, and replace defective part.
- Vigorously operate the manual operator to dislodge dirt, etc. Check contact continuity. — OK / Not OK
- Take cover off contact block. Check for worn, burned, pitted, or misaligned contacts. — OK / Not OK
- Replace or straighten contacts.
- Check for broken mechanical coupling between coil and contact blocks or between contact block assemblies. — OK / Not OK
- Replace defective parts.
- Replace defective coil spring.
- Slip bond paper between contacts and close contacts on paper. With contacts closed, pull paper out. Check contact continuity. — OK / Not OK
- Check for broken or annealed contact springs. — OK / Not OK
- Replace contact block assembly.
- Replace defective springs.
- Reassemble relay. Check to see that enclosure doors are closed and enclosure is clean, to avoid future contact trouble.

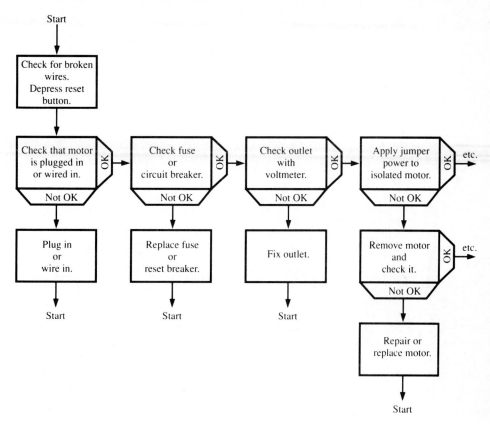

FIGURE 3–26
Blower Motor Troubleshooting Chart

If, however, the company uses diagrams to communicate with customers and parts suppliers, the diagrams may need to conform to industry and government standards. Then the company, the customer, and governmental, regulatory, and industrial organizations will all have the same base of reference. Such standardization is also occurring internationally.

Two examples of listings of standards are shown in Figure 3–27. One set of standards is a small partial listing of military standards applicable to U.S. government work. The other listing is of some of the standards issued by the American National Standards Institute (ANSI). Not all of the standards listed apply to electrical diagrams. Standards can apply to mechanical and design parameters as well as electrical ones.

Who issues standards? The federal government issues the military standards. Those standards and also listings by subject are available from various governmental sources, including the National Bureau of Standards (NBS) in

Military Standards

MIL-STD-12	Abbreviations for use on drawings and in technical-type publications
MIL-STD-15-2	Electrical wiring equipment symbols for ships' plans—Part 2
MIL-STD-17	Mechanical symbols
MIL-STD-27	Designations for electric switchgear and control devices
MIL-STD-32	Status notes for use on military sheet form standards
MIL-STD-100	Engineering drawing practices
MIL-STD-106	Mathematical symbols
MIL-STD-108	Definitions of and basic requirements for enclosure for electric and electronic equipment
MIL-STD-167	Mechanical vibrations of shipboard equipment
MIL-STD-196	Joint electronics type designation system
MIL-STD-242	Electronic equipment parts, selected standards
MIL-STD-280	Definition of item levels, item exchangeability, models, and related terms
MIL-STD-409	Alloy nomenclature and temper designation system for magnesium base alloys
MIL-STD-412	Alloy designation system for titanium
MIL-STD-429	Printed-wiring and printed circuits terms and definitions
MIL-STD-454	Standard general requirements for electronic equipment
MIL-STD-455	Alloy designation system for wrought copper and copper alloys
MIL-STD-681	Identification coding and application of hook-up and lead wire
MIL-STD-710	Synchros, 60 and 400 hertz
MIL-STD-736	Unitized equipment design
MIL-STD-1132	Switch and associated hardware, selection and use of
MIL-STD-1313	Microelectronic terms and definitions
MIL-STD-1378	Requirements for employing standard hardware program modules

ANSI Standards

Surface roughness, waviness, and lay	B46.1
Alloy and temper designation system for wrought aluminum	H35.1
Abbreviations for use on drawings and in text	Y 1.1
ANSI drafting manual	
Drawing sheet size and format	Y14.1
Line conventions, sectioning, and lettering	Y14.2
Multi and sectional view drawings	Y14.3
Pictorial drawing	Y14.4
Dimensioning and tolerancing for engineering drawings	Y14.5
Screw threads	Y14.6
Gears, splines, and serrations	Y14.7
Gear drawing standards	Y14.7.1
Forging	Y14.9
Metal stampings	Y14.10
Plastics	Y14.11
Mechanical assemblies	Y14.14
Electrical and electronics diagrams	Y14.15
Graphic symbols for electrical and electronics diagrams	Y32.2
Graphic symbols for logic diagrams (Two-state devices)	Y32.14
Reference designations for electrical and electronics parts and equipments	Y32.16

FIGURE 3–27
Standards Applicable to Diagrams

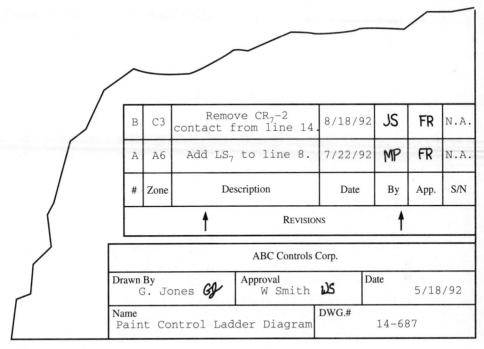

#	Zone	Description	Date	By	App.	S/N
B	C3	Remove CR_7-2 contact from line 14.	8/18/92	JS	FR	N.A.
A	A6	Add LS_7 to line 8.	7/22/92	MP	FR	N.A.

REVISIONS

ABC Controls Corp.		

Drawn By
  G. Jones *GJ*
	Approval
	  W Smith *WS*
		Date
		    5/18/92

Name Paint Control Ladder Diagram	DWG.# 14-687	

FIGURE 3–28
Typical Diagram-Revision System

Washington. Another listing of standards for commercial work is by ANSI, a clearinghouse for a number of trade and technical societies. It determines overall standards for different organizations. Some other sources of standards are the Electronic Industries Association (EIA), the American Standards Institute (ASI), the American Institute of Electrical and Electronic Engineers (AIEEE), and the National Machine Tool Builders Association. The Machine Tool Society has a Joint Industries Council (JIC), which issues standards. Many other organizations, such as the American Instrument Society and architectural societies, are also involved in setting up standards.

There are standards and codes applicable to areas other than drawings. Some such standards, such as the National Electric Code (NEC), are concerned with safe and proper installation of electrical equipment. Other standards concern safety, maintainability, and other areas. The NEC code and some other commonly used codes and standards will be discussed later in the book.

3–11   DIAGRAM REVISION SYSTEMS

Does the electrical drawing for an installation actually match the wiring of the equipment or device? Usually yes, but sometimes no. If they do not match, there are two possible reasons. First, you may not have a copy of the latest drawing. Check your copy against a new copy from the master file. Second, the drawing may be the latest in the master file, but the equipment could have been changed without the drawing having been brought up to date. The latter is often the case, especially in a multishift operation.

The lack of drawing change control can cause havoc in a manufacturing plant. Also, the lack of control can cause some very expensive field problems, including large lawsuits.

With incorrect drawing information, troubleshooting can be difficult. For example, suppose you are working to fix a punch press that will not punch. You observe two switches on the control panel that are not on the drawing. Also, you cannot find relay CR_5 in the equipment, even though it is shown on the drawings. You need an accurate, up-to-date drawing to effectively proceed with troubleshooting.

A major procedural step to prevent the mismatch of drawing and equipment is an effective drawing-revision system. Figure 3–28 shows the title block and revision system of a typical drawing. Note that the revision system can apply to drawings of all types, as well as to such documents as parts lists. The entire drawing is not shown in the figure.

The drawing was first issued on 5/18/92, as designated. Subsequently, two drawing revisions were made, on 7/22/92 and on 8/18/92. The figure shows how the revisions were recorded and designated. Note that units manufactured before the drawing changes were made still have the old configuration. Serial numbers or manufacturing dates will be needed to identify the older units.

The revision blocks in the figure include the following information:

☐ the zone of the diagram in which the revision is located (if the drawing is large and is divided into zones)
☐ a description of the change (as complete as possible)
☐ the effective change date
☐ the person making the change
☐ approval authorizing the change
☐ the serial number of the first unit in which the change was incorporated (optional)

There are some basic considerations for an effective drawing-revision system. The equipment should not be changed unless the drawing is changed. Additionally, a control must be in effect so that not just anyone can revise a drawing. In computer and CAD systems, code access keys must be used to prevent unauthorized computer access. Revision approval must be made by an appropriate supervisor.

After the drawing has been revised, copies of old drawings must be removed from the system. New, revised drawings must then be issued to the persons and departments that need them. As previously stated, the serial number or the date of change of the products produced must be remembered. After a change, there can effectively be two models of a product in the field.

SUMMARY

- □ All electronic and electrical components have symbols, which may be used as shorthand on control diagrams.
- □ It is important to use the correct type of diagram for every situation requiring reference to a diagram.
- □ Schematic diagrams are the basic diagrams of electronic circuits.
- □ Ladder diagrams are the basic diagrams of electrical control circuits.
- □ Wiring diagrams are used for construction and installation purposes. They are made up from a corresponding schematic or ladder diagram.
- □ Printed-circuit diagrams are based on electronic schematic diagrams. They are used for PC board manufacture and for component placement on the boards.
- □ Block diagrams are constructed by breaking down a device or system into labeled functional blocks. The blocks are then connected with lines and arrows, which describe the flow of information.
- □ Pictorial diagrams are similar to wiring diagrams. They are more accurate dimensionally. Instead of symbols, actual drawings or photographs of the individual components are shown.
- □ Troubleshooting diagrams are a help for organized, rapid analysis of malfunctions. They are constructed in block form, with lines and arrows showing the order of action for analysis.
- □ Many situations require that drawings and diagrams be constructed according to special specifications and codes.
- □ An organized revision system for diagrams and drawings is usually required. The system ensures that diagrams always match actual parts, systems, and construction.

EXERCISES

1. List 20 electrical or electronic devices. Draw their symbols.
2. List the major types of diagrams. How is each type best used?
3. Describe two processes with discrete inputs and outputs.
4. Describe two processes with analog inputs and outputs.
5. Describe a process with both discrete and analog inputs or outputs.
6. Refer to another textbook, and look at the symbols used. Which are the same as those listed in this book, and which are different?
7. Make a ladder diagram for the following sequence:
   When SW_1 is closed, CR_1 goes ON.
   After CR_1 goes ON, SW_2 can turn CR_2 ON.
   When CR_2 goes ON, PL_1 goes OFF.
8. Make up a line cross-reference system for Exercise 7.
9. Make up a sequence listing for the ladder diagram in Figure 3–29.
10. What is incorrect in the ladder diagram of Figure 3–30?
11. Convert the ladder diagram of Figure 3–31 to a wiring diagram.
12. Convert the wiring diagram of Figure 3–32 to a ladder diagram.

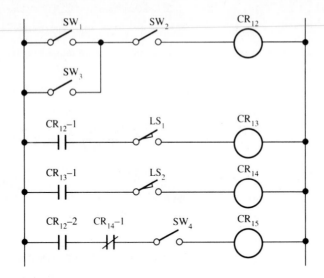

FIGURE 3–29
Ladder Diagram from Which to Make a Sequence

13. Convert the schematic in Figure 3–33 to a printed-circuit-board layout diagram.

14. From a reference book, obtain a simple printed-circuit-board layout diagram. Draw a schematic from it.

15. Construct a block diagram for a stereo system. The system has four speakers, two for each channel. It also has a CD player, a tape player, AM and FM stereo, a turntable, and a TV input.

16. Make an interconnection diagram for Exercise 15. Include the number of wires or the type of cable used.

17. Sketch a pictorial diagram of the circuit in Figures 3–8 and 3–14.

18. Make up a troubleshooting diagram for a fluorescent light fixture with two bulbs. A fluorescent light has a ballast and a starter associated with it.

19. Discover some other diagram standards by referring to other textbooks.

20. Your company makes a control assembly (model MH) for a robot. The drawing was first issued on 2/25/92. The following changes were subsequently made:

   An inductor, PN 45637, was added on 6/15/92.
   A resistor was changed from 50 Ω to 65 Ω (PN 6785) on 7/16/92.

How would the title and revision blocks appear on the schematic diagram?

21. How would you distribute drawings for Exercise 20? Assume both suppliers and one customer are involved.

22. Five of the products of Exercise 20 are returned. How can you determine which version of the drawing to refer to for product analysis?

23. Are there other types of diagrams in other textbooks? One type not covered in this chapter is the sequence diagram. List and describe any other types you find.

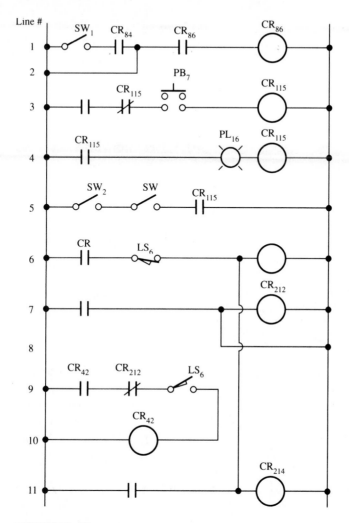

FIGURE 3–30
Questionable Ladder Diagram

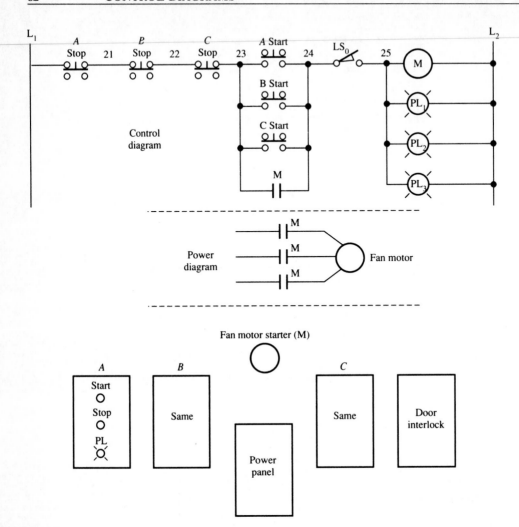

FIGURE 3–31
Ladder Diagram for Conversion to a Wiring Diagram

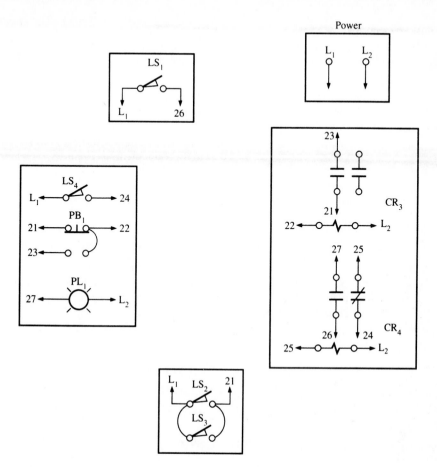

FIGURE 3–32
Wiring Diagram to Be Converted to a Control Ladder Diagram

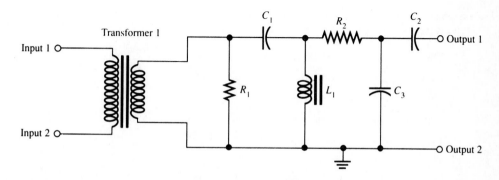

FIGURE 3–33
Schematic to Be Converted to a Printed-Circuit-Board Layout

APPLICATION ASSIGNMENT ANSWER

In addition to tools, meters, and other equipment, you would take the following diagrams and information:

1. an electrical ladder diagram and a sequence listing—for analysis of the operation
2. a wiring diagram—to make sure the unit is wired properly and to look for loose wires (The wiring diagram will show you the location of all relays, timers, etc.)
3. a parts list—in case you need to determine which part needs replacement (The list would be associated with one of the equipment diagrams.)
4. a block diagram—useful in discussions with operating personnel regarding malfunction states
5. troubleshooting diagrams, charts, and listings, as available—for aid in locating the malfunction
6. installation instructions (not discussed in the chapter)—to determine whether the customer installed the machine properly and whether the electrical service is proper.

One final word: Make sure you have the latest possible diagrams for the unit installed at the customer's plant. If the printing press wiring and controls do not match the diagrams you have, analysis is much more difficult.

4

Solid-State Devices

CHAPTER OUTLINE

4–1 Introduction □ **4–2** Diodes: Control, Power, and Zener □ **4–3** Transistors: Bipolar and FET □ **4–4** Integrated Circuits (ICs): Linear and Digital

LEARNING OBJECTIVES

□ Draw symbols for and describe the major types of semiconductor devices.
□ Describe how diodes are used in electronic circuitry.
□ Describe *pnp* and *npn* transistor operation and applications.
□ Describe IC chips, their application, and their basic construction.

APPLICATION ASSIGNMENT

An electronic device containing a large number of diodes is malfunctioning. You have been assigned to check the diodes for proper function. What are some of the procedures you would follow?

4–1   INTRODUCTION

In this chapter, which is a broad review of general semiconductor types, the fundamentals of diodes, transistors, and integrated circuits (ICs) will be covered. Listed in the bibliography are books that cover those devices in more detail. These books may be referred to if needed. There are also selected reference data sheets for solid-state devices in the appendix.

4–2   DIODES: CONTROL, POWER, AND ZENER

The *semiconductor diode* is a two-layer, two-terminal device. The basic function of the semiconductor diode is to pass current in one direction and not in the other. Semiconductor diodes have taken the place of tube diodes and selenium rectifiers, which are found in some older equipment. Tube diodes wear out, as do all tubes. Also, tube diodes are inefficient because of the power required to heat their anode terminals for operation. The older selenium rectifiers are also inefficient, as they do pass some current in the reverse direction. Also, selenium rectifiers have low current ratings.

Semiconductor diodes—or *junction diodes,* as they are sometimes called—are used extensively for voltage and current rectification. Rectification is the process of conversion of alternating current to direct current. Other applications of semiconductor diodes include circuit protection, waveform modification, voltage regulation, and use in logic circuits.

Figure 4–1 illustrates the appearance, connections, and construction of the semiconductor diode. The diode is constructed by joining two different types of material at a junction as shown. A terminal is connected to each of the materials. The *p*, or positive-end, connection is called the *anode*. The *n*, or negative, connection is called the *cathode*. The two layers of material making up the diode are called *p* and *n*. At the *p-n* junction, electrons can flow one way but not the other. Therefore, the diode is a rectifier. One of the diodes shown in the illustration has a band circling it. The band indicates the positive end. It is important to observe this polarity when replacing or installing a diode. Reversed polarity can damage the diode and possibly other components in a circuit.

There are two important diode characteristics to observe when choosing or replacing a semiconductor diode. The first is current rating. The higher the current-carrying capacity required, the larger and more costly the diode will be. The diode chosen must be able to carry the current required by the circuit. The second important characteristic is the peak inverse voltage (PIV). Exceeding the rated peak inverse (reverse) voltage (PIV) will cause the diode to break down. After breakdown the diode effectively becomes a short circuit. Once the diode breaks down, it can become permanently damaged after the offending voltage is removed. If it becomes a short circuit, other components can also be adversely affected by the resulting excessive current and voltage changes.

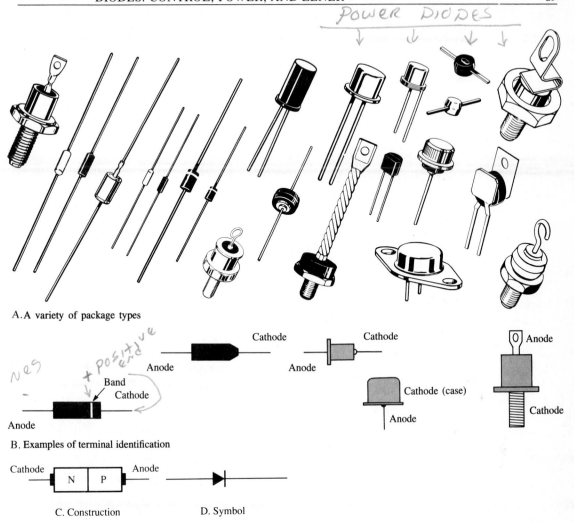

A. A variety of package types

B. Examples of terminal identification

C. Construction   D. Symbol

FIGURE 4–1
Semiconductor Diodes

Figure 4–2 shows a typical operating curve for a basic diode. In the forward direction, the current flows at a very low voltage. The diode does have a small resistance in the forward direction. The diode resistance can be ignored in most applications, as it is very small. Note that applying supply voltage directly across the diode in the forward direction will cause high current and probable diode burnout. An appropriate load resistor is needed in series with the diode.

In the reverse direction, the diode conducts very little when the voltage is below the breakdown value. In circuit operation, the reverse leakage of current below breakdown voltage is not exactly zero. Leakage does occur, but it is in

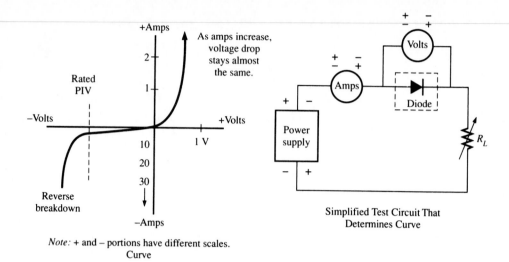

FIGURE 4–2
Characteristic Curve for a Typical Diode

microamps and can normally be neglected. As previously stated, the diode will normally be destroyed by high current when excess reverse voltage is applied to it.

Diodes are normally made of silicon or germanium as a base material. Their *p* and *n* sections have small amounts of an additive (impurity). The additive materials result in the unilateral operation of the diode. The two types of diodes, silicon and germanium, are similar in operation. Details of the materials, construction differences, and operational details can be found in textbooks listed in the bibliography.

Power diodes have a different physical appearance from that of circuit diodes. They are the larger diodes on the right side of Figure 4–1A. Additionally, many power diodes have heat sinks attached to them to dissipate heat and provide cooler operation. Greater heat dissipation results in a device with a higher current rating.

One special application of diodes is voltage regulation. The diodes used as voltage regulators are called zener diodes. A regular diode will not work effectively as a zener diode, and vice versa. The previously discussed regular diodes are operational in the plus direction. The zener diode, on the other hand, is used in the circuit to function in the negative direction. Typical zener diode curves are shown in Figure 4–3. The curves are similar to those of the regular diode.

A simplified typical zener regulating circuit is shown in Figure 4–4. Assume that the zener diode is rated at 18.0 V. The zener's function is to limit the voltage across the output resistor, R_L, to nearly 18 V—if the supply voltage is 18 V or more.

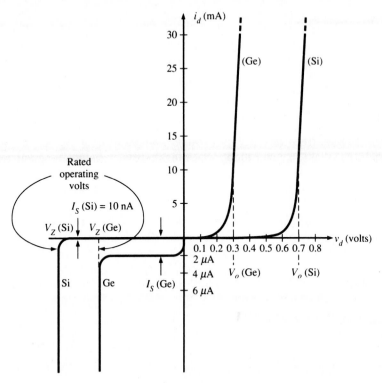

FIGURE 4–3
Typical Zener Diode Characteristic Curves

With a supply voltage below 18 V, the zener will not conduct. The output voltage will then be equal to the supply voltage minus the R_S voltage drop. If the supply voltage is turned up to exceed 18 V, the zener will start to conduct. The zener current will add to the load current going through R_S. Resistor R_S will develop enough voltage drop to keep the output voltage at 18 V. Ohm's law will be

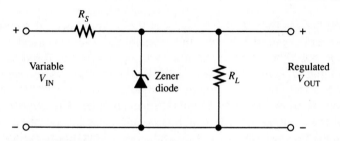

FIGURE 4–4
Simplified Zener Voltage Regulator Circuit

used to illustrate how the regulation is accomplished. Assume that R_L is 1500 Ω, R_S is 200 Ω, and the input voltage is 26.5 V. The output voltage will be kept to 18 V by the zener. Current in R_L is 18.0/1500, or 12 mA. Note that the R_L voltage is not 26.5 V. The voltage drop across R_S must therefore be 26.5 − 18.0, or 8.5 V. The current through R_S is thus 8.5/200, or 42.5 mA. Therefore, the current in the zener is 42.5 − 12, or 30.5 mA. The added zener current in R_S lowers the voltage to 18 V but no lower.

This zener illustration is greatly simplified. In actual operation, the voltage varies slightly from 18 V as input voltage and load current vary. The reason for the variation is that the zener diode has some internal resistance, which was neglected in this analysis. Also, a detailed analysis involves high and low limits on current and voltage. Other textbooks, listed in the bibliography, cover zener analysis in detail.

Zener diodes are rated for the voltage at which they operate. The voltage rating in the example just discussed was 18.0 V. Another key characteristic (often neglected) is the zener reverse-current rating. (There may also be a rating in watts.) In the example, the reverse-current rating would have to be more than 30.5 mA. A zener rated at 25 mA would not hold up, as the current rating would be exceeded. Therefore, the key ratings for a zener are for voltage and *reverse current*, not forward current.

4–3   TRANSISTORS: BIPOLAR AND FET

Transistor is short for *tran*sfer re*sistor*. There are textbooks devoted to transistor construction, theory, and application. Only the basics of transistors will be covered in this section.

One major type of transistor is the bipolar transistor. It is a three-layer device with three leads. The bipolar transistor's construction and appearance are shown in Figure 4–5. The three terminals of the device are called the collector, C, the emitter, E, and the base, B. There are two possible material configurations: *pnp* and *npn*. To differentiate between *pnp* and *npn,* the symbols differ as shown in the figure.

Transistors are used primarily as amplifying devices. A bipolar transistor amplifies a small electrical signal applied between two of its terminals. An amplified, or modified, output electrical signal is then found between two other terminals. Since there are only three terminals, one of the terminals must be common to input and output. The common terminal can be any of the three terminals. The three possible connection schemes for transistors are shown in Figure 4–6.

What are the key characteristics of bipolar transistors? A complete set of specifications is shown in the appendix. Normally a transistor is replaced with a new one of the same kind. Often such replacement is not possible. A substitute transistor must be used. There are numerous books and charts giving equivalency ratings. If no charts are available, a comparison analysis must be made. The major characteristics to match when comparing two transistors are as follows:

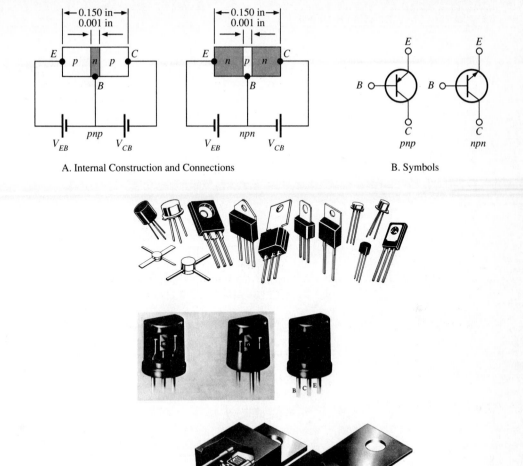

A. Internal Construction and Connections

B. Symbols

D. Terminal Connections

FIGURE 4–5
Bipolar Transistor Construction, Symbols, and Appearance

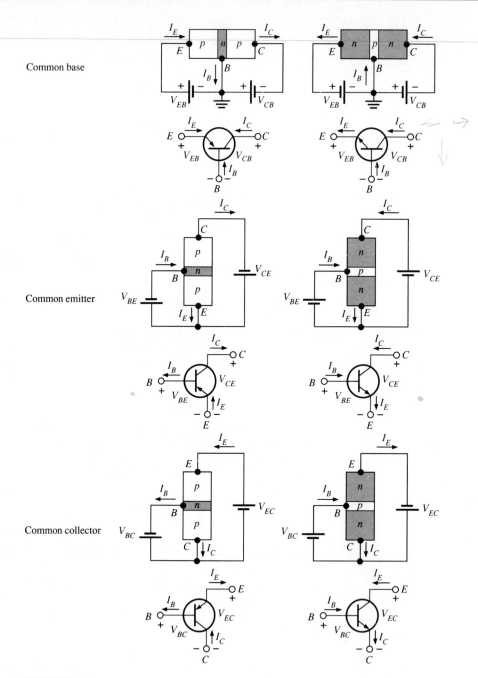

FIGURE 4–6
Bipolar Transistor Connection Schemes

type (*pnp* or *npn*)

terminal-to-terminal voltage rating

current ratings

wattage rating

current transfer rating, the effective gain

Other characteristics to consider include the following:

maximum storage temperature

operating temperature range

frequency limitations

breakdown voltages

One common use of transistors in control circuits is the transistor switch. Two types are shown in Figure 4–7, one for *pnp* and one for *npn*. As the voltage is raised to a certain point on the base, the transistor will begin conducting. This type of switch might be used where many low-current circuits must be turned ON and OFF. The transistor switch is small and inexpensive, and it never wears out, as it has no moving parts.

The second major type of transistor is the field-effect transistor, or FET. The FET is also a three-terminal device but is constructed differently from the bipolar type. The three terminals are called the source (S), the drain (D), and the gate (G). There are two subtypes of FET transistors. They are the junction FET (JFET) and the metal-oxide semiconductor FET (MOSFET).

The construction and symbols for the JFET are shown in Figure 4–8. Also shown in the figure is a typical circuit connection. The *p* and *n* material may be reversed, as in the bipolar transistor. Therefore there are two types of JFETs, called *n*-channel and *p*-channel. As an increasing voltage is applied to the JFET gate, the flow of electrons or current is increasingly limited. If enough voltage of the proper polarity is applied to the gate, the source–drain flow is stopped com-

FIGURE 4–7
The Transistor Switch

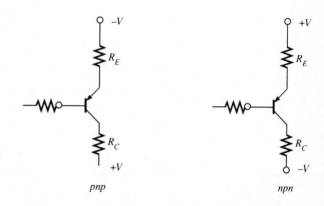

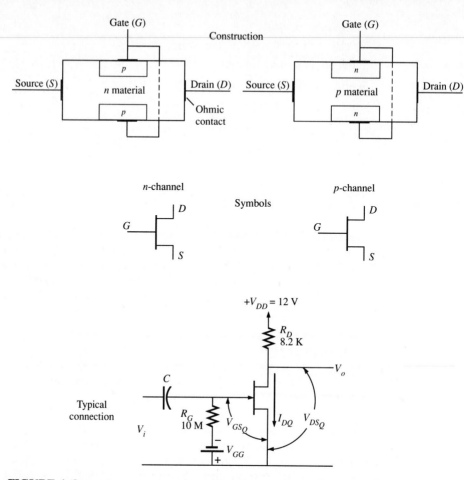

FIGURE 4–8
JFET Construction, Symbols, and Typical Connection

pletely. Some of the key ways in which the JFET differs from the bipolar transistor include:

   very high input impedance—typically 100 MΩ
   less operational variation with temperature variation
   less noise in communications applications
   smaller operating-frequency bandwidth
   greater fragility

   The construction, the symbols, and a typical circuit for the MOSFET are shown in Figure 4–9. Again, by reversing the p and n materials, two types are

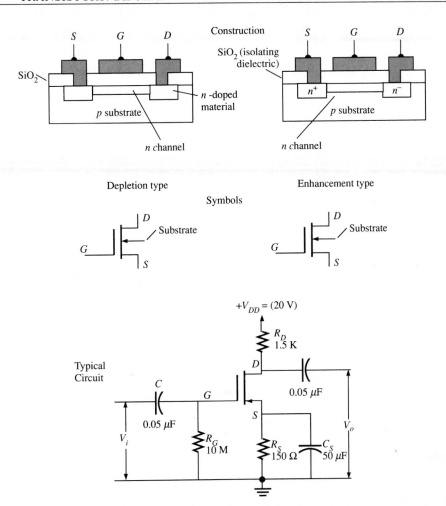

FIGURE 4–9
MOSFET Construction, Symbols, and Typical Circuit

created: p-channel and n-channel. The two types of MOSFET construction, both shown in the figure, are depletion and enhancement. The major differences of the MOSFET compared with the JFET are as follows:

much higher input resistance

much higher gain

much greater fragility

susceptibility to destruction by a small charge of static electricity

4–4   INTEGRATED CIRCUITS (ICs): LINEAR AND DIGITAL

The *integrated circuit* (*IC*) is a very small electronic device that includes miniaturized electronic components. The IC is used instead of a printed circuit board with many discrete components. The IC is made up to include diodes, transistors, capacitors, and resistors.

There are many forms of ICs. Some of the forms are shown in Figure 4–10. The most common configuration is a flat wafer with terminals along each side. Some wafer ICs are shown in the diagram. Wafer ICs are often called ''chips.'' Actually, the *chip* is the small internal semiconductor wafer to which the IC package terminals are connected.

Integrated-circuit chips are reliable and low in cost, and they come in small packages. They work at very low voltage levels. Therefore, to function in most applications, they must have input and output stages for voltage conversions.

There are many electronic categories of ICs. Two major types are *linear* (or analog) and *digital* (or discrete).

A typical linear chip is one used as an amplifier. A signal or signals are fed into the IC. The IC produces an appropriate output or outputs. Both the inputs and the outputs are continuous electrical signals between two limiting values. A typical linear IC is used in audio amplifiers. The input and output signals of a simple linear IC are shown in Figure 4–11.

Digital ICs fall into two major subcategories. One type contains a number of logic gates. Inputs and outputs are ON or OFF. The second type receives, responds to, and puts out digital pulses. The pulses are of different frequencies and pulse lengths. Figure 4–12 illustrates the operations of these two types of digital ICs.

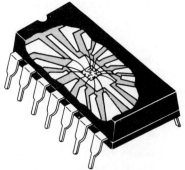

FIGURE 4–10
Typical Integrated Circuit (IC) Packages

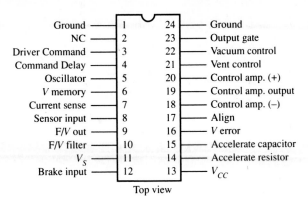

A. Terminal assignment

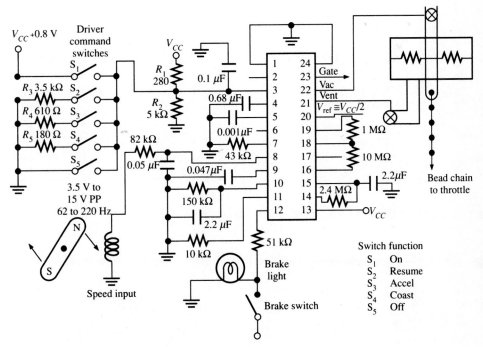

B. Typical Connection Diagram

FIGURE 4–11
Linear IC Operation

Integrated circuits are continually being developed that contain thousands of transistors and other components in an increasingly small space. The cost of developing an IC for manufacture can run to seven figures. Therefore, ICs are not practical for small production runs. If a manufacturer is going to make 500,000 clock radios, however, it is reasonable to make an IC chip to operate the radio.

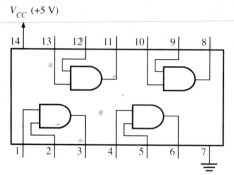

A. Digital Logic — Quad AND Gate

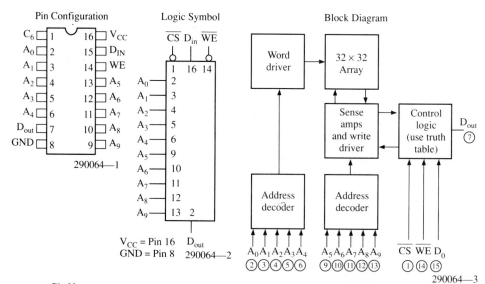

B. Digital Signals — Static RAM

FIGURE 4–12
Digital IC Operation

Better yet, the chip required for the application may already be in production. Many stock ICs can be purchased for less than a dollar. Applications of linear and digital ICs include the following:

operational amplifiers   function generators
counters   TV modules
timers   memories
oscillators   buffers and latches
coders and decoders   voltage regulators
complete radios   communication converters

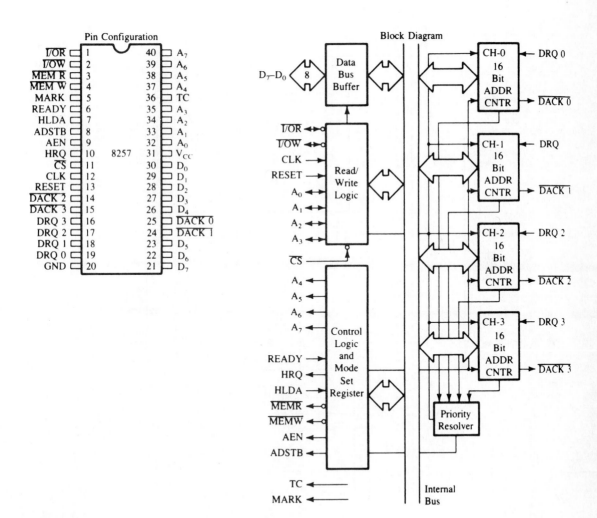

FIGURE 4–13
Microprocessor IC and Associated IC (Courtesy of Intel Corporation)

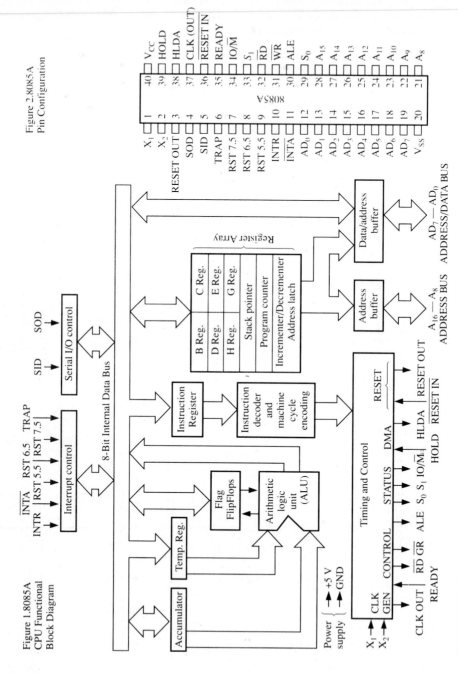

Figure 2. 8085A
Pin Configuration

Figure 1. 8085A
CPU Functional
Block Diagram

FIGURE 4–13
(continued)

The ultimate IC so far is the microprocessor IC chip. It contains all the components contained in the central processing unit of a larger computer. It has made the small, portable computer possible at a low cost. Figure 4–13 illustrates a microprocessor IC. The other ICs shown are associated with the microprocessor chip. They complete the computer circuitry. These other chips are for control, clock, input, output, and so on.

How is an IC manufactured? Other textbooks explain the process in detail. Some general principles of construction will be given here. The chip is constructed in layers. Layers of *p* material, *n* material, insulation, metal, and connections are put on in sequence. The first layer is deposited on the base and appropriately etched. Other layers are then put on and appropriately etched. Note that there are layer modification processes other than etching. Figure 4–14 shows how a chip might be made of 40 diodes in four layers. A transistor IC is made by

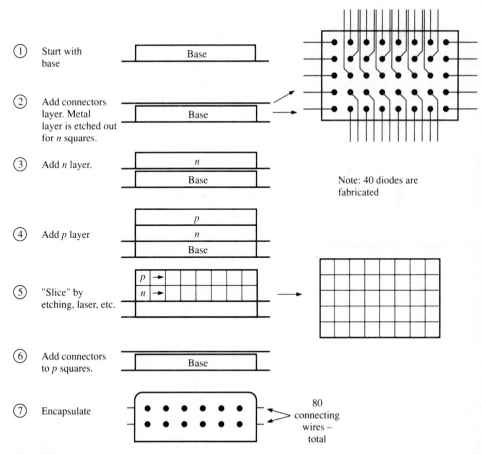

FIGURE 4–14
IC Construction for Diodes

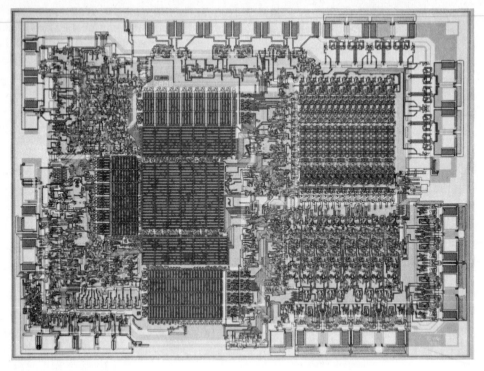

FIGURE 4–15
IC Chip Magnified 5000 Times (Courtesy of Intel Corporation)

creating three layers and appropriate connections. A capacitor is made with three layers, two of metal and one of dielectric insulation, plus connections. Resistors are made by using appropriate lengths of resistive material. Other semiconductors can be produced on a chip in a similar manner. There are no inductors in a typical chip, as there is no practical way to fabricate them. Actual ICs include combinations of diodes, transistors, resistors, and capacitors.

How does a typical IC look under a microscope? Figure 4–15 is a view of a typical IC.

SUMMARY

□ Semiconductor diodes are two-layer, two-terminal devices. They are used primarily to change varying AC voltages to unidirectional DC voltages. They are also used for many other purposes, such as wave shaping and circuit protection. Power diodes are physically larger than circuit control diodes.

□ Zener diodes are used for voltage regulation at a rated value. They use the reverse voltage of the diode characteristic for operation.

□ Bipolar transistors are three-layer, three-terminal devices. They are of two types, *pnp* and *npn*. They are used in many applications, such as amplification and switching.

☐ Field-effect transistors are used where high input impedance is necessary to reduce electrical loading of the input circuit. They are of two general types, the JFET and the MOSFET.

☐ Integrated circuits, or chips, are made up of micro layers of semiconductor material. The layers are etched, cut, and connected to form the equivalent of resistors, capacitors, and semiconductors. An IC chip can contain thousands of equivalents of components such as transistors.

EXERCISES

1. A semiconductor diode conducts electricity in only one direction. True or false? Explain your answer.

2. A semiconductor diode is rated at 200 V PIV. At 195 V reverse voltage it conducts 25 μA. In the forward direction, it has a forward resistance under load of 1 Ω. Sketch its characteristic curve.

3. What happens if you place two identical diodes in each of the following configurations?
   a. two in series with the same polarity
   b. two in series with opposite polarity
   c. two in parallel with the same polarity
   d. two in parallel with opposite polarity

4. Use Figure 4–4. The zener diode is rated at 14.5 V. Resistor R_L is 800 Ω; R_S is 50 Ω. Input voltage varies from 12 V to 26 V. Plot the voltage across R_L as the input voltage is varied through its range. What wattage should R_S have? What should be the current rating (reverse) of the zener?

5. List the major differences between bipolar and JFET transistors.

6. What are the major differences between the JFET and the MOSFET?

7. Refer to one of the textbooks on semiconductors in the bibliography. What are the characteristics, advantages, and disadvantages of the three types of transistor circuitry? Refer to Figure 4–6.

8. What is the advantage of an FET voltmeter?

9. Refer to Figure 4–14. How could you make up an IC chip for
   a. 20 capacitors?
   b. 15 transistors?
   c. 10 SCRs?

APPLICATION ASSIGNMENT ANSWER

First, you would visually inspect for burned or broken diodes that need to be replaced. Then you would check the rest of the diodes for resistance both ways. Resistance should be high one way and low the other. High readings both ways indicate an open diode. Low readings both ways indicate a shorted diode. Note that if you get a bad reading, you may have to remove one end of each diode from the circuit for rechecking. Other parallel connected circuit devices may be involved in the reading, making it lower. When you find a bad diode, replace it with one of the same or of a greater rating. Diode values are not normally inscribed on the small case. You will have to refer to the device parts list to determine the diode part number and characteristics.

5

Transducers and Sensors

CHAPTER OUTLINE

LEARNING OBJECTIVES

□ Classify the types of thermal transducers and sensors, and describe the characteristics of each type.

□ Classify the types of optical transducers and sensors, and describe the characteristics of each type.

□ Classify the types of magnetic transducers and sensors, and describe the characteristics of each type.

□ Classify the types of electromechanical transducers and sensors, and describe the characteristics of each type.

□ Given a particular application that requires monitoring of some physical variable, choose a device that would provide a satisfactory response.

APPLICATION ASSIGNMENT

In a large warehouse, automatic overhead door openers are being installed to allow forklift trucks and shuttle vehicles to pass through. Sensors are required to detect approaching vehicles so that the door openers can be actuated. It is essential that the vehicles be detected early enough not to have to slow down. The vehicle path is 70 ft wide. Although the sensors will be in a clean environment, they will be exposed to loud noises and various levels of ambient light.

As a field application engineer for a large photosensor manufacturer, you are asked to recommend an appropriate sensor for the application. Upon completion of this chapter, you should be able to make the appropriate recommendation. The solution is given at the end of the chapter.

5-1   INTRODUCTION

If an industrial process is to be truly controlled, it is essential that certain system variables be monitored. That is, conditions within the system must be constantly measured and converted to representative electrical signals. These signals will ultimately determine how the controlled process must be modified to accommodate the present conditions within the system. For example, to control the temperature of a piece of metal as it is being machined, the current temperature of the metal must be constantly measured. With this information, the speed of the machine tool can be adjusted to keep the temperature of the workpiece within an acceptable range. Therefore, integral parts of any industrial process control system are those elements or subsystems that are capable of sensing the system conditions. These elements can be classified into two groups: *input transducers* and *sensors*. There is such a fine distinction between an input transducer and a sensor that the two words are often used interchangeably. In many cases, treating the two groups as one is acceptable. Nevertheless, in this book they will be considered functionally different components.

An electronic process control system must acquire information—in the form of electrical signals—about process variables, make decisions based on that information, and make appropriate changes in the process by electronically adjusting the output actuators. Process variables such as temperature, light intensity, and position are not electrical in nature. Likewise, output adjustments such as heat, motion, and fluid flow bear no resemblance to electrical signals. Some method of conversion is required at both ends of the control system. The devices that make such conversions are called transducers.

Technically, a transducer is a device capable of transmitting energy from one system to another (usually converting the energy form along the way). Since this book is concerned with electronic control systems, the focus here is on electrical transducers. Furthermore, this book will concentrate on the input section of the control system and will consider only devices that change their electrical characteristics in accordance with a change in the physical condition to which they are sensitive. For example, the resistance of certain electrical transducers varies as their temperature changes. Other transducers produce potential differences when they are subjected to forces along certain axes.

Since only the electrical characteristics of a transducer change, support circuitry is required to detect or amplify this change. Therefore, an electrical input transducer alone is not a complete sensing subsystem.

Sensors, on the other hand, are integrated assemblies that usually include one or more transducers, along with all of the necessary signal-conditioning circuitry. The output of the sensor may be a physical contact, a semiconductor switch, an analog voltage or current, or even a binary number. Some sensors are sophisticated enough not only to acquire information about conditions within the system, but also to format and transmit those data to a remote computer.

Input transducers and sensors are classified according to the physical conditions to which they are sensitive. The most common (and most readily available) devices fall into the following categories: thermal, optical, magnetic, and electro-mechanical. This chapter will examine the principles of operation of the various devices in each category.

5–2   THERMAL TRANSDUCERS AND SENSORS

Components that change their electrical characteristics in accordance with a change in temperature fall into three categories: *thermistors, resistance temperature detectors (RTDs)*, and *thermocouples*.

Thermistors

The conductivity of a material depends upon the number of free charge carriers that exist, and the frequency with which those charge carriers collide. As the temperature of a material rises, the number of free charge carriers and the frequency of collisions between these carriers both increase. If the chemical composition of a material is such that few free charge carriers exist at low temperatures, then the additional charge carriers created by an increase in temperature cause the material resistance to decrease. If, on the other hand, a great number of free charge carriers exist at low temperatures, then the additional charge carriers created by an increase in temperature only serve to increase the frequency of collisions between charge carriers, increasing the resistance of the material.

Negative temperature coefficient (NTC) thermistors are constructed by bonding wire to various chemical compounds (usually semiconductors or metal oxides) that exhibit a decrease in resistance as their temperature increases. These materials can be formed into virtually any shape, but they are most often formed into beads, discs, or cylinders (as shown in Figure 5–1). Typically, an NTC thermistor may have a resistance range of 100 kΩ to 100 Ω over a usable temperature range of $-50°$ to 200°C. The fact that thermistor resistance changes so drastically over such a narrow temperature range (see Figure 5–2) makes the thermistor the most sensitive thermal transducer. In addition, since thermistors can be made very small, they can respond quickly to temperature variations.

Positive temperature coefficient (PTC) thermistors are constructed of barium titanate–based semiconductors that exhibit an increase in resistance as their temperature increases. The response of a PTC thermistor is quite different from that of an NTC thermistor. The resistance of a PTC thermistor actually decreases slightly with temperature until the temperature reaches the *transition temperature* (sometimes called the switching or Curie temperature). At this point the temperature coefficient switches from negative to positive. As the temperature increases beyond the transition temperature, the resistance rapidly increases (see Figure 5–2). Although the chemical composition of the thermistor allows the transition temperature to be set anywhere from about $-20°C$ to $+300°C$, it is typically around 50°C. The sensitivity of a PTC thermistor is even higher than that of an

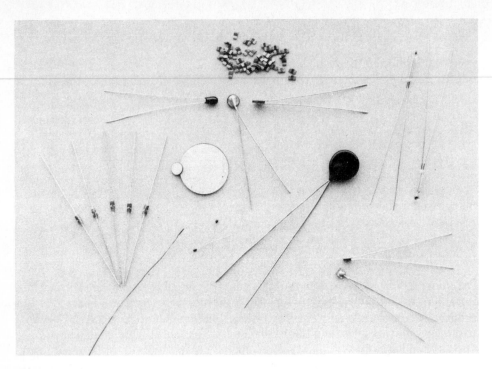

FIGURE 5–1
Typical Commercially Available Thermistors (Courtesy of Fenwal Electronics) and
Thermistor Symbol

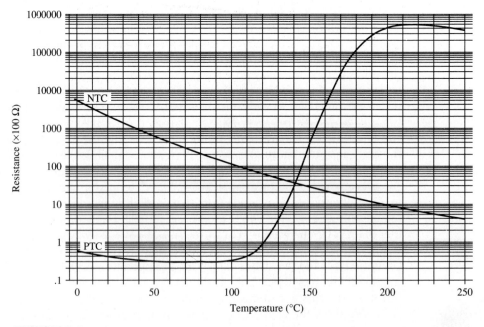

FIGURE 5–2
Typical Thermistor Response Curve

NTC thermistor. In addition, the maximum resistance and usable temperature of a PTC thermistor are usually much higher than those of an NTC thermistor.

In spite of all their favorable attributes, there are drawbacks to using thermistors. The resistance of a thermistor does not change linearly with changes in temperature. If the temperature doubles, the resistance of a thermistor may change by a factor of five. In addition, depending on the present temperature of the thermistor, the rate of resistance change will vary. This fact is demonstrated by the response curve shown in Figure 5–2. If a thermistor is to be put to practical use, some sort of linearization circuit must be employed. Because of variation in materials and manufacturing processes, thermistor curves have not been standardized to the extent that curves for thermocouples and RTDs have. The typically large resistance values of thermistors cause two additional problems. First, even a small amount of current through this large resistance can heat the thermistor enough to cause an erroneous measurement (self-heating). Second, thermistors are prone to permanent decalibration when subjected to high temperatures for extended periods. That is, a thermistor with a resistance of 30 kΩ at 25°C may, after being subjected to high temperatures, develop a different resistance at 25°C.

Resistance Temperature Detectors

If an enormous number of free electrons already exist in a material (as they do in metals), the addition of thermally generated free electrons will not significantly increase the conductivity. Rather, the increased thermal motion of the atoms within the material will cause more collisions between these electrons, making it harder for them to drift through the material under the influence of an electrical force. Thus, an increase in thermal energy will actually decrease the conductivity of the material. Although most metals exhibit this phenomenon, only a select few are used in the construction of RTDs.

Resistance temperature detectors are constructed by the winding of coils of wire on ceramic or glass forms or by the deposit of metal films on ceramic substrates. An RTD has a positive temperature coefficient. That is, as the temperature increases, the resistance of the device also increases. Some commercially available RTDs are shown in Figure 5–3. Although RTDs may be constructed of copper, nickel, and sometimes even tungsten, platinum is most commonly used, for it is easily purified and is inherently chemically stable. The resistance of the device is seldom over 1 kΩ at 0°C, and it is typically closer to about 100 Ω. Whereas thermistors have usable temperature ranges that are typically 0°C to 100°C, RTDs can generally be used from about -200°C to around 700°C. In addition to the wider temperature range, an RTD has a much more linear response than a thermistor (see Figure 5–4). In fact, for narrow ranges of temperature, the response may be considered linear. Since most RTDs are made of platinum, the characteristics are well standardized. Most manufacturers produce RTDs that follow the response curves set by the Deutsches Institut für Normung (DIN), a German agency that sets engineering standards. These curves specify that the temperature coefficient for a platinum RTD should be 0.00385 $\Omega/\Omega/$°C. With that

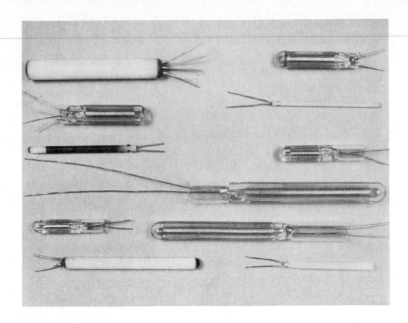

FIGURE 5–3
Typical Wire-Wound RTD Elements (Photo provided by Omega Engineering—An Omega Technologies Company) and RTD symbol

temperature coefficient, an RTD that has a resistance of 100 Ω at 0°C will change its resistance by 0.385 Ω for each degree of temperature change. If the response of an RTD is very linear (as it is for platinum), the following equation can be used to determine the resistance at temperatures other than 0°C:

$$R_T = R_0(1 + \alpha T) \qquad \textbf{(5.1)}$$

where

R_T = resistance at desired temperature

R_0 = resistance at 0°C

α = temperature coefficient of the RTD

T = desired temperature

Thus, for a temperature range from 0°C to 700°C, the resistance of the RTD will vary from 100 Ω to 369.5 Ω. If a thermistor could operate over such a large temperature range, its resistance range would be several hundred megohms.

Thus, the primary drawback of using RTDs for temperature measurements is the small resistance deviation. For a circuit to use an RTD to produce an accurate temperature measurement, the circuit must be extremely sensitive to minute changes in resistance. Aside from the response, the cost of the RTD can be an

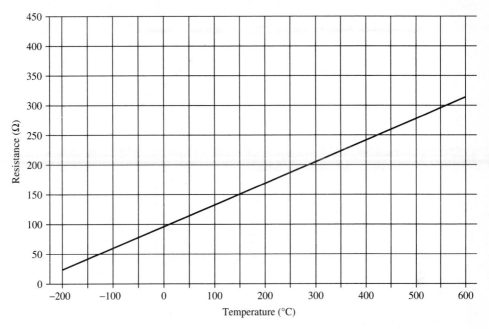

FIGURE 5–4
Typical RTD Response Curve

important factor when deciding which temperature transducer to use. An RTD has usually three to five times the cost of an equivalent-quality thermistor.

Thermocouples

The operation of a thermocouple is based on a principle called the *Seebeck effect* (named after Thomas Seebeck). This principle states that if a closed loop of wire is constructed by joining two dissimilar metal wires at both ends, a continuous current will flow through the loop as long as the junctions are at different temperatures. Furthermore, for small changes in temperature, the current will be proportional to the temperature difference between the junctions. If the loop is opened, a voltage will appear across the opening that will likewise be proportional to the junction temperature difference (see Figure 5–5). If the temperature of one of the junctions is known, this voltage can be used to calculate the temperature of the other junction. Normally, one of the junctions, called the *hot junction*, is subjected to the temperature that is to be measured, while the other junction, called the *cold junction,* is held at a constant known temperature.

At first glance, it seems that two inherent problems will be encountered when thermocouples are implemented in practical applications. First, one of the junctions must be held at a constant, known temperature. A convenient tempera-

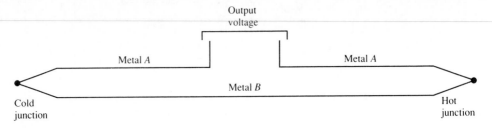

FIGURE 5–5
Open Thermocouple Loop

ture reference to use is 0°C, for this is the ice point for water. As a matter of fact, most of the standardized thermocouple response curves assume that the temperature of the cold junction is 0°C. It would be, at the very least, inconvenient to keep one of the junctions immersed in an ice bath. The second problem arises from the making of electrical connections to the thermocouple. Each connection point constitutes a new thermocouple junction, introducing an error into the final output signal (see Figure 5–6).

Both of these problems, although very real, are easily resolved. As shown in Figure 5–7, two thermocouples sharing a common intermediate metal (metal *B* in this case), can be reduced to a single thermocouple, as long as the two original thermocouples are at the same temperature. In other words, the need for the intermediate metal is eliminated. This is exactly the situation that exists in Figure 5–6. Therefore, the thermocouple circuit of Figure 5–6 can be replaced by the equivalent circuit of Figure 5–8, as long as the terminal block is held at a constant 0°C (the same temperature as the cold junction). The terminal block connections in Figure 5–8 have effectively become the cold junction of the thermocouple. The

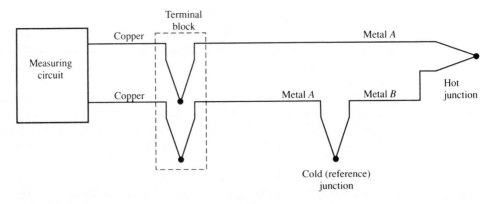

FIGURE 5–6
Connection of Measuring Apparatus to a Thermocouple Loop

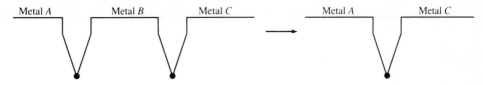

FIGURE 5–7
Reduction of Two Thermocouple Junctions to a Single Junction

error voltage produced by these connections can easily be compensated for by an electronic circuit that will produce an opposing voltage exactly equal in magnitude to the error voltage at all temperatures. Such circuits are commercially available and are called *cold junction compensators*. Additionally, this circuit can incorporate functions that will produce the same voltage that the cold junction would normally produce at 0°C. Such functions will eliminate the need to regulate the temperature of the terminal block at the ice-point temperature of water. Such devices are also commercially available and are called *electronic ice-point references*. Since different compensation voltages are required for different thermocouple types, the compensation circuits must be matched to the thermocouple that is being used. The result is an arrangement such as that shown in Figure 5–9.

As mentioned earlier, the Seebeck effect occurs for metals that are chemically different. Technically, any two different metals can be used. However, the characteristics of a few have been standardized, and those metals are therefore in common use throughout industry. The American National Standards Institute (ANSI) has identified and standardized certain thermocouple types and has given each an identifying letter designation. For each type, it defines such things as types of metal used, usable temperature range, and color codes. The table of

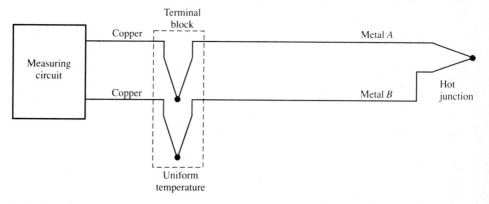

FIGURE 5–8
Elimination of Cold Junction

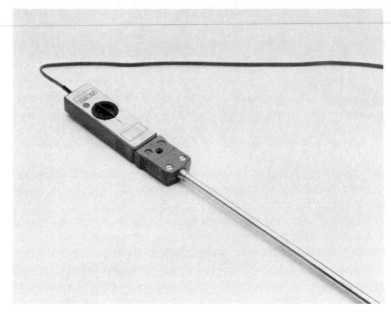

FIGURE 5–9
Commercial Thermocouple Probe with Compensation Circuit (Photo provided courtesy of Omega Engineering—An Omega Technologies Company)

Figure 5–10 shows the characteristics of each standard type. The responses of several ANSI standardized thermocouple types are shown in Figure 5–11.

Temperature Sensors

Among the temperature sensors that are currently available commercially, three stand out in their widespread application. *Integrated solid-state temperature sensors, temperature transmitters,* and *infrared pyrometers* are rapidly becoming commonplace in industrial settings.

A solid-state temperature sensor is an IC that contains a thermal transducer (usually a forward-biased *pn* junction is used for this), along with calibration and linearization circuitry that will provide a voltage or current proportional to the temperature being sensed. Some commercial IC temperature sensors are shown in Figure 5–12. Typically, the output for such devices is 1–10 mV/°C or 1–10 μA/°C. These devices are fast, accurate, and inexpensive. In addition, they have the advantage of being PC-mounted. This is especially important in applications that involve measuring the temperatures of other PC board components for the purpose of regulation and protection.

A temperature transmitter (see Figure 5–13) is similar to a solid-state temperature sensor in that it produces a linear output voltage or current that is

ANSI T/C	Symbol Single	Generic and Trade Names	Color Coding – Single	Color Coding – Overall T/C Wire	Color Coding – Overall Extension Grade Wire	Magnetic Yes	Magnetic No	Maximum Usful Temp. Range	EMF (MV) Over Useful Temp. Range	Average sensitivity µV/°C	Environment (Bare Wire)
T	TP	Copper	Blue	Brown	Blue		X	°F -328 – 662	-5.602 – 17.816	40.5	Mild oxidizing, reducing. Vacuum or inert. Good where moisture is present.
	TN	Constantan, Cupron. Advance	Red				X	°C -200 – 350			
J	JP	Iron	White	Brown	Black	X		°F 32 – 1382	0 – 42.283	52.6	Reducing. Vacuum, inert. Limited use in oxidizing at high temperatures. Not recommended for low temps.
	JN	Constantan, Cupron. Advance	Red				X	°C 0 – 750			
E	EP	Chromel, Tophel, T¹ Thermokanthal KP	Purple	Brown	Purple		X	°F -328 – 1652	-8.824 – 68.783	67.9	Oxidizing or inert. Limited uses in vacuum or reducing.
	EN	Constantan, Cupron. Advance	Red				X	°C -200 – 900			
K	KP	Chromel, Tophel, T¹ Thermokanthal KP	Yellow	Brown	Yellow		X	°F -328 – 2282	-5.973 – 50.633	38.8	Clean oxidizing and inert. Limited in Vacuum or reducing.
	KN	Alumel, Nial T² Thermokanthal KN	Red			X		°C -200 – 1250			
S	SP	Platinum 10% rhodium	Black		Green		X	°F 32 – 2642	0 – 14.973	10.6	Oxidizing or inert. Atmos.
	SN	Pure platinum	Red				X	°C 0 – 1450			
R	RP	Platinum 13% rhodium	Black		Green		X	°F 32 – 2642	0 – 16.741	12.0	Oxidizing or inert. Atmos. Do not insert in metal tubes. Beware of contaminations.
	RN	Pure platinum	Red				X	°C 0 – 1450			
B	BP	Platinum 30% rhodium	Gray		Gray		X	°F 32 – 3092	0 – 12.426	7.6	
	BN	Platinum 6% rhodium	Red				X	°C 0 – 1700			
C*	CP*	Tungsten 5% rhenium	White/red trace		White/red trace		X	°F 32 – 4208	0 – 37.066	16.6	Vacuum, inert, hydrogen atmospheres. Beware of embrittlement.
	CN*	Tungsten 26% rhenium	Red				X	°C 0 – 2320			
G*	GP*	Tungsten	White/blue trace		White/blue trace		X	°F 32 – 4208	0 – 38.564	16.0	
	GN*	Tungsten 26% rhenium	Red				X	°C 0 – 2320			
D*	DP*	Tungsten 3% rhenium	White/yellow trace		White/yellow trace		X	°F 32 – 4208	0 – 39.506	17.0	
	DN*	Tungsten 25% rhenium	Red				X	°C 0 – 2320			

* Not ANSI symbol

FIGURE 5–10
Characteristics of Standard Thermocouple Types

117

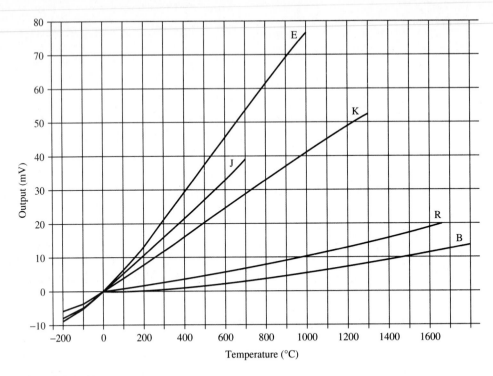

FIGURE 5–11
Response of Several Popular ANSI Thermocouple Types

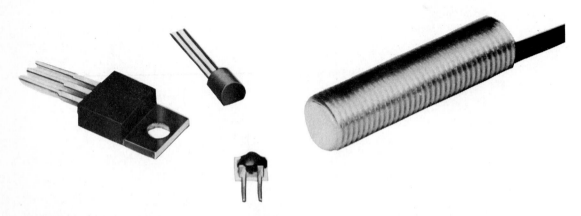

FIGURE 5–12
Commercially Available IC Temperature Sensors (Courtesy of Micro Switch, A Honeywell Division)

FIGURE 5-13

A Popular Commercially Available Temperature
Transmitter (Courtesy of Newport Electronics
Inc.)

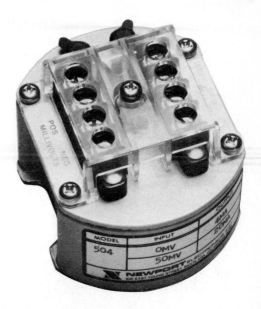

proportional to the temperature at the input. The input in this case, however, is
usually a remote thermocouple or RTD. Typically, a temperature transmitter has
two input connections (to which an RTD or thermocouple is attached) and two
output connections. The power for the device is usually supplied along the same
lines as the output. The power supply and the load are connected in series with the
output, and current through the load is modulated by the transmitter in accordance with the signal from the thermocouple or RTD at the input. Figure 5-14
should help illustrate this concept.

Infrared pyrometers (or optical pyrometers) operate on the principle that
thermal energy is transmitted in the form of electromagnetic radiation (EMR). For
energy levels perceived as heat, the frequency of the EMR is in the infrared area
of the frequency spectrum (just below visible red light). An infrared pyrometer is
shown in Figure 5-15. The transducer involved is an optical device, usually a
phototransistor or photodiode, onto which the radiation is focused. This probe
assembly is connected to a remote module that amplifies and linearizes the signal
from the probe. The result is a linear voltage or current that is proportional to the

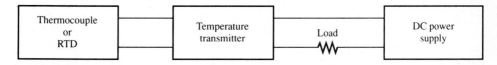

FIGURE 5-14

Application of a Temperature Transmitter

FIGURE 5–15
Typical Optical Pyrometer (Photos provided courtesy of Omega Engineering—An
Omega Technologies Company)

measured temperature. The obvious advantage of this type of measurement is that
it requires no physical contact. These sensors find application in environments
that are dangerous, delicate, or electrically unsafe and in temperature measurement of moving objects. Since the infrared pyrometer is a noncontact sensor, it
can be used to measure extremely high temperatures. A typical temperature range
may be as wide as 0°C–3000°C.

5–3 OPTICAL TRANSDUCERS AND SENSORS

Components that change their electrical characteristics in accordance with a change in the intensity or frequency of incident EMR are called optical transducers. The word *optical* tends to connote visible light; however, many of the optical transducers that are currently in use are sensitive to frequencies outside the visible range. (This explains why infrared pyrometers are sometimes called *optical* pyrometers.) Generally, optical transducers fall into one of two categories: *photoconductive* (those that produce a change in resistance in accordance with EMR intensity) and *photovoltaic* (those that produce a current or voltage in accordance with EMR intensity).

It is important to understand the difference between the *spectral response* and the *sensitivity* of an optical transducer. Sensitivity of an optical transducer is a measure of the amount by which the electrical characteristic of the transducer changes with respect to changes in illumination at a particular frequency of EMR. Spectral response describes how the sensitivity of the transducer changes with respect to frequency of EMR.

While many people have a general feel for physical variables such as temperature, position, and velocity, most of us have trouble quantifying optical phenomena. The distinction between the terms used to describe optical quantities is often quite subtle. For example, although luminance, luminous intensity, illumination, and luminous flux all sound quite similar, each has a specific, different definition. In addition, many of the older, obsolete terms are still in widespread use. Further confusion results from the fact that there are different terms used to describe the same optical quantity, depending upon whether the measurement applies to radiant energy across the entire electromagnetic spectrum (radiometry) or whether the measurement applies only to the area of the spectrum in the immediate vicinity of visible light (photometry). In general, the optical transducers are the ones that are sensitive to electromagnetic frequencies near or within the visible spectrum.

As a point of reference, consider the following. Luminance is a photometric measure of optical power output whereas illumination is the photometric measure of optical power input. Although footlamberts, candelas/m², and stilbs are all units that are used to quantify luminance, candelas/m² are the recommended SI standard. Similarly, although footcandles, lux, and phots are all used to quantify illumination, lux is the recommended SI standard. An ordinary candle has an approximate luminous intensity of one candela. Suppose that a sphere having a radius of one meter surrounds the candle. The resulting illumination on one square meter of the spherical surface will be one lux.

Photoconductive Cells

Photoconductive cells, or *photoresistors,* are devices that vary their conductivity in accordance with variations in the amount of EMR that strikes them. Their operation is best understood if the nature of electrical conduction is first dis-

FIGURE 5–16
Typical CdS Photoconductive Cell (Courtesy of Hamamatsu Corp.) and Photoconductive Cell Symbols

cussed. For electrons to move freely through a material, they must be relatively free from the influence of their parent atoms. That is, each electron must acquire enough energy to break free from the attractive force exerted on it by the atom's nucleus. The amount of energy required varies from material to material. In addition, for any given material, an exact amount of energy is required; too much or too little simply will not do. To help illustrate this point, imagine standing at the bottom of a staircase, holding a tennis ball. If you wish to bounce the ball onto the third step, you must bounce it with a certain amount of force. If too much force is applied, the ball will bounce off the side of the fourth step. If too little force is applied, the ball will bounce off the edge of the third step. Although the principle involved at the atomic level is not exactly the same, the concept is quite similar. Electromagnetic radiation carries energy along with it. As the frequency of the EMR increases, the amount of energy carried along increases (that is why X rays are more dangerous than the visible light emitted by a lightbulb, even though both are EMR). Therefore, exposing any material to EMR of the proper frequency should make that material conduct.

A typical photoconductive cell is shown in Figure 5–16. Commercial photoconductive cells are constructed by depositing a film of photoconductive material on a ceramic substrate. Most are tailored to be sensitive to EMR in the visible frequency range, and all of them become more conductive when exposed to the EMR. The amount of change in resistance is proportional to the intensity of the EMR, but as shown in Figure 5–17, the relationship is not linear. Some of the more common materials used in the construction of photoconductive cells are cadmium sulfide (CdS), cadmium selenide (CdSe), lead sulfide (PbS), and lead selenide (PbSe). Each material has a sensitivity that peaks at a particular frequency. CdS cells have a peak sensitivity to frequencies in the blue-green region of the visible spectrum; CdSe cells have a peak sensitivity to frequencies in the orange-red region of the visible spectrum; and both PbS and PbSe cells are most sensitive to frequencies in the infrared region.

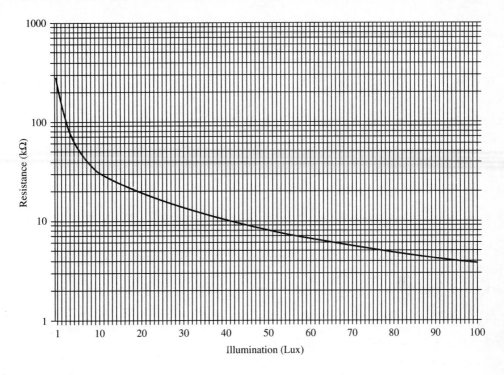

FIGURE 5–17
Typical Photoconductive Cell Response

One important factor to consider when using photoconductive cells is their *light history*. This term refers to the response time, dark resistance, and illuminated resistance, which will all vary depending upon the light conditions to which the cell has previously been exposed. In general, when a photoconductive cell is kept in darkness for a long period of time, its resistance will be lower than that of an identical cell that has been kept at a bright level of light. In addition, cells kept in darkness exhibit slower response times than cells kept at a bright light level.

Photovoltaic Transducers

A photovoltaic transducer is a semiconductor *pn* junction across which a voltage is induced by exposure to EMR. When a *pn* junction is first created, there is an initial diffusion of charge carriers across the junction, leaving a region devoid of free charge carriers (the *depletion region*), and a minute potential difference across this region. This process occurs at any *pn* junction, whether it is to be used as a photovoltaic transducer, a rectifier diode, or even the base-emitter junction of a transistor. When EMR of the proper frequency strikes an electron in the *depletion region,* the electron gains enough energy to drift away from its parent atom,

FIGURE 5–18
Encapsulated Photovoltaic Cells (Courtesy of Hamamatsu Corp.) and Photovoltaic Cell
Symbols

and it is accelerated across the junction by the minute potential difference that
exists. At the same time, the "hole" left behind by the liberated electron is
accelerated across the junction in the opposite direction. As this action continues,
the net potential difference across the junction decreases until incident EMR will
produce no further current. Therefore, there is a maximum voltage that a photovoltaic cell will produce. As an example, silicon photovoltaic cells will produce
approximately 0.6 V in direct sunlight. The same phenomenon prevents the output
voltage of a photovoltaic cell from being a linear function of the input.

A photovoltaic cell that is used for direct energy conversion (from EMR to a
voltage) is often referred to as a *solar cell,* indicating that the sun supplies the
input energy. Such a device is constructed by diffusing a thin layer of *p*-type
material onto a layer of *n*-type material. The *p*-type material is made thin to allow
EMR to pass through and penetrate the depletion region. In addition, the surface
area of the device is maximized to allow maximum exposure to the EMR. Oftentimes a solar cell is manufactured as a large, flat disk. A selection of photovoltaic
cells is shown in Figure 5–18.

Photodiodes

A photodiode is literally a small photovoltaic cell. Instead of the flat, wafer appearance that a solar cell has, a photodiode usually has a transparent TO-92 or
TO-18 case. Unlike solar cells, photodiodes are used for detection of EMR rather
than for direct energy conversion.

A photodiode can be used in a forward or reverse mode. In the forward
mode, the diode operates in exactly the same manner as a solar cell, producing a
voltage output for an EMR input. This signal can then be amplified and linearized
to suit the needs of the particular control system being implemented.

In the reverse mode of operation, current through the diode is due primarily to leakage. That is, thermally generated minority carriers are accelerated across the junction by the source voltage. Usually, this leakage current is on the order of a few tens of microamperes. If the number of minority carriers increases, by any means, the reverse current through the diode likewise increases. Electromagnetic radiation of the proper frequency will produce minority carriers in the depletion region of the diode, which will in turn increase the current through the diode. This current may then be converted to a voltage, amplified, and linearized. In this mode the photodiode operates more like a photoconductive cell than like a photovoltaic device.

Phototransistors

A phototransistor operates according to the same principle as a reverse-biased photodiode. As EMR strikes the collector-base junction, minority carriers are injected into the base region and accelerated across the base-emitter junction by the forward bias that exists. The advantage of this device over the photodiode is the inherent amplification that takes place. For the same amount of illumination, the current through a phototransistor is much greater than that through a photodiode. Naturally, EMR must be permitted to strike the junction of the transistor, so these devices, like photodiodes, are usually packaged in transparent or partially transparent cases. Three connections are provided on the package—one each for emitter, collector, and base; however, the base is seldom connected. Since the EMR essentially produces the base current, there is no need for a base connection. However, sometimes the transistor is configured as an amplifier, and the EMR is allowed to modulate the base current that is a result of the biasing network.

A selection of photodiodes and phototransistors is shown in Figure 5–19.

Optical Sensors

Integrated optical sensors are used to detect the presence or position of objects. Some commercially available optical sensors are shown in Figure 5–20. Generally, an integrated optical sensor consists of two subsystems: an optical transmitter and an optical receiver. Sensing is accomplished by detection at the receiver of an optical signal originating at the transmitter. Detection of the object takes one of three forms: *opposed sensing* (*through-beam sensing*), *retroreflective sensing,* or *diffuse sensing* (*proximity sensing*). The basic principle of operation is the same for all of these methods. The difference lies in the relative positions of the transmitter, the receiver, and the object to be sensed.

In an opposed-sensing optical system, shown in Figure 5–21, the transmitter and receiver are set in opposition, so that the signal is transmitted directly between them. If an object passes between the transmitter and the receiver, the signal is blocked, and the receiver detects the absence of the signal. For this type of sensing, the transmitter and receiver must be separate units. Because this

FIGURE 5–19
Typical Photodiodes and Phototransistors (Courtesy of Hamamatsu Corp.), Photodiode
Symbols, and Phototransistor Symbols

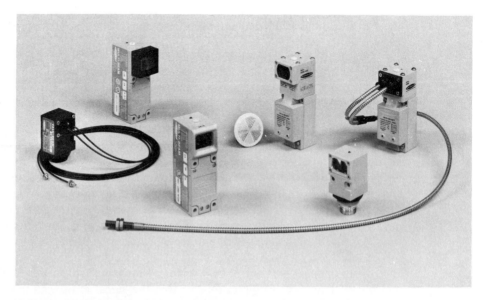

FIGURE 5–20
Self-Contained Optical Sensors (Courtesy of Banner Engineering Corp.)

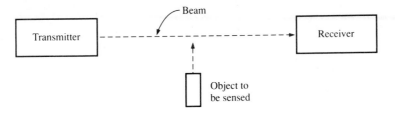

FIGURE 5–21
Opposed (Through-Beam) Optical Sensing Arrangement

method involves direct transmission, the detection range may be as far as several hundred feet.

Retroreflective sensing, shown in Figure 5–22, is similar to opposed sensing in that the beam is interrupted by the sensed object. However, instead of transmitting the optical signal directly to the receiver, the transmitter directs the signal at a reflector. This reflector bounces the signal back to the receiver. This method allows the transmitter and the receiver to reside in the same package, thus making the sensor more compact and the wiring layout simpler. Since the beam is reflected, there is a certain amount of loss due to diffusion and beam spreading. Therefore, the detection range for this method is seldom more than 30 ft.

Diffuse sensing, in turn, is similar to the retroreflective method. The object itself, however, acts as the reflector. Obviously, the object must be reflective. To a certain degree, transparent or translucent objects do reflect and so are detectable by this method. Diffuse sensing also lends itself to situations that prohibit the placing of receivers or reflectors on the side of the object opposite the transmitter. Because the sensed object acts as the reflector, the sensing distance for diffuse sensing is usually less than 10 ft. A diffuse-sensing system is illustrated in Figure 5–23.

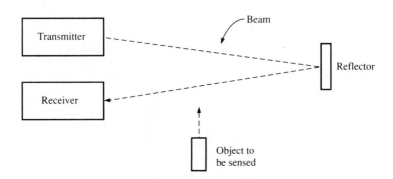

FIGURE 5–22
Retroreflective Optical Sensing Arrangement

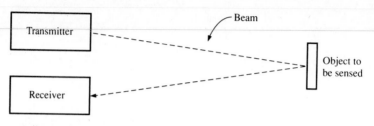

FIGURE 5–23
Diffuse (Proximity) Optical Sensing Arrangement

The signal passed from transmitter to receiver may be modulated or unmodulated. For modulated systems an infrared-emitting diode (IRED) is pulsed at a predetermined frequency in the transmitter. The receiver contains a filter that allows recognition of only that particular signal frequency. Such systems are of particular value in environments where the ambient light may interfere with normal operation. Unmodulated systems do not require an oscillator in the transmitter, nor do they require a filter in the receiver. Such systems, consequently, are considerably less expensive than their modulated counterparts.

Optical sensors of the types described here are quite easily implemented. The transmitters generally require only two connections (for power) and will operate from DC supplies at 3–30 V or from AC supplies up to 250 V. The output of the sensor is taken from the receiver and usually takes the form of an open-collector transistor (for DC outputs) or a triac (for AC outputs). The load and the supply for the load are connected in series with the output, much in the same way as for the temperature transmitter shown in Figure 5–14.

Optoisolators

Based on the same principle as optical sensors, optoisolators (or *optocouplers*) are used extensively in circuit protection and industrial computer interfacing. As will be discussed in Chapter 7, an optoisolator, in its simplest form, is an LED and a phototransistor, set in opposition, in a permanently sealed package. Often a dual-in-line package (DIP) is used so that the devices will be package-compatible with other ICs on the same PC board. The primary function of an optoisolator is to provide electrical isolation between the devices connected to the transmitter and those connected to the receiver. Information is transmitted through an optical link. If a computer, operating at 5 V DC, were directly connected to circuitry controlling a 240 V AC load, for example, the possibility would exist that a circuit failure could result in the high voltage being applied directly to the computer. Isolating the computer from the high-voltage circuit by using an optoisolator prevents the permanent damage that such a circuit failure could cause, since there is no physical connection between the transmitter and the receiver inside the optoisolator. Figure 5–24 shows such an arrangement.

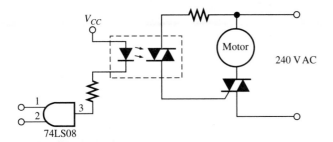

FIGURE 5–24
An Optoisolator Used to Protect a Logic Gate from a 240 V AC Line

Optoisolators have become much more sophisticated in recent years. Though standard simple optoisolators are still available today, the many improvements have increased the already widespread application of optoisolators. Input control-logic circuitry has been added to determine under what conditions the IRED should be activated. Such output devices as photo-Darlington transistor configurations, light-activated SCRs and triacs, and even constant-current output amplifiers have replaced the single phototransistors. Figure 5–25 shows some of the various package styles that are currently available.

FIGURE 5–25
Commercially Available Optoisolators (Courtesy of Hamamatsu Corp.)

Optointerrupters

An *optointerrupter,* shown in Figure 5–26, is a variation of an optoisolator and, in fact, is identical to an optoisolator in every way except application. The primary difference between the two is that an optointerrupter is not a sealed system. A slot exists between the IRED and the phototransistor, into which a perforated card, disk, or tape can be inserted. Whenever a hole in the card passes the IRED, light passes through to the phototransistor. Therefore, the position or velocity of the card (and of whatever is attached to it) can be accurately monitored.

5–4   MAGNETIC TRANSDUCERS AND SENSORS

Magnetic transducers are devices that change their electrical characteristics in accordance with the presence of or a change in a surrounding magnetic field. Such devices are useful in device current monitors, solid-state switches, and countless position and velocity measurement applications. For the most part, these devices fall into three categories: *induction devices, reluctance devices,* and *Hall-effect devices.*

To fully comprehend the operation of magnetic devices, a few basic principles must be understood. First of all, by convention, magnetic lines of force (*magnetic flux*) are defined to flow from north to south. Consequently, the direction of the magnetic field is from north to south. Second, any current-carrying conductor has a circularly polarized magnetic field surrounding it. An easy way to remember the direction of this field is to point the thumb of your right hand in the direction of current (conventional flow) through the conductor, and the fingers of your right hand will wrap around the conductor in the direction of the magnetic field. Last, certain materials (often called ferromagnetic materials) conduct magnetic flux very well. In fact, such materials actually enhance, or strengthen, the field.

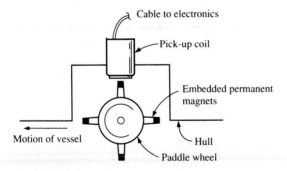

FIGURE 5–27
Example of an Induction Transducer (From Berlin and Getz, *Principles of Electronic Instrumentation and Measurement*. Merrill/Macmillan, 1988.)

Induction Transducers

Induction devices operate on the principle that if there is relative motion between a magnetic field and a conductor, a current can be induced in the conductor. This is true as long as some component of the motion is perpendicular to the length of the conductor. Usually, the conductor is a wire, and usually the wire is wound on some type of form to produce a coil. As a magnetic field passes the coil, it induces in the coil a voltage that is proportional to the strength of the field, the velocity with which it passes, and the number of turns of wire on the coil.

The motion of the field can be produced by physical motion of a magnet or by a current change in a nearby conductor. An important point to remember is that the same effect can be obtained with a stationary field and a moving coil. In either case, the induced voltage is small and of short duration. Therefore, some type of conditioning is required to make the signal usable in a control system. Figure 5–27 shows one possible arrangement. Each time one of the magnets passes the coil, a pulse of voltage is induced in the coil and applied to the input of an amplifier. This amplified pulse may be further conditioned by a noninverting buffer to be compatible with system logic. The built-in hysteresis of the buffer tends to eliminate the slow rise and fall times of the input signal.

Variable Reluctance Transducers

A similar effect can be obtained by using the principle of variable reluctance. Reluctance in magnetic circuits is the equivalent of resistance in electric circuits. A low-reluctance path is a good magnetic conductor. If a ferromagnetic material is brought in close proximity to a permanent magnet, the strength of the field surrounding the magnet increases. In addition, the flux is redirected to pass through the material. If the coil in Figure 5–27 were wrapped around a permanent magnet rather than a ferromagnetic core, and if the magnets on the shaft were not magnets at all, but simply extensions of the shaft, the same output would be obtained from the buffer. Each time one of the teeth passed the magnet, the field would be altered, and a pulse of current would be induced in the coil.

Hall-Effect Transducers

Although inductive and reluctance coils are still sometimes used as transducers, they are rapidly being replaced by devices that are smaller, cheaper, and more versatile. These devices are called Hall-effect transducers. The Hall effect is the principle whereby moving charge carriers in a conductor experience a force due to a magnetic field. This effect is especially pronounced in semiconductors. As shown in Figure 5–28, as charge carriers pass through the material, they are deflected, producing a potential difference across the material as long as the magnetic field is present. This potential difference is called the Hall voltage. If the magnetic field is increased, the Hall voltage will likewise increase. If the current or magnetic field direction is reversed, the polarity of the Hall voltage will reverse.

The advantage of Hall transducers is that, since they are made of semiconductor material, it is very easy to implement signal conditioning circuitry on the same piece of material. Therefore, it is possible to manufacture a device that will detect a magnetic field and provide a linear output voltage proportional to the strength of the field, all on a piece of silicon no larger than a transistor.

Technically, a Hall transducer requires four connections—two to provide current through the device and two to provide the output voltage. Generally, however, the negative-current supply line and the negative output line are internally connected together, producing a three-terminal device that looks like a transistor.

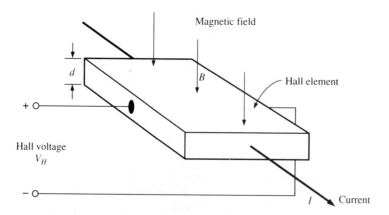

FIGURE 5–28
The Hall Effect in a Semiconductor (From Berlin and Getz, *Principles of Electronic Instrumentation and Measurement*. Merrill/Macmillan, 1988.)

FIGURE 5–29

Typical Commercially Available Current Transformer (Courtesy of Square D Electro-
magnetics) and Current Transformer Symbol

Current Transformers

Current transformers (see Figure 5–29) are transducers that are used to monitor
and regulate conductor current. They differ from potential transformers (voltage
transformers) in that the primary winding is in series with the current to be
measured. The current induced in the secondary winding is proportional to the
ratio of the number of primary windings to the number of secondary windings. For
example, if there are 100 secondary windings for every primary winding, the
current induced in the secondary will be approximately 1/100 of that flowing in the
primary. An important fact to realize is that a current transformer is designed to
operate with the secondary shorted by some sort of load. This load may be a
resistor, a meter, a solenoid, or even a motor. If the secondary is open, the device
may act as a step-up potential transformer, producing voltages so high that the
transformer may be destroyed. This is especially true of transformers designed to
carry hundreds of amperes of primary current. One other important consideration
is that for a current to be induced in the secondary, the primary current must be
changing. Consequently, current transformers find little application in DC circuits.

For larger current transformers, it is an established convention to design the
secondary so that either 1 A or 5 A flow when the primary is at its rated current.
This convention facilitates and standardizes the production of devices that are to
be connected in the secondary circuit. Certain transformers are especially designed to operate with the secondary open. In those devices the output voltage is
proportional to the primary current. Figure 5–30 illustrates both of these concepts. Two types of current transformers are characterized by the manner in
which primary current is applied. *Wound current transformers* have the primary

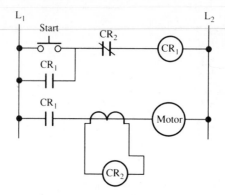

A current transformer may be implemented as a protection device. In this case, when the primary (motor) current is too high, CR_2 interrupts current to the motor.

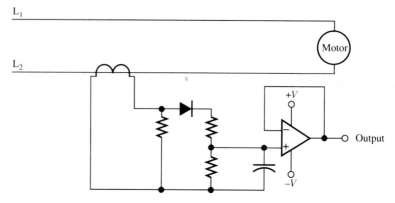

A current transformer may also be used as a linear current monitor. In this case a portion of the signal from the transformer is rectified and amplified to provide a voltage proportional to motor current.

FIGURE 5–30
Two Uses for Current Transformers

and secondary conductors wound around the same square or toroid core. Connections for the primary and the secondary are provided on the transformer case. *Through-primary current transformers* have only the secondary conductor wound on the core. The primary conductor is meant to pass directly through the center of the core. Connections are only for the secondary. Large, high-current transformers are constructed almost exclusively as through-primary devices. In fact, a bar of copper or aluminum is usually permanently connected through the center of the core, and screw lugs are provided for connecting the primary in series with the load.

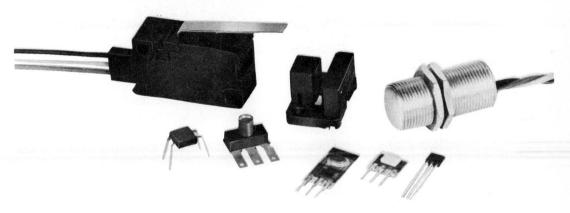

FIGURE 5–31
Typical Integrated Hall-Effect Sensors (Courtesy of Micro Switch, A Honeywell Division)

Integrated Magnetic Sensors

Magnetic sensors are simply Hall transducers coupled with integrated signal-conditioning circuitry and packaged in a way that facilitates implementation. Usually, they are three- or four-terminal devices (see Figure 5–31) that provide one of two types of output—either a digital output that indicates the presence (or absence) of a magnetic field or a linear output that indicates the relative strength of the field.

One interesting variation of the magnetic sensor is the *vane sensor* (Figure 5–32). In application it is very similar to an optointerrupter. A permanent magnet and a digital-output Hall sensor are set in opposition, with a slot between them. When a metal vane (a toothed metal card) is passed through the slot, the teeth shunt the magnetic flux away from the Hall sensor. So instead of the presence or absence of light, the vane sensor detects the presence or absence of a magnetic

FIGURE 5–32
Typical Magnetic Vane Sensor (Courtesy of Micro Switch, A Honeywell Division)

field. Just as with an optointerrupter, the position or velocity of the vane can be accurately monitored.

Solid-state switches also make use of integrated magnetic sensors. Externally, such a switch looks no different from its mechanical counterpart. Internally, however, things are quite different. Connected to the switch plunger is a small permanent magnet. When the plunger is depressed, the magnet comes in close proximity to a Hall sensor, which in turn closes a semiconductor switch. The obvious advantage is that there are no mechanical contacts to wear or become pitted from arcing. In addition, all mechanical switches "bounce." That is, when a mechanical switch is opened or closed, the contacts make and break thousands of times before becoming stable. Bounce becomes a matter of particular concern when the switch is interfaced to digital circuitry. Solid-state switches virtually eliminate switch bounce. One drawback of solid-state switches is that the internal Hall sensor requires a DC supply. Therefore, each switch requires two extra control wires. The added life and reliability, though, are usually enough to offset this problem.

5–5   ELECTROMECHANICAL TRANSDUCERS AND SENSORS

Oftentimes it is necessary to convert mechanical variables, such as motion, acceleration, and pressure, into equivalent electrical signals. This section will cover five of the most common types of devices in current use: *limit switches, potentiometers, linear variable differential transformers, force transducers, flow sensors,* and *level sensors.*

Limit Switches

One of the simplest electromechanical transducers available today is the limit switch. Limit switches, shown in Figure 5–33, differ from ordinary switches only in their construction and application. In every other way, they are simply switches. Limit switches are integral parts of the control system, and they are usually actuated by some mechanism that is also part of the system. Their purpose, when actuated, is to start, stop, speed up, slow down, or in some other way change some process that is part of the control scheme.

Limit switches consist of three parts: the housing, the actuator assembly, and the internal contacts. The housing of the switch protects the internal contacts from the environment and often serves to make the switch oiltight or explosion-proof. Usually, the housing can be opened, so that the internal components can be serviced (limit switches are usually repaired rather than replaced). The actuator (the mechanism that opens or closes the internal contacts) may be a rotary type, a plunger, or any of a variety of levers, and it often has a roller attached to prevent excessive wear. The internal contacts are often separate assemblies that may be removed and replaced. Although the simplest limit switches have one NO contact and one NC contact, many options are available. Several contacts may be speci-

FIGURE 5–33
Typical Limit Switches (Courtesy of Micro Switch, A Honeywell Division)

fied, and each may be maintained or momentary, NO or NC—even Hall-sensor contacts can be obtained.

Potentiometers

One of the simplest methods of representing position or motion with an electrical signal is through the use of a potentiometer. Remember from Chapter 2 that the movable contact (the wiper) on a potentiometer may be actuated in a linear manner or a rotary manner. In addition, the resistance may vary linearly or logarithmically. Rotary potentiometers are available as single-turn units or multiturn units (with up to 20 turns), and linearly actuated potentiometers are available with wiper travels of up to 4 ft.

For a potentiometer to indicate anything about the position or motion of an object, some mechanical linkage must exist between the object and the potentiometer. This can be accomplished through the use of cams, rollers, pulleys, gears, or levers.

An example of a potentiometer used as a transducer is shown in Figure 5–34. If a particular amount of slack must be maintained in the material sheet, the speed of the rollers must be controlled. For the speed to be controlled, however, the amount of slack currently in the material must be known. This information is provided by the potentiometer. As the material slack varies, the freewheeling rider rises or falls, changing the position of the wiper shaft and so also the voltage

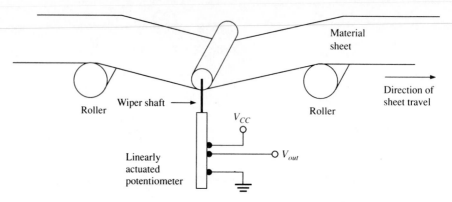

FIGURE 5–34
Typical Application of a Potentiometric Transducer

available at V_{out}. This voltage can then be used to determine if and how a roller speed adjustment must be made.

An important consideration is the resolution of the transducer. Two factors affect resolution in a potentiometric transducer. One is the degree of travel of the wiper; the other is the precision with which the mechanical linkage is constructed. Although the mechanical aspects will not be considered in this textbook, suffice it to say that the precision of the transducer can be no better than the precision of its mechanical linkage. Consider, however, a rotary potentiometer used to indicate the position of a workpiece. If the potentiometer is a single-turn unit, it will rotate approximately 300°. If 10 V are applied across the outside terminals, the voltage available at the wiper will vary by 33.3 mV per degree of shaft rotation. If a 20-turn unit is used instead, the shaft will rotate a total of approximately 7200°, and the resolution will increase to about 1.4 mV per degree of shaft rotation. In other words, if a given movement of the workpiece could cause full rotation of either potentiometer, a smaller movement would be measured much more precisely by the 20-turn unit.

Linear Variable Differential Transformers

A linear variable differential transformer (LVDT) is a transformer that consists of a primary winding and two identical secondary windings positioned symmetrically on both sides of the primary. The primary and the secondaries are wound on a hollow plastic or ceramic tube, into which a ferrous cylinder (core) can be placed. The core may be threaded to screw into the tube, or it may be smooth, to slide in freely. The position of the core with respect to the two secondary windings determines the output of the device. Therefore, LVDTs are used to measure physical displacement, both linear and rotary. A typical LVDT is shown in Figure 5–35.

FIGURE 5–35
A Typical LVDT (Courtesy of Kavlico Corp.)

The secondaries of an LVDT are usually connected in a series-opposing arrangement, as shown in Figure 5–36, so if the voltage outputs from both secondaries are equal, the net output from the LVDT is zero. The factor that affects the secondary output is the amount of magnetic flux that is coupled from the primary

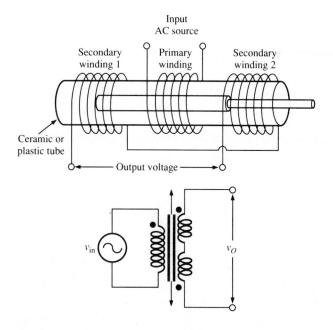

FIGURE 5–36
Structure of an LVDT (From Berlin and Getz, *Principles of Electronic Instrumentation and Measurement*. Merrill/Macmillan, 1988.)

to the secondary by the core. If the core is positioned so that both of the secondaries are equally coupled, the secondaries will produce identical outputs, 180° out of phase, resulting in an output of 0 V. If, however, the core moves to a position that results in more magnetic flux coupled through one of the secondaries, that particular secondary winding will produce a larger output than the other. This difference will result in a measurable output from the LVDT. Since the secondaries produce outputs that are 180° out of phase, the dominating secondary determines the phase of the LVDT output. The amount of displacement of the core determines the amplitude of the output. Oftentimes this output is rectified in such a way that the DC voltage indicates the amount of displacement, and the polarity of this DC voltage indicates the direction of the displacement.

Rotary variable differential transformers (RVDTs) are also available. The movable core in these devices takes the form of an elliptically shaped cam that is attached to a shaft. As the shaft is rotated, the relative amount of flux coupled from the primary to each of the secondaries changes in accordance with the position of the cam lobe. The output of the transformer is therefore proportional to the angular position of the shaft.

The requirement of an AC supply is an obvious drawback to implementation of LVDTs or RVDTs. To remedy this, some manufacturers have developed DC-operated devices. These implement a DC-driven solid-state oscillator that provides the AC necessary to operate the transformer. The output of the transformer is rectified and filtered to provide a DC signal that is proportional to the position of the shaft.

Force Transducers

It is often necessary to accurately measure such quantities as weight, pressure, and acceleration. Measurement of any of these quantities requires measurement of force. Virtually all force transducers fall into two categories: *piezoresistive* and *piezoelectric*.

Piezoresistivity is a property that all materials exhibit to a certain degree. However, it is most prominent in certain semiconductors. The resistivity of those materials changes in response to applied stress or strain. In fact, piezoresistive transducers are usually called *strain gauges*. Of these, two types are in common use: wire or foil strain gauges and silicon strain gauges. Both operate according to the same principle: That is, if the atomic lattice structure of the material is deformed, it will be more difficult for charge carriers to drift through. Hence, the resistance of the material will increase as it is deformed.

A wire (or foil) strain gauge consists of resistance wire (or foil), laid out in a serpentine fashion on a carrier such as paper or plastic (see Figure 5–37). Often several gauges are provided on the same carrier, positioned at various angles with respect to one another. Such arrangements eliminate the problem of precisely aligning several individual gauges when attaching them to the device under test. Solder pads are provided for connecting the gauge to the measuring circuit. Strain gauges must be permanently affixed to the device being measured. Therefore,

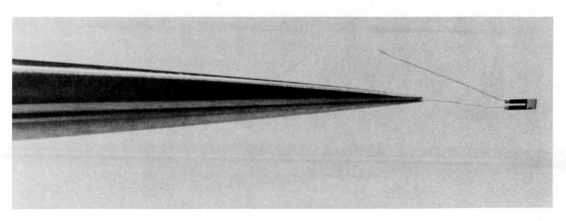

FIGURE 5–37
Commercially Available Strain Gauge (Photo courtesy of Entran Devices, Inc.)

strain gauges are not reusable transducers. Special adhesives are available for bonding gauges to other materials.

The gauge sensitivity, or *gauge factor,* is the ratio of the change in the gauge resistance to the change in the gauge length. In equation form, this is stated as follows:

$$\text{Gauge factor} = \frac{\%\ \text{Resistance change}}{\%\ \text{Length change}}$$

Typically, wire and foil strain gauges have gauge factors of around 2. The actual factor is determined by testing a sample of a particular production lot and marking the gauges with the exact value measured. Typical unstrained resistance values range from about 50 Ω to 3 kΩ. The most commonly used values, however, tend to be below 500 Ω. The amount of compression or elongation that most gauges will withstand is usually around 5%. This, coupled with the low gauge factor, means that the gauge resistance will change by only about 10% from an unstrained state to a fully strained state. This low level of change is a marked disadvantage, since temperature variations alone may change the resistance by several percent. Therefore, temperature compensation is always a consideration when implementing wire strain gauges as transducers. Usually RTDs or thermistors are used for this task. Since the resistance changes by such a small amount, sensitive measuring circuitry is required, to accurately detect strain in the materials being measured.

The usual method of application of wire strain gauges is to arrange them in a Wheatstone bridge configuration, as shown in Figure 5–38. If the bridge is perfectly balanced (all four resistors equal), the output will be zero. If any one of the resistors changes (the strain gauge is the variable), the bridge will produce some measurable output. Sometimes this output is directly measured and is propor-

FIGURE 5–38
Strain Gauge Arranged in a Wheatstone Bridge
(From Berlin and Getz, *Principles of Electronic
Instrumentation and Measurement.* Merrill/
Macmillan, 1988.)

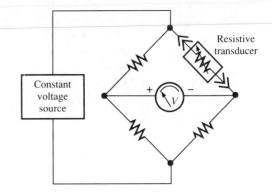

tional to the applied strain. An alternative method is to have, as one of the bridge
legs, a variable resistor. When the bridge produces an output (due to strain), this
resistor is adjusted until the bridge again produces a zero output.

The process of bonding a strain gauge to a workpiece and calibrating the
bridge to produce a realistic, measurable, and repeatable output is by no means a
trivial procedure. In general, one would not go through this process each time that
a force measurement is required. Rather, integrated assemblies called load cells
are used for routine force measurements. These devices consist of metal blocks,
cylinders, rings, and beams (see Figure 5–39) with strain gauge bridges permanently affixed to them. Most load cells have full-scale outputs that are on the order
of 2 mV per volt of bridge excitation voltage (supply voltage). That is, a load cell
that is rated for a maximum load of 100 lb., with a rated output of 2 mV/V and a
bridge supply voltage of 10 V, will produce an output of 20 mV with a load of 100

FIGURE 5–39
Commercially Available Load Cells (Courtesy of Transducer Techniques)

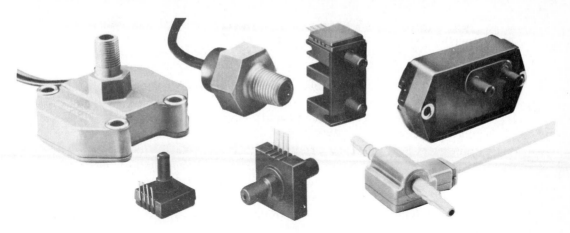

FIGURE 5–40
Commercial Pressure Sensors Containing Integrated Semiconductor Strain Elements
(Courtesy of Micro Switch, A Honeywell Division)

lb. Load cells are available that have maximum load ratings ranging from as little as less than 1 lb. to as much as 100,000 lb.

Semiconductor strain gauges are usually constructed of silicon; however, germanium is sometimes employed. Although the principle of operation is identical to that of wire strain gauges, these gauges are seldom employed as single, discrete units. Instead, they are more often integrated into pressure-sensing assemblies. Usually a monocrystalline section of silicon is implanted with piezoresistive areas by the process of diffusion. The transducers become an inseparable part of the silicon, eliminating the need for any type of bonding adhesive. Moreover, integrated Wheatstone bridges can be diffused into a single silicon chip. If this chip is then used as a diaphragm to which the pressure is applied, the result is an integrated pressure sensor. All that is required is a package and four wires—two for the bridge supply and two for the output. Some typical commercial pressure sensors of this kind are shown in Figure 5–40.

One of the primary advantages of a semiconductor strain transducer is its gauge factor, which can be as much as one hundred times that of a wire gauge. Additionally, since the resistivity of silicon is high, the gauges can be made very thin and very small. Integrated pressure sensors are largely temperature independent; that is, the resistance of the gauge will not change appreciably with temperature variations. However, at temperatures much above 200°C, the integrated gauges tend to break down and become indistinguishable from the monocrystalline substrate into which they have been implanted.

Certain crystalline materials exhibit a property called piezoelectricity. That is, if a crystal is deformed, a potential difference across the crystal will be produced. Depending on the type of material, the size of the crystal, and the applied force, this potential difference can be thousands of volts. Quartz and barium titanate are common materials employed for this property, and they are used most commonly as ultrasonic transducers. Because such high potentials are produced, these transducers are not suitable for normal strain measurements. They are capable, however, of detecting minute changes in gas or liquid density caused by acoustic signals. This capability makes them ideal for use as microphones and ultrasonic sensors.

A relatively recent commercial development in piezoelectric transducers is polyvinylidene fluoride (PVDF) polymer films (commonly called *piezo film*). The transducers look like clear plastic film that is anywhere from 10 μm to 100 μm thick, with metallization on each side. Although the voltages produced are not as high as those produced by the more rigid quartz and ceramic transducers, piezo film has the advantages of flexibility and high mechanical strength. Piezo film is used in switches, accelerometers, vibration sensors, medical monitors, and audio and ultrasonic sensors.

Flow Sensors   *on TEST DOPPlER only)*

Flow sensors measure the rate of fluid flow, usually indirectly, by measuring the velocity of the fluid. Many contain intelligent (microprocessor-based) signal conditioners that produce analog or digital signals proportional to the rate of flow. Regardless of whether the sensor contains a microprocessor, the output signals are usually a proportional analog voltage or current (often 4–20 mA or 1–5 V), a digitally compatible output switch (an *npn* current-sinking transistor or a *pnp* current-sourcing transistor), or a pulse train that varies in frequency with the flow rate. Virtually all of the available flow sensors require that the fluid flow be interrupted and the sensor inserted in line. Because sensors of this type are so sophisticated and complex, they are quite expensive.

Generally, flow sensors can be classified as either penetration or nonpenetration sensors. Penetration flow sensors are those that must physically penetrate the sensed medium to obtain the measurement. That is, the transducer must be placed in line with the fluid. Of these sensors, three types are in common use today: turbine, or paddle-wheel; vortex-shedding; and thermal.

As shown in Figure 5–41, a turbine flow sensor incorporates a free-spinning metal rotor, housed in a nonmagnetic body. The rotating blade is sensed by a magnetic transducer, and the resultant frequency is either converted to a proportional voltage or current or converted to a digitally compatible train of pulses.

Vortex-shedding flow sensors incorporate a nonstreamlined object in the path of the fluid. A series of eddies or vortices (swirling currents in the fluid) are produced downstream from the object when the fluid flows past (see Figure 5–42). The production of these vortices is directly proportional to the velocity of the fluid. Pressure fluctuations associated with the vortices are often sensed with piezoelectric transducers, and the resultant signal may be conditioned in a manner similar to the frequency conversion of a turbine sensor.

A thermal flow sensor consists of a heating element and a thermal transducer, housed in a probe and inserted into the fluid. As the probe is heated to a

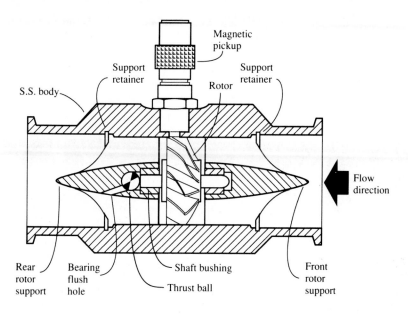

A. Construction

B. Appearance

FIGURE 5–41
Turbine Flow Sensor (Photo courtesy of The Foxboro Company)

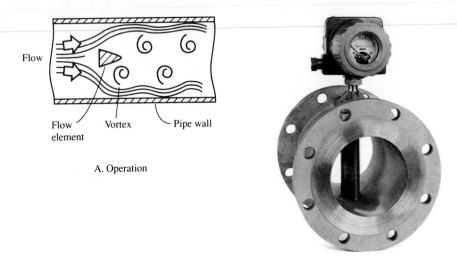

A. Operation

B. Commercial Model

FIGURE 5–42
Vortex Flow Sensor (Photo courtesy of The Foxboro Company)

stable temperature, the thermal transducer produces an output proportional to the temperature of the probe. A decrease in the probe temperature is an indication that fluid is flowing past the probe, carrying away some of the heat. The amount of temperature decrease is proportional to the fluid flow rate. Again, the signal obtained from the transducer may be conditioned to provide a linear output with respect to the flow rate.

Penetration sensors are not always practical. Sometimes, especially for slurries or viscous liquids, noncontact (or nonpenetration) sensors are required. Generally, such sensors employ either *magnetic sensing* or *ultrasonic Doppler sensing*. The primary advantage of nonpenetration sensing is that the transducer does not interfere with the fluid flow.

Operation of magnetic flow sensors, illustrated in Figure 5–43A, is based on Faraday's law, which states that relative motion between a conductor and a magnetic field induces a potential difference in the conductor. In magnetic flow sensors, the flowing liquid, which serves as the moving conductor, is surrounded by a magnet. Electrical contacts on both sides of the liquid detect the potential difference, which is then amplified and conditioned to provide a linear, proportional output signal. The operation of a magnetic flow sensor is contingent on the conductivity of the fluid being sensed. Nonconductive fluids will not provide the induction required.

The Doppler effect is responsible for the successful operation of ultrasonic Doppler flow sensors. The Doppler effect states that if EMR or acoustic energy is

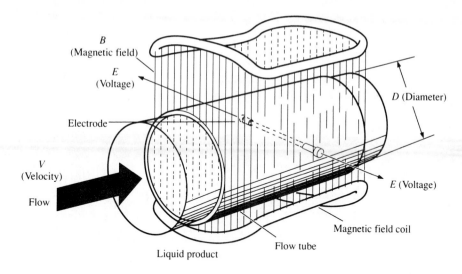

$$E \propto VBD$$

E = Induced voltage
V = Average liquid velocity
B = Magnetic field
D = Distance between electrodes (pipe I.D.)

A.

B.

FIGURE 5–43
Magnetic Flow Sensor (Photos courtesy of The Foxboro Company)

directed at a moving object (in this case, the fluid), the energy will be reflected back at a different frequency—higher if the object is moving toward the transmitter and lower if the object is moving away. As shown in Figure 5–44*A*, acoustic energy is directed upstream, and an ultrasonic transducer senses the reflected energy. The frequency difference, which is proportional to the flow rate, is calculated and converted to a proportional linear signal.

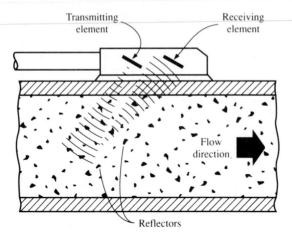

A. Operation

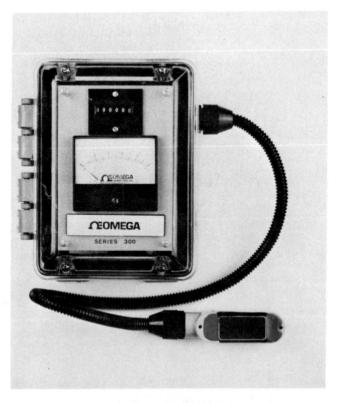

B. Commercial Model

FIGURE 5–44
Ultrasonic Doppler Flow Sensor (Photo provided courtesy of Omega Engineering—An Omega Technologies Company)

Level Sensors

Level sensors are used to determine the level, or height, of liquid or granulated solid material in tanks and hoppers. Some level sensors simply determine whether or not the material has reached a specified level, but others produce a linear analog signal proportional to the height. The output signal available from most flow sensors is usually either a proportional analog voltage or current (often 4–20 mA or 1–5 V) or a digitally compatible output switch (an *npn* current-sinking transistor or a *pnp* current-sourcing transistor).

Level sensors are classified as either contact or noncontact. Contact sensors, which must come in physical contact with the sensed medium, are generally capable of sensing only liquids; noncontact sensors are suitable also for granulated solids.

Typical contact level sensors are shown in Figure 5–45. Of the contact level sensors currently available, three types are most common: *conductive probe, float,* and *ultrasonic.*

A conductive probe level sensor detects the presence of the liquid by the change in conductivity between two stainless steel probes. When conductivity is established, a semiconductor switch is activated. Usually this is an open-collector transistor that will either source or sink current. Obviously, application of this type of sensor is limited by the conductivity of the liquid.

A float sensor detects the presence or level of the liquid by displacement of a float. Usually the float is a magnetic toroid situated around a tube. Inside the tube is a reed switch or a Hall sensor. As the liquid level increases and causes the float to rise, the reed switch opens, or the Hall sensor produces a signal that indicates motion of the float.

An ultrasonic contact sensor incorporates an ultrasonic transmitter and a receiver, set in opposition and separated by an air gap. When liquid fills the gap, the ultrasonic acoustic energy is transmitted to the receiver, resulting in closure of a switch (usually a semiconductor switch).

Typical noncontact level sensors are shown in Figure 5–46. Three categories of noncontact level sensors exist: *ultrasonic, capacitive proximity,* and *inductive proximity.*

Ultrasonic sensors operate on a retroreflective principle. Ultrasonic acoustic energy is directed at the surface of the sensed medium and is reflected back toward the receiver. The distance between the surface and the sensor can be calculated electronically from the time it takes the energy to return from the surface. A proportional analog signal is then provided as an output from the sensor. This signal indicates the relative height of the surface of the sensed medium. Usually the signal is a current between 4 mA and 20 mA or a voltage between 1 V and 5 V. The high cost of these sensors (several hundred to several thousand dollars) is offset by their long sensing distance, which may be 20 ft or more.

A capacitive proximity level sensor incorporates an oscillator, a detector, an output switch, and a sensing plate. The sensing plate is an extension of the

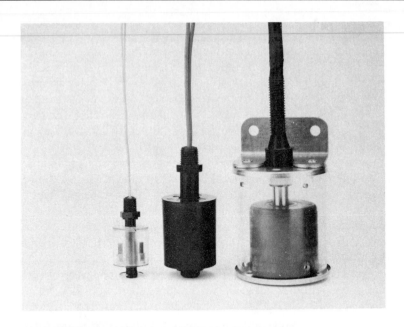

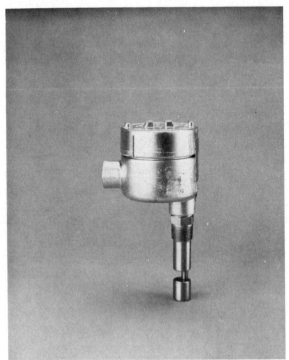

FIGURE 5–45
Contact Level Sensors (Photos provided courtesy of Omega Engineering—An Omega Technologies Company)

A.   B.

FIGURE 5–46
Noncontact Level Sensors (Photo A. Courtesy of Electro Corporation. Photo B. Courtesy of Turck Multiprox Inc.)

oscillator capacitor. If a material passes within the electric field of this plate, the effective capacitance will change, causing a corresponding shift in oscillator frequency. If the frequency change is great enough, the detector will send a signal to the output switch, activating it. As in most sensors, the output switch consists of an *npn* current-sinking transistor, or a current-sourcing *pnp* device. Although a linear output, proportional to distance, is a possibility with this type of sensor, it is uncommon. In fact, most capacitive proximity sensors are actually called capacitive proximity switches. Generally, the sensing distance increases with an increase in plate surface area. Even so, the range of most capacitive sensors is limited to just a few centimeters. Capacitive sensors are sensitive to most materials, including metals, liquids, plastics, and paper products. However, metals and other conductors provide the greatest sensing distance.

An inductive proximity level sensor is very similar to its capacitive counterpart. Instead of a sensing plate, however, it incorporates a sensing coil. When a metallic object passes into the sensing area, energy is drawn from the oscillator, causing oscillations to cease. The detector then causes the output switch to be activated. Again, the standard transistor switch arrangements are usually available. As with capacitive sensors, the sensing distance is limited to a few centimeters. The advantage to using an inductive method of sensing (as opposed to a capacitive method) is that an inductive sensor tends to ignore nonconductive objects. It should be noted that proximity level sensors can be employed in *any* application in which object proximity must be sensed.

SUMMARY

☐ Electrical input transducers and sensors are classified according to the physical variables to which they are sensitive. The four main categories are thermal, optical, magnetic, and electromechanical. The following tables give a brief summary of the operating characteristics of the transducers and sensors discussed in this chapter.

Thermal Transducers

	Operation	Advantages	Disadvantages
Thermistor	Device resistance changes inversely with variations in temperature.	Large nominal R Large R variation Inexpensive Fast	Nonlinear Narrow temperature range Requires power Self-heats
RTD	Device resistance changes directly with variations in temperature.	Linear R variation Most stable Most accurate	Expensive Small R variation Requires power
Thermocouple	Device produces voltage or current proportional to temperature.	Nearly linear output Wide temperature range Used at high temperatures Self-powered	Low output Least sensitive Requires reference

Temperature Sensors

	Operation	Advantages	Disadvantages
IC Temperature Sensor	Device produces voltage or current numerically equivalent to absolute temperature.	Linear output Small packages Inexpensive Fast	Low temperature range Low output
Temperature Transmitter	Device accepts thermocouple or RTD input. Output current is proportional to temperature.	Linear output Facilitates remote sensing	Accepts only thermocouples or RTDs as inputs
Optical Pyrometer	Focuses infrared energy on IC temperature sensor in probe. Output current is proportional to temperature.	Linear output Used at high temperatures Noncontact	Expensive

Optical Transducers

	Operation	Advantages	Disadvantages
Photoconductive Cell	Device resistance changes inversely with variations in EMR intensity.	Large *R* variation	Slow response Requires power Temperature sensitive
Photovoltaic (Solar) Cell	Converts EMR into electrical current. Current is proportional to EMR intensity.	Self-powered Linear output current	Large surface area Inefficient Slow response Nonlinear output voltage
Photodiode	Converts EMR into electrical current. Current is proportional to EMR intensity.	Small Fast	Low output Requires external power supply
Phototransistor	Converts EMR into current, which is injected into the base. Device then responds as typical transistor.	Sensitive High output Fast	Requires external power supply

Optical Sensing Methods

	Operation	Advantages	Disadvantages
Opposed Sensing	EMR beam is directed from transmitter to receiver. Object is detected when beam is interrupted.	Long sensing range	Transmitter and receiver must be precisely aligned. Separate receiver and transmitter required
Retroreflective Sensing	EMR beam is directed from transmitter to a reflector and back to receiver. Object is detected when beam is interrupted.	Transmitter and receiver may be housed in the same unit.	Sensing range limited to about 30 ft Requires remote reflector

NOT ON TEST

Optical Sensing Methods (*continued*)

	Operation	Advantages	Disadvantages
Diffuse Sensing	EMR beam is directed at sensed object, which acts as reflector. Object is sensed when beam is reflected back to receiver.	Transmitter and receiver may be housed in the same unit. No reflector required	Sensing range limited to less than 10 ft Sensed object must be reflective.

Magnetic Transducers

	Operation	Advantages	Disadvantages
Inductive	Relative motion of magnetic field and coil induces voltage in coil.	Requires no power supply	Detects motion only
Reluctive	Proximity of a low-reluctance object alters magnetic field surrounding device and induces voltage in coil.	Requires no power supply	Detects motion only
Hall-Effect	Presence of a magnetic field produces voltage proportional to strength of field.	Senses stationary fields Inexpensive	Requires external power supply

Electromechanical Transducers

	Operation	Advantages	Disadvantages
Limit Switch	Electrical contact is maintained or broken according to position of actuator.	Usually requires no external power supply Available in a variety of styles	Must contact the sensed object
Potentiometer	Resistance (or voltage) between wiper and either terminal changes as shaft is turned.	Durable	Requires external power supply Requires some sort of gearing or linkage

Electromechanical Transducers (*continued*)

	Operation	Advantages	Disadvantages
Wire Strain Gauge	Device resistance changes as gauge is compressed or elongated.	Durable	Requires external power supply Low gauge factor
Semiconductor Strain Gauge	Device resistance changes as gauge is compressed or elongated.	High gauge factor	Requires external power supply

Flow Sensors

	Operation	Advantages	Disadvantages
Turbine	Fluid flowing past turbine causes blades to spin. Rotation is detected by magnetic sensor.	Relatively inexpensive	Impedes fluid flow
Vortex	As fluid flows past a vortex shedder, vortices are produced downstream. Resultant vibrations are sensed piezoelectrically.	Minimal moving parts Accurate	Expensive Impedes fluid flow
Thermal	Probe is heated to constant temperature. Thermal transducer measures rate of cooling, which is proportional to flow rate.	No moving parts Relatively inexpensive	Slow response Impedes fluid flow Fluid must be thermally conductive.
Ultrasonic	Acoustic energy is transmitted upstream into fluid, at a given frequency. Frequency of reflected energy is proportional to fluid velocity.	No moving parts Flow may be sensed through pipe walls. Does not impede fluid flow	Most expensive
Magnetic	Voltage is induced in fluid as it passes through magnetic field.	No moving parts Relatively inexpensive	Fluid must be electrically conductive.

Flow Sensors (*continued*)

Operation	*Advantages*	*Disadvantages*
Electrodes sense voltage, which is proportional to flow.	Does not impede fluid flow	

Level Sensors

	Operation	*Advantages*	*Disadvantages*
Float	When rising liquid level reaches float, float rises and actuates either a magnetic or a mechanical switch.	Inexpensive	Senses liquids only
Conductive Probe	When rising liquid level contacts the probes, current flows and is available at output.	No moving parts Inexpensive	Senses liquids only Liquid must be electrically conductive.
Ultrasonic Contact	Acoustic energy is directed across a gap toward receiver. Liquid within gap completes transmission path.	No moving parts Senses any liquid	Expensive
Ultrasonic Noncontact	Acoustic energy is transmitted toward surface to be sensed. Reflection time determines surface distance.	No moving parts Surface contact not required Measures absolute height	Most expensive
Proximity	Frequency change in oscillator indicates object proximity. Output is activated when oscillator changes frequency.	No moving parts	Sensing range limited to about 10 cm

EXERCISES

1. Explain how the electrical characteristics of a thermistor change as its temperature varies.

2. Is the response of a thermistor considered linear?

3. Give a typical room-temperature resistance for:
   a. an NTC thermistor
   b. a PTC thermistor

4. Which of the following temperature ranges would not provide an appropriate environment in which to implement a thermistor? (Choose one or more.)
   a. $-20°C-+100°C$
   b. $1000°C-1500°C$
   c. $0°C-200°C$
   d. $2500°C-3500°C$

5. Explain how the electrical characteristics of an RTD change as its temperature varies.

6. Is the response of an RTD considered linear?

7. Which of the following temperature ranges would not provide an appropriate environment in which to implement an RTD? (Choose one or more.)
   a. $-20°C-+100°C$
   b. $0°C-700°C$
   c. $0°C-200°C$
   d. $2500°C-3500°C$

8. Give a typical room-temperature resistance for an RTD.

9. Explain the operation of a thermocouple.

10. Is the response of a thermocouple considered linear?

11. Explain why compensation circuits are required when implementing thermocouples.

12. Which of the following temperature ranges would not provide an appropriate environment in which to implement a thermocouple? (Choose one or more.)
   a. $-20°C-+200°C$
   b. $2500°C-3500°C$
   c. $0°C-800°C$
   d. $5000°C-8000°C$

13. Give an example of a situation in which an optical pyrometer would be the only choice for measuring temperature.

14. Explain how the output of a typical IC temperature sensor changes as its temperature varies.

15. Give an example of a situation in which the use of temperature transmitters would be an attractive alternative.

16. Explain how the electrical characteristics of a photoconductive cell change with variations in light intensity.

17. Are photoconductive cells sensitive to all colors of light? Explain.

18. Explain how the electrical characteristics of a photovoltaic cell change with variations in light intensity.

19. Is the response of a photovoltaic cell considered linear?

20. Explain the two modes of operation of a photodiode.

21. Explain the operation of a phototransistor.

22. Compare the sensitivity of photodiodes to that of phototransistors.

23. If a phototransistor were implemented as a small signal amplifier, with the base lead used to bias the device, how would light affect the output signal?

24. Explain why modulated light is often used in integrated optical sensors.

25. Give an example of a situation in which each of the following optical sensor arrangements would be required:
   a. opposed
   b. retroreflective
   c. diffuse

26. Explain the operation of a typical optoisolator. Why are optoisolators used so often in industrial electronics?

27. Give a typical isolation voltage for an optoisolator.

28. Explain the construction and application differences between optoisolators and optointerrupters.

29. Can an inductive or reluctive coil sense a stationary magnetic field? Explain.

30. What physical variable (e.g., temperature, pressure, or displacement) are coils usually used to measure?

31. Explain the operation of a Hall-effect transducer.

32. What is the difference between a Hall transducer and a Hall sensor?

33. What types of output are usually available from a Hall sensor?

34. Explain the difference between a mechanical switch and a solid-state switch.

35. Explain the operation of a current transformer.

36. What physical variable are current transformers used to measure?

37. Give one important caution to observe when applying current transformers.

38. Explain the construction and application differences between standard mechanical switches and limit switches.

39. Give an example of a situation in which a potentiometer is used as a transducer.

40. Explain the operation of an LVDT.

41. What physical variable are LVDTs used to measure?

42. What type of output does an LVDT produce?

43. Explain the difference between piezoresistive and piezoelectric transducers.

44. Compare the sensitivity of wire or foil strain gauges to that of semiconductor strain gauges.

45. Explain what *gauge factor* means. Give a typical gauge factor for a wire or foil strain gauge and for a semiconductor strain gauge.

46. Explain the difference between penetration and nonpenetration flow sensors.

47. If the flow rate of a dry, nonconductive gas is required, which of the following sensors is most appropriate?

    a. turbine
    b. vortex
    c. magnetic
    d. ultrasonic
    e. thermal

48. If the level of plastic pellets in a hopper must be controlled, which of the following sensors is most appropriate?
    a. float
    b. probe
    c. ultrasonic proximity
    d. capacitive proximity

APPLICATION ASSIGNMENT ANSWER

Of the three optical sensing arrangements discussed in this chapter, only one is appropriate in this situation. Two key elements determine the choice: the required sensing range and the exposure to ambient light.

Since the vehicles must be detected at a great distance from the door, diffuse proximity sensing will not provide adequate performance. The sensing beam must be directed across the vehicle path perhaps several hundred feet from the door. Either an opposed arrangement or a retroreflective arrangement will provide the required response. However, since the width of the vehicle path exceeds the maximum sensing distance for retroreflective sensors, an opposed arrangement must be used. In addition, since the sensors will be in an environment of variable ambient light, modulated transmitters and receivers should be used.

II

SIGNAL CONDITIONING AND POWER CONTROL

6

Signal Conditioning

CHAPTER OUTLINE

6–1 Introduction □ **6–2** Operational Amplifiers □ **6–3** Instrumentation Amplifiers □ **6–4** Bridge Circuits □ **6–5** Frequency to Voltage Conversion □ **6–6** Digital Signal Conditioning

LEARNING OBJECTIVES

□ Given a control system containing several operational amplifiers, identify the basic circuit configurations.

□ Given an inverting or noninverting amplifier, determine the circuit gain.

□ Given a voltage comparator and the necessary input conditions, determine the state of the output.

□ Given the graph of the signal input to an integrator, determine the graph of the output signal.

□ Given the graph of the signal input to a differentiator, determine the graph of the output signal.

□ Given a differential amplifier with signals applied to each input, determine the circuit gain and the resultant output.

□ Given a control system containing one or more instrumentation amplifiers, identify the characteristics of the circuit, and describe the output.

□ Given a bridge circuit, along with characteristics of the resistive transducer, determine the bridge voltage.

□ Determine the resolution of an analog to digital converter.

□ Given an analog to digital converter, along with the characteristics of the input signal, determine the output number.

APPLICATION ASSIGNMENT

As a technician at an industrial manufacturing facility, it is your responsibility to implement the instrumentation that will provide an indication of the temperature of a boiler. It has been decided that a thermocouple will be used to monitor the temperature. The output of the thermocouple will vary from 5 mV to 30 mV within the anticipated temperature range. The meter that will be used, however, requires an input between 1 V and 10 V. Your first task, then, is to design an amplifier that will maintain compatibility between the thermocouple output and the meter input.

After studying this chapter, you should be able to select the proper amplifier configuration and component values. One possible solution is given at the end of the chapter.

6-1   INTRODUCTION

Many input transducers exhibit a change in resistance; many offer minute output voltages or currents; and many provide nonlinear outputs. In order for many of the transducers introduced in Chapter 5 to be of significant use, their response must be *conditioned* or changed to a way that will make them compatible with the other circuits that make up an electronic control system. This chapter focuses specifically on those circuits that are used to modify the response of electrical input transducers. Since operational amplifiers are an integral part of most signal conditioners, a review of their characteristics and basic circuit configurations is offered first.

6-2   OPERATIONAL AMPLIFIERS

Operational amplifiers (op amps) find widespread application in every branch of electronics, and are especially important in the implementation of electronic control systems. It is therefore imperative that students become fluent in the operational characteristics and application of op amps as integral system components.

Although op amps may be implemented in such diverse applications as amplifiers, oscillators, signal generators, filters, and power supplies, emphasis will be on applications associated specifically with control systems. In addition, the op amp will be presented as an ideal amplifier. That is, such topics as input and output impedance, slew rate, and frequency stabilization will not be considered. The most commonly used circuit configurations will be presented and explained. The chapters that follow will provide numerous applications.

Fundamentals

The basic op amp symbol, shown in Figure 6–1, is characterized by a triangular shape and usually has five standard connections. Most op amps require a bipolar supply, so there are connections for $+V$ and $-V$. Ground is indirectly connected to the device through the external input and output circuits, which all share the same common reference point as the $+V$ and $-V$ supplies. For devices that can be operated from a single supply, the $-V$ input connection may be replaced by a connection to ground.

The output of the op amp is taken from the apex of the triangle, and the two inputs are connected to the opposite base. The inverting and noninverting inputs

FIGURE 6–1
Standard Op Amp Connections

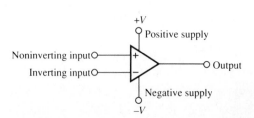

are so named because the polarity of the output is always the same as that of the noninverting input. Likewise, the polarity of the output is always opposite that of the inverting input. That is, if the inverting input is positive with respect to the noninverting input, the output will be driven negative with respect to ground. If the noninverting input is positive with respect to the inverting input, the output will be driven positive with respect to ground.

Op amps gained initial popularity as analog computation devices, the output voltage of which was the result of a mathematical operation performed on the input voltages (hence the name *operational amplifier*). Therefore, it may be thought that an exceptional grasp of mathematics is required to put op amps to use in practical applications. On the contrary, the characteristics of virtually every op amp circuit configuration can be understood by applying the following guideline:

> Given any input conditions, the output of an op amp will change in such a way as to force (or attempt to force) the inverting input to the same potential as the noninverting input.

Voltage Follower

One of the most straightforward illustrations of an op amp in an amplifying configuration is the *voltage follower* (sometimes called a *buffer* or *isolation amplifier*). A voltage follower provides a gain of 1, while maintaining the inherent high input impedance and low output impedance of the device. The unity gain configuration makes the voltage follower of little use as a voltage amplifier. However, because of the high input impedance of the circuit, a voltage follower is able to capture weak signals from high-impedance outputs, with very little signal loss. Moreover, since the output impedance is so low, the voltage follower is able to provide virtually all of the signal to low-impedance loads.

A voltage follower is nothing more than an op amp with 100% of its output fed back to the inverting input (see Figure 6–2). The input signal is applied directly to the noninverting input. If the input signal rises to +1 V with respect to ground, the output, in an attempt to force the inverting input to the same potential as the noninverting input, will likewise rise to +1 V. In fact, the output will rise to a level that may differ from the input by a few tens of microvolts. For all practical purposes, however, the input and output voltages may be considered identical.

FIGURE 6–2
Voltage Follower

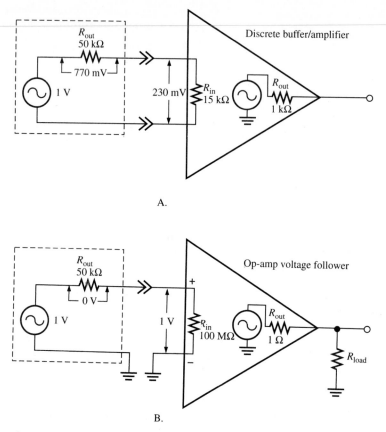

A.

B.

FIGURE 6–3
Practical Application of a Voltage Follower

Figure 6–3 illustrates a practical application of a voltage follower. Shown in Figure 6–3A is a signal source with an output resistance of 50 kΩ, connected to an amplifier that has an input resistance of 15 kΩ. If the applied signal is 1 V, only 230 mV will appear at the input of the amplifier. The remaining 770 mV will be dropped across the output resistance of the signal source.

If, however, the same signal source is connected, as in Figure 6–3B, to a typical op amp voltage follower (with an input resistance of 100 MΩ), virtually all of the input signal will appear at the op amp input. In addition, since the output resistance of the voltage follower is so low, any load connected to its output (including the amplifier shown in Figure 6–3A) will receive the full, unattenuated signal.

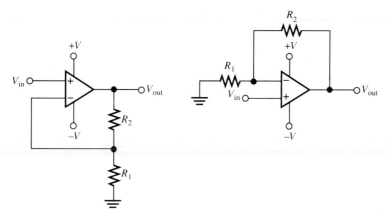

FIGURE 6–4
Noninverting Amplifier

Noninverting Amplifier

The *noninverting amplifier,* like the voltage follower, will provide a high input impedance, while maintaining a low output impedance. In addition, however, this configuration will provide a gain that is continuously adjustable from 1 to several hundred. In fact, the only factors that limit the gain of the amplifier are the open-loop gain of the op amp and the stability of the resultant circuit.

This configuration differs from the voltage follower in that only a portion of the output voltage is fed back to the inverting input (see Figure 6–4). The op amp will still attempt to force the differential input voltage to 0. However, in this case, the output will have to rise to a higher potential to do so.

Suppose that V_{in} in Figure 6–4 is +1 V. In addition, assume that the values of R_1 and R_2 are 1 kΩ and 9 kΩ, respectively. For the inverting input to attain a potential of +1 V, the output of the op amp must force 1 mA through R_1. Because of the high input impedance of the op amp, virtually no current flows into the inverting input. Therefore, all of the current that flows through R_1 must also flow through R_2. If the output provides a current of 1 mA through a total resistance of 10 kΩ ($R_1 + R_2$), the output potential must be +10 V. Thus, since the input is +1 V and the output is +10 V, the circuit provides a gain of 10. It should be noted that for an input that is positive with respect to ground, the output is also positive with respect to ground. That is, the output is in phase with the input (hence the name *noninverting amplifier*).

The ratio of R_2 to R_1 determines the closed-loop gain of the circuit (the gain is termed *closed-loop* because feedback exists from output to input). If the value of R_1 is decreased, the op amp will be forced to produce a higher current to attain the same voltage drop. The higher current will result in a higher output voltage and a correspondingly higher gain. The same result may be had by increasing the value of R_2. The exact relationship between gain and the values of R_1 and R_2 is given by equation 6.1:

$$A_{CL} = \frac{R_1 + R_2}{R_1} = 1 + \frac{R_2}{R_1} \tag{6.1}$$

Voltage-to-Current Converter

Since op amps are primarily voltage amplifiers, most have limited output current capability. Nevertheless, many control actuators and indicators require a regulated variable-current input. If op amps are to be used in these applications, some means of current amplification and control must be implemented.

The circuit shown in Figure 6–5 will provide an output current proportional to the input voltage. Close inspection will reveal that this configuration is nothing more than a noninverting amplifier. The feedback resistor, in this case, is made up of R_1 and the base-emitter junction of Q. As the op amp output voltage increases, base current is forced through Q, resulting in a proportional collector current through Q, the load, and R_2. When the resultant voltage across R_2 is equal to the input voltage, the output will stabilize. The magnitude of the load current, therefore, may be controlled by varying the input voltage or the value of R_2.

It should be recognized that this circuit is limited to positive values of input voltage. The exact relationship between load current, input voltage, and the value of R_2 is given by the following formula:

$$I_{\text{load}} = \frac{V_{\text{in}}}{R_2} \tag{6.2}$$

The value of R_1 must be sufficient to protect the base-emitter junction of Q. The transistor, Q, serves only to carry the higher load current that the op amp cannot provide. The characteristics of the transistor are not critical. It should be capable

FIGURE 6–5
Voltage-to-Current Converter

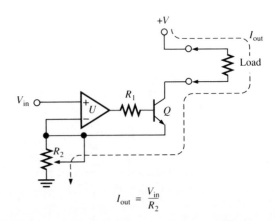

$$I_{\text{out}} = \frac{V_{\text{in}}}{R_2}$$

of carrying the required load current, and it should have a beta of at least 100, to ensure that it does not require more base current than the op amp can provide.

Suppose that R_2 in Figure 6–5 is adjusted to 500 Ω. If an input voltage of +1 V is applied to the noninverting input of the op amp, the output voltage will increase until the voltage across R_2 is also +1 V. This condition will exist when 2 mA flows through the load. As long as the input is maintained at +1 V, the load current will be maintained at 2 mA.

Note that +V must be of sufficient magnitude to force 2 mA through the load. If the load resistance were 50 kΩ, +V would have to be +101 V to produce a current of 2 mA!

Naturally, the ratio of input voltage to load current can be adjusted simply by changing the value of R_2. If, for example, the value of R_2 is increased to 1 kΩ, the same +1 V will result in a load current of only 1 mA.

EXAMPLE 6–1 Given the circuit shown in Figure 6–5, suppose that an input voltage-to-output current ratio of 2 V/mA is desired. To what value must R_2 be adjusted?

Rearranging equation 6.2 will yield the required value of R_2:

$$R_2 = \frac{V_{\text{in}}}{I_{\text{load}}} = \frac{2 \text{ V}}{1 \text{ mA}} = 2 \text{ k}\Omega$$

Inverting Amplifier

Another op amp configuration capable of providing voltage gains far greater than 1 is the *inverting amplifier*. As the name implies, a signal at the input will result in a signal of opposite polarity at the output. That is, the output voltage is 180° out of phase with the input. The output signal is usually larger than the input signal. As shown in Figure 6–6, a portion of the output is still fed back to the inverting input. The input signal, however, is also applied to the inverting input. The noninverting input is usually grounded, although technically it could be held at any potential other than ground.

To understand the operation of this configuration, consider the circuit of Figure 6–7. If +1 V is applied to the input, the output of the op amp will attempt to

FIGURE 6–6
Inverting Amplifier

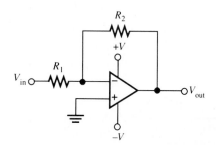

FIGURE 6–7

Operation of an Inverting Amplifier

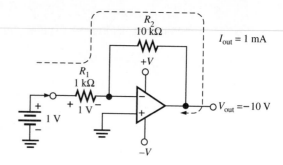

maintain the inverting input at 0 V (the same potential as that of the noninverting input). If that attempt is successful, the voltage drop across R_1 will have to be 1 V, with the polarity as shown. A voltage of 1 V across 1 kΩ is due to a current of 1 mA. Since (because of high input impedance) no current enters or exits the inverting input, this 1 mA must be provided by the output. Any current flowing through R_1 must also flow through R_2 (they are in series). Consequently, R_2 will drop 10 V. Note, however, that to produce the proper polarity across R_1, the output potential must be negative with respect to ground. Since the inverting input is virtually at the same potential as ground, the voltage across R_2 is also the output voltage. Thus, with an input of $+1$ V and an output of -10 V, the amplifier provides a gain of -10 (the minus sign simply indicates a phase inversion).

Note that the output of the op amp in Figure 6–7 will attempt to provide 1 mA regardless of the value of R_2. Naturally, if the value chosen for R_2 is too large (or if the value chosen for R_1 is too small), the op amp output may saturate before attaining the proper magnitude. By changing the value of R_2, it is possible to vary the output voltage for any given input and, therefore, the gain. Likewise, changing the value of R_1 will also vary the gain. The exact relationship between the gain and the values of R_1 and R_2 is as follows:

$$A_{CL} = -\frac{R_2}{R_1} \tag{6.3}$$

Summing Amplifier

The circuit of Figure 6–8 may be used to obtain an output that is the algebraic sum of the magnitudes of its inputs, multiplied by the gain of the amplifier. Although only two inputs are shown, the circuit may be easily expanded, simply by adding a resistor in parallel with R_1 and R_2 for each additional input required.

The similarity of this circuit to the inverting amplifier is no mere coincidence. Operation is, in fact, identical. The output of the op amp will strive to maintain the inverting input at the same potential as the noninverting input, which is at ground potential. To accomplish this result, the output will have to force enough current through R_1 and R_2 to produce voltage drops equivalent to each of

FIGURE 6–8
Summing Amplifier

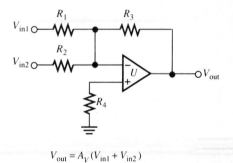

$$V_{out} = A_V(V_{in1} + V_{in2})$$

their respective signal inputs. Since all of this current must flow through R_3, the gain may be controlled by adjusting its value. That is, adjusting the value of R_3 forces the output to attain a higher or lower voltage in order to produce the appropriate amount of current through R_1 and R_2.

If the values of all input resistors are equal, the amplifier gain is given by the following formula:

$$A_V = -\frac{R_3}{R_1} \tag{6.4}$$

The amplifier output may then be described by the following equation:

$$V_{out} = A_V(V_{in1} + V_{in2}) \tag{6.5}$$

EXAMPLE 6–2 Given the summing amplifier shown in Figure 6–8, suppose that $R_1 = R_2 = 2.2\ k\Omega$, and that $R_3 = 47\ k\Omega$. If $V_{in1} = +0.84\ V$, and $V_{in2} = -1.2\ V$, what output will be produced?

The output can be determined directly from equation 6.5:

$$V_{out} = -\frac{47\ k\Omega}{2.2\ k\Omega}\ (0.84\ V - 1.2\ V) = +7.69\ V$$

Voltage Comparator

If an op amp is operated in an open-loop mode (that is, with no feedback from output to inverting input), the circuit is called a *comparator*. Basically, the output of a comparator indicates which of its inputs is at a higher potential. If the inverting input is positive with respect to the noninverting input, the output will be driven to negative saturation (to within about 1 V of $-V$). If the inverting input is negative with respect to the noninverting input, the output will be driven to positive saturation (to within about 1 V of $+V$). Note that unlike the outputs of linear amplifying configurations, which have continuously variable output volt-

FIGURE 6–9
Voltage Comparator Circuit

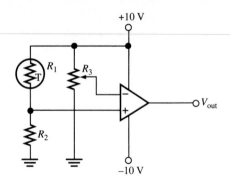

ages, the output of a comparator has only two stable states—positive saturation and negative saturation.

Whether the output of the comparator saturates positive or negative depends on which of its inputs is at a higher potential. If, for example, a signal of +2 V DC is applied to the inverting input while the noninverting input is held at +3 V DC, the output will saturate positive. As the input circuitry of the op amp detects the voltage difference, a signal is sent to the output amplifier that will result in an adjustment. Since the potential at the inverting input is less than that at the noninverting input, the output will be driven positive (in an attempt to force the inverting input to the same potential as the noninverting input). Obviously, no change will occur at the inverting input, since no feedback exists. Nevertheless, the input circuitry will continue to request a higher output voltage. The magnitude of the output voltage will continue to increase until it reaches the physical limit of $+V$.

A practical application should give a more realistic perspective. Consider the circuit of Figure 6–9. Note that the inverting input can be adjusted to any potential between 0 V and +10 V. Suppose that it is now set at +2 V. If the thermistor has a resistance range from 100 kΩ when cold to 5 kΩ when hot, then the noninverting input will have a cold-to-hot voltage range of +0.91 V to +6.7 V. When the temperature sensed by the thermistor is cold, the output of the comparator will saturate negative (the output will attempt to force the inverting input down to +0.91 V). As the temperature increases, the magnitude of the voltage at the noninverting input will likewise increase. Eventually, this voltage will exceed +2 V, and the output will be driven to positive saturation. This output level may indicate that the temperature of the thermistor has risen to an appropriate level.

Comparator with Hysteresis

The output of a comparator may be driven from one saturation level to another as the input varies about the reference voltage by only a few microvolts. In many situations this response is unacceptable, as it makes the device too sensitive to input changes.

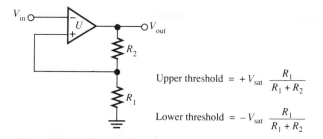

Upper threshold $= +V_{sat}\,\dfrac{R_1}{R_1 + R_2}$

Lower threshold $= -V_{sat}\,\dfrac{R_1}{R_1 + R_2}$

FIGURE 6–10
Op Amp with Positive Feedback

With feedback applied to the noninverting input, it seems as though the circuit of Figure 6–10 will oscillate. In fact, this configuration is a comparator with two stable output levels ($+V_{sat}$ and $-V_{sat}$). The effect of applying a portion of the output to the noninverting input is to change the switching threshold for the comparator. Assume, for example, that the output of U in Figure 6–10 is saturated at its positive level. Since R_1 and R_2 form a voltage divider, the amount of output signal that is fed back to the noninverting input is

$$+V_{sat}\left(\frac{R_1}{R_1 + R_2}\right)$$

To force the output to negative saturation, the magnitude of the voltage at the inverting input must exceed this value. As soon as the output switches, however, the feedback voltage changes to

$$-V_{sat}\left(\frac{R_1}{R_1 + R_2}\right)$$

To force the output back to positive saturation, the magnitude of the voltage applied to the inverting input must become more negative than this value.

The voltage divider network formed by R_1 and R_2 creates two separate thresholds—a high threshold (positive feedback voltage) and a low threshold (negative feedback voltage). These may be simultaneously adjusted by changing the R_1/R_2 ratio. The overall effect of two switching thresholds is to decrease the sensitivity of the comparator to input variations. For example, suppose that the saturation levels of U in Figure 6–10 are $+15$ V and -15 V and that the values of R_1 and R_2 are 1 kΩ and 10 kΩ, respectively. To force the output to switch from $-V_{sat}$ to $+V_{sat}$, the input will have to change from $+1.36$ V to -1.36 V, a total variation of 2.73 V. With a fixed threshold of $+1.36$ V, a standard comparator without feedback would require that the input vary from perhaps $+1.368$ V to $+1.352$ V, a total variation of only 0.016 V.

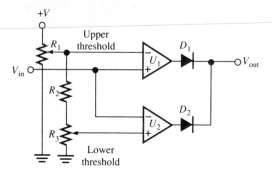

$$V_{out} = 0 \text{ V if Lower threshold} < V_{in} < \text{Upper threshold}$$

FIGURE 6–11
Window Comparator

Window Comparator

Figure 6–11 shows a circuit that offers operation similar to that of Figure 6–10. The main difference is that the circuit of Figure 6–11 allows independent adjustment of the upper and lower thresholds. As long as the magnitude of the input voltage is between the two thresholds (within a given voltage ''window''), the output will be 0 V. Otherwise, the output voltage will be at the positive saturation level (minus one diode voltage drop) of U_1 and U_2. For this reason the circuit is called a *window comparator.*

The upper and lower thresholds are set by adjusting the potentiometers R_1 and R_3. Connecting R_2 and R_3 to the wiper of R_1 ensures that the upper threshold voltage will always be higher than the lower. This condition is essential for proper operation. To avoid loading the lower half of R_1, R_2 and R_3 should have considerably higher values than R_1.

Suppose that the upper and lower thresholds are set at +8 V and +5 V, respectively. If the magnitude of the input voltage is +3 V (below both thresholds), the output of U_2 will be saturated positive, forward-biasing D_2 and providing current to the load. The output of U_1 will be at negative saturation, and the presence of D_1 will therefore prevent any load current from U_1. If the input voltage increases to +6 V, which is between the two thresholds, the outputs of both U_1 and U_2 will be driven to negative saturation. The diodes D_1 and D_2 will prevent any load current, and the resultant output will be 0 V. If the input voltage increases even further, to perhaps +10 V, which is above both thresholds, the output of U_1 will be driven to positive saturation, and load current will be provided through D_1. The output of U_2, however, will remain at negative saturation, holding D_2 reverse-biased.

The output of the comparator may be used to drive another piece of hardware, such as a relay, a solenoid valve, or even another circuit. The signal may also be sampled by a microprocessor-based system such as a programmable controller or computer. Additionally, since the potential at the inverting input is adjustable, the temperature at which the comparator output switches is also adjustable.

Integrators

If a capacitor is connected in the feedback loop of an op amp, the circuit output will be a function not only of the input signal, but also of time. Such a circuit configuration is called an *integrator* because of the mathematical operation (integration) performed on the input signal. Very simply, the output of an integrator will reflect not only the magnitude of the input signal, but also how long the signal has been present at the input. This information is necessary in many control systems, where the duration of an error signal is a factor used to help bring the system back within an acceptable operating range.

Charge, voltage, and capacitance are related by the following equation:

$$Q = CV \tag{6.6}$$

Since charge (Q) can also be described as the product of current and time, the equation can also be written as follows:

$$It = CV \tag{6.7}$$

According to equations 6.6 and 6.7, if a potential difference of 1 V exists across the plates of a 1 F capacitor, there must be a 1 C charge differential between the two plates. Likewise, if a current of 1 A is transferred between the plates of a 1 F capacitor for 1 s, the resultant potential difference will be 1 V.

If switch S in Figure 6–12 is closed, the voltage across capacitor C will follow the charge curve that is characteristic of this circuit. From equations 6.6 and 6.7, one would expect that after two equal time increments, the capacitor

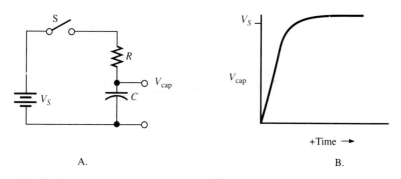

A.   B.

FIGURE 6–12
Characteristic Charge Curve for a Capacitor Charged by a Constant Voltage Supply

FIGURE 6–13
Op Amp Integrator Circuit

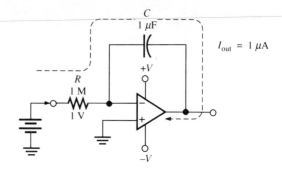

voltage would double. One look at the charge curve reveals that this simply is not true. It seems as though the circuit of Figure 6–12 does not obey the equations. In fact, it does. The difference is that the capacitor in Figure 6–12 is being charged from a constant voltage supply. As the capacitor accumulates more and more charge, the charging current decreases. Equations 6.6 and 6.7 assume a constant charging current. Therefore, for the capacitor voltage to be directly proportional to the length of time that a signal is applied, the capacitor must be charged with a constant current.

An op amp configured as an inverting amplifier inherently provides a constant current through its feedback path (for a constant input voltage). If the feedback resistor is replaced by a capacitor, the op amp will attempt to maintain a constant current through the input resistor, and so also through the capacitor.

Consider the circuit of Figure 6–13. With an input signal of +1 V DC, the output will change to a potential that will force 1 μA through R. Naturally, since R and C are in series (remember that no current flows into the op amp inputs), 1 μA must also flow through C. As C begins to charge, the output will be forced to a higher potential (a higher negative voltage) in order to maintain this current. Since the inverting input is forced to the same potential as ground, the voltage across the capacitor is also the output voltage.

If C in Figure 6–13 is initially uncharged, the output may be described by the following relationship:

$$V_{out} = -\frac{It}{C} \qquad\qquad (6.8)$$

Using this relationship, one can construct a table (Figure 6-14) that describes the output of the integrator in Figure 6–13 at various points in time (after the +1 V DC has been applied to the input).

FIGURE 6–14
Integrator Output Signal in Response to
Constant DC Input

Time (seconds)	V_{out} (volts)
0.5	−0.5
1.0	−1.0
1.5	−1.5
2.0	−2.0
2.5	−2.5

Notice that V_{out} represents the accumulated value of the input current multiplied by the amount of time that the input was applied. This property is true regardless of the values chosen for R and C.

Figure 6–14 shows the output signal in response to a constant DC input. If the input signal is changing, the output is not so easily calculated. The general relationship between input and output is given by the following equation:

$$V_{out} = -\frac{1}{RC}\int_0^T V_{in}\, dt + V_0 \qquad\qquad (6.9)$$

where V_0 represents the initial capacitor voltage. That is, the output voltage is given by the integral of the input voltage, plus any initial capacitor voltage. For the case in which the input signal is a constant DC voltage, equation 6.9 simplifies to the following equation:

$$V_{out} = -\left(\frac{V_{in}}{RC}\right) t + V_0 \qquad\qquad (6.10)$$

Shown in Figure 6–15 are some typical input graphs with the resulting output graphs, which may better demonstrate the implications of equations 6.9 and 6.10. From these graphs another useful characteristic should be recognized. That is, the rate of change of the output of an integrator is proportional to the magnitude of its input. If the magnitude of the input increases, the output changes at a greater rate.

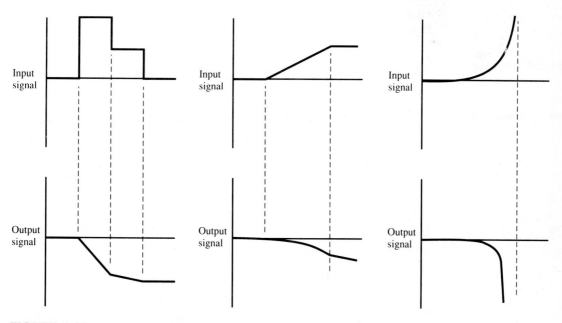

FIGURE 6–15
Typical Integrator Input and Output Signals

EXAMPLE Given the integrator shown in Figure 6–13, suppose that the capacitor is initially
6–3 uncharged. If a voltage of −2.5 V is applied to the input of the op amp for a period
of 3 s, what final output will result?
Equation 6.10 yields:

$$V_{out} = - \left(\frac{-2.5 \text{ V}}{1 \text{ sec}} \right) 3 \text{ s} + 0 \text{ V} = +7.5 \text{ V}$$

Differentiators

If a capacitor is connected in series with the input of an op amp, the output is a
function of the rate of change of the input signal. Such a circuit is called a
differentiator because of the mathematical operation (differentiation) performed
on the input signal. Very simply, the output of a differentiator is proportional to
the rate at which the input signal is changing. Like the integrator, this circuit
provides information that is necessary in certain types of control systems, to help
bring the system back within an acceptable operating range.

As in the integrator, there must be a constant charging current to the capacitor for the circuit to be useful. Notice that the differentiator differs from the
integrator in that the positions of *R* and *C* have been reversed (see Figure 6–16).
This seemingly subtle circuit change drastically changes the manner in which the
output responds to input signals.

If *S* in Figure 6–16 is closed, the inverting input will instantaneously be
driven to +1 V DC, because the voltage across a capacitor cannot be instantly
changed. Since the noninverting input is at ground potential, the op amp output
will be driven to negative saturation. Consequently, a charging current will flow
through *R* and *C* long enough to charge *C* to 1 V DC. When the capacitor is fully
charged (to 1 V DC), the current through *R* will fall to zero. Since there will no
longer be a voltage across *R,* the output voltage will also be zero. Furthermore,
the output will remain at zero as long as there are no further changes at the input.

Suppose, however, that the input signal is changing at a linear rate (that is,
during equal time increments, the input changes by equal amounts). Equation 6.7

FIGURE 6–16
Op Amp Differentiator Circuit

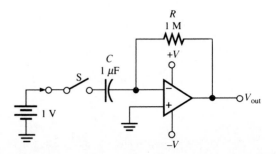

can be rearranged to show that if the capacitor voltage increases linearly, the current will be constant:

$$I = C\left(\frac{V}{t}\right) \qquad (6.11)$$

For example, suppose that the input voltage (and so, also, the capacitor voltage) in Figure 6–16 is changing at a rate of 1 V/s. Substituting this value into equation 6.11 will yield a constant current of 1 μA. Since this current flows through R, the voltage developed across R (and so, also, the output voltage) will be a constant 1 V. If the input signal changes at a greater rate, or if the value of R is increased, the constant voltage output will be greater. Thus, the output of a differentiator is proportional to the rate of change of the input signal.

If the input signal is changing at a nonlinear rate, the feedback current will not be constant. The output, therefore, will also be changing. The general relationship between input and output is given by the following equation:

$$V_{out} = -RC\left(\frac{dV_{in}}{dt}\right) \qquad (6.12)$$

To better illustrate the implications of equation 6.12, some typical input graphs and the resulting output graphs have been provided in Figure 6–17.

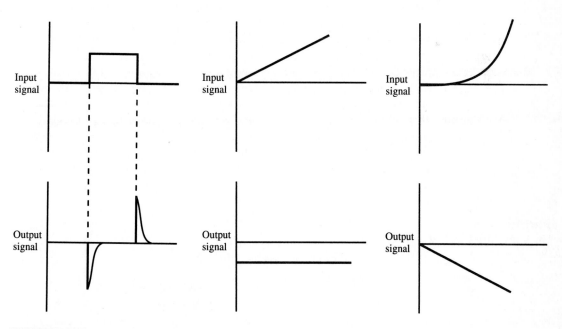

FIGURE 6–17
Typical Differentiator Input and Output Signals

EXAMPLE 6–4 Given the differentiator shown in Figure 6–16, suppose that the op amp is provided with an input signal that is increasing at a constant rate of 4 V/ms. If R = 100 kΩ and C = 0.015 μF, what output voltage will result?

Since the input is changing at a constant rate, the output will be driven to a constant value given by equation 6.12:

$$V_{out} = -(100 \text{ k}\Omega)(0.015 \ \mu\text{F})\left(\frac{4 \text{ V}}{1 \text{ ms}}\right) = -6 \text{ V}$$

Differential Amplifiers

To measure accurately the output from a transducer whose output varies between 50 μV and 250 μV, the signal must be amplified. At first glance, amplifying the signal seems to be a straightforward task. After all, the inverting and noninverting amplifiers are both easy to construct, and each does a fine job of amplifying whatever is present at its input. That is precisely the problem. Along with the desired signal, noise (spurious unwanted signals) will be induced in the wires connected to the inputs. Noise is most often due to changes in nearby electromagnetic fields, and it can range from several hundred microvolts to several tens of millivolts. Since the noise will be amplified along with the signal output from the transducer, the desired signal will be totally obliterated. What is required in such cases is an amplifier capable of rejecting the noise and amplifying only the desired signal. Such circuits exist and are called *differential amplifiers*.

As shown in Figure 6–18, a differential amplifier is really a combination of an inverting amplifier and a noninverting amplifier. If input 1 is grounded, the circuit takes on the appearance of a noninverting amplifier, with only a portion of the signal from input 2 appearing at the input. If, however, input 2 is grounded, the circuit becomes an inverting amplifier, with the noninverting input connected to ground through a resistor. If $R_1 = R_3$ and $R_2 = R_4$, the inverting gain equals the noninverting gain. Consequently, the output will reflect the difference between the two input signals, amplified by the differential gain R_2/R_1 or R_4/R_3. Note that the resistors must be precisely matched. In addition, to adjust the gain, the ratios

FIGURE 6–18
Op Amp Differential Amplifier

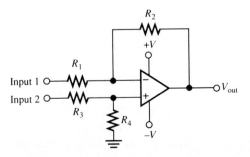

of R_2 to R_1 and R_4 to R_3 must be adjusted simultaneously and equally. Such adjustment is difficult to accomplish, and so differential amplifiers are usually of fixed gain.

Since only the difference signal at the input is amplified, noise, which usually appears equally at each input, is automatically rejected. One specification that is important in situations that warrant use of a differential amplifier is the *common-mode rejection ratio* (*CMRR*) of the op amp. Basically, this is the ratio of differential gain to common-mode gain, expressed in decibels. *Common-mode gain* is the gain by which signals appearing equally at both inputs are amplified. In situations where high-precision measurements are required, an amplifier with a high CMRR would be a wise choice. The CMRR is calculated according to the following formula:

$$\text{CMRR} = 20 \log \frac{A_{\text{DIFF}}}{A_{\text{CM}}} \tag{6.13}$$

A CMRR of 90 (typical of many op amps) indicates that the differential signals are amplified almost 32,000 times more than the common-mode signals.

Op Amp Compensation

Although the characteristics of modern-day op amps closely approximate those of an ideal amplifier, no electronic circuit is perfect. Circuitry must be implemented to compensate for the nonideal characteristics that most op amps possess. In control applications in which high-precision measurements are required, it is essential that any op amp output voltage be due only to variations at the input. In reality, imbalances in the internal circuitry of the op amp itself may contribute to the output voltage. Such imbalances, of course, must be eliminated if accurate response is required.

The input of every op amp consists of a transistor differential amplifier. If the transistors are bipolar junction transistors (BJTs) or junction field effect transistors (JFETs), a slight amount of bias or leakage current will flow into or out of each input. As shown in Figure 6–19A, the bias currents flow through different equivalent resistances. If the output resistance of the signal sources is ignored, the bias current for the inverting input flows through a resistance made up of the parallel combination of R_1 and R_2. The bias current for the noninverting input, however, flows directly to ground. The bias current, therefore, will produce different voltage drops at each input. Although the difference in potential may be very slight, it will be amplified by the open-loop gain of the amplifier and will result in an initial output offset voltage. To compensate for this error, all that is required is a resistor at the noninverting input, as shown in Figure 6–19B. The value of this resistor should be equivalent to the parallel combination of R_1 and R_2 so that the bias current will produce identical voltage drops at both inputs.

The compensation resistor may not be all that is required to eliminate the output offset voltage. Aside from the fact that the input bias currents may not be

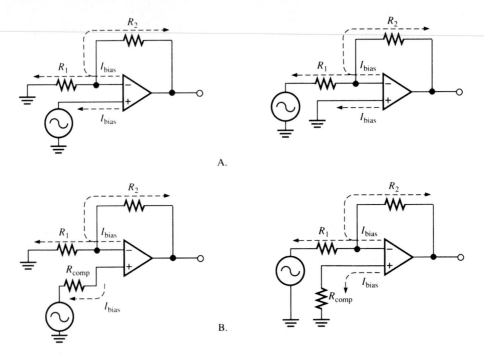

FIGURE 6–19
Addition of Bias Compensation Resistor to Op Amp

equal, the transistors that make up the input differential amplifier may have differing characteristics. To compensate for such differences, the bias of the differential amplifier must be adjusted. Many op amps provide pins specifically intended for this function. Usually, the pins are labeled "null adjust." For most op amps, nulling the device requires that a potentiometer be connected between the null-adjustment pins. The wiper of the potentiometer is then connected to the positive or negative supply. Because a variety of methods are employed, individual device specifications should be consulted to determine the proper arrangement and procedure. If null-adjustment pins are not provided, the bias current can be adjusted externally by using the circuits shown in Figure 6–20.

6–3   INSTRUMENTATION AMPLIFIERS

An instrumentation amplifier is made up of several monolithic op amps, which can be positioned into two distinct functional groups—a *differential amplifier* and a *buffered differential input*.

The circuit configuration shown in Figure 6–21 serves two functions. The voltage followers provide the required high impedance at both inputs, and the

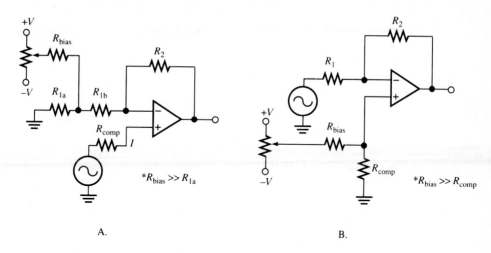

A.   B.

FIGURE 6–20
External Null Adjustments for (*A*) a Noninverting Amp and (*B*) an Inverting Amp

three resistors provide a means of adjusting the magnitude of the differential output voltage. If $R_1 = R_2$ in Figure 6–21, the resultant voltage drop across R_g is

$$V_{in2} - V_{in1}$$

The voltage drop is such because for each of the voltage followers, the inverting inputs are driven to the same potentials as the noninverting inputs. Since the

FIGURE 6–21
Differential Input Buffer

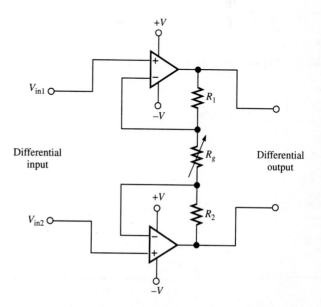

differential output voltage is due to the current through all three resistors, varying R_g will have the effect of changing the current that flows through R_1 and R_2 and, therefore, the voltage drop across the entire network. If $R_1 = R_2 = R$, the voltage across this network is given by:

$$(V_{in2} - V_{in1})\left(1 + \frac{2R}{R_g}\right) \tag{6.14}$$

If we connect the circuit of Figure 6–21 to that of Figure 6–18, the resultant circuit is what is referred to as an instrumentation amplifier, a high-input-impedance differential amplifier with an adjustable gain (see Figure 6–22). Since R_g varies the amount of differential signal input to the differential amplifier, it effectively controls the overall circuit gain. Terminals are usually provided on an instrumentation amplifier package so that R_g can be provided externally by the user. If R_3, R_4, R_5, and R_6 are all equal, the differential amplifier will have unity gain, and the output voltage of the instrumentation amplifier will be given by equation 6.14.

In addition to the connections for an external gain-adjust resistor, most instrumentation amplifiers have connections called *sense* and *reference*. Figure 6–23 will help illustrate the need for these. If one side of R_6 is grounded, and one side of R_4 is connected directly to the output (as in Figure 6–22), the effective resistance of R_6 will be increased by the resistance in the ground path from the load. The effective resistance of R_4, however, will not change. Even though this

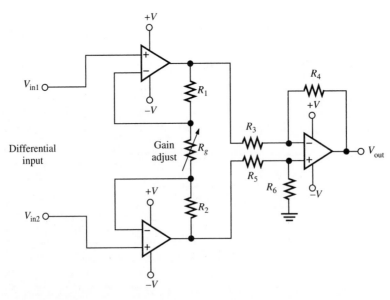

FIGURE 6–22
Basic Instrumentation Amplifier

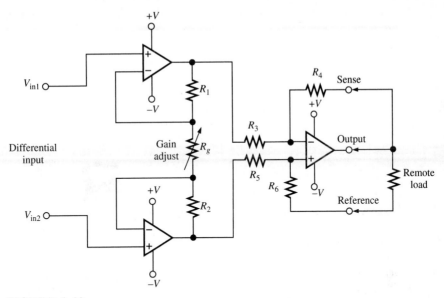

FIGURE 6–23
Typical Instrumentation Amplifier Connected to a Remote Load

additional resistance may seem negligible, it will upset the gain balance and so, also, the accuracy of the differential amplifier. If all three wires shown in Figure 6–23 are connected to the load, any additional resistance in the R_6 path will be offset by the same amount of resistance in the R_4 path.

The block diagram of a commercial instrumentation amplifier, the Analog Devices AD624, is shown in Figure 6–24. The AD624 provides integrated laser-trimmed resistors that allow preset gains of 1, 100, 200, or 500 to be selected simply by shorting pin 3 to pin 11, 12, or 13. Other preset gains can be obtained by also shorting pin 16 to pins 11, 12, or 13. In all, there are 14 preset gains that may be selected. Of course, *any* gain can be obtained by connecting an external resistor between pins 3 and 16. If this latter option is chosen, the gain is given by:

$$A_v = 1 + \frac{40 \text{ k}\Omega}{R_g} \tag{6.15}$$

A typical application of the AD624 is shown in Figure 6–25. Here a signal from a compensated type J thermocouple is amplified by a factor of 200. Note that R_5 has been provided to null the input circuitry of the amplifier. Two separate nulling circuits have been implemented in the AD624. Pins 4 and 5 balance the input circuitry, while pins 14 and 15 are used to balance the output. Output offset errors are independent of gain and are usually nulled only for low-gain applications. Any offset error generated at the input, however, will be amplified along with the desired signal. For this reason, input offset nulling is usually employed for applications that require gains in excess of 10.

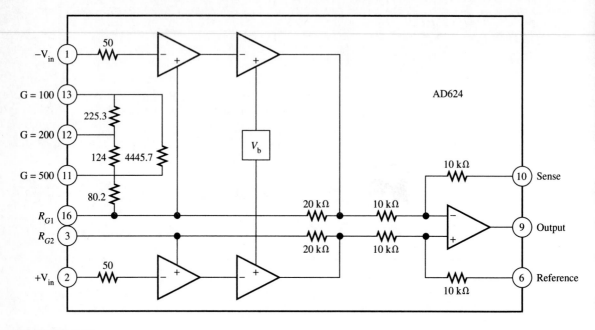

FIGURE 6–24
The Analog Devices AD624 Instrumentation Amplifier

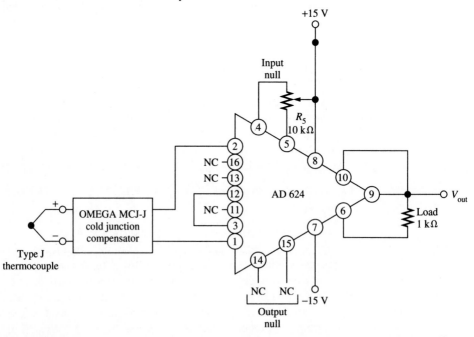

FIGURE 6–25
Typical AD624 Application

6-4   BRIDGE CIRCUITS

Many transducers exhibit a resistance change with respect to changes in the physical variable that they are used to measure. The most common way of converting this change in resistance to a change in voltage or current is through the use of a resistance bridge. Although various configurations are possible, most bridges employ four resistors (see Figure 6–26A).

For the arrangement shown in Figure 6–26A, the values of R_1 and R_2 are fixed and often identical. The difference between the voltage across R_3 and the voltage across the transducer is referred to as the *bridge voltage*. If the value of R_3 is adjusted to match the resistance of the transducer, the bridge voltage will be zero (the bridge will be nulled). As the resistance of the transducer changes, the

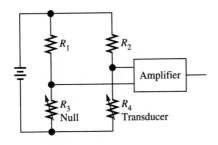

A. Typical resistive transducer
bridge arrangement

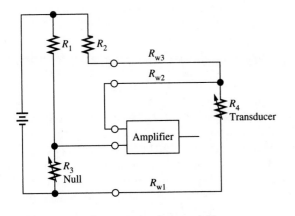

B. Connection of a remote transducer to a bridge

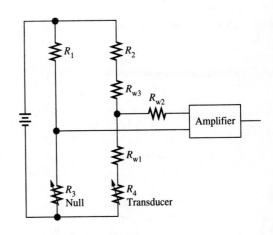

C. Remote transducer bridge equivalent circuit

FIGURE 6–26
Typical Resistive Transducer Bridge Arrangement, Connection of a Remote Transducer to a Bridge, and Remote Transducer Bridge Equivalent Circuit

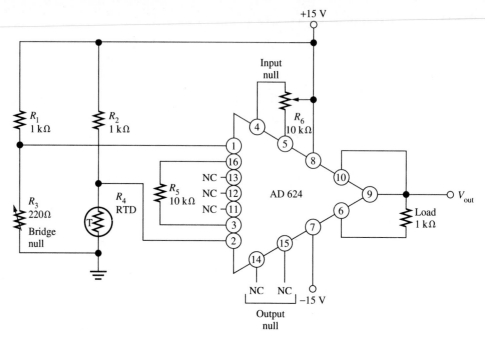

FIGURE 6–27
Connection of an Instrumentation Amplifier to a Bridge

bridge voltage will change. It is this bridge voltage that will be used to gauge how the process variable is changing. For example, suppose that R_4 in Figure 6–26A is an RTD with a resistance of 100 Ω at 0°C. If the value of R_3 is adjusted to 100 Ω, the bridge voltage will be 0 V at 0°C. At temperatures other than 0°C, the RTD resistance and the bridge voltage will change. The bridge voltage can then be measured to indirectly measure the temperature of the RTD. Instrumentation amplifiers such as the one shown in Figure 6–27 are usually employed to amplify the often small bridge voltage variations.

In many cases, the transducer must be located a great distance from the other bridge resistors. The additional resistance added by the connecting wires will upset the bridge balance. As shown in Figures 6–26B and 6–26C, however, the use of a three-wire cable to the transducer will eliminate the error. Note that if all three wires are the same length and type, the resistance of each length will be identical. Since R_2 and the resistance of the transducer will both be increased by the same amount (the connection cable resistance), the voltage across the transducer will not change. In addition, if the input impedance of the amplifier is very high, virtually no current will flow through R_{W2}. The end result is that regardless of the length of the wires connecting the transducer to the bridge, the use of a three wire cable will eliminate any error introduced by the added cable resistance.

Figure 6–27 shows a typical application of a transducer bridge connected to an instrumentation amplifier.

EXAMPLE 6–5 Suppose that the value of R_3 in Figure 6–27 has been adjusted to 100 Ω. In addition, suppose that the RTD has a temperature coefficient of 0.00385 Ω/Ω/°C and a resistance of 100 Ω at 0°C. Draw a response curve of temperature versus amplifier output over a temperature range of 0°C to 200°C.

The response curve will be a straight line, since the resistance of an RTD varies linearly with temperature. Therefore only 2 points need be plotted. The first point will occur at 0°C and 0 V (at 0°C, the RTD resistance is 100 Ω and the bridge voltage is 0 V). In order to determine the second point, the bridge voltage at 200°C must be obtained.

The RTD resistance at 200°C can be calculated by using equation 5.1:

$$R_{200} = 100 \ \Omega[1 + (0.00385 \ \Omega/\Omega/°C)200°C] = 177 \ \Omega$$

The voltage across R_3 is:

$$V_{R3} = 15 \ V \left(\frac{100 \ \Omega}{1.1 \ k\Omega}\right) = 1.3636 \ V$$

The voltage across the RTD is:

$$V_{RTD} = 15 \ V \left(\frac{177 \ \Omega}{1.177 \ k\Omega}\right) = 2.2557 \ V$$

With the AD624 gain set at 5, the output voltage at 200°C is:

$$V_{out} = 5(2.2557 \ V - 1.3636 \ V) = 4.461 \ V$$

Plotting these two points produces the response curve shown in Figure 6–28.

FIGURE 6–28

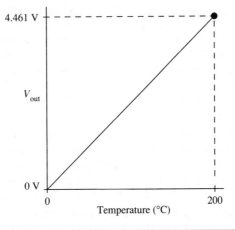

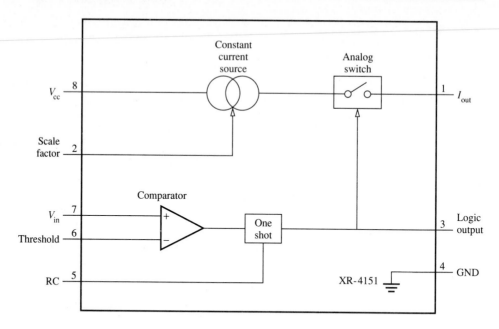

FIGURE 6–29
The EXAR XR-4151 Frequency to Voltage Converter

6–5   FREQUENCY TO VOLTAGE CONVERSION

Inductive, reluctive, and Hall effect transducers as well as certain flow sensors and proximity switches produce switched outputs. That is, the output is normally held at some positive DC value until the physical variable is sensed, at which time the output falls to 0 V. When these transducers and sensors are used to measure motion, the result is a variable frequency pulse output. In order for these signals to be of use, the frequency must be converted to a proportional voltage or current. The circuits that perform this function are called frequency to voltage converters and are available as commercial integrated circuits.

Figure 6–29 shows the block diagram of the EXAR XR-4151 frequency to voltage converter. Although this diagram describes the internal circuitry of a particular device, it is typical of the many similar devices that are commercially available. Note that there are four main components in the XR-4151—a comparator, a one-shot, a constant current source, and an analog switch. The principle of operation is quite simple. The threshold input is usually biased at some voltage midway between 0 V and V_{CC}. When the signal at V_{in} exceeds the threshold, the output of the comparator switches to $+$SAT and triggers the one-shot. The one-shot then produces a pulse of duration determined by an external RC circuit. The pulse enables the analog switch, which allows a short-duration current pulse (of about 140 μA) to pass through to the output. These output pulses can then be

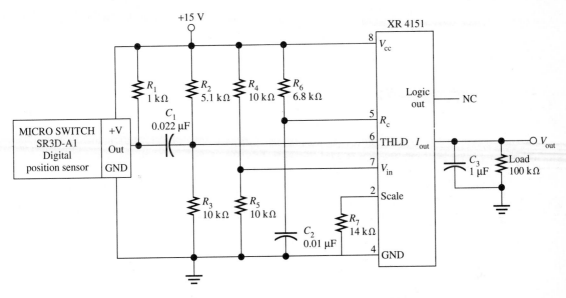

FIGURE 6–30
Typical Application of the XR-4151

filtered to produce a steady DC voltage. The constant current source and one-shot make the device output independent of the amplitude and duty cycle of the input signal.

A practical application of the XR-4151 is shown in Figure 6–30. In this circuit, V_{in} and the threshold are biased at +7.5 V and +9.9 V respectively. This holds the output of the comparator at 0 V. As the SR3D-A1 position sensor detects a magnetic field, its output falls to 0 V. Since the sensor's output is capacitively coupled to the threshold input of the XR-4151, each time that a magnetic field is detected, the threshold input is momentarily pulled down to 0 V. This action causes the one-shot to trigger and produce a 75 μs ($1.1R_6C_2$) pulse of current to the output of the device. The magnitude of the output voltage can be adjusted by changing the value of the filter capacitor or load resistor, or by changing the duration of the pulse. In general, however, this is arbitrary since the output can be further amplified or attenuated with additional circuitry.

6–6   DIGITAL SIGNAL CONDITIONING

The price of microprocessors and microcontrollers is decreasing as fast as their power and complexity are increasing. Many controllers now contain microprocessors. This means that many controllers now require that process variable signal information be provided in a digital format. Whether a particular microprocessor accepts data in an 8-bit, 16-bit, or a 32-bit wide format, in general it views its input data in one of two ways. The data may be viewed as 8 (or 16, or 32) individual, possibly unrelated bits (*binary digits*). It may also be viewed as a binary *number* made up of those 8, 16, or 32 bits. In either case, the data must first be made digitally compatible.

Digital Threshold Detection

There are many instances in which the actual magnitude of a process variable signal is unimportant. Rather, it is the magnitude of the signal with respect to some threshold that is of concern. The op-amp comparators discussed earlier provide exactly this function. They compare a process variable signal to another threshold signal. The problem is that the output signals available from most op amps are incompatible with the input requirements of most microprocessors. Since most microprocessors require input signals that switch between 0 V and +5 V, the output of any circuit that is providing threshold-based data must be converted to switch between these levels. A signal that exceeds +5 V or falls below ground by even a few volts can destroy a microprocessor interface.

Figure 6–31A shows a typical application of a level shifting circuit. As the temperature of the thermistor increases, the voltage at the noninverting input of the op amp will increase. The voltage at the wiper of R_3 determines the temperature at which the output of the op amp will switch from $-$SAT to $+$SAT. When the output of the op amp is at $-$SAT, D_1 is reverse-biased and blocks current through the IRED at the input of the optoisolator. With no current through the IRED, the output transistor will be at cutoff and the output voltage will be approximately +5 V (logic 1). As soon as the output of the op amp switches to $+$SAT, current will flow through D_1, R_4, and the IRED. The output transistor will saturate, and the output voltage will fall to approximately 0 V (logic 0). That is, a logic 0 will indicate to the microprocessor system that the temperature of the thermistor is above the threshold set by R_3. Note that the output from the circuit shown in Figure 6–31A constitutes only one bit of data. A typical microprocessor would be able to simultaneously sample the output of eight such circuits.

Aside from shifting the voltage level, the optoisolator in Figure 6–31 provides electrical isolation between the op amp and the microprocessor system. This isolation is absolutely essential. Even though ±12 V may be considered low voltage, it will destroy a microprocessor system interface just as effectively as 120 V. In situations that involve higher voltages (such as the one shown in Figure 6–31B), isolation becomes an even more important issue.

Analog to Digital Conversion

The circuits shown in Figure 6–31 provide a limited amount of information to the microprocessor. For example, the output of the circuit of Figure 6–31A does not indicate the exact temperature, nor does it indicate how far above (or below) the threshold the input signal is. Adding additional thresholds (as with a window comparator) will provide more information. Generally, however, if more resolu-

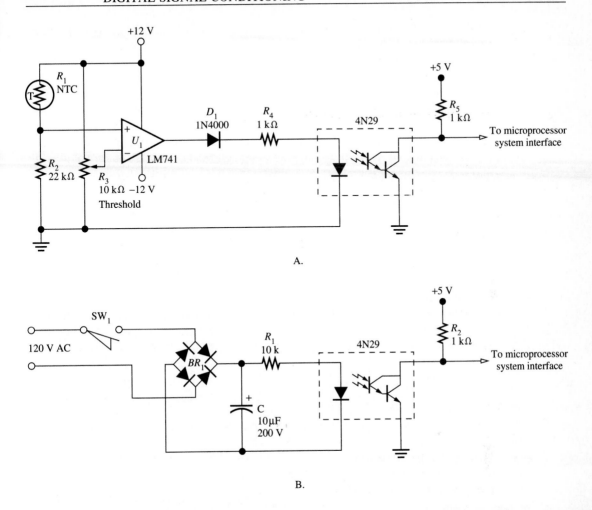

FIGURE 6–31
Interfacing Noncompatible Signals to a Microprocessor System

tion is required, an analog to digital converter (ADC) is employed. An ADC converts an analog signal to a proportional binary number.

There are many techniques used to convert analog signals into binary numbers. A common method, successive approximation, is presented here. From the perspective of the microprocessor, the exact method of conversion is unimportant. What is critical is the dialogue that must take place between the microprocessor and the ADC in order for the conversion to be performed.

ADCs provide outputs that are 8, 10, 12, 14, or 16 bits wide. Most will accept input signals that vary between 0 V and +5 V. Keep in mind that the output of an

ADC is a binary number. The output range for an 8-bit ADC, expressed in hexa-decimal, is 00H to FFH. One important figure of merit for an ADC is the resolution. Resolution is directly related to the number of output bits that the converter provides. It is not uncommon for an 8-bit converter to be said to have 8-bit resolution. While technically correct, this statement is a bit misleading, for resolution also depends upon the input signal dynamic range (the amount by which the signal varies). Generally, the smaller the input voltage change that the ADC is capable of detecting, the higher the resolution is for that particular ADC. Resolution voltage for an ADC (the smallest voltage that the device is capable of detecting) is given by:

$$V_{res} = \frac{V_{ref}}{2^N - 1} \qquad\qquad (6.16)$$

where V_{ref} = the reference voltage—the highest voltage that the ADC is capable of detecting

N = the number of output bits that the ADC provides.

Note that even though the number of output bits does not change, if the maximum input voltage for the ADC is decreased, the resolution increases (V_{res} becomes smaller). An 8-bit ADC will produce only 255 different output numbers other than zero. This is true regardless of the input voltage range. If the ADC input voltage range is 0 V to 4 V (a dynamic range of 4 V), the smallest voltage that the ADC will be capable of detecting will be 15.7 mV (4 V/255). If the maximum input voltage for the ADC is cut in half, the resolution voltage will be 7.8 mV (2 V/255). The ADC with a dynamic range of 2 V will measure input signals more precisely, but it will not be able to measure signals over as large a range as the ADC with a dynamic range of 4 V. It is important to note that an ADC cannot measure an input voltage that exceeds its reference voltage. In addition, since most ADCs provide an adjustable reference voltage input, the dynamic range (and so also the measurement precision) for the ADC can be changed.

EXAMPLE 6–6   An 8-bit ADC has a reference voltage set at 5 V. The input voltage for the ADC is provided by a temperature sensor that produces a linear output voltage that varies from 0 V to +5 V as the temperature varies from 0°C to 100°C. What is the smallest temperature change that can be detected by the ADC?

Since the output of the sensor never exceeds the ADC reference voltage, the ADC will be able to sense temperatures across the entire range of the sensor. The ADC will resolve the input voltage (and so also the temperature) into 255 different numbers (excluding zero). The temperature resolution, therefore, can be determined by using equation 6.16:

$$Temp_{res} = \frac{Temp_{max}}{255} = \frac{100°}{255} = 0.392°$$

That is, the temperature must change by at least 0.392° in order for the ADC output to change by one digit.

As shown in Figure 6–32, the block diagram of a successive approximation ADC consists of three main blocks. The successive approximation register (SAR) is responsible for approximating the output number, while the comparator is responsible for determining whether the current approximation is too high or too low. The digital to analog converter (DAC) will produce an output voltage that is proportional to its binary input. Each input bit has a particular value or weight. The most significant bit (MSB) always has a weight of $V_{ref}/2$; the next most significant bit has a weight of $V_{ref}/4$; the next has a weight of $V_{ref}/8$, and so on. Suppose that the reference voltage and input voltage of the 8-bit successive approximation ADC shown in Figure 6–32 are +4 V and +1.4 V, respectively. The resulting bit weights of the DAC are shown in Figure 6–32. When the conversion process starts, the SAR produces its first approximation: the binary number 10000000_2. This results in a voltage of +2 V at the output of the DAC. With +1.4 V at the input of the ADC, the output of the comparator will be at +SAT. This indicates to the SAR that the current approximation is too high. The SAR then makes its next approximation: the binary number 01000000_2. Note that the SAR has cleared the MSB, since it resulted in an approximation that was too high. The result of this next approximation produces +1 V at the DAC output, and −SAT at the output of the comparator. This indicates to the SAR that the current approximation is too low. When the SAR makes its third approximation, it retains the logic 1 from the second approximation and adds to it an additional logic 1 in the next most significant bit position. This process continues (as shown in Figure 6–32) for eight total approximations. At the conclusion of this process, the SAR will yield a number that is its best digital approximation of the analog input signal. Note that the output will never be an exact measurement. However, examination of Figure 6–32 will show that the output of the DAC has replicated the input signal within 9.4 mV. The final output number for this example is 59H (89_{10}). To the microprocessor system that reads this number from the converter, 59H indicates that +1.4 V is present at the input of the ADC. Of course, it is not necessary to perform this type of analysis each time that a conversion is to be done. For any given input conditions, the output number (in base 10) from an 8-bit ADC can be determined by the following equation:

$$ADC\ output = 255 \left(\frac{V_{in}}{V_{ref}} \right) \tag{6.17}$$

EXAMPLE 6–7   Given an 8-bit ADC with a reference voltage of +5 V, determine the number that will be produced for an input voltage of +3.78 V.

The output number is obtained by using equation 6.17:

$$ADC\ output = 255 \left(\frac{3.78\ V}{5\ V} \right) = 192.8_{10} = C1H$$

In order for an ADC to be microprocessor-compatible, the microprocessor must be able to initiate a conversion. In addition, the ADC must have the capabil-

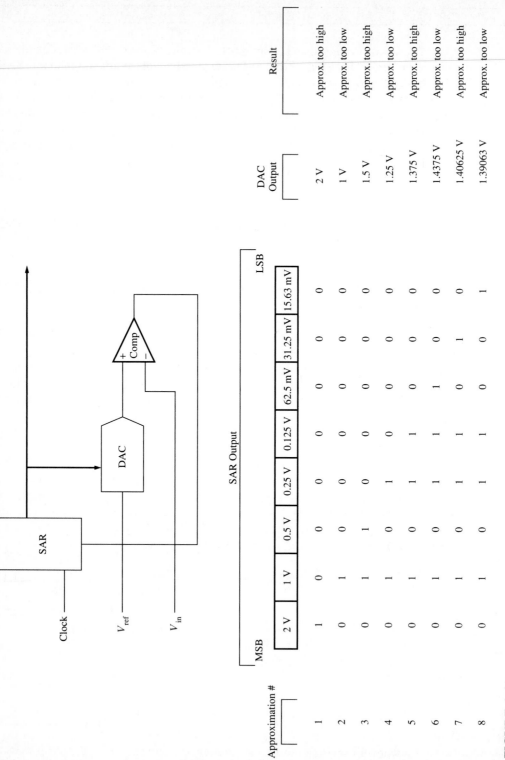

Approximation #	SAR Output MSB							LSB	DAC Output	Result
	2 V	1 V	0.5 V	0.25 V	0.125 V	62.5 mV	31.25 mV	15.63 mV		
1	1	0	0	0	0	0	0	0	2 V	Approx. too high
2	0	1	0	0	0	0	0	0	1 V	Approx. too low
3	0	1	1	0	0	0	0	0	1.5 V	Approx. too high
4	0	1	0	1	0	0	0	0	1.25 V	Approx. too low
5	0	1	0	1	1	0	0	0	1.375 V	Approx. too high
6	0	1	0	1	1	1	0	0	1.4375 V	Approx. too low
7	0	1	0	1	1	0	1	0	1.40625 V	Approx. too high
8	0	1	0	1	1	0	0	1	1.39063 V	Approx. too low

FIGURE 6–32

Successive Approximation Method of Analog to Digital Conversion

ity to inform the microprocessor that the current conversion is complete. The National Semiconductor ADC0804 is a popular 8-bit successive approximation ADC with these capabilities. Figure 6–33 shows the pin diagram and the necessary control timing. The ADC0804 accepts signals from a differential analog input. Neither input should be brought below ground or above +5 V, since this will destroy the device. The reference voltage is provided externally by the user. An integrated gain stage allows the reference to be set at one half of the desired

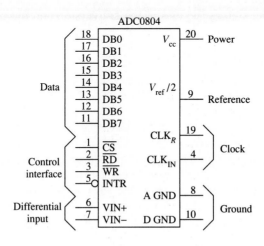

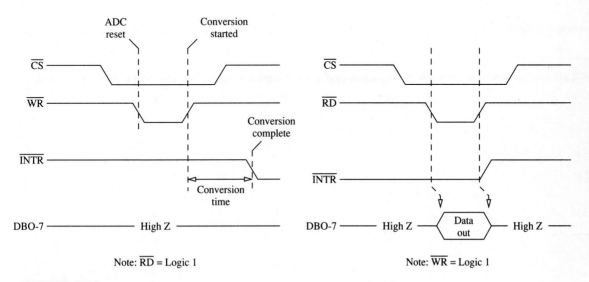

FIGURE 6–33
The National Semiconductor ADC0804

reference level. That is, if the voltage presented to pin 9 of the ADC is +2 V, the internal reference voltage will be +4 V. Since the supply voltage for the ADC0804 is +5 V, the voltage applied to pin 9 should never exceed +2.5 V.

In order to cause the ADC0804 to start the conversion process, the CS input (pin 1) and the WR input (pin 3) must be simultaneously brought to a logic 0 level. Technically, the conversion is started when CS is at a logic 0 and the WR input changes from a logic 0 to a logic 1. The amount of time required for the conversion depends upon the RC circuit connected to the clock inputs on the device. When the conversion is complete, the INTR (pin 5) changes from a logic 1 to a logic 0. This transition signals the end of the conversion and indicates that the ADC has assembled a binary number that is its best approximation of the analog input. In order to obtain the number from the ADC, the microprocessor system must simultaneously bring the CS and RD (pin 2) inputs to a logic 0. When both of these signals are at a logic 0 level, data are transferred from the internal circuitry to the output (pins 11–18). When the RD signal changes back to a logic 1, the INTR signal also returns to a logic 1.

A practical application of the ADC0804 is shown in Figure 6–34. An Omega PX510-100S 5 V pressure sensor provides an input voltage that varies from 0 V to 5 V as the pressure varies from 0 psi to 100 psi. The ADC reference voltage can be adjusted to any value between 0 V and 5 V, but it should be set at 5 V in order to sense the entire pressure range of the sensor. When the pressure is 100 psi, the output of the sensor will be 5 V and the ADC will yield 255_{10} (FFH). The ADC output for any pressure from 0 psi to 100 psi can be determined by using equation 6.17.

EXAMPLE 6–8

Suppose that the voltage at pin 9 of the ADC in Figure 6–34 is 2.5 V. If the most recent conversion produced an output of 7AH, what was the pressure at the input of the sensor?

With 2.5 V at pin 9 of the ADC, the reference voltage will be 5 V. Expressing 7AH as a decimal number yields 122_{10}. Rearranging equation 6.17 yields the input voltage that produced that 7AH output:

$$V_{in} = \frac{(V_{ref})(\text{ADC output})}{255} = \frac{(5\ V)(122)}{255} = 2.39\ V$$

Since the output of the pressure sensor is linear, the pressure can be determined for any given output voltage by setting up a simple proportion:

$$\frac{100\ psi}{5V} = \frac{pressure}{2.39\ V}$$

$$Pressure = \frac{(2.39\ V)(100\ psi)}{5\ V} = 47.8\ psi$$

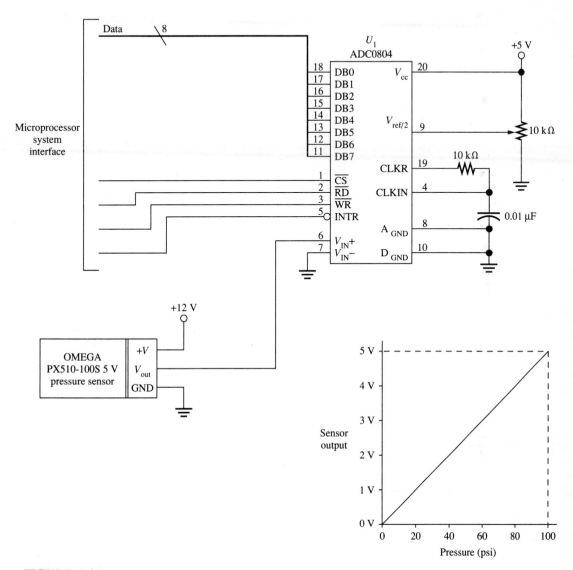

FIGURE 6–34
Typical Application of the ADC0804

SUMMARY

☐ An operational amplifier is a general-purpose amplifier that provides high gain, high input impedance, and low output impedance. In addition, its operational characteristics can be controlled almost entirely by external components.

□ The main operational characteristic of an op amp is that the output changes in a manner that forces the inverting input to the same potential as the noninverting input.

□ A voltage follower provides a gain of 1, while maintaining the high input impedance and low output impedance of the op amp. Often this circuit is used to match impedances or to isolate two circuits from each other.

□ The noninverting amplifier provides a gain that is set by the ratio of the two resistors, R_2 and R_1, accounting to the following equation:

$$A_{CL} = 1 + \frac{R_2}{R_1}$$

□ The inverting amplifier provides a gain that is set by the ratio of the two resistors, R_2 and R_1, according to the following equation:

$$A_{CL} = \frac{R_2}{R_1}$$

□ A voltage comparator is characterized by the absence of feedback from output to inverting input. Additionally, the output is, at all times, saturated either positive or negative. If the inverting input is positive with respect to the noninverting input, the output will be negative. If the inverting input is negative with respect to the noninverting input, the output will be positive.

□ The output of an integrator reflects the amount of time that an input signal has been present. This property is due to the accumulating charge of the feedback capacitor. For a static DC input, the output increases over time. If the input is increasing, the output increases at an even greater rate. It is sometimes helpful to recognize that the rate of change of the output voltage for an integrator is proportional to the magnitude of the input voltage.

□ The output of a differentiator reflects the rate at which an input signal is changing. The output voltage is proportional to the rate of change of the input voltage. If the input is not changing, the output will be zero.

□ An instrumentation amplifier is basically a differential amplifier. It provides precise, adjustable gain and rejects any signals that are common to both inputs. The gain is set by a single external resistor.

EXERCISES

1. Suppose that, in Figure 6–4B, $R_1 = 22$ kΩ and $R_2 = 1.1$ MΩ. What is the gain for this amplifier? Will the output signal be in phase with the input?

2. Suppose that a gain of 100 is desired from a noninverting amplifier. If a feedback resistor of 470 kΩ is chosen, what values of R_1 should be used?

3. If the amplifier used in Exercise 1 is powered with a bipolar 12 V supply, what is the maximum signal input for the circuit if the op amp is not to be driven into saturation?

4. Consider an inverting amplifier with a 500 mV signal at the input. Describe the output of the amplifier if the connection from the output to the feedback resistor is open.

5. For the circuit of Figure 6–9, assume that the thermistor has a cold-to-hot resistance range of 100 kΩ to 5 kΩ. If the value of R_2 is changed to 47 kΩ, and the inverting input is adjusted to +2 V DC, describe how circuit operation will be affected.

6. For the circuit of Figure 6–9, suppose that the output is required to change in response to a higher temperature. Should the voltage at the inverting input be increased or decreased?

7. Given an integrator with $R = 2.2$ MΩ, and $C = 0.5$ μF. If a $+1$ V DC signal is connected to the input, what will the output voltage be after 3 s?

8. What is the maximum integration time for an integrator with $R = 1$ MΩ, $C = 4.7$ μF, and $V_{in} = +1$ V if the op amp is powered by a bipolar 10 V supply?

9. If the value chosen for the input resistor for an integrator is too large, how will the circuit response be affected?

10. A differentiator is required that will produce an output signal of -2 V DC for a linear input change of 1 V/s. If an input capacitor of 5 μF is chosen, what value of R should be used?

11. If the value chosen for the input capacitor for a differentiator is too small, how will the circuit response be affected?

12. For the differential amplifier of Figure 6–18, if $R_2 = R_4 = 220$ kΩ and $R_1 = R_3 = 10$ kΩ, what is the circuit differential gain?

13. For the circuit of Figure 6–21, suppose that $V_{in1} = 25$ mV and that $V_{in2} = 15$ mV. If $R_1 = R_2 = 5$ kΩ, and if $R_g = 1$ kΩ, what is the output voltage?

14. To increase the circuit gain of the instrumentation amplifier of Figure 6–23, should R_2 be increased or decreased?

15. For the circuit of Figure 6–6, suppose that the values of R_2 and R_1 are 100 kΩ and 22 kΩ, respectively. What value of compensation resistor should be chosen?

16. For the circuit of Figure 6–10, suppose that the values of R_2 and R_1 are 10 kΩ and 1 kΩ, respectively. If U saturates at ± 15 V, what will the switching thresholds be?

17. To decrease the sensitivity of the comparator in Figure 6–10, should the value of R_1 be increased or decreased?

18. For the circuit of Figure 6–11, can the upper threshold be adjusted independently of the lower? Can the lower threshold be adjusted independently of the upper?

19. Suppose that the values of R_1 and R_2 in Figure 6–8 are both 1 kΩ. If the value of R_3 is 100 kΩ, what output will be produced for inputs of 25 mV and 72 mV at inputs V_{in1} and V_{in2}?

20. Suppose that the value of R_2 in Figure 6–5 is adjusted to 2.2 kΩ. What is the voltage-to-current conversion ratio?

21. If the value of R_2 in Figure 6–5 is too low, what will result?

22. Given the AD624 instrumentation amplifier shown in Figure 6–24, what gain will result if a 2.1 kΩ resistor is connected between pins 3 and 16?

23. Given the AD624 instrumentation amplifier shown in Figure 6–24, what gain will result if pin 3 is connected to pin 12 and pin 11 is connected to pin 16?

24. Using the graph shown in Figure 5–11 of Chapter 5, determine the output of the instrumentation amplifier shown in Figure 6–25 when the thermocouple temperature is 300°C.

25. Given the amplified RTD bridge shown in Figure 6–27, determine the output voltage when the RTD temperature is 165°C. Assume that the bridge has been nulled at 0°C.

26. Explain how the operation of the circuit shown in Figure 6–31A would be affected if the positions of R_1 and R_2 were accidentally reversed.

27. Given an 8-bit ADC with a reference voltage of 4 V, what output would be obtained if the input voltage was 3.2 V?

28. What is the resolution voltage for the ADC in Exercise 27?

29. Explain how the operation of the circuit shown in Figure 6–34 would be affected if the voltage at pin 9 of the ADC were reduced to 1 V.

APPLICATION ASSIGNMENT ANSWER

You must consider two factors when contemplating a solution to this problem. First, you must decide on an appropriate gain for the amplifier. Second, you must choose an appropriate amplifier configuration.

A gain of 200 would produce an output of 1 V for an input of 5 mV. A gain of 333 would produce an output of 10 V for an input of 30 mV. As a compromise, you could choose a gain that is somewhere between these. A gain of 300 will result in an output range of 1.5 V to 9.0 V for an input range of 5 mV to 30 mV. This range is certainly within the guidelines and should provide satisfactory results.

Since the input signals will be of such low amplitude, it is important that induced noise be rejected by the amplifier configuration that you choose. Remember that noise amplitudes may be several hundred millivolts. A differential amplifier, therefore, is a good choice. Figure 6–35 shows one possible solution. Note that R_3/R_1 and R_4/R_2 are both equivalent to 300.

FIGURE 6–35

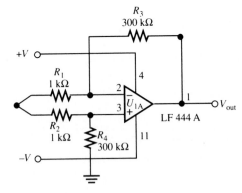

7

SCRs, Triacs, and Other Thyristors

CHAPTER OUTLINE

7–1 Introduction □ **7–2** Thyristors: General Makeup and Nomenclature □ **7–3** The Silicon-Controlled Rectifier (SCR) □ **7–4** The Unijunction Transistor (UJT) □ **7–5** The Diac and the Triac □ **7–6** Thyristor Voltage and Power Calculations □ **7–7** Other Major Thyristors □ **7–8** Light-Activated Semiconductors and Optoisolators

LEARNING OBJECTIVES

□ Recognize the nomenclature codes for thyristors.
□ Describe the SCR, its application, and its specifications.
□ Describe the UJT and its relation to the SCR.
□ Show how diacs and triacs are used in power circuits.
□ Calculate thyristor voltage and power outputs, using trigger angle.
□ Describe other major thyristor and thyristor-related semiconductors and their applications.
□ Describe the major photosemiconductors, and tell how they are used.
□ Describe the use of optoisolators for circuit applications and protection.

APPLICATION ASSIGNMENT

You are a technician assigned to analyze an erratically operating speed control. Occasionally the motor being controlled speeds up and then slows back to required speed. Where would you start the analysis?

7–1   INTRODUCTION

This chapter will discuss thyristors, which are multilayer solid-state devices. Thyristors are used extensively in industrial control. One of the major uses of thyristors is in changing a fixed AC voltage directly into a variable AC or DC voltage. Variable DC voltage is needed for variable speed control of processes, particularly DC drive motors.

First, this chapter will cover thyristors in general, their physical makeup, and their nomenclature. Next, it will cover the most common thyristor, the SCR, which is the major type of thyristor used to convert AC to DC. SCR characteristics, including turn-on methods and firing angle control, will be shown. In the next sections the UJT, which is often used to trigger SCRs, and the diac and triac for full-wave AC to DC rectification will be described. The method of calculating thyristor firing angles for varying voltage and power levels will also be discussed, as well as the other major types of thyristors in common use.

The final subjects to be covered are light-activated semiconductor devices and optoisolators. A light-activated semiconductor has a window into which varying light levels are inputted. The varying light levels change the characteristics of the semiconductor device. Optoisolators are used extensively to transmit voltages and signals between electronic stages. Optoisolators have the advantage of electrically isolating stages while transmitting the electrical signals.

Two sections in the appendix are applicable to thyristors. One is a table of major thyristors and their characteristics. The other is a table of resistance checks for determining good and bad devices.

7–2   THYRISTORS: GENERAL MAKEUP AND NOMENCLATURE

Semiconductor diodes, discussed in Chapter 4, are two-layer devices. One layer of p material and one of n material make up a diode. Transistors, also discussed in Chapter 4, are three-layer devices, either pnp or npn. Thyristors, in general, are four-layer devices of alternating p and n material. A thyristor can have two, three, or four terminals, depending on the device.

One of the earliest four-layer thyristor devices developed was the *silicon-controlled rectifier* (*SCR*). It was developed to replace the thyratron and ignitron vacuum tubes. Thyratrons, ignitrons, and SCRs efficiently change AC voltage to a variable DC voltage. The SCR was more efficient and reliable than the two tube devices in performing this rectification. Thyratron and ignitron tubes became obsolete as SCR technology improved.

Other thyristors have been developed over the years. The other devices are used in a variety of applications, not just for rectification. New devices are being created each year in the thyristor family. The next six sections will show the construction and application of the major types of thyristors in use today.

Each letter in the name or abbreviation for a thyristor stands for a word. For example, in the abbreviation *SCR, S* is for *silicon, C* is for *controlled,* and *R* is for *rectifier*. The following is a key for letters in the names and abbreviations for thyristors:

A   asymmetrical (having different characteristics in forward and reverse operation) *or* activated

B   bilateral (operating electronically in both directions)

C   controlled *or* complementary (In a complementary device *p* and *n* materials are reversed in construction, and, consequently, device polarity is reversed.)

D   diode

E   emitting

G   gate

L   light

LA   light-activated (turned ON by a light pulse)

P   programmable *or* photo

R   rectifier (a device that changes AC power to DC power)

S   silicon *or* switch *or* semiconductor

T   transistor

TO   turn-OFF

TR   triode

U   unilateral (operating in one direction only)

UJ   unijunction

VAR   variable

This list is a general guide to identifying and understanding thyristor names.

7–3   THE SILICON-CONTROLLED RECTIFIER (SCR)

Suppose that you are assigned the task of constructing a variable-voltage DC supply for 30 A. The input voltage is 220 V single-phase AC. The output is to be continuously adjustable from 0 V to 120 V DC. You have the three construction options shown in Figure 7–1. Among other options would be a three- or four-unit motor generator system (not shown).

For many years the construction options shown in Figure 7–1*A* and *B* were the only methods for the conversion of fixed AC voltage to continuously variable DC voltage. The drawback of the first option (Figure 7–1) is very low efficiency, due to power losses in the rheostat. The second option is more efficient than the first. However, it is costly and bulky, and the transformer's moving brushes require maintenance.

In the 1960s two vacuum-tube devices became available. These two tubes, the ignitron and the thyratron, could function in the manner of the SCR in Figure 7–1*C*. These two types of tubes were successful in only a few applications. They had low efficiency, low power capabilities, high cost, and short life.

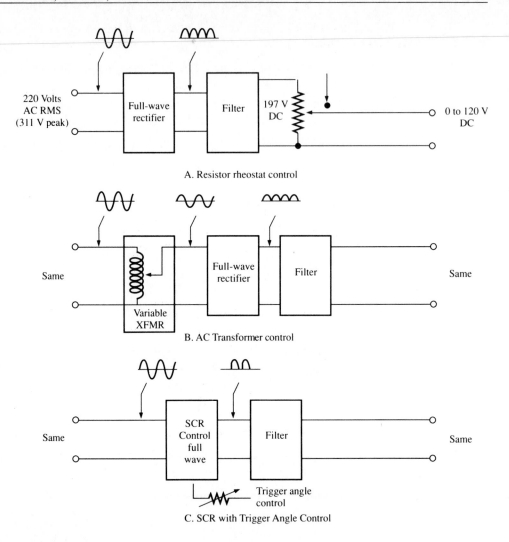

FIGURE 7–1
AC-to-DC Conversion Methods

The SCR has now almost completely replaced the options in Figure 7–1*A* and *B* and the tube option of Figure 7–1*C*. The SCR, used as shown in Figure 7–1*C,* is a high-efficiency, high-power, long-life device. The SCR is also less costly and less bulky, and it gives off a relatively small amount of heat. Also, SCRs have been found to have applications other than power conversion. They are used as control devices in many electronic circuits. Note that low-power AC-to-DC power converters (rectifiers) will be covered in detail in Chapter 8.

The SCR is a four-layer, three-lead thyristor semiconductor. Some forms of the SCR are illustrated in Figure 7–2. Also, the construction and the symbol are included in the figure.

Some of the key characteristics of the SCR, with rated anode-to-cathode voltage applied, are as follows:

☐ With positive voltage applied to the gate, the SCR acts as a diode and is turned ON.

☐ With negative voltage applied to the gate, the SCR will not turn ON.

☐ With no voltage applied to the gate, the SCR will not normally conduct.

☐ Once the SCR is turned ON, it will stay ON under full load, even after the positive gate voltage is removed.

☐ Applying positive voltage to the gate during conduction has no effect.

☐ Applying negative voltage to the gate during conduction will not turn the SCR OFF.

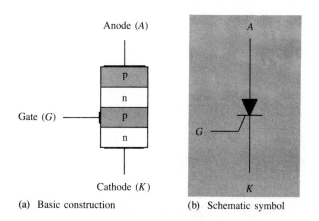

(a) Basic construction   (b) Schematic symbol

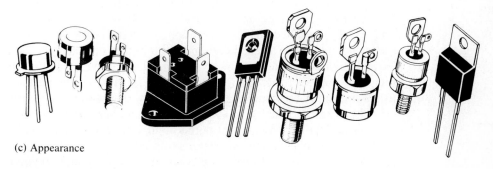

(c) Appearance

FIGURE 7–2
SCR Construction, Symbol, and Appearance

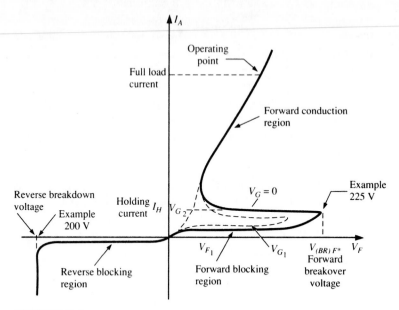

FIGURE 7–3
SCR Characteristic Curve

These characteristics are easily explained by reference to the typical SCR characteristic curve in Figure 7–3. The curve is quite similar to the diode curve with the
addition of the "knee" in the blocking region. In the reverse direction, as in the
diode curve, there is little current flow until the breakdown voltage (200 V) is
reached. At reverse breakdown voltage, the SCR becomes a reverse short circuit.

In the forward direction, with no gate voltage applied, 225 V must be applied
to cause appreciable conduction. The point at which such conduction occurs is
called the breakover point. Breakover voltage is much higher than the working
voltage. Breakover voltage is not normally used for firing the SCR.

With rated voltage applied from A to K, to get the SCR to fire, a gate
voltage, V_{G_2}, is applied. When V_{G_2} is applied, the operating curve is the one shown
as a dashed line. Forward conduction therefore takes place in the same manner as
it does in a diode. Note that for an intermediate gate voltage, V_{G_1}, the breakover
voltage is reduced. Increasing gate voltage further will move the knee of the
dashed-line curve farther to the left, eventually almost to the V_{G_2} curve's position.

The characteristic curve also illustrates another important factor, the holding current. The holding current is the minimum current that will keep the SCR
conducting when gate voltage is removed. If, for example, the rated holding
current of the SCR is 7.4 mA, 5.0 mA will not keep conduction going. Any value
of 7.4 mA or more—for example, 8.5 mA—will keep conduction going, even if
gate voltage is removed.

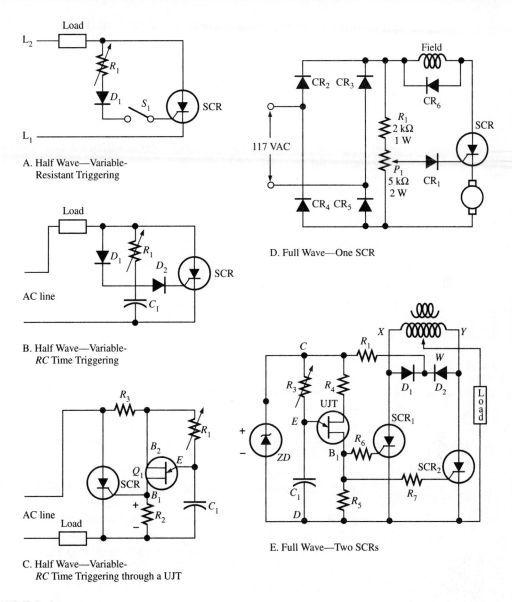

A. Half Wave—Variable-
Resistant Triggering

B. Half Wave—Variable-
RC Time Triggering

C. Half Wave—Variable-
RC Time Triggering through a UJT

D. Full Wave—One SCR

E. Full Wave—Two SCRs

FIGURE 7–4
SCR Connection Schemes

Figure 7–4 shows some typical circuits in which the SCR is used. There are five schemes shown for triggering the gate. In all cases, the SCR or SCRs are connected in series with the load and AC voltage supply. In all cases, the variation of the firing angle is controlled by a variable resistor. Variations of the firing angle result in variations of average DC voltage and effective AC power to the load device. These circuits illustrate the line synchronization of the gate pulse voltage. The gate voltage must pulse consistently at the same angle with respect to the line voltage. This constant angle is necessary for proper operation of the circuit.

Normally, to turn the SCR ON with anode-to-cathode voltage applied, rated gate voltage is applied to the gate with respect to the cathode. The following ways of turning the SCR ON are to be avoided:

1. turning A-to-K voltage up beyond breakover (This method is used with some other comparable thyristors but not with an SCR.)
2. high temperature (SCRs are heat sensitive.)
3. a voltage value below rated gate voltage but rising at a high pulse-rise rate (This effect is called the *dv/dt* effect.)

There is a special SCR called the LASCR, which is turned ON by a light pulse. It will be covered in Section 7–7.

To turn the SCR OFF when it is conducting, take one of the following actions:

1. Open the A-to-K circuit.
2. Reduce the operating current below the holding current value.
3. Use commutation: Parallel the SCR load with a low-value shunt resistor. The shunt must be of a value that will reduce the SCR flow below holding current.

There is a modified SCR called the silicon-controlled switch (SCS). It has an extra gate, and the SCR turns OFF when an appropriate voltage is applied to that extra gate. The SCS will be discussed in Section 7–6.

As previously stated, controlling the firing time of the gate, with respect to the A–K voltage, permits control of the output voltage. The later the gate pulse is in the cycle, the later the sine wave output is turned ON. The later the triggering ON of the SCR, the lower the average DC output voltage. This concept is shown in Figure 7–5 for one SCR, half wave, and for two SCRs back to back, full wave. The mathematical relationship between gate firing angle and output value for half and full wave will be explained in Section 7–6. Note that as the triggering occurs later in the cycle, the DC output voltage decreases.

For reference, a typical SCR specification sheet is shown in the appendix. Note that values for characteristics discussed in this section are included, along with other specified values and charts.

7–4   THE UNIJUNCTION TRANSISTOR (UJT)

The unijunction transistor (UJT) is used primarily as a triggering device. A triggering device is one that goes from stable ON to stable OFF very quickly. The UJT is used extensively to trigger, or fire, SCRs and triacs. The UJT is also used in timers, oscillators, and function waveform generators. The UJT is not strictly a

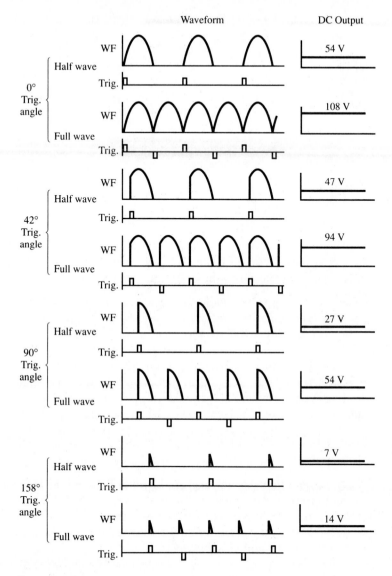

FIGURE 7–5
SCR System DC Output Voltage Control

thyristor. It is a special type of transistor. It is included at this point in the chapter because it is associated with thyristors.

Physically, the UJT looks like a three-lead transistor. Its symbol and construction are shown in Figure 7–6. It has two base terminals and one emitter terminal. In applications the two base terminals are not interchangeable. Terminals B_1 and B_2 must be connected to the proper individual circuit terminals.

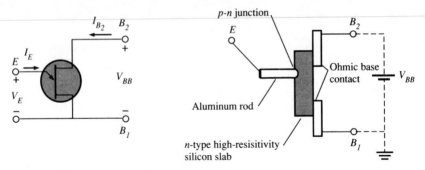

FIGURE 7–6
UJT Symbol and Construction

The characteristic curve for the UJT is illustrated in Figure 7–7. Some of the key characteristics of the UJT, shown by this curve, are as follows:

- There is a very small amount of leakage current in the reverse direction with reverse polarity applied. It is a few microamps.
- The UJT fires at a voltage V_P.
- Once the UJT fires, current flows readily from the emitter to B_1.
- As current increases after the firing point is reached, less voltage is needed to maintain the current. This area is called the negative-differential-resistance region. This means an increasing current is produced by a decreasing voltage. This negative-resistance characteristic makes the UJT usable for the regenerative action required in oscillator circuits.
- Once the valley point is reached, the voltage increases slightly as current is increased. Essentially, the voltage becomes constant at V_E (sat).

An equivalent circuit for the UJT is shown in Figure 7–8. Typical values for the equivalent internal component parts are shown. After the voltage at the emitter reaches V_P, the UJT fires. The R_{B_1} value decreases from thousands of ohms to a few ohms: 5–25, typically. The resistance change is abrupt. Resistor R_{B_1} has one value or the other and nothing in between.

How is the firing point described and calculated? It is related to a ratio called the intrinsic stand-off ratio (ISTOR), which is denoted by the Greek letter eta (η). Its value is the ratio of R_{B_1} to R_{BB} for the UJT in the unfired state. Its numerical value is given in every UJT's specifications. The firing point, V_P, is determined by the following formula:

$$V_{\text{firing}} = V_P = V_D + (\eta \times V_{\text{supply}})$$

For the circuit of Figure 7–8, the η would be $R_{B_1}/R_{BB} = 4.95 \text{ k}\Omega/8.70 \text{ k}\Omega = 0.57$. For a supply voltage of 12.5 V,

$$V_P = 0.65 + (0.57 \times 12.5) = 7.775 \text{ V}$$

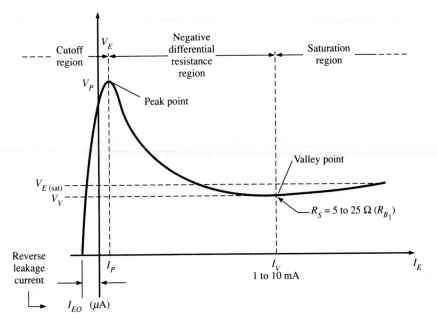

FIGURE 7–7
UJT Characteristic Curve

Note an important characteristic of the UJT. It does not fire at a fixed value. It fires at a voltage that depends on a ratio of the supply voltage. As supply voltage goes up, the value of V_P also goes up.

    An oscillator circuit that utilizes a UJT is shown in Figure 7–9. The variable DC power supply is set at 12.5 V DC. The UJT equivalent to the circuit in Figure 7–8 is used. When the DC power is turned on, the capacitor charges according to the standard RC rise curve shown in Figure 7–10. The capacitor voltage rises to the UJT firing voltage, 7.775 V. Resistance R_{B_1} then changes from 4.95 kΩ to a low value—say, 20 Ω. The capacitor then quickly discharges through R_{B_1} of the UJT

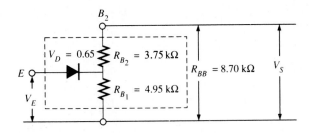

FIGURE 7–8
UJT Equivalent Circuit

FIGURE 7-9
UJT Oscillator Circuit

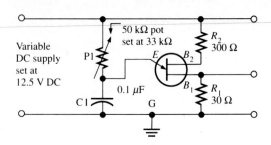

and through R_2. When the capacitor is discharged, R_{B_1} goes back to 4.95 kΩ. The charging process of the capacitor then starts over again.

Some calculations can be made for the oscillator circuit. With a supply voltage of 12.5 V as before, the firing voltage will be 7.775 V for the UJT. The standard RC rise curve shown in Figure 7–10 can be used to determine the time to reach 7.775 V. The time can be calculated mathematically, but that will not be done here. The time constant, T, is determined by multiplying the value of resistance by the value of the capacitance in the circuit.

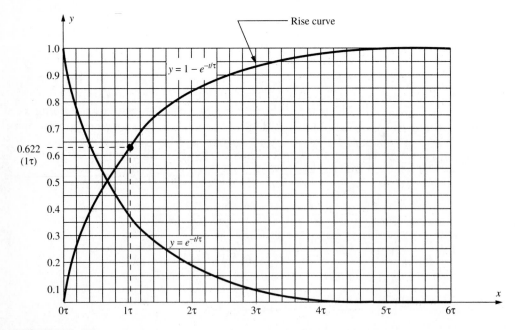

FIGURE 7-10
Standard RC Rise Time Curve

$$T = R \times C = 33 \text{ k}\Omega \times 0.1 \times 10^{-6} = 3.3 \text{ ms}$$

This constant defines the horizontal axis scale of Figure 7–10. The point on the vertical scale is determined by $7.775/12.5 = 0.622$. This point is shown on the graph. The corresponding point on the horizontal axis is at 0.85 of one time constant, 3.3 ms. The time for firing is therefore 3.3 ms.

The waveforms for this oscillator are shown in Figure 7–11. The discharge time is so short as to appear as a straight vertical line. Actually it is $5 \times RC$ time.

$$\text{Discharge time} = 5 \times (30 + 20) \times 0.1 \times 10^{-6} = 25 \ \mu\text{s}$$

The emitter-to-ground, V_{EG}, curve follows the standard rise curve until the UJT fires. Then it goes back almost to 0 and starts over. The V_{B_1G} curve shows the short-duration high voltage (and current) during the firing time. The V_{B_2G} curve is shown for reference. The V_{EG} voltage can be used for waveform generation. This V_{EG} curve approaches a sawtooth form. The V_{B_1G} curve can be used for triggering other devices, such as an SCR. Only a short-duration pulse is needed to fire an SCR, as previously discussed.

Figure 7–12 shows an SCR triggered by a UJT. The circuit is best explained with a look at its waveforms shown in Figure 7–13. On the positive AC cycle, voltage is applied through the diode to the load and the SCR. The SCR will not

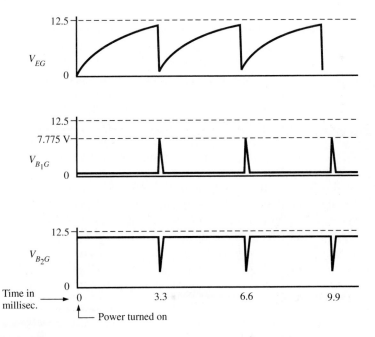

FIGURE 7–11
UJT Oscillator Waveforms

FIGURE 7–12
SCR Triggered by a UJT

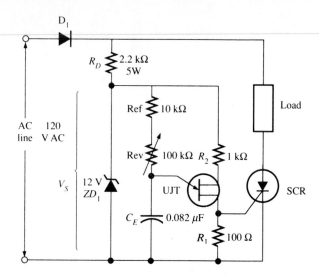

conduct until it is fired by the UJT. The firing has to be synchronized with the SCR voltage, as is shown by the waveform curves.

Note the importance of the line-synchronized nature of the UJT firing pulse. The firing angle is always the same with respect to the zero-voltage, zero-angle point of the AC sine wave. This characteristic is inherent in the circuit of Figure 7–12.

The synchronized triggering pulses are supplied by the zener diode and the UJT circuitry. The angle of delay can be adjusted by varying potentiometer Rev in the RC rise circuit. The waveform for V_{AC} is the standard half-wave rectifier curve. The 120 V AC RMS has a peak value of $120 \times 1.414 = 170$ V.

The voltage across the zener, V_{ZD_1}, only reaches the zener voltage, assumed to be 12 V for this illustration. The waveform approaches a 12-V-peak square wave. The resulting voltage at the emitter of the UJT is the rise curve across C_E of the RC circuit. The UJT is assumed to fire at 8 V. When the UJT fires, the voltage across R_1 follows the burst of the triggering current through the UJT. The resulting burst of voltage in turn fires the SCR.

As the value of Rev is adjusted, the RC time changes accordingly. The RC curve changes its slope, and consequently the rise time will change. As the rise time changes, the time to reach 8 V will change. This change directly varies the firing time of the SCR. The variation of SCR firing time results in the variation of the average DC output voltage to the load, as previously discussed.

7–5   THE DIAC AND THE TRIAC

The *diac* is a two-way breakover triggering device. The *triac* is essentially a two-way SCR with one gate. The diac and the triac are used together in a variety of applications, such as the room incandescent-light dimmer switch. A diac-triac combined package, called a *quadrac,* is sometimes used.

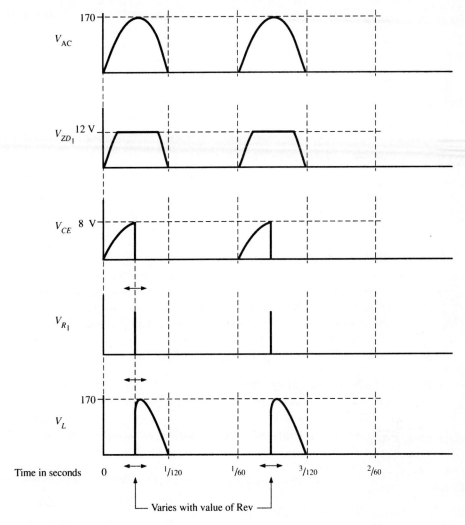

FIGURE 7–13
Waveforms for Figure 7–12

The diac symbols (two are possible), construction, and characteristic curve are shown in Figure 7–14. Physically the diac looks like a regular semiconductor diode. As the voltage is increased from 0 in either direction, a small amount of leakage current occurs, as shown by the characteristic curve. When V_{BR} is reached in either direction, the diac fires. There is then a negative-differential-resistance region, as with the UJT. After the breakover voltage is reached, the diac conducts current easily and has very little internal resistance.

The diac looks like a semiconductor diode. It is not coded for polarity because it acts the same in both directions. It is rated by current and voltage. The

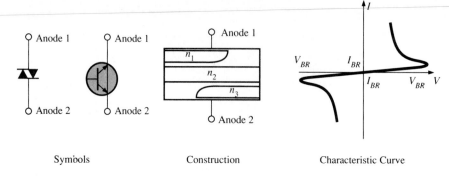

FIGURE 7–14
Diac Symbols, Construction, and Characteristic Curve

voltage will have a tolerance specified, for example, 22 V ± 1 V. The two break-over voltages, forward and reverse, for a given diac will be very close. A word of caution: Some diacs are asymmetrical. For example, a rating of 16/12 is possible. In one direction breakdown is 16 V, and in the other it is 12 V. A circle is sometimes used on this type of diac to identify the higher breakdown direction.

The diac can be used in an oscillator circuit in the same way as a UJT can. Figure 7–15 shows a relaxation-oscillator circuit for the diac. The figure also shows the approximate resulting waveforms that the circuit produces. The diac is used in only one direction for this application. As R_2 is varied, the frequency of oscillation varies as in the UJT oscillator circuit.

The triac symbol, construction, and characteristic curve are shown in Figure 7–16. The triac is a bilateral device. Where two or four SCRs must be used for full-wave AC operation, only one triac is needed. The triac, however, is limited in

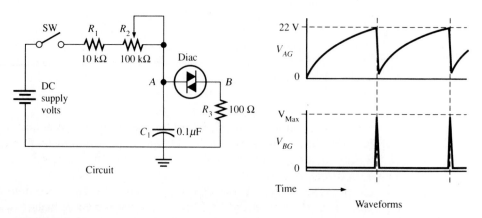

FIGURE 7–15
Diac Relaxation-Oscillator Circuit and Waveforms

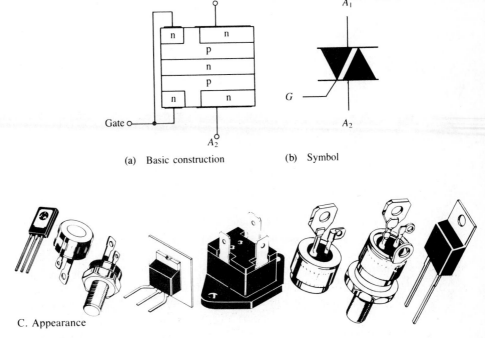

(a) Basic construction   (b) Symbol

C. Appearance

FIGURE 7–16
Triac Construction, Symbol, and Appearance

power-handling capability. The triac is also limited in frequency range capability to near the 60 Hz line frequency. In high-power, rectified DC systems, the triac is not used, because of its limitations. Unlike the diac, the triac needs holding current to keep conducting when the gate current is removed. (The same is true of the SCR.)

Physically the triac resembles the package of the transistor. Its major rating factors are voltage and current capabilities and gate-firing voltage. The triac is turned ON and OFF by the same means as discussed for the SCR.

The characteristic curve for a typical triac is shown in Figure 7–17. Note that it is essentially two SCR curves "back to back."

A basic AC power control circuit for a diac-triac combination is shown in Figure 7–18. The approximate circuit waveforms are also shown. The output voltage to the load is controllable, as that of the SCR circuit is, by varying a resistor. Instead of the SCR half-wave power, there is full-wave control for twice the power output. The circuit in Figure 7–18 contains an *RC* combination on the right to absorb back-surges of voltage from inductive loads, which might destroy the triac. Also, it is necessary to connect anodes 1 and 2 correctly. They are not interchangeable as might be assumed.

FIGURE 7–17
Triac Characteristic Curve

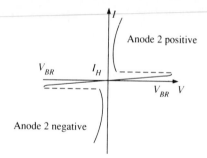

The output power is varied by varying the setting of R_1. As with the SCR power circuit, the R_1 variation changes the voltage rise time across the capacitor. The capacitor voltage reaches the firing voltage of the diac at a time set by the RC rise time. When the diac fires, it in turn fires the triac.

What polarity of gate voltage is needed to fire the triac? Normally a positive voltage is used in the positive direction (quadrant I) and a negative voltage in the negative direction (quadrant III). A typical specification sheet for a triac is shown in the appendix. It includes a quadrant table. The polarity operational capability is specified by listing the operating quadrants. The following is what is meant by each quadrant:

Quadrant	Polarity A_1–A_2	Gate Polarity
I	+	+
II	+	−
III	−	−
IV	−	+

If the quadrant is specified, the triac will conduct with that positive-negative combination. As previously stated, quadrants I and III are normally used. The triac will not conduct or function properly in an unlisted quadrant. The quadrants must be considered when substituting one triac for another in a given application.

7–6   THYRISTOR VOLTAGE AND POWER CALCULATIONS

How do the output voltage and the output power vary with the triggering angle for the SCR and the triac? Figure 7–19 illustrates the curves that show the relation of triggering angle to average voltage and average power output. These curves may also be used for other thyristors that are controlled by triggering.

For illustration, the input voltage is assumed to be 67 V AC RMS, fed into a trigger-controlled thyristor circuit. Also, the output of the thyristor being controlled is assumed to be fed to a resistance of 16.5 Ω. The circuit and input voltage waveforms for full and half wave are shown in Figure 7–20. The average DC values for the input voltages are calculated as in any rectifier circuit. The 67 V AC

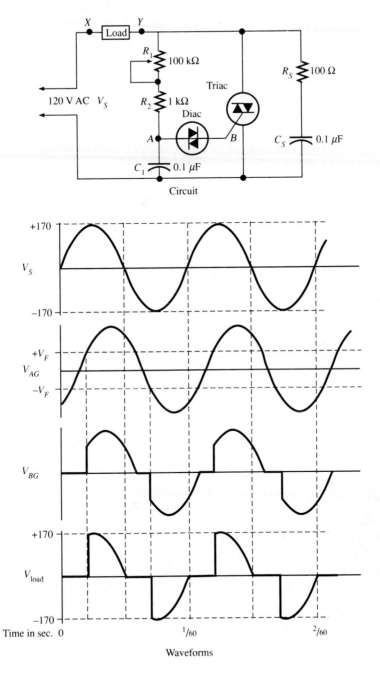

FIGURE 7–18
Diac-Triac Power Control by Phase Control

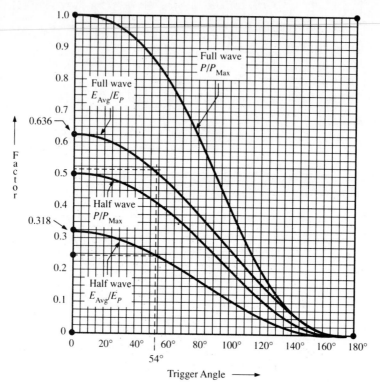

FIGURE 7–19
Voltage and Power Curves for Triggering Angle

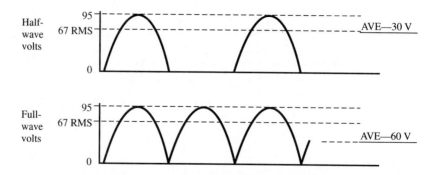

FIGURE 7–20
AC-Input Rectified Waveforms

has a peak AC value of 67 × 1.414 = 95 V. The average DC for half wave is 95 × 0.636 × 1/2 = 30.2 V. For full wave the voltage is 95 × 0.636 = 60.4 V.

As the triggering angle is increased from 0° through 180°, the output voltage will decrease. Suppose that the triggering angle is 54°. For voltage analysis the E_{avg}/E_P curve is used in Figure 7–19. On the graph for half wave, find the 54° point on the curve as shown. Then go across to the factor scale as shown. The factor scale shows a value of 0.255. Therefore, the average DC voltage output will be 95 × 0.255 = ~24 V. For full wave, the factor from the curve is 0.51. The average full wave DC voltage will be 95 × 0.51 = ~48 V. Voltages for other angles are calculated in a similar manner: From the trigger angle, use the curve to find the factor, and multiply accordingly. Note that the full-wave curve starts at the 0.636 point for 0°. The half-wave curve starts at 1/2 × 0.636, or 0.318. These curves, of course, are valid for sine waves only.

Another way to use the curves is in the inverse manner. Suppose that the input voltage is again 67 V AC, or 95 V peak. The following procedure can be used to find the triggering angle needed for an output voltage of 21 V half wave: First, determine the factor. It would be 21/95 = 0.22. Next, start at 0.22 on the factor scale, look across to find the corresponding point on the curve, and then look down at the angle scale. A triggering angle of about 66° is required for 21 V. The same procedure, carried out on the full-wave graph, will show that a full-wave angle of 106° is needed for 21 V DC output.

The curves in Figure 7–19 may also be used for power calculations. The power curves are labeled "P/P_{max}." Note that these curves are only for loads that are purely resistive. Figure 7–21 shows the standard power curves for AC power. The peak power is determined by the product of peak volts and peak amps. The effective power, maximum power, is determined by using effective voltage and effective current in appropriate formulas.

As with voltage, if triggering takes place after 0°, the average DC output power is less than full power. The power decreases as the trigger angle increases and may be determined by using the curves. To illustrate, the same voltage and

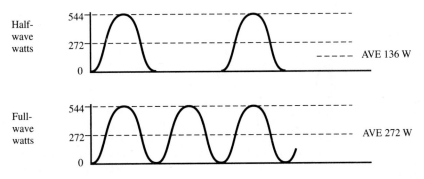

FIGURE 7–21
AC Power Waveforms

load resistance as before are used: 67 V AC RMS input and a resistive load of 16.5 Ω. For average maximum power, $P = (V \times V)/R = (67 \times 67)/16.5 = 272$ W (average).

The power curves take into account the variation of the RMS factor from 0.707. Once the voltage waveform stops being a pure sine wave, the value of 0.707 is no longer valid. The variation from 0.707 is considered in the construction of the two curves shown. The 0.707 consideration does not affect the calculations.

A triggering angle of 105° is assumed for illustration of the use of the two power curves in Figure 7–19. Use the power curve in the same manner as the voltage curve. The factor for half wave is 0.11. For full wave the factor is 0.22. The power for half wave is $0.11 \times 272 = 30$ W. For full wave the power is $0.22 \times 272 = 60$ W.

Suppose 185 W is needed from this circuit. The factor is $185/272 = 0.68$. For half wave this factor is impossible. The half-wave factor is above the curve. For full wave, the angle would have to be about 73°.

7–7   OTHER MAJOR THYRISTORS

Figure 7–22 illustrates the symbols and characteristic curves of two thyristors: complementary UJT (CUJT) and the programmable UJT (PUT). Both of these semiconductors operate as triggering devices.

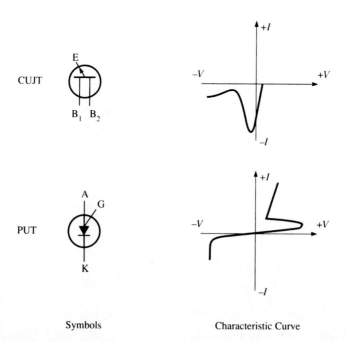

Symbols   Characteristic Curve

FIGURE 7–22
CUJT and PUT Symbols and Characteristic Curves

The CUJT is built exactly the same way as the UJT except that its *p* material and its *n* material are reversed. It is sometimes called a *p*-channel UJT, with the regular UJT designated *n*-channel. The operational voltage polarity to the bases is reversed. The trigger voltage applied between emitter and base for the CUJT is negative. The characteristic curve is therefore a third-quadrant mirror image of the UJT first-quadrant curve. The CUJT is used where its reverse polarity is required in a circuit.

The PUT is constructed similarly to the SCS, which will be covered shortly, but it has no anode gate. As in the UJT, a gate voltage pulse produces a large increase in anode-to-cathode current. The PUT can trigger at a lower current than the UJT—for example, at 0.1 μA, compared with 10 μA for the UJT. Why is the PUT designated a programmable device? The operational characteristics of the PUT may be varied by varying the ratio of external resistors in its associated circuitry. The PUT works well at lower frequency and in long-duration timer applications.

Figure 7–23 shows five thyristor devices not previously discussed. The first is the silicon-controlled switch (SCS). The SCS is a special SCR with an added gate, which can turn the SCS OFF. The SCS is fired by gate 1, just as an SCR is. When a negative voltage is applied to gate 2, the conducting SCS is turned OFF, provided gate 1 is not active. The SCS is used where a sharp turn-OFF of current is desired. The SCR turns OFF fairly fast, but it does take some time to do so. The rate of turn-OFF is called *OFF slew rate*. Note that SCRs are also rated for ON slew rates. The SCS ON slew rate is like the SCR rate. On the other hand, the SCS OFF slew rate is many times faster than the SCR rate.

The other four thyristors shown in Figure 7–23 are triggering devices. The silicon unilateral switch (SUS) is essentially a small SCR that is used as a unilateral triggering device. The SUS is triggered by breakover voltage. In some cases the breakover-voltage operating value is controlled by supplying a limited amount of gate voltage also.

The silicon bilateral switch (SBS) is a bilateral SUS. It can be used to fire two or four full-wave SCRs or higher-powered triacs. It replaces two SUSs.

The silicon asymmetrical switch (SAS) and the asymmetrical SBS (ASBS) are similar to the SBS. They differ in that they have two different firing voltages—one forward and one reverse—for example, 8.5 V and 3.0 V or 1 V and 15 V. One application of these devices is to reduce hysteresis in a circuit. Hysteresis, the effect of different rise and fall curves, causes unwanted heating of components. For 1 V/15 V triggering, operation is mostly in the first quadrant, so there is very little reversing. Polarity reversing produces hysteresis, as in magnetic-device circuits.

Three other thyristor devices are shown in Figure 7–24. The first is the gate turn-OFF switch (GTO). It is a small SCR that can be turned OFF as well as ON from a single gate. The GTO has one advantage over the SCR. The turn-ON and turn-OFF times, the slew rates, are 10–20 times as fast as for an SCR. There are some limitations on the use of GTOs. The GTO requires a higher triggering current than the SCR, 20 mA versus 20 μA for the SCR. Also, the GTO is limited to 5 A and 25 W, whereas the SCR can handle thousands of watts.

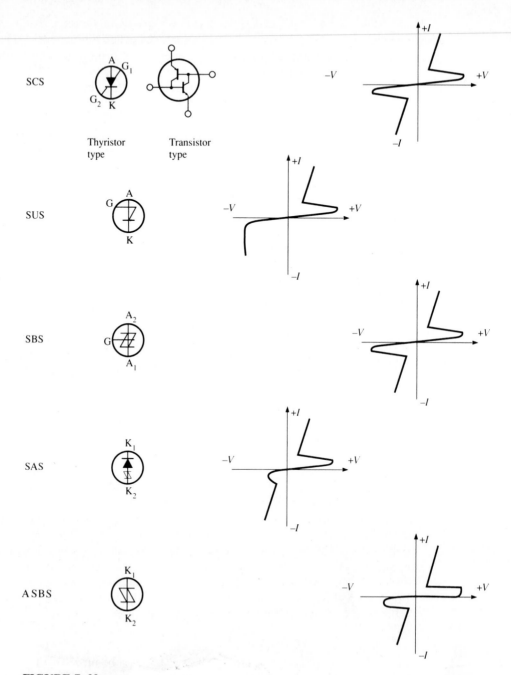

FIGURE 7–23
Thyristor Switching Device Symbols and Characteristic Curves

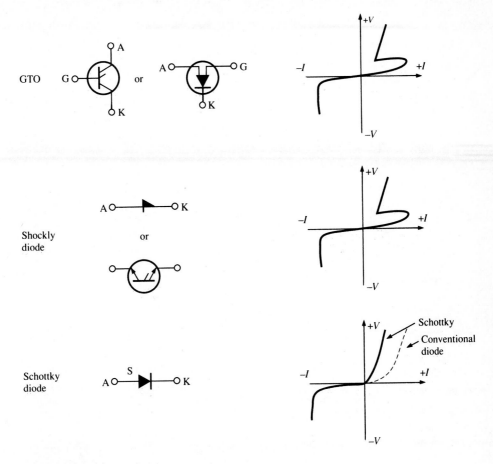

FIGURE 7–24
GTO and Thyristor Diode Symbols and Characteristic Curves

The Shockley diode shown in the figure is named for one of the inventors of the transistor. It is essentially a one-way diac. It is fired by breakover voltage in its applications.

The Schottky diode of Figure 7–24 is a four-layer device also. It has a better forward current-carrying capability than a conventional diode. This capability is shown by the characteristic curve. The four layers of semiconductor material give it the capability of an avalanche effect.

IMPORTANT
READ
ON TEST!

7–8   LIGHT-ACTIVATED SEMICONDUCTORS AND OPTOISOLATORS

Many semiconductor devices utilize light for operation. Six of those devices are shown in Figure 7–25. Each device has a small window in its top or side. Light of

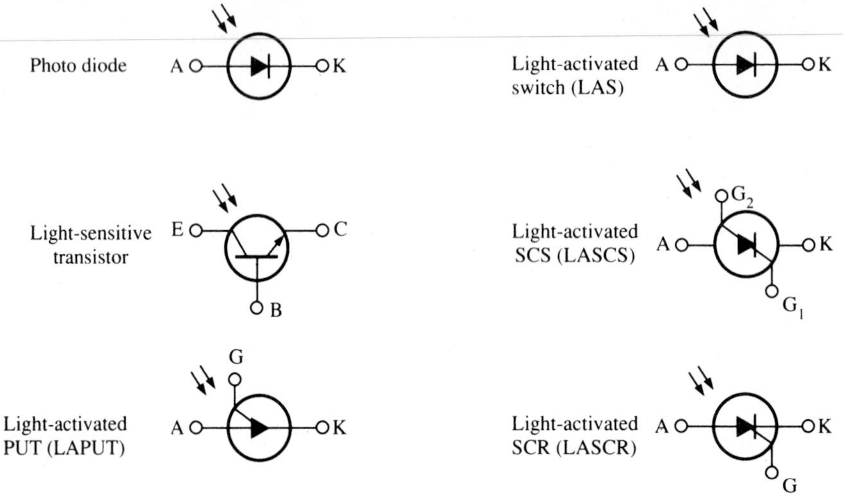

FIGURE 7–25
Light-Activated Semiconductor Symbols

varying intensity is applied through the window, usually from a light-emitting diode. The components and circuitry of photoactivated devices must be enclosed so that varying room light will not affect them. As the light input varies to these devices, the device characteristic curve varies. Within each device, the varying light is received at a PN junction or a light-sensitive area.

Two kinds of varying light input are used. One is input light that goes from OFF to ON. For example, the LASCR is triggered by a light pulse. One advantage of this arrangement is electrical isolation from the trigger circuit. Another advantage is that a low input current is required for the input light. This prevents circuit "loading" of the triggering circuit.

The other kind of variable light input is one in which varying levels of light are used, not just ON and OFF status. Suppose four levels of input light are used in a phototransistor. A single transistor can then behave like four different transistors, depending on the light level. In a circuit requiring switching between four transistors, varying the light for one transistor would eliminate the need for three other transistors.

Optoisolator symbols are shown in Figure 7–26. The optoisolator package is a flat, rectangular package with three or more leads—usually four. The input to the optoisolator is a varying voltage applied to a light-emitting diode (LED) inside the package. As the input voltage is varied, the internal LED emits a varying light, usually nearly directly proportional to the input voltage. The varying signal is therefore changed to a varying amount of emitted light. The LED-emitted light is then received by another internal semiconductor device.

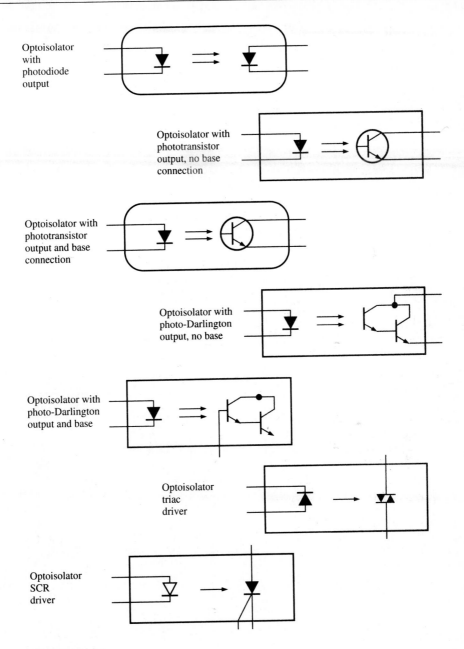

Optoisolator
with
photodiode
output

Optoisolator with
phototransistor
output, no base
connection

Optoisolator with
phototransistor
output and base
connection

Optoisolator with
photo-Darlington
output, no base

Optoisolator with
photo-Darlington
output and base

Optoisolator
triac
driver

Optoisolator
SCR
driver

FIGURE 7–26
Optoisolator Types and Symbols

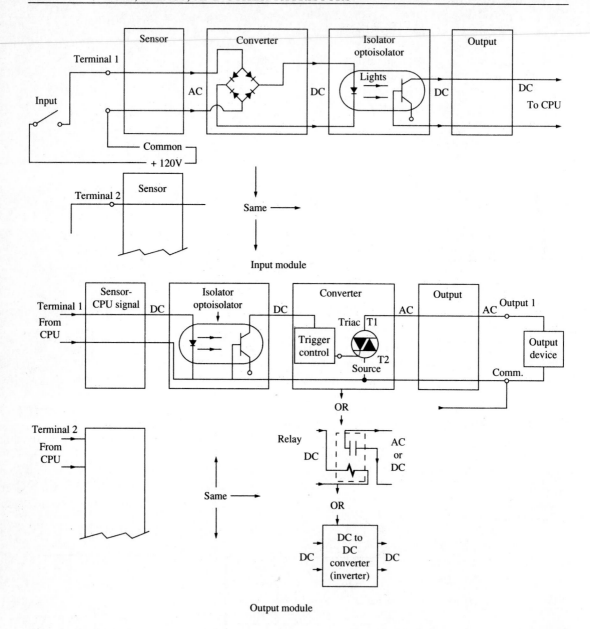

FIGURE 7–27
Use of Optoisolators in a Programmable Logic Controller

In the figure, seven of the many possible types are shown. As the internal varying light is received internally, the internal output devices are correspondingly activated. Outputs of optoisolators may be ON/OFF or varying, as described for light-activated devices. The triac and SCR drivers shown have a pulsed, ON/OFF triggering arrangement. Other optoisolators shown in the figure are analog in nature. An output audio amplifier would use an analog optoisolator. The varying input audio signal is then electrically isolated from the varying output audio signal.

An example of the use of optoisolators is in the input and output modules of a programmable controller. The programmable logic controller is a user-friendly computer that is connected to the "outside world" by interfacing modules. The "outside world" is usually 110 V AC. The central processing unit (CPU) works on a transistor-transistor logic (TTL) voltage of ±5 V DC. What if the system failed and the 110 V got into the CPU? The CPU would most probably be damaged. A layout of the programmable controller input and output modules is shown in Figure 7–27. Note that the optoisolators electrically isolate the CPU from possible damage by the much higher external voltages.

SUMMARY

- □ Most thyristors are referred to by their abbreviations or by names that are shortened versions of the terms that describe them. Learning the words for which certain letters usually stand can help you understand thyristor abbreviations that are new to you.
- □ Thyristors are, in general, four-layer devices of p or n material or both.
- □ The SCR is a four-layer, three-terminal device. The major type of thyristor, it converts fixed AC voltage directly to variable AC voltage. This conversion is accomplished by varying the triggering angle to the SCR gate.
- □ The UJT, a major type of triggering device, is often used to trigger an SCR. It is essentially a trigger pulse amplifier.
- □ The triac is a bidirectional SCR. It has some operational limitations in comparison with the SCR.
- □ The diac is used to trigger triacs. It is bidirectional.
- □ There is a chart that relates output power and voltage to trigger angle. Calculations of voltage and power values for a given trigger angle can be made by utilizing the chart.
- □ There are many other types of thyristors. They perform circuit functions similar to those of the SCR, the UJT, the diac, and the triac.
- □ Most semiconductor devices, including thyristors, have equivalent photosensitive devices. Such devices actuate or change internal characteristics as varying light levels are applied to them.
- □ The optoisolator is used to electrically and electronically isolate one circuit section from another. Optoisolators are available in a number of different configurations.

EXERCISES

1. According to the list in Section 7–2, which of the following abbreviations could be the names of real thyristors?
   a. CSCR   **g.** PGTO
   b. BSCR   **h.** ASUS
   c. ABUJT   **i.** UTRIAC
   d. CTRIAC   **j.** LSUS
   e. CDIAC   **k.** BSCS
   f. UDIAC

2. What is the significance of the intrinsic stand-off ratio (ISTOR)? Does it apply to the SCR? Why or why not?

3. Explain the ways to turn an SCR ON, with reference to the curve of Figure 7–3.

4. Explain the ways to turn an SCR OFF, with reference to Figure 7–3.

5. Explain how the circuits of Figure 7–4A, B, and D function to produce variable DC. What happens if V_S, AC linevolts, is increased 10% for Figure 7–4B, half wave?

6. Find the ISTOR and the firing voltage for a UJT circuit that has the following values:

$$V_S = \text{either } 21 \text{ V or } 28 \text{ V}$$
$$R_{B_1} = 5.66 \text{ k}\Omega$$
$$R_{B_2} = 3.86 \text{ k}\Omega$$
$$V_D = 0.58 \text{ V}$$

7. For the UJT in Exercise 6, when the 21 V is applied in the reverse direction, what is the back-leakage current value?

8. What would the waveforms look like for each voltage for the UJT of Exercise 6? Use the circuit of Figure 7–9.

9. Explain how the circuits of Figure 7–4C and E function to produce variable DC. What happens if V_S is increased 10% for a given variable resistor setting?

10. Refer to Figure 7–9. How would the waveforms change if the pot was reset to 20 kΩ with a 12.5 V supply voltage?

11. Describe how the circuit in Figure 7–12 functions.

12. How will the curves of Figure 7–15 change as the supply voltage is increased? How will the curves change as R_2 is increased in value?

13. Refer to Figure 7–17. Could a triac be used as an SCR, with proper circuitry? How do characteristic curves for the triac and the SCR differ?

For the next three problems refer to Figure 7–19.

14. Input voltage to an SCR (half wave) is 120 V RMS. The same voltage is applied to a triac (full wave). Fill in the following table for the firing angles given:

Angle (°)	SCR Voltage, Half Wave	Triac Voltage, Full Wave
0		
55		
90		
110		
180		

15. Input is 240 V RMS. Find the firing angles, half and full wave, to produce the voltages given.

DC Output Voltage (V)	Trigger Angle, Half Wave	Trigger Angle, Full Wave
240		
171		
120		
95		
55		

16. Input is 440 V. Load resistance is 2.5 kΩ. Find the power for half and full wave at the angles given.

Trigger Angle (°)	Watts, Half Wave	Watts, Full Wave
0		
40		
90		
140		
180		

17. From memory, name six of the ten devices in Figures 7–22, 7–23, and 7–24. Draw the symbols and characteristic curves for each of the six devices. Describe a possible application for each device.

18. Describe some applications for the devices illustrated in Figure 7–25.

19. For the devices of Figure 7–26, how important is it for the output to be linear with respect to the input? In what type of applications is this linearity important?

APPLICATION ASSIGNMENT ANSWER

The best place to start your analysis would be to observe the waveform (or waveforms) of the power from the electronic drive output to the motor. Is it consistent? What happens to it when the motor speeds up? If the waveform changes when the motor speeds up, the next step is to check the waveform of the voltage triggering the SCR, the SCRs, or the triac—whichever is the drive thyristor. Is the triggering erratic? Once these steps have been taken, further analysis can be made as needed.

8

Solid-State Power Control

CHAPTER OUTLINE

LEARNING OBJECTIVES

- □ Describe the operating characteristics of AC and DC power switches.
- □ Describe the acceptable methods of triggering AC and DC power switches.
- □ Show the proper method of connecting loads to AC and DC power switches.
- □ Given a linear analog or digital amplifier, along with an appropriate input signal, determine the resultant load voltage or current.
- □ Given a pulse width modulation power control system, along with an appropriate input signal, determine the duty cycle and average load voltage or current.
- □ Given a linear ramp or microprocessor-based AC power control system, along with the appropriate operating parameters, determine the firing delay angle and the load power.

APPLICATION ASSIGNMENT

An older section of a factory has been undergoing a continuing process of modernization. As part of this plan, the control for some 240 V AC electric heaters has been changed from electromagnetic to solid-state relays. Ever since the changeover, however, the solid-state relays have been a constant source of trouble. Often, the outputs are found shorted, leaving the heaters on continuously. The plant manager is ready to abandon the solid-state relays in favor of the older more "reliable" electromagnetic relays. Before she makes her decision, however, she asks you to investigate the source of the trouble. What would you look for?

8–1   INTRODUCTION

If an electronic control system is to maintain control of a process or operation, it must have the capability of controlling the amount of power that is supplied to the system output actuator(s). After all, it is the output actuator that actually effects the changes in the process. Unfortunately, the power requirements of output actuators, such as motor-driven solenoids, pumps, fans, heaters, solenoid valves, lamps, and similar devices, are beyond the output capability of the op amps, logic gates, and computer I/O (input/output) ports that comprise the outputs of most electronic control systems today.

This chapter will deal with the circuits that interface delicate control system electronics with the process-correcting devices that keep the system under control. These circuits are divided into two broad categories—those that simply switch the output actuator ON or OFF and those that supply a variable amount of power to the actuator.

8–2   POWER SWITCHES

Many control system applications require only that power to the load be switched on and off. The electronic circuits that accomplish this task are classified as either DC power switches (often called DC output modules) or AC power switches (often called solid-state relays). Although each of these can be constructed from discrete components, they are more often found as integrated modules.

DC Output Modules

Many companies provide integrated DC power switches called DC output modules. One such device is shown in Figure 8–1. The function of these modules is to provide the capability of switching high currents and voltages to a load while using very low voltages and currents as the control signals. That is, they operate much like an electromechanical relay. Since the output of a DC output module is polarized, however, these devices are *not* generally referred to as solid-state relays.

Aside from the output polarity issue, DC output modules offer several distinct advantages over their electromechanical counterparts. The DC output module is generally smaller than an equivalent electromechanical relay; the input is noninductive; there are no moving parts that will wear under heavy use; and the device is inherently explosion-proof. The input circuitry of a DC output module can be made sensitive enough to be driven directly from the output of a programmable controller, a computer I/O port, an op amp, or even a logic gate. In contrast, however, DC output modules are generally more expensive; they are only (currently) available as single contact outputs; and they generally fail in a shorted condition.

The three circuits shown in Figure 8–2 characterize typical equivalent circuits of many of the DC output modules that are commercially available today.

FIGURE 8–1
A Typical DC Output Module

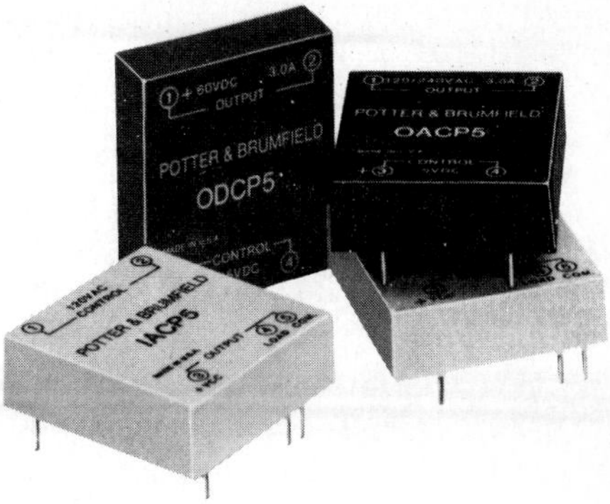

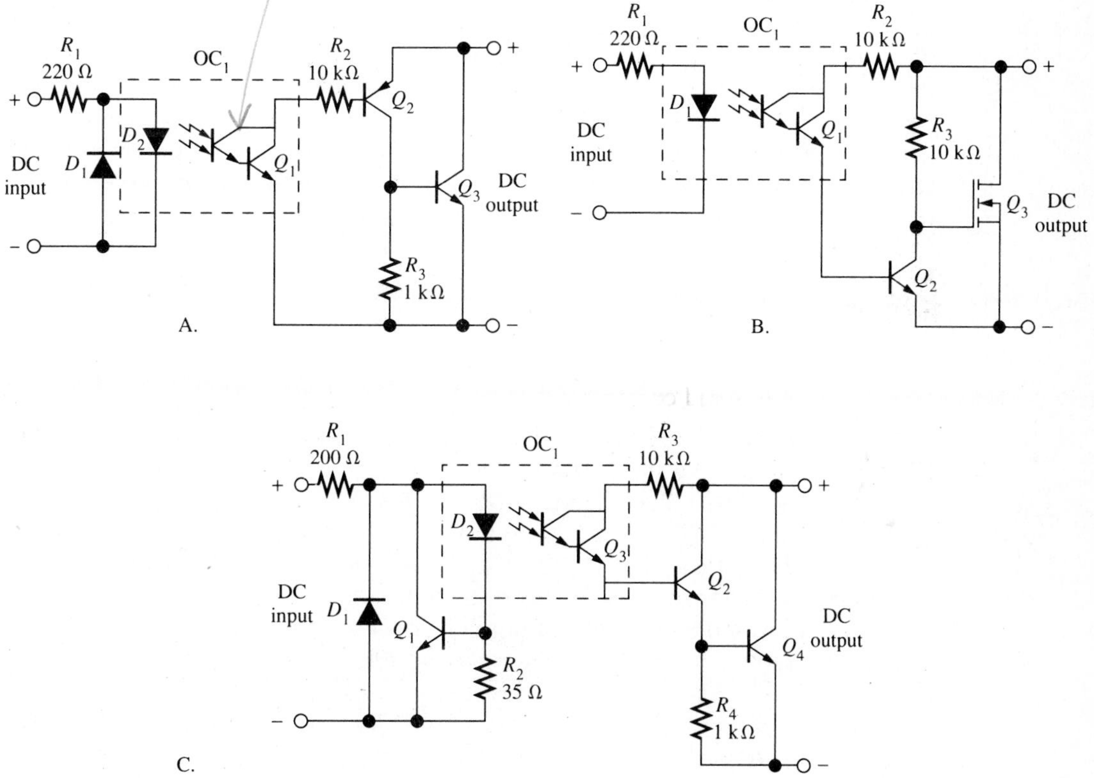

OPTOISOLATER

A.

B.

C.

FIGURE 8–2
Equivalent Circuit Configurations of DC Output Modules

Note that each circuit is nothing more than an optically isolated transistor switch. Each, however, is slightly different in both the transistor arrangement and the manner in which the optoisolator is driven. The purpose of the optoisolators is to prevent output faults from electrically coupling high load voltages into the circuitry driving the module.

The circuits shown in Figures 8–2A and 8–2B have simple arrangements for driving the infrared-emitting diode (IRED) at the input of the optoisolator. The 220 Ω resistor, R_1, limits the current to a safe level. Due to this simple arrangement, however, the range of input voltages that will safely and reliably switch the output into conduction will be quite narrow. In fact, these modules would probably be rated for a fixed input voltage of perhaps 5 V. The diode, D_1, in the circuit of Figure 8–2A provides protection in the event that the input voltage is accidentally reversed. Most infrared-emitting diodes have peak inverse voltages of about 3 V. If D_1 is forward-biased (by reversing the input voltage), the reverse IRED voltage will be clamped to only 0.7 V. For DC output modules that do not employ such protection, the entire module can be destroyed with as little as 5 V applied to the input with the reverse polarity.

The input circuit of the DC output module shown in Figure 8–2C is a bit more elaborate. In addition to the current-limiting resistor and reverse voltage protection diode, this module also employs a simple current regulator (Q_1 and R_2) that allows the input to be driven from a much wider range of voltages. Input current passes through R_1, the IRED, and R_2 as long as the input voltage is low enough to keep the input current below about 20 mA. At a current of 20 mA, the voltage dropped across R_2 will be 0.7 V—just enough to forward-bias the transistor Q_1. As Q_1 begins to conduct from collector to emitter, it shunts current away from the IRED. The input voltage can therefore increase considerably beyond the level that would have destroyed the IREDs in the modules of Figures 8–2A and 8–2B. It is quite typical for a module of this type to be rated for an input voltage range of 3 V to 32 V.

Although Figure 8–2 shows only three examples, there are dozens of possible transistor switching configurations that can make up the output section of a DC output module. The outputs of the modules shown in Figures 8–2A and 8–2C are quite similar. When sufficient current passes through the IRED, the phototransistor saturates, providing a current path through the 10 kΩ resistor and the base-emitter junction of Q_2. In turn, Q_2 saturates, providing enough base current to saturate the output transistor (Q_3 in Figure 8–2A and Q_4 in Figure 8–2C). Since the output transistor remains off until an input voltage is applied to the module, these devices would be regarded as normally open switches. The circuit of Figure 8–2B operates as a normally closed power switch. Note that with no IRED current, the phototransistor in the optoisolator will be at cutoff. Since this interrupts the base current for Q_2, Q_2 will also be at cutoff. Any voltage impressed across the output of the module, therefore, will be dropped across Q_2. This action will produce enough gate-to-source voltage to force Q_3 into saturation. In order to turn Q_3 off, all that is required is sufficient IRED current to saturate Q_2. This will essentially short the gate of Q3 to its source, forcing it into cutoff.

As long as the input and output specifications for a particular DC output module are known, it is not necessary to show the details of the internal circuitry. Instead, a schematic symbol such as that used in Figure 8–3 is used. The source of all output currents is an external DC power supply. The load (the device being switched ON and OFF) is placed in series with the external supply and the output transistor (see Figure 8–3). It is interesting to note that DC output modules do not *output* anything; they merely provide a path for the load current.

The input current for the module can be provided in a number of different ways. Figure 8–3 shows four possible configurations. The particular configuration chosen depends upon the device that will be used to provide input current for the module. For example, the output of the U1A gate shown in Figure 8–3A is capable of sinking about 16 mA to ground when its output is at a logic LOW. When the output is at a logic HIGH, however, it is only capable of sourcing about 0.5 mA (not enough to trigger the module). The LM324 op amp, on the other hand can both source *and* sink the amount of current required by the module input circuitry. Remember that the objective is to provide a current path from the positive terminal of the voltage supply, through the module input circuitry, and out to the

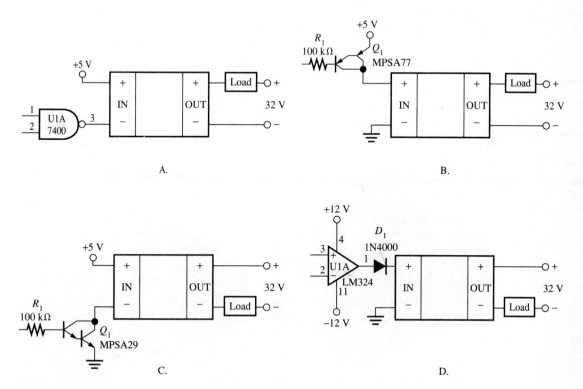

FIGURE 8–3
Acceptable Methods of Providing Necessary Input Current for DC Output Modules

negative terminal of the voltage supply. Any method that accomplishes this is a valid manner in which to turn on the module.

Commercially available DC ouput modules typically have input voltage ranges of 3 V to 32 V, output current ratings of 3 A to 5 A, and maximum output voltage ratings of up to about 60 V. These ratings, of course, vary from manufacturer to manufacturer. For example, some modules will only reliably switch with input voltages that vary from 12 V to 20 V, and some will handle output currents of up to 10 A. Although most output modules are optically isolated, this is not a requirement. Similarly, certain modules may not have input protection against reverse polarity. There is only one way to be certain about the module operation: always consult the device specifications before implementing it in a control system.

Solid-State Relays

Solid-state relays are similar to DC output modules in both appearance (see Figure 8–4) and application. The primary difference is that solid-state relays are intended specifically to switch power to AC loads. The output voltage ratings are typically 120 V AC or 240 V AC. In addition, the output current capacities are considerably higher than those of DC output modules. Current ratings of 10 A to 50 A are quite common, and solid-state relays are available that will handle up to 100 A of load current. As in the DC output modules discussed earlier, the load is placed in series with the output of a solid-state relay. Similarly, any of the DC output module trigger circuits shown in Figure 8–3 can also be employed to trigger solid-state relays that have DC input trigger voltages.

Solid-state relays have the same advantages *and* disadvantages as the DC output modules discussed earlier. Due to the high voltages and currents associ-

FIGURE 8–4
A Typical Solid-state Relay

ated with solid-state relays, however, there are two precautions that must be observed. All solid-state switches have a nominal ON-state voltage drop. Typically, this voltage is 1.5 V. With a load current of 20 A, the device power dissipation will be 30 W. Therefore, as with all semiconductor devices, the maximum current ratings only apply when the device is attached to a heat sink with enough surface area to carry away the heat. In addition, the OFF-state leakage current that flows through the load can be as high as 20 mA. At a potential of 120 V AC or 240 V AC this current is high enough to pose a safety hazard.

The input characteristics of solid-state relays tend to be a bit more standardized than those of DC output modules. Commercially available solid-state relays with DC inputs generally have input trigger voltage ranges of 3 V DC to 32 V DC. Those that have AC inputs generally trigger with input voltages from 90 V AC to 280 V AC. Although these ranges have not been formally established by agencies such as ANSI (American National Standards Institute) or DIN (Deutsches Institut für Normung), they have evolved because of the widespread use of solid-state relays.

Regardless of the output voltage or current rating, the outputs of all solid-state relays are classified as either random-trigger or zero-voltage trigger. A relay with a random-trigger output switches current to the load at the instant that the input trigger current requirement is met. (Actually, there is a turn-on delay of up to a few milliseconds.) This can cause several problems, especially when switching current through loads that normally have large inrush currents associated with them (such as heating and incandescent lighting loads). Suppose, for example, that a solid-state relay will be used to turn ON and OFF an ordinary 120 V AC, 100 W incandescent lamp. According to Ohm's law, the resistance of the lamp should be 144 Ω. In fact, the filament resistance is 144 Ω only when the filament is hot and emitting light. When cold, the filament resistance will be, perhaps, 10 Ω. If a random-trigger solid-state relay is used, and if, by chance, the trigger voltage is applied at the instant the line voltage is at its peak value (approximately 170 V), current through the lamp will be 17 A. This can happen even with a conventional mechanical switch, and it explains why bulbs usually burn out when they are first turned on. This situation is undesirable because the excessive current is damaging to the solid-state relay and to the device being controlled (in this case, the lamp). In addition, the electromagnetic interference (EMI) associated with this high current can disturb the operation of nearby electronic circuits.

The solution to this problem is a zero-voltage-trigger solid-state relay. In order for this type of relay to switch current through the load, two criteria must be met: the required trigger voltage must be present at the input *and* the load supply voltage must be less than about 20 V. If either of these conditions is violated, the output of the solid-state relay will remain in an OFF state. This action is demonstrated in Figure 8–5. Shown is a DC input solid-state relay along with its associated input and output voltage waveforms. Note that the input trigger voltage occurs at the point when the load supply voltage is crossing its peak value. Notice the difference between the load voltage that would result from the use of a random-trigger solid-state relay and that which would be produced from the use of a

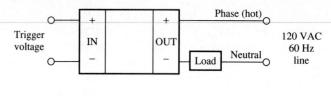

A.

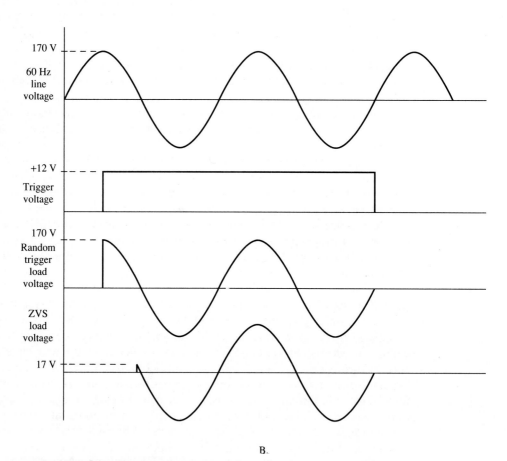

B.

FIGURE 8–5
Providing the trigger voltage at the instant the AC line crosses peak produces drastically different load voltages, depending upon which type of solid-state relay is used.

zero-voltage-trigger solid-state relay. The random-trigger relay switches current to the load almost instantly, producing an instantaneous load voltage of 170 V (along with the associated high current and EMI). Even though the trigger voltage was present at 90° into the cycle, the zero-voltage-trigger relay does not switch current to the load until nearly 180°—not until the load supply voltage falls below the threshold (which, for this relay, appears to be 17 V). Note that after the current has been switched to the load, the load voltages produced by the two relay types are indistinguishable. In addition, after the trigger voltage falls to zero, both relays switch off at the next zero-crossing point of the AC line.

It may seem that a zero-voltage-trigger solid-state relay is the obvious choice for all applications. In fact, it may be the worst possible choice for switching current through inductive loads such as transformers, motors, and contactor coils. When a voltage is applied to an inductor, a counter electromotive force (CEMF) is produced in the inductor. The magnitude of this CEMF is proportional to the applied voltage and opposes current flow through the inductor. Since the voltage applied to a load by a zero-voltage-trigger solid-state relay will always be low, the CEMF produced by an inductive load will likewise be low. This may allow the inductor current to increase to the point where the inductor core saturates. If this happens, even though the current through the inductor continues to increase, the resulting magnetic field density does not. Since the inductance of a device depends upon changing magnetic field densities, the inductance may drop to near zero. The resulting current through the inductor may then rise to many times more than that which is expected. This condition may last for several cycles and may produce the same problems encountered in switching current to incandescent loads using random-trigger solid-state relays. As a rule, zero-voltage trigger relays should be used only in situations where the resistance characteristics of the load result in undesirably high turn-on currents.

The internal circuitry of both solid-state relay types is shown in Figure 8–6. The random-trigger relay shown in Figure 8–6A is quite simple in operation. As soon as sufficient current flows through R_1 and D_2, Q_2 provides a path for the gate current to the triac, Q_3 (note the similarity to the input circuitry of Figure 8–2C). The bridge rectifier in front of R_1 and D_1 allows the relay to be triggered by AC voltages. The zero-voltage-switching relay shown in Figure 8–6B is a bit more complex. Gate current for the triac, Q_5, must flow through the bridge rectifier, BR_1, and the SCR, Q_4. Note, however, that Q_4 must be triggered into conduction by Q_2. Normally, all that is required to trigger the SCR is sufficient current through R_1 and D_2. If, however, the load supply voltage is high enough to break over the zener diode, D_3, the transistor, Q_3, will saturate and shunt the SCR trigger current away from the gate, preventing the SCR (and so also the output triac) from triggering.

Both of the solid-state relays shown in Figure 8–6 have RC snubbing networks connected across the output terminals. The purpose of these networks is to prevent the output triac of the solid-state relay from breaking over into conduction in response to the high-frequency voltage spikes that are sometimes induced on the AC line. If the values for R and C are chosen properly, the spikes will be shunted away from the relay output. Although some relays include these internally, most require that the networks be attached externally. This allows the user to choose values for R and C that are appropriate for their particular application. Typical snubbers have R and C values that are 33 Ω and 0.1 μF respectively.

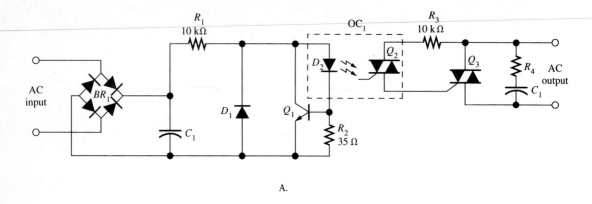

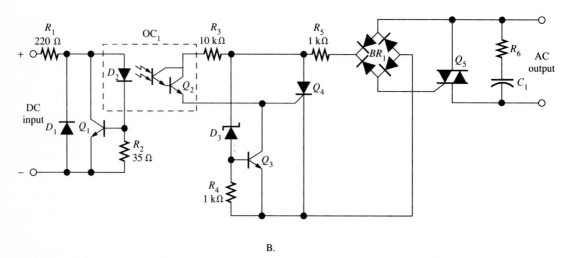

FIGURE 8–6
Equivalent Circuit Configurations of (A) Random-trigger Solid-state Relay and (B)
Zero-voltage-switching Solid-state Relay

8–3   PROPORTIONAL DC POWER CONTROL

In many applications simply switching load power ON and OFF is not acceptable.
Controlling the speed of a DC motor or the position of a proportional solenoid
valve are examples of situations in which a variable amount of direct current is
required. Both analog and digital means of accomplishing this task are discussed
in this section.

 Linear Analog DC Amplifiers

Operational amplifiers do a fine job of amplifying DC signals. Unfortunately, the
output of most op amps is limited to currents of less than 100 mA and voltages of
less than ±20 V. Still, it would be wise to take advantage of the simplicity of

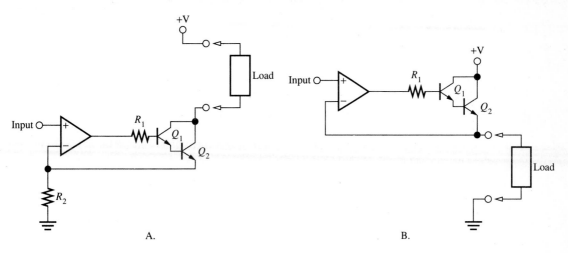

FIGURE 8-7
Operational Amplifiers Can Be Employed in High-power Linear Amplifiers in Both
Constant-current and Constant-voltage Applications.

design that op amps afford. Both of the circuits shown in Figure 8–7 employ op amps to control the power supplied to the load. The circuit of Figure 8–7A maintains a constant current through the load. The magnitude of the current depends upon the op-amp input voltage and the value of R_2. Notice that the op amp in this circuit is configured as nothing more than a noninverting amplifier, with R_1, and the base-emitter junctions of Q_1 and Q_2, comprising the feedback resistor. Remember that the op-amp output will change in such a manner as to force the noninverting input to the same potential as the noninverting input. If the input voltage is +1 V, the output voltage of the op amp will rise until Q_1 and Q_2 conduct enough to allow enough current to flow through R_2 to produce a voltage drop of +1 V. The current that flows through R_2 must first flow through the load and Q_2. It should, therefore, be apparent how the value of R_2 can affect the load current. If the value of R_2 is 1 Ω, then 1 A must flow through R_2 (and so also the load) in order for the voltage at the inverting input of the op amp to be +1 V. With a value of 1 Ω for R_2, 1 A of load current will be produced for each volt input to the op amp. That is, if the input voltage is 5 V, the load current will be maintained at 5 A. The value of R_1 is not critical. It should be low enough to allow the op amp to drive the Q_1/Q_2 darlington configuration into saturation (if required), yet high enough to protect the base of Q_1 from excessive current. The exact value depends upon the overall current gain of the darlington and the maximum allowable base current for Q_1.

It is important to note that the load current for the circuit of Figure 8–7A remains constant even if the load changes. There are, of course, limitations. If the resistance of the load is too high, the op amp may not be able to achieve the required load current even if its output is driven to saturation. Furthermore, the power rating of R_2 must be chosen to withstand the heat that will be generated.

For example, if the value of R_2 is 1 Ω, and input voltage is 5 V, R_2 will dissipate 25 W (5 V × 5 A). Similarly, the continuous collector current and maximum power dissipation ratings for Q_2 should be well in excess of the maximum anticipated load current and transistor power dissipation.

     The circuit shown in Figure 8–7B operates in a very similar manner to that of Figure 8–7A. The difference is that R_2 has been replaced by the load itself. This subtle substitution changes the circuit from a load *current* regulator to a load *voltage* regulator. If, for example, the input voltage is +5 V, the output voltage of the op amp will rise until enough current flows through Q_2 and the load so that the load voltage is +5 V. If the load resistance changes, the load current will change, but the output voltage of the op amp will adjust to ensure that the load voltage remains at +5 V. This circuit has one very obvious limitation. The output voltage of the op amp must be at least 1.4 V higher than the load voltage (in order to maintain the required forward bias on Q_1 and Q_2). This limits the maximum load voltage to less than the maximum op-amp output voltage.

EXAMPLE 8–1

Suppose that the load in Figure 8–7A has a nominal resistance of 20 Ω. Generate a response curve that plots input voltage versus load current for R_2 = 5 Ω as the input voltage varies from 0 V to 10 V.

     Since the response of the circuit is linear, only two points need to be plotted. One of these points occurs where the input voltage is 0 V and the resulting load current is 0 A. To determine the other point, choose any voltage between 0 V and 10 V and calculate the resulting load current. If the input voltage chosen is 10 V, the output of the op amp will increase until the base current through Q_1 is high enough to produce a 10 V drop across R_2. With 10 V across R_2, the load current can easily be calculated (the load current must pass through R_2).

     The plotting of these two points results in the response curve shown in Figure 8–8.

$$I_{\text{load}} = \frac{V_{R2}}{R_2} = \frac{10\ \text{V}}{5\ \Omega} = 2\ \text{A}$$

FIGURE 8–8

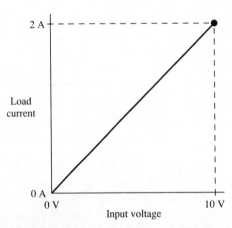

FIGURE 8–9
Typical R/2R Ladder Network

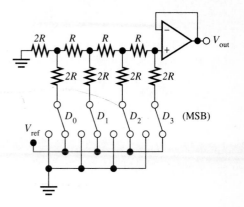

DAC = DIGITIAL TO ANALOG CONVERTER

Linear Digital DC Amplifiers

The circuits shown in Figure 8–7 operate superbly as long as there exists an analog signal with which to drive the op amp. More and more, however, microprocessors and microcontrollers are being implemented in control applications. A typical output port on one of these devices may consist of eight signal lines, each of which will be at a potential of either +5 V (logic 1) or 0 V (logic 0). Together, the signal lines represent an 8-bit (*binary-digit*) number that can take on a value that ranges from 0_{10} to 255_{10} (00 to FF hexadecimal). In order to vary the power that is applied to a load, these binary numbers must be converted into an analog equivalent. The circuit that performs this function is called a digital to analog converter (DAC).

The heart of virtually every commercial DAC is a resistive network called an *R/2R* ladder (see Figure 8–9). This network requires only two different resistor values—one twice the value of the other. Each switch in the network has the capability of connecting one of the resistors to either 0 V or some positive voltage called the *reference voltage*. There will be one switch for each input bit. That is, a DAC that is capable of converting 8-bit numbers will have 8 switches in the network (the DAC shown in Figure 8–9 is only capable of converting 4-bit numbers). With 4 switches, and 2 positions for each switch, there are 16 possible switch combinations. Each combination produces a unique resistor network configuration, resulting in a different voltage at the noninverting input of the op amp. The purpose of the op amp in Figure 8–9 is to isolate the ladder network from the DAC load. Consider Figure 8–10*A*, where the switches shown in Figure 8–9 have been set so that D_0, D_1, and D_2 are grounded (logic 0) and D_3 is connected to V_{ref} (logic 1). Given the equivalent circuit for this combination, it should be easy to see why the network produces a voltage that is equal to $V_{ref}/2$. In the same manner, if switch D_2 is the only one connected to a logic 1, the equivalent circuit yields a voltage of $V_{ref}/4$. As a matter of fact, the same procedure can be used to show that D_1 contributes a voltage of $V_{ref}/8$ when connected to a logic 1, and D_0 contributes a voltage of $V_{ref}/16$ when connected to a logic 1. For an input combination where D_0, D_1, and D_3 are all at a logic 1 state, the resulting output voltage will be $V_{ref}/16 +$

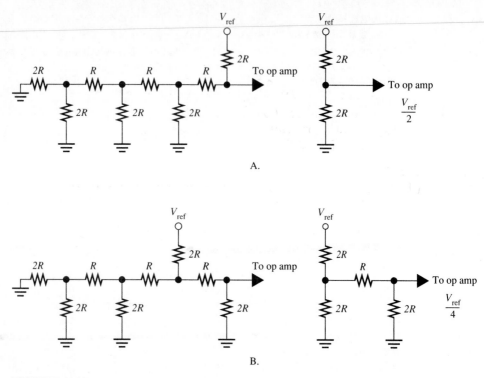

FIGURE 8-10
A. Ladder network of Figure 8-9 with 1000_2 applied (along with its equivalent circuit)
B. Ladder network of Figure 8-9 with 0100_2 applied (along with its equivalent circuit)

$V_{ref}/8 + V_{ref}/2$. This pattern holds true for DACs with any number of inputs bits: the most significant bit (MSB) always contributes $V_{ref}/2$; the next most significant bit always contributes $V_{ref}/4$; the next bit contributes $V_{ref}/8$, and so on. The gain of the op amp (if any) scales the output, but the inputs are always *weighted* in a binary fashion.

Commercial DACs are integrated circuits in which the switches shown in Figure 8-9 are actually transistors. The minimum number of inputs is usually eight. Some DACs have integrated op amps and precision voltage references, some have on-chip circuitry that facilitates microprocessor interfacing, and some even offer multiple, selectable output channels. For any binary input number presented to a DAC, the output voltage can be expressed as:

$$V_{out} = \text{Input}\left(\frac{V_{ref}}{2^N - 1}\right) \qquad \textbf{(8.1)}$$

where *Input* is the magnitude of the binary number expressed as its base 10 equivalent, and N is the number of binary inputs to the DAC. The expression $V_{ref}/$

$2^N - 1$ is referred to as the *resolution voltage* for the converter. It represents the smallest nonzero voltage that can be produced at the output. Note that any signal that may be produced at the output of a DAC must be constructed of whole-number multiples of the resolution voltage. Therefore the greater the number of inputs, the smaller the resolution voltage and the more faithfully an analog signal can be produced. For example, if a particular 8-bit DAC uses a reference voltage of 10 V, the resolution voltage will be (10 V/255) or 39.2 mV. Therefore it is impossible to produce an output voltage of *exactly* 3 V. It is possible to produce an output of 2.98 V (76 × 39.2 mV) or 3.02 V (77 × 39.2 mV), but not 3.00 V.

EXAMPLE 8–2   A given 8-bit DAC has a reference voltage of 5 V. Determine the resolution voltage. In addition, determine the input number that would be required in order to produce an output of 4 V.

The resolution voltage can be determined by evaluating the right-hand term of equation 8.1 for the given reference voltage and input bit-width (number of input bits).

$$V_{res} = \frac{V_{ref}}{2^N - 1} = \frac{5 \text{ V}}{255} = 19.6 \text{ mV}$$

Equation 8.1 can then be rearranged to determine the input number required to produce an output of 4 V.

$$\text{Input} = \frac{V_{out}}{V_{ref}} (2^N - 1) = \frac{4 \text{ V}}{5 \text{ V}} (255) = 204_{10} = \text{CCH}$$

An example of a commercially available DAC, the Analog Devices AD558, is shown in Figure 8–11. Note that in addition to the $R/2R$ ladder network, the device contains a data latch, an integrated amplifier, and a tracking reference. The device operates over a supply voltage range of 4.5 V to 16 V. Regardless of the supply voltage, however, the tracking reference ensures that the output from the ladder network will vary from 0 V to approximately 400 mV as the binary input to the DAC varies from 00000000_2 (00H) to 11111111_2 (FFH). Internal laser-trimmed resistors are provided so that the gain of the output amplifier may be changed without the requirement of external components. The output (V_{out}) is always connected to V_{out} Sense. This provides feedback for the amplifier. The gain is controlled by connecting V_{out} Select directly to either V_{out} or ground. If V_{out} Select is connected to V_{out}, then the op-amp feedback resistor will be the equivalent of a 14 kΩ resistor and a 40 kΩ resistor in parallel. This will result in a gain of approximately 6.2 (low-gain configuration). If, instead, V_{out} Select is connected to ground, the feedback resistor will remain at 40 kΩ, and the input resistor will be the equivalent of 14 kΩ and 2 kΩ in parallel, resulting in a gain of approximately 23.9 (high-gain configuration). The internal resistor values have been selected so that in the low-gain configuration the output voltage will vary from 0 V to 2.56 V.

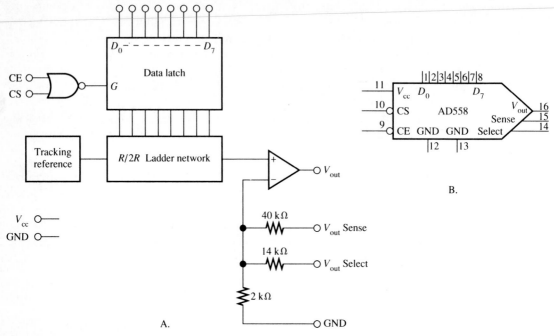

FIGURE 8–11
A. AD558 DAC Block Diagram B. AD558 Schematic Symbol

Additionally, if the DAC is connected in the high-gain configuration, the output will vary from 0 V to 10 V (as long as the DAC supply voltage is between 11.5 V and 16 V).

In order to change the output voltage of the AD558, data must be *clocked* into the data latch. That is, data must be present at the inputs D_0–D_7 while the CE and CS inputs are both brought to a logic 0. Data will then propagate through the latch and be presented to the inputs of the $R/2R$ ladder network. If CE or CS is brought back to a logic 1, further input changes at D_0–D_7 will be ignored by the device, and the input data will have been captured (latched). The output voltage can then be determined according to equation 8.1. With this device, however, the magnitude of V_{ref} depends upon the AD558 configuration. If the high-gain configuration is used, V_{ref} will be 10 V. If the low-gain configuration is chosen, V_{ref} will be 2.56 V.

Digital to analog converters are not designed to produce high output voltages or currents. They merely provide an interface between analog and digital circuitry. If a computer or other similar digital circuit is to truly control the power supplied to a load, an analog amplifier will still be required. Since a DAC *does* produce an analog output signal, it should be compatible with the input of any analog power amplifier. The circuit of Figure 8–12 shows an AD558 DAC driving

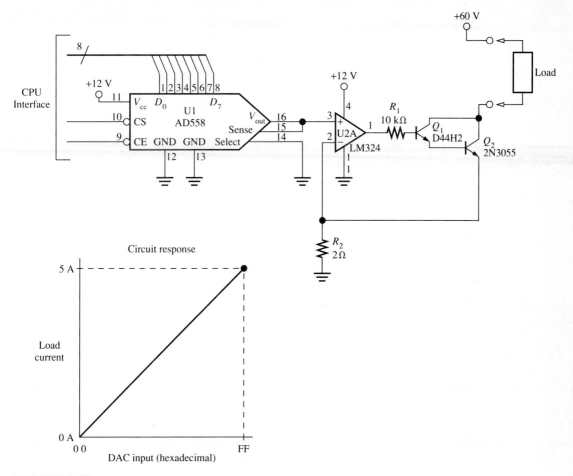

FIGURE 8–12
AD558 DAC Interfaced to Linear Amplifier of Figure 8–7A

the input of the power amplifier from Figure 8–7*A*. Naturally, additional circuitry will be required in order to properly interface the DAC with the CPU.

The DAC in Figure 8–12 is configured to provide an output that may be varied from 0 V to +10 V as the DAC input is varied from 00H to FFH. Furthermore, each volt that is applied to pin 3 of U2A results in 500 mA of load current. (This is due to the fact that the value of R_2 has been chosen to be 2 Ω.) The overall operation of the circuit is described by the response curve that accompanies the circuit in Figure 8–12. That is, the load current can be varied from 0 A to 5 A merely by changing the DAC input number from 00H to FFH. Equation 8.2 may be used in order to determine the required DAC input number (in base 10) for any given desired load current.

$$\text{DAC Input} = \frac{(255)I_{\text{load}}}{5 \text{ A}} \qquad \qquad \textbf{(8.2)}$$

As with any analog/digital interface, one important consideration is the resolution. Since an 8-bit DAC is used in Figure 8–12, the load current may be varied only in increments of 5 A/255, or approximately 19.6 mA. With this circuit properly interfaced to a microprocessor-based controller (such as a personal computer), the load current can be varied under software control, based upon data that are obtained through human intervention or electronic input sensors.

EXAMPLE 8–3 Given the circuit shown in Figure 8–12, determine the number that the CPU would have to send to the AD558 in order to produce a load current of 1 A.

When equation 8.2 is used, the result is:

$$\text{DAC Input} = \frac{(255) \ (1 \text{ A})}{5 \text{ A}} = 51_{10} = 33\text{H}$$

Pulse-Width Modulation

$PWM =$

Providing power to a load using a linear amplifier has one primary disadvantage: low efficiency. Consider, for example, the circuit of Figure 8–12. If the resistance of the load were 5 Ω, and the load current were 2.5 A, the load would dissipate approximately 31 W. At the same time, however, Q_2 would dissipate approximately 75 W! That is, in order to provide 31 W of power to the load, the power supply must actually provide at least 106 W (31 W + 75 W). The 75 W dissipated by Q_2 results in no useful work being done and is therefore wasted power. In addition to the inefficiency, there is a more tangible consequence to consider. All of the additional heat generated by Q_2 must be carried away via some heat-sinking structure. This requires additional space. The system power supply must be larger in order to be able to provide the additional power that will be dissipated by Q_2. In addition, Q_2 itself must be capable of sustaining the high temperature that accompanies a power dissipation of this magnitude.

An attractive alternative to linear amplification is pulse-width modulation (PWM). The concept is quite simple. Instead of operating the output transistor in its linear region, it is operated in a switched mode. The transistor is either fully saturated or fully at cutoff. Under these conditions, the load is subjected to voltage pulses of an amplitude nearly equal to the supply voltage. By controlling the duration (width) of the pulses, and by properly filtering these pulses, the average amount of power supplied to the load can be controlled.

The primary advantage of PWM power control is its high efficiency. When the transistor is saturated, the voltage across it will be, at most, a few volts. Even with a high current through the transistor, its power dissipation will be low. When the transistor is at cutoff, the current through it will be negligible. Even though the full source voltage is dropped across the transistor, its power dissipation will again be low.

A PWM power control system consists of four main components: a triangle (or ramp) wave oscillator, a comparator, a power switch, and an output filter. In Figure 8–13 an LM324 op amp is utilized as the comparator; Q_1 and Q_2 form the power switch; and D_1, L, and C comprise the output filter. It should be understood that in most cases the frequency of the oscillator is generally in the neighborhood of 20 kHz, and that the control voltage varies quite slowly with respect to this frequency. As the oscillator output voltage varies above and below the level of the control voltage, DC pulses are generated at the output of U1A. These pulses switch Q_1 and Q_2 alternately from saturation to cutoff at a rate equal to the frequency of the oscillator. The duration (and, therefore, the duty cycle) of each pulse is determined by the level of the control voltage. The output filter averages the pulses to yield a voltage that is approximately equal to the product of the pulse duty cycle and the DC supply voltage.

In order to fully understand the operation of the circuit of Figure 8–13, consider the input and output signals shown in Figure 8–14. The input signals for the LM324 are the control voltage and the triangle-wave oscillator output. For convenience the exact value of $0.5V_P$ has been chosen, although the control voltage may vary anywhere from 0 V to V_P. Since the control voltage is connected to the noninverting input on U1A, the output of U1A will only be driven to positive saturation (+SAT) when the magnitude of the control voltage exceeds the voltage at the output of the oscillator. With the control voltage set at one half of the peak oscillator voltage, the output of U1A will be at +SAT for the first one fourth of the oscillator cycle time. After the oscillator output has exceeded the magnitude of the control voltage, the output of U1A will be driven to 0 V (−SAT). The output of U1A will remain at −SAT until the oscillator output voltage again drops below the

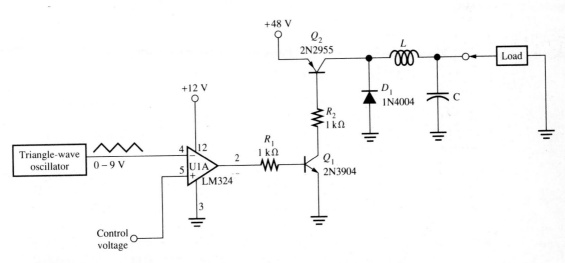

FIGURE 8–13
Typical PWM Power Controller

FIGURE 8–14
PWM Output with Respect to Control
Voltage

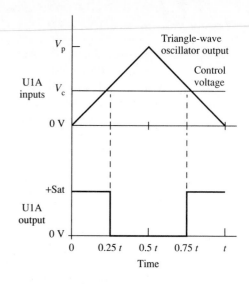

magnitude of the control voltage. With the control voltage set at exactly $0.5V_P$, the output of U1A will be at +SAT for a total of one half of the total oscillator cycle time. Given this relationship, it should be easy to understand why the time that the output of U1A remains at +SAT each cycle increases as the control voltage increases. For this particular circuit, the general relationship between the control voltage, the peak oscillator voltage, and the output duty cycle for U1A is:

$$\text{Duty cycle} = \frac{V_C}{V_P} \qquad (8.3)$$

Where:

V_C = Control voltage

V_P = Peak oscillator output voltage

duty cycle is expressed as a decimal

If the inputs to U1A in Figure 8–13 were reversed, the U1A output waveform shown in Figure 8–14 would be inverted. Similarly, as the control voltage increased, the time that the output of U1A remained at +SAT each cycle would decrease. In this case, the duty cycle would be described by:

$$\text{Duty cycle} = 1 - \frac{V_C}{V_P} \qquad (8.4)$$

Where:

V_C = Control voltage

V_P = Peak oscillator output voltage

duty cycle is expressed as a decimal

There is one point that is very important to understand. From an electrical standpoint, *duty cycle* is a general term used to describe the duration of electrical pulses. As such, it is usually calculated by dividing the ON time of the pulse by the period of the cycle (ON time + OFF time). In the case of PWM power control systems, the ON time to be concerned with is the time that current flows through the load. This time does *not* necessarily correspond to the time during which the pulse is at its highest voltage level. For example, in the circuit of Figure 8–13, current *does* flow through the load when the output of U1A is at +SAT. The load voltage duty cycle, therefore, is described by equation 8.3. Suppose, however, that Q_1 and Q_2 were arranged in such a way that a voltage level of −SAT at the output of U1A produced current through the load. As the pulse width at the output of U1A decreased, the load power would actually be increasing. The load voltage duty cycle for this arrangement would then be described by equation 8.4. Simply dividing the ON time by the cycle time, in this case, would yield erroneous results.

In order to decide which equation describes the load voltage duty cycle for any given circuit, analyze the relationship between the control voltage and the power delivered to the load. If the load power *increases* as the control voltage increases, equation 8.3 describes the duty cycle. If the load power *decreases* as the control voltage increases, equation 8.4 describes the duty cycle.

Applying high voltage pulses directly to a load may not produce the expected response. Most PWM power control systems employ some type of output filter that acts to smooth out the high-frequency transitions produced at the output. Figure 8–13 uses a simple LC averaging filter to produce a voltage that is proportional to the duty cycle. For this filter the load voltage can be expressed in terms of the duty as:

$$V_{\text{load}} = \text{Duty cycle} \times V_{\text{supply}} \qquad\qquad \textbf{(8.5)}$$

Where:

duty cycle is expressed as a decimal

In order to understand the operation of this filter, consider the circuits shown in Figure 8–15. When the output of U1A is at +SAT (as in Figure 8–15*A*), Q_1 saturates, providing enough base current to also saturate Q_2. This provides a current path from the +48 V supply, through Q_2, L, and the load. Note that during this time D_1 is reverse-biased and C charges to the load voltage. When the output of U1A falls to −SAT (0 V), Q_1 and Q_2 are both deprived of base current. With no current flowing through Q_2, the magnetic field surrounding L collapses, producing the polarity shown in Figure 8–15*B*. Due to the polarity produced, D_1 is forward-biased, clamping the negative side of L to ground and completing a conduction path to the negative side of the load. As the magnetic field collapses, L forces current through the load in the same direction as in Figure 8–15*A*. As this current decreases, it is supplemented by the discharge of C. If the values of L and C are properly chosen, the load voltage will not decrease much before the next +SAT

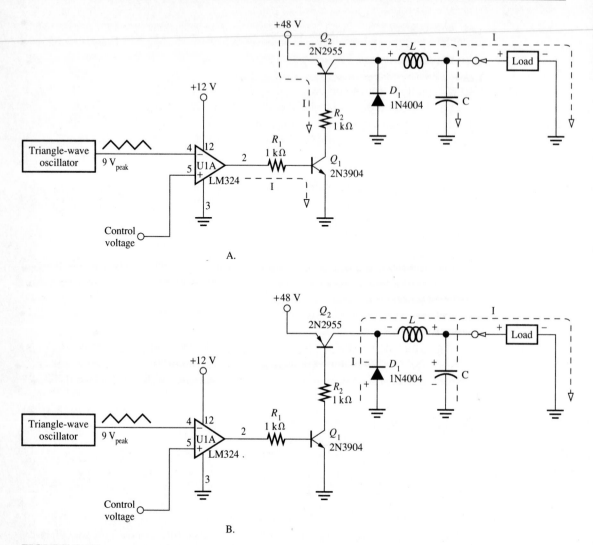

FIGURE 8–15
A. Load Current While Q_1 Is Conducting B. Load Current While Q_1 Is Not Conducting

pulse from U1A is produced. The relationship between the control voltage, the oscillator output, the output of U1A, and the load voltage is shown in Figure 8–16. Note how the average load voltage increases as the duty cycle of the pulses increases.

The control voltage in a PWM power control system can be provided by something as simple as the wiper of a potentiometer to something as complex as the output of a DAC such as the one shown in Figure 8–17. In this particular circuit, the CPU provides binary numbers to the DAC, which, in turn, provides an

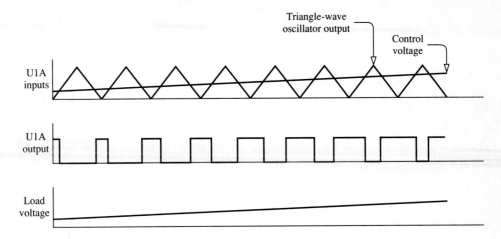

FIGURE 8–16
Load Voltage with Respect to Control Voltage and Duty Cycle in a PWM Power Control System.

analog control signal to the pulse-width modulator. Note that the AD558 is configured to provide an output of 0 V to 10 V. Since the output of the triangle-wave oscillator peaks at 9 V, maximum power will be applied to the load when the DAC output is just above 9 V (or when the DAC input is somewhat less than FFH). Rearranging equation 8.1 and solving for the DAC input produces the number E6H. The response curve in Figure 8–17 describes the relationship between the DAC input and the average load voltage.

The advantage of the circuit of Figure 8–17 is that the load power can be automatically controlled based upon decisions made within a software program. Furthermore, if the required load power control scheme changes, all that is required is modification of the software.

EXAMPLE 8–4 Given the circuit shown in Figure 8–13, determine the control voltage that would result in an average load voltage of 32 V.

According to equation 8.5, the required duty cycle is;

$$\text{Duty cycle} \approx \frac{V_{\text{load}}}{V_{\text{supply}}} = \frac{32 \text{ V}}{48 \text{ V}} = 0.67 = 67\%$$

From the response curve shown in Figure 8–14, it is clear that the load power increases as the control voltage increases. Equation 8.3 can therefore be used to determine the control voltage. Rearranging the equation results in:

$$V_C = V_P \text{ (Duty cycle)} = 9 \text{ V}(0.67) = 6.03 \text{ V}$$

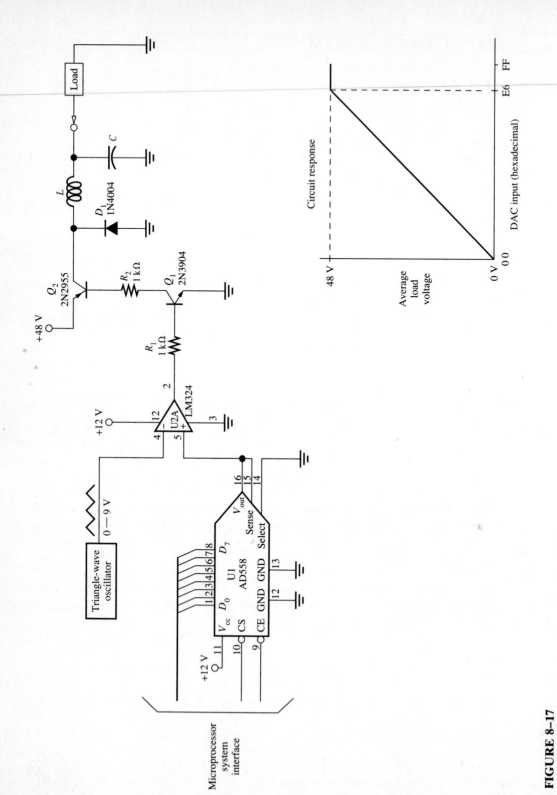

FIGURE 8–17
AD558 DAC Interfaced to PWM Power Controller of Figure 8–13

8-4   PROPORTIONAL AC POWER CONTROL

Systems that vary the amount of power supplied to AC loads usually involve the use of thyristors triggered by a capacitor-based timing circuit (see Chapters 7 and 14). In these circuits the firing delay angle of the thyristor (and so also the load power) is controlled by varying the charging resistance in the triggering circuit. It is often more convenient to be able to control the firing delay angle with a voltage that is proportional to some process variable such as temperature or pressure. In this section two methods of AC power control will be examined that are alternatives to the more traditional RC triggering circuits.

Linear Ramp Control

Traditional thyristor triggering circuits operate in a manner that is quite similar to DC PWM power control systems. A reference signal is compared to an input control signal, and the output device is switched into conduction when the magnitude of the reference signal increases beyond the magnitude of the control signal (or vice versa). The primary difference is that in AC power control systems, the reference signal is a sine function; and the control signal is usually either a sine function derived from the reference signal or an exponential function provided by an RC circuit. This fact makes calculations as well as implementation and calibration of the system difficult. The solution is to derive a *linear* DC reference signal from the AC line. If this signal increases linearly from 0 V at the beginning of the AC cycle to its maximum value after 180° of the 360° cycle, then the magnitude of the reference signal will be directly proportional to the amount of time that has elapsed during each half cycle. That is, the reference signal will have reached one fourth of its maximum value after 45° of the cycle, one half of its maximum value after 90° of the cycle, and three fourths of its maximum value after 135° of the cycle. Any trigger signal that is based upon this reference signal, therefore, will also be proportional to time (and angle). The circuit shown in Figure 8–18 provides this function. The circuit of Figure 8–18 is functionally divided into three distinct parts: a zero-crossing detector, a ramp generator, and an AC switch. The output of the ramp generator (an integrator) starts from near 0 V and increases linearly toward +SAT. The maximum value of the ramp is determined by the value of R_4. Synchronization to the AC line is achieved through the use of the zero-crossing detector. Pins 2 and 5 of U1 are connected to ground and to one side of the transformer secondary. As soon as point A rises a few millivolts above ground, the output of U1A will be driven to +SAT. This voltage will be immediately applied to R_2 and the anode of D_1. Since this voltage will forward-bias D_1, current will flow through R_3, forcing Q_1 into saturation. This will short C_3, resetting the output of the ramp generator to 0 V. After about 500 μs, C_1 will be fully charged, D_1 will no longer be forward-biased, and Q_1 will have turned off, allowing the output of U2A to begin ramping toward +SAT again. As soon as point A falls a few millivolts below ground, the output of U1B will be forced to +SAT, resulting in the same ramp generator reset action. At the same time, the output of U1A

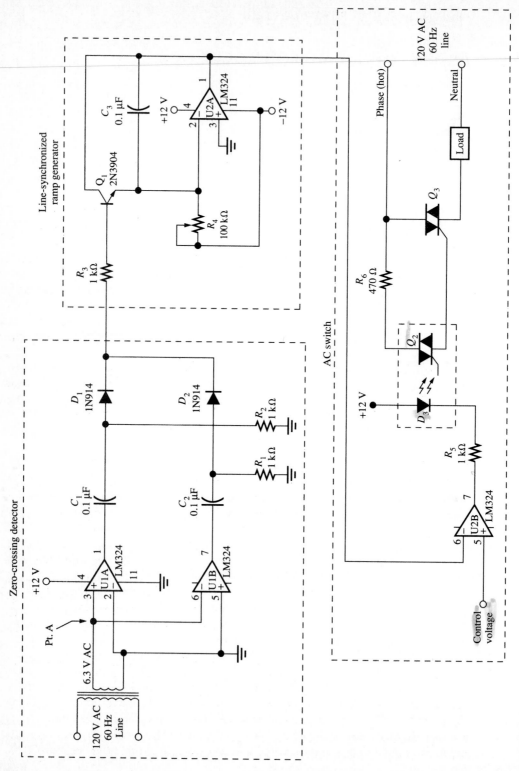

FIGURE 8–18

Operation of Linear Ramp AC Power Control System, Similar to the Operation of a PWM Power Control System

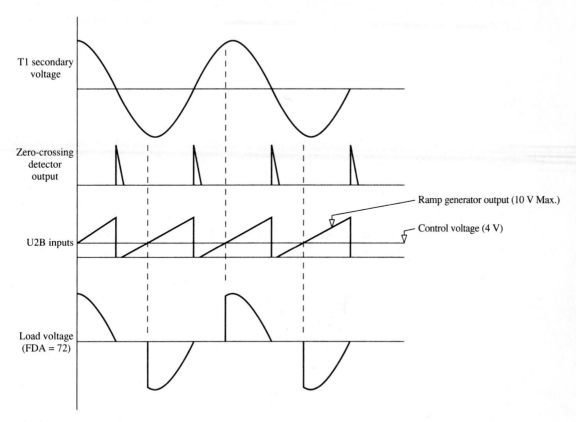

FIGURE 8-19
Load Voltage with Respect to Control Voltage in a Linear Ramp AC Power Control
System

will be forced to −SAT (0 V), discharging C1 so that another reset pulse can be
produced at the next zero-crossing point of the AC line. In this manner U1A and
U1B produce reset pulses on alternate zero-crossing points of the AC line. Since
the ramp generator is reset upon receipt of either of the pulses, the output of U2A
begins ramping at a zero-crossing point and reaches its maximum value at the next
zero-crossing point (180° later). The relationship between the reset pulses and the
ramp generator output is shown in Figure 8–19.

The output of the ramp generator in Figure 8–18 is compared to the control
voltage by U2B. When the output of the ramp generator exceeds the control
voltage, the output of U2B is driven to −SAT, providing current through the
IRED, D_3. Light emitted by D_3 latches Q_2 into conduction, resulting in gate
current for Q_3 and, ultimately, load current. At the next zero-crossing point of the
AC line, Q_3 turns off. At the same time, the ramp generator is reset. Current will
not flow through the load again until the output of the ramp generator again

exceeds the control voltage. If the magnitude of the control voltage is fixed, load current will be switched on at the same point into each half cycle. If the control voltage is varied, the point at which the load current turns on will also vary. That is, if the magnitude of the control voltage is increased, more time will elapse before Q_3 switches current through the load (the firing delay angle will increase). The firing delay angle can be expressed in terms of the ramp and control voltages as shown in equation 8.6. For the values given in Figure 8–19, the resultant firing delay angle is 72°.

$$\text{FDA} = \frac{(V_{control})180°}{V_{ramp_{max}}} \qquad (8.6)$$

In order to determine the power dissipated by the load, it is necessary to first determine the RMS voltage that is impressed across the load. For resistive loads, the load voltage can be expressed in terms of the firing delay angle as:

$$V_{RMS} = \frac{V_{peak}}{\sqrt{2\pi}} \sqrt{\pi - \theta + \frac{1}{2}\sin 2\theta} \qquad (8.7)$$

Where θ = the firing delay angle expressed in radians (1 radian = 57.3°)

This voltage can also be determined by using the graph shown in Figure 7–19 in Chapter 7.

EXAMPLE 8–5 Given the circuit shown in Figure 8–18, determine the control voltage that would result in a firing delay angle of 40°. Assume that R_4 is adjusted so that the ramp reaches a peak voltage of 8 V before being reset. What is the resultant load voltage?

Rearranging equation 8.6 results in:

$$V_{control} = \frac{(\text{FDA})(V_{ramp_{max}})}{180°} = \frac{(40°)(8\text{ V})}{180°} = 1.78\text{ V}$$

Forty degrees is equivalent to 0.7 radian. Substituting this into equation 8.7 yields:

$$V_{load} = \frac{170\text{ V}}{\sqrt{2\pi}} \sqrt{\pi - 0.7 + \frac{1}{2}\sin 1.4} = 116\text{ V}$$

Microprocessor-based Control

The advantage of integrating microprocessors into electronic control systems consists in the fact that circuitry such as timers, arithmetic circuits, and logic circuits can all be replaced by program instructions (software). This greatly simplifies hardware design, implementation, and troubleshooting. In addition, the

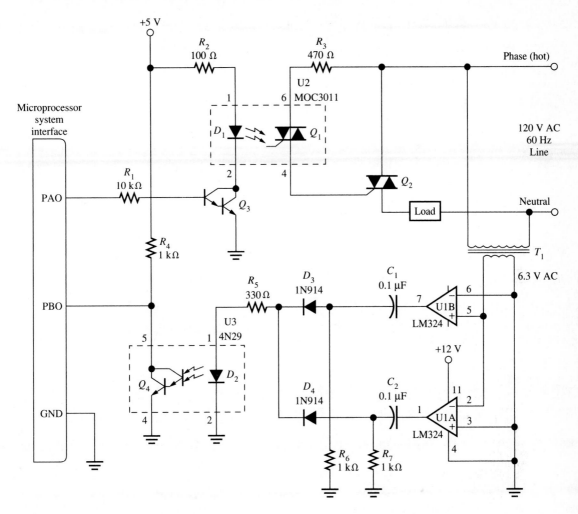

FIGURE 8–20
A Microprocessor-based AC Power Control System Relies on Program Instructions to
Accomplish its Control Scheme.

ease with which software can be changed offers great flexibility. Evidence of this
is shown in Figure 8–20. Although this circuit performs essentially the same
function as that of Figure 8–18, it actually requires less hardware (note the absence of the ramp generator). Of course the microprocessor system itself constitutes a large amount of hardware. Keep in mind, however, that this microprocessor system can control (virtually simultaneously) many other circuits in addition
to the one shown in Figure 8–20.

The zero-crossing detector in Figure 8–20 is virtually identical to the one
shown in Figure 8–18. Each time that the AC line voltage crosses 0 V, a positive

FIGURE 8–21
Software Control Flowchart for the
Power Interface of Figure 8–20

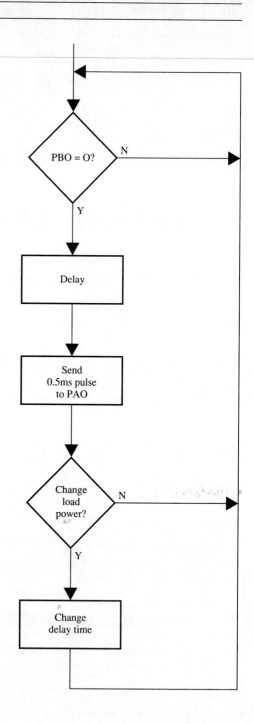

voltage pulse is produced at the intersection of D_3, D_4, and R_5. This voltage pulse produces current through D_2, forcing Q_4 into saturation. The result is a logic 0 (nearly 0 V) at PB0 (Port B, bit 0). By monitoring the logic level present at PB0, the CPU can determine the point at which each half cycle of the AC line starts.

The AC switch in Figure 8–20 is controlled by PA0 (Port A, bit 0). If the voltage present at PA0 is at a logic 1 (near +5 V), current flows through R_1, saturating Q_3 and providing current through D_1. Light emitted by D_1 provides the gate current necessary to latch Q_1 and Q_2 into conduction. Current then flows through Q_2 and the load. Naturally, if the voltage present at PA0 is at a logic 0 level, load current will be blocked by Q_2.

Although details of the operation of the microprocessor interface will not be discussed here, it will be assumed that through the proper software instructions the CPU *can* determine the logic level present at PB0 and control the logic level sent to PA0. The possible schemes that can be used to control the load power in Figure 8–20 are numerous. One such method is shown in Figure 8–21. According to this flowchart, the CPU monitors the logic level present at PB0. After PB0 has fallen to a logic LOW (the AC line is crossing 0 V), the CPU generates a trigger pulse at PA0, switching AC line current through the load. This trigger pulse is generated after a software delay that may be changed at the user's request. As long as the delay time is between 0 s and about 8 ms, the firing delay angle can be controlled between 0° and 180°. If, for example, the delay were set at 4 ms, the firing delay angle would remain at about 90° (4 ms into each half cycle). The delay time change requests, for this example, could be provided by keyboard input. With the addition of an ADC, the CPU could sample an analog control signal provided by some process-variable-sensing circuitry and modify the delay time (and firing delay angle) accordingly.

SUMMARY

☐ Power switches are classified as either DC output modules or solid-state relays. DC output modules typically require DC trigger voltages of 3 V to 32 V and generally provide from 1 A to 5 A to loads powered from voltage supplies of up to about 60 V. Solid-state relays typically require trigger voltages of either 3 V DC to 32 V DC (for DC triggered relays) or 90 V AC to 280 V AC (for AC triggered relays) and generally provide from 10 A to 100 A to loads powered from the 120 V AC or 240 V AC line.

☐ Linear amplifiers can be employed to provide a variable amount of voltage or current to a load. Although these circuits are simple in concept, construction, and operation, they are relatively inefficient. They are therefore used for low-power applications only.

☐ Digital to analog converters (DACs) can be used to provide microprocessor-controlled power interfaces. Voltage can be provided to the load only in increments of the DAC resolution voltage, as given by:

$$V_{res} = \frac{V_{ref}}{2^N - 1}$$

☐ The relationship between the DAC input number and its output voltage can be expressed as:

$$V_{out} = \text{Input} \left(\frac{V_{ref}}{2^N - 1} \right)$$

☐ Pulse-width modulation systems control load power by controlling the rate at which the full supply voltage is switched across the load. The load voltage duty cycle is given by:

$$\text{Duty cycle} = \frac{V_C}{V_P}$$

for cases where load voltage increases as V_C increases and by:

$$\text{Duty cycle} = 1 - \frac{V_C}{V_P}$$

for cases where load voltage decreases as V_C increases.

☐ If properly filtered, the load voltage in a PWM power control system is given by:

$$V_{load} \approx \text{Duty cycle} \times V_{supply}$$

☐ An AC equivalent of PWM control can be achieved by varying the firing delay angle of a triac according to the magnitude of a DC control signal. This is done by comparing the control signal to a ramp signal that is synchronized to each half cycle of the AC line, and by using the output of the comparator to trigger the triac. This results in a firing delay angle and output voltage given by:

$$\text{FDA} = \frac{(V_{control})\ 180°}{V_{ramp_{max}}}$$

$$V_{RMS} = \frac{V_{peak}}{\sqrt{2\pi}} \sqrt{\pi - \theta + \frac{1}{2} \sin 2\theta}$$

☐ Microprocessors are ideally suited to control the firing delay angle of thyristors. As long as the CPU can detect the zero-crossing points of the AC line, the firing delay can be controlled by software delay routines.

EXERCISES

1. List three advantages that solid-state relays have over electromagnetic relays.

2. List three disadvantages of solid-state relays.

3. Show two methods (other than those shown in Figure 8–3) of triggering a DC output module that has an input trigger voltage range of 3 V DC to 32 V DC.

4. Show two methods of triggering a solid-state relay that has an input trigger voltage range of 90 V AC to 280 V AC.

5. Suppose that a solid-state relay is to be used to control a 1200 W, 120 V AC heater. What important characteristics should the relay have?

6. Given the circuit shown in Figure 8–7*A*, suppose that the value of R_2 is 10 Ω. What load current will result from an input voltage of 2 V?

7. Given the circuit shown in Figure 8–7*A*, suppose that the load supply voltage is 32 V. If $R_2 = 1$ Ω and the load is 10 Ω, what is the maximum attainable load current?

8. Given the circuit shown in Figure 8–7*B*, suppose that the op-amp supply voltage is ± 12 V. If the collector supply voltage for Q_1 and Q_2 is 32 V, what is the maximum attainable load voltage?

9. An 8-bit DAC has a reference voltage of 5 V. What is the resolution voltage? What input number would result in an output of 1.75 V?

10. Given the circuit shown in Figure 8–12, what DAC input number would result in a load current of 1.5 A?

11. How would the response curve shown in Figure 8–12 change if the value of R_2 were changed to 3 Ω?

12. Given the circuit shown in Figure 8–13, what control voltage will produce a duty cycle of 63%?

13. If the control voltage in Figure 8–13 is 3 V, what average load voltage will result?

14. Suppose that the inputs to the op-amp in Figure 8–13 were reversed. If the control voltage were 7 V, what duty cycle and average load voltage would result?

15. Given the circuit shown in Figure 8–17, determine the DAC input that would result in an average load voltage of 20 V.

16. Suppose that the control voltage in Figure 8–18 is 6 V. If the maximum ramp generator output is 10 V, what firing delay angle will result? What RMS load voltage will result?

17. Explain what would happen if the value of R_4 in Figure 8–18 were adjusted so that the output of U2A saturated within 4 ms.

18. Given the circuit shown in Figure 8–20, along with the software flowchart shown in Figure 8–21, draw the resulting load voltage waveform for a delay time of 2 ms.

APPLICATION ASSIGNMENT ANSWER

Heating loads invariably produce high inrush currents upon start-up. If random-trigger solid-state relays are being used, they would very likely be destroyed on a regular basis. Zero-voltage-switching solid-state relays should be installed in their place. In addition, factory environments often have electrically *noisy* power distribution systems. That is, short-duration voltage spikes are often superimposed onto the system power bus by various pieces of electrical equipment. Snubber networks should be installed across the output terminals of each solid-state relay.

III

MOTORS AND CONTROLS

9

Industrial-Use Motors

CHAPTER OUTLINE

9–1 Introduction □ **9–2** Magnetics-Motor Principle □ **9–3** Industrial Motor Construction □ **9–4** Choosing the Correct Industrial Motor for an Application □ **9–5** DC Motors □ **9–6** Three-Phase AC Induction Motors □ **9–7** Single-Phase AC Motors □ **9–8** Other Common AC Motors

LEARNING OBJECTIVES

- □ Describe the magnetics-motor principle.
- □ List the major parts of an industrial motor.
- □ Describe the general construction of an industrial motor.
- □ List the operational characteristics of the three major categories of industrial motors.
- □ Explain the operation of the DC motor.
- □ Describe the electrical connection schemes for DC motors.
- □ Choose the type of DC motor needed for different applications.
- □ Describe basic speed-control systems for DC motors.
- □ Explain the relationship of counter electromotive force (CEMF) to motor operation.
- □ Explain the operation of a three-phase AC motor.
- □ Define and calculate synchronous speed and slip for AC motors.
- □ Choose by code letter the type of three-phase AC motor needed for an application.
- □ Define *frame size*.
- □ Explain the operation of a single-phase AC motor.
- □ List and describe the connections for the four major types of single-phase AC motors.
- □ Describe the operation of the motor-internal centrifugal switch.
- □ Show how dual rotation, speed, and voltage of motors are accomplished.
- □ Describe the operation of other major types of AC motors.
- □ Differentiate among the three types of repulsion motors.
- □ Define *commutator, slip ring, short-circuiter,* and *interpole*.

APPLICATION ASSIGNMENT

You are a technician in charge of replacing old motors on a machine. There are three small blower motors, used for cooling three small electronic modules. There is also a 2 horsepower (HP) grinder motor with variable speed. The working speed of the grinder must vary between 0 and 4000 RPM. A 20 HP motor drives a hydraulic pump.

What types of new motors would you choose, and why? What type of incoming electrical power would be needed?

9–1   INTRODUCTION

The industrial motors discussed in this chapter are used in commercial, agricultural, and residential applications as well as in manufacturing plants. Over 50% of all power produced in the United States is used to run motors. The millions of motors that use the power convert electrical energy to many types of mechanical energy.

The chapter begins with a discussion of magnetics-motor principles and a description of the makeup and construction of motors. The description of the construction is background for discussion of the operation of various types of motors in this chapter and in Chapter 10.

Next, a general description of the major industrial motor categories is included in a discussion of choosing the proper motor for a given application.

The first type of specific motor covered is the DC motor, of which there are various types. Next, the three-phase AC motor is discussed, along with basic motor parameters.

Finally, there are two sections on the characteristics of different types of single-phase motors. The first of the two sections covers the most commonly used single-phase AC motors. The other section, the final section of the chapter, covers some of the more common special types of single-phase AC motors.

For precise position control and rotational control, motors other than these industrial motors are used. Such control motors will be discussed in Chapter 10. Fundamental motor control will be introduced in the present chapter. Detailed motor control systems will be described in Chapters 11 and 12.

9–2   MAGNETICS-MOTOR PRINCIPLE

Electric motors operate on the principle that two unlike magnetic fields attract each other (and like fields repel). The magnetic field attraction causes a rotor to turn to seek another magnetic field produced by outer, stationary coil and current. Most motor magnetic fields are produced by electrical current flowing through coils of wire wound on magnetic iron or steel. Figure 9–1 illustrates the basics of this principle, which will be expanded on throughout the chapter when applied to specific motor types. Figure 9–1A shows the polarity, north or south, produced in a magnetic iron core by current flow. The magnetic field polarity depends on two factors: the direction of current flow and the direction in which the coil is wound. In motors magnetic polarity is reversed by reversing the current flow in order to reverse motor rotation. In some cases for smaller motors, magnetic fields are produced by permanent magnets. Figure 9–1B shows the DC motor rotation principle. The rotor field is offset a few degrees by the manner in which the electric current is fed into the rotor. The rotor turns as the rotor field is attracted by the fixed field. Figure 9–1C show the AC motor principle. A rotor magnetic field is attracted to one outer field at one instant. Another outer field then occurs a time interval later, when the original field decreases in value. The field in the rotor is then attracted to the new field causing the rotor to rotate as shown. This principle will be more completely illustrated in the next sections of this chapter.

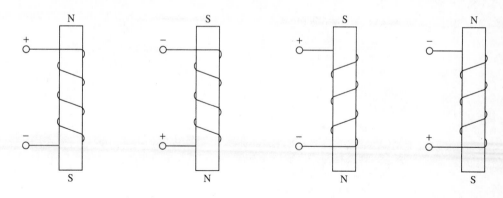

A. Magnetic Principle

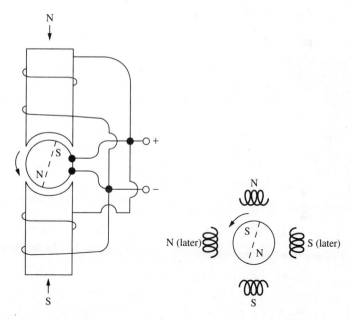

B. DC Motor Rotation Principle   C. AC Motor Rotation Principle

FIGURE 9–1
Magnetics-Motor Rotation Principle

9–3   INDUSTRIAL MOTOR CONSTRUCTION

All industrial motors (actually, practically all types of motors) have four basic parts: the *stator;* the rotor; and two end shields.

A finished stator for an AC motor, typical of all motors, is shown in Figure 9–2. The stator consists of three subparts: the outer shell, stacked punchings

FIGURE 9–2
Typical Motor Stator

called *laminations,* and coils of insulated wire wound through insulated slots in the lamination stack.

The stator outer shell may be a casting or a rolled and welded piece of flat metal stock. The laminations, all of which are exactly identical, are aligned, stacked, and pinned or welded together. The stacked laminations are welded to or pressed inside the shell. Laminations are used rather than a solid piece of metal because they reduce motor heating by reducing core heating due to eddy currents. The individual laminations are heat-treated with an oxide coating so that they are essentially insulated from each other. Therefore, eddy currents are reduced, since they can flow only in one plane.

Lamination iron or steel has varying electrical characteristics and varying motor-torque-producing characteristics. The cost also varies, increasing with per-formance capabilities. Which iron or steel is used in a given motor model depends on the motor's design parameters.

The wire coils in the stator are made of multiple turns of insulated wire. The wire used for the coils is coated with a flexible insulating varnish before winding. After winding, the coils are therefore insulated from each other. Coils are inserted in an insulated jacket in the stator lamination stack. After the coils have been inserted in stator slots, the coils and the lamination stack are dipped in varnish. The extra varnish gives better insulation qualities, as well as mechanical stability.

Each turn of wire must be insulated from the others and from the lamination stack. Otherwise, winding burnout will ultimately occur. Note that the slot insula-tion jacket prevents the metal outer shell from reaching a hazardous potential. If the electrically energized coils touched the stator, the metal outer shell could become electrically "hot." That situation would pose an electrical hazard. Often insulation strips are also added between windings of different potential—for ex-ample, between main winding coils and start winding coils.

Note that each coil group is made of multiple coils with different spans. This distribution of coils gives a smoother turning torque throughout each turn of the rotor through the coil's magnetic field. The coil ends are brought out and connected to regular wire, which is, in turn, brought out to power the motor. Electrical power connection to the external wires is accomplished directly or is made in terminal boxes or, occasionally, on terminal strips.

Figure 9–3 shows the end shields for some typical industrial motors. The end shields may have either sleeve bearings or ball bearings to support the rotor inside the stator. Ball bearings are used mainly in larger motors. In most single-phase motors, one end shield includes a stationary, axially acting switch, as shown.

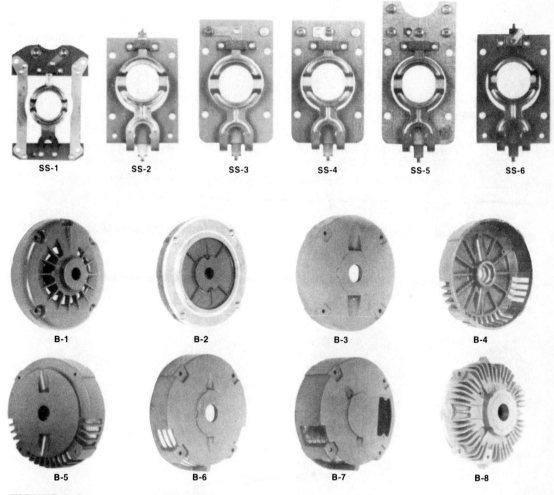

FIGURE 9–3
Typical Motor End Shields and Stationary Switches (Courtesy of Marathon Electric)

Two major types of rotors are shown in Figure 9–4. The first is a DC wound rotor with a commutator. The second is an induction rotor for a three-phase motor. Some rotors, not shown, are wound rotors with slip rings.

The wound rotor with a commutator (Figure 9–4A) is used for all DC motors and for the AC universal motor. Wires are wound in coils in the slots of the rotor. The slots are in a stack of laminations. The laminations are processed and stacked in the same manner as stator laminations. Like stator slots, rotor slots have insulation in them. Also, the rotor is varnish-dipped in the same manner as the stator windings. The rotor laminations are pressed onto the shaft, in a stack of specified length, before the windings are installed. After the windings have been installed, the coil ends are brought out. Coil ends are connected to various segments of the commutator.

The second rotor illustrated (Figure 9–4B) is an induction rotor. It is constructed similarly to the DC wound rotor except that it has no wires or commutator. The laminations are first pressed onto a metal pin or dowel. Then, aluminum is "squirted" by die casting through holes in the laminations. When die casting is

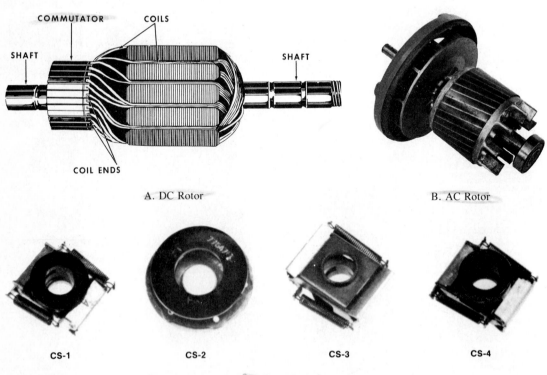

A. DC Rotor   B. AC Rotor

CS-1   CS-2   CS-3   CS-4

C. Centrifugal Switch

FIGURE 9–4
Typical Motor Rotors and Centrifugal Switches

complete, there are aluminum bars between two end caps. After some cooling, the pin is removed, and the rotor core is pressed onto the shaft while still somewhat hot. As the rotor core cools, it contracts and grips the shaft tightly. The induction rotor is used in AC single-phase and three-phase motors of all types and sizes.

For single-phase motors, centrifugal switches are pressed onto the shaft. These have axial moving bobbins, which contact the stationary switch. The moving bobbin opens and closes the stationary switch at an appropriate speed. Typical centrifugal switches are shown in Figure 9–4*C*.

Figure 9–5 shows the outlines of a typical rotor lamination and a typical stator lamination.

An optional cooling fan can be cast with an induction rotor. Separate sheet metal fans can be pressed onto the shafts of other rotors. The rotors are balanced after assembly by adding weight to the core end or the fan blades. Sometimes weight is subtracted by nipping out a section of the fan. Note that the rotor slots are not straight axially, as the stator slots are. The angle at which the slots are arranged is called *skew*. The skew gives the rotor better and smoother operating characteristics. Skew varies from motor to motor. Rotors of different skews are not interchangeable, even if their diameters and lengths are the same.

All motors operate best when the air gap between the rotor and stator faces is at a minimum. However, the smaller the air gap, the greater the possibility of the rotor rubbing against the stator. Motors are designed and manufactured with this trade-off in mind. Therefore, after taking a motor apart, make sure to reassemble it in the same manner, with all parts in the same orientation. Otherwise, the required concentric circles might not be lined up properly. The motor could rub and then stall, and the rotor-stator interference could burn out the windings.

Motor size increases as horsepower increases. Motor sizes conform to NEMA standards. The frame sizes are given numerical designations, such as 48 frame and 120 frame.

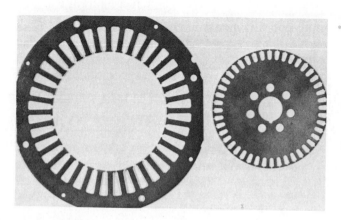

FIGURE 9–5
Typical Rotor and Stator Laminations

9–4   CHOOSING THE CORRECT INDUSTRIAL MOTOR FOR AN APPLICATION

When choosing an electric motor for a given application, the first step is to choose the general type to be used. The main consideration is the type and speed of mechanical or rotary torque required. Other considerations are the type of electrical supply available, the precision of speed control needed, and, of course, the cost. In Figure 9–6 industrial motors are divided into four general categories. Of the four classes of motors, three will be covered in this chapter. The fourth, indicator and control motors, will be covered in Chapter 10. Each of the three classes to be discussed in this chapter has special application capabilities and characteristics.

The first type, the direct-current (DC) motor, is used when full torque is needed at widely varying speeds. In contrast, alternating-current (AC) motors are essentially constant-speed devices. (However, as will be seen in Chapter 12, there are electronic drives that can make the three-phase AC motor operate as a variable-speed motor.) As a trade-off for the desirable characteristic of varying speed, DC motors are more costly and more complicated than AC motors. Additionally, DC motors require variable DC current, which must be obtained from a rectification system at additional cost. They also have carbon brushes, which ride on the rotating commutator. This brush-commutator combination increases maintenance cost and possible downtime.

The second type is the three-phase AC motor. By far the most commonly used subtype is the induction type. Of all the motors, the three-phase induction motor has the lowest cost per horsepower and the smallest size per horsepower. Three-phase AC motors are built in ratings up to hundreds of horsepower. They are relatively easy to manufacture, and their maintenance is the least of any type of motor. They use induction rotors, so there are no windings in the rotor, no brushes, and no commutator. One major requirement, of course, for these motors is that three-phase power be available. Three-phase power is not available every-

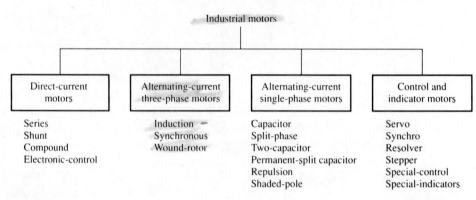

FIGURE 9–6
General Industrial Motor Classes

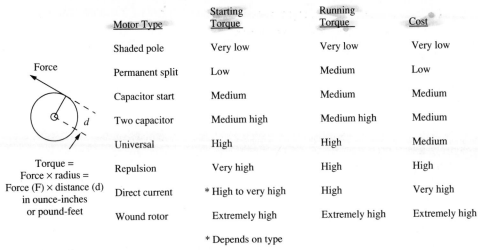

Motor Type	Starting Torque	Running Torque	Cost
Shaded pole	Very low	Very low	Very low
Permanent split	Low	Medium	Low
Capacitor start	Medium	Medium	Medium
Two capacitor	Medium high	Medium high	Medium
Universal	High	High	Medium
Repulsion	Very high	High	High
Direct current	* High to very high	High	Very high
Wound rotor	Extremely high	Extremely high	Extremely high

* Depends on type

Force

Torque =
Force × radius =
Force (F) × distance (d)
in ounce-inches
or pound-feet

FIGURE 9–7
Relation of Torque Level to Cost in Various Motor Types

where. For example, residences and many commercial establishments have only single-phase power. Residential or commercial three-phase power is normally very costly to have installed.

The third classification is the single-phase AC motor. There are many types of single-phase AC motors. They are used when three-phase power is not available and where the horsepower rating needed is small—commonly less than 2 HP, although there are single-phase AC motors rated as high as 10 HP/220 V. Single-phase motors are often called fractional-horsepower motors. Actually, only motors under 1 HP are true fractional motors.

In contrast to the three-phase motor, the single-phase motor requires an internal starting system. The starting system consists of a centrifugal device on the rotor, a stationary switch on one end shield, and additional windings for starting. Single-phase motors have induction rotors as do three-phase motors. The four basic types of single-phase AC motors will be discussed in Section 9–7. Other common AC motors will be covered in Section 9–8.

Throughout this chapter, the torque capabilities of different types of motors will also be discussed. Figure 9–7 provides a look at the relative torque-level capabilities of different types of motors. It also illustrates that the cost of the motor goes up with the relative torque level of the motor. It therefore is necessary to consider this trade-off upon selecting a motor type for a given application.

Figure 9–7 also defines torque mechanically. Torque is the product of the rotary force produced and the radius at which the force is acting. As the loading of the motor goes up, the motor torque capabilities must increase to match the load. Higher torque is attained by using a larger, higher horsepower motor or by using a different type of motor.

Information is given for both starting and running torque in Figure 9–7. Starting torque refers to the turning ability of a motor starting from standstill. Running torque refers to the ability of a motor to keep a load turning once the motor has gotten up to running speed.

9–5   DC MOTORS

Figure 9–8 is a cutaway view of a DC motor. The locations of the various parts are shown in the figure. Both the stator and the rotor have wires wound in slots in their faces. The inserted wire windings are coils made up of multiple turns of prevarnished wire, as previously described.

Note that each DC motor has specifically designed rotors and stators. Rotors and stators of one model of DC motor are not interchangeable with rotors and stators of another model. Interchanging parts improperly will cause lowered performance and even motor burnout.

Two-pole DC motors have two stator winding groups to produce the required electrical north and south poles. Other DC motors have four winding groups or more, in multiples of two. Some large DC motors have extra, auxiliary pole windings called *interpoles*. Interpoles are positioned halfway between the

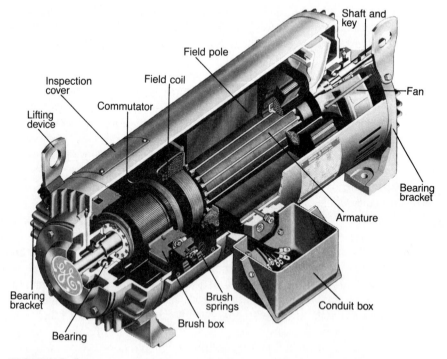

FIGURE 9–8
DC Motor—Cutaway View Showing Components and Construction (Courtesy of General Electric)

main poles. They are used to "straighten out" the magnetic field to reduce brush-to-commutator arcing during rotation and improve motor performance.

Permanent-magnet DC motors have no field coils. The stator has permanently magnetized field poles for operation. The size of a permanent-magnet motor is limited by its field poles' magnetic capability. The magnetic-field-generating capability of a permanent magnet is relatively small compared with that of a coil.

Wires attached to the coils of the DC motor stator are brought to the outside of the motor, often in a terminal box. The coils from the rotor are connected to various segments of a commutator during rotor assembly, as previously shown in Figure 9–4. The electrical path connection from the rotor commutator is then made through brushes that ride on the commutator as it rotates. The brushes ride axially in and out, by spring pressure, in fixed brush holders that are attached to the stationary motor stator frame. The brushes, in turn, are connected to the motor power source, directly or through a terminal box. The brushes will come in contact with different commutator segments as the rotor rotates. As different active commutator segments are contacted by the brushes, different coils in the rotor are electrically activated.

Typical DC motor connection schemes are shown in Figure 9–9. In the figure, the stator, or *armature,* is designated by the circle labeled "A." The simple conditions for series, shunt, and permanent-magnet motors are illustrated. There are four possible connections for the compound DC motor. Figure 9–9 shows the compound possibilities for combinations of long or short with cumulative (fields helping) or differential (fields opposing).

The DC motor rotation is accomplished by the interaction of the electrical fields of the stator and the rotor. As the commutator moves under the brushes, different segments of the commutator are electrically contacted. Consequently, different rotor coils are sequentially energized. The armature field is kept in the proper relative position for continuous operation. There are various configurations for connecting DC motor fields, as shown in Figure 9–10. A field coil with a few turns of large-diameter wire is used as a series field. A different type of field coil, the shunt field, has many turns of a smaller-diameter wire. A DC motor may be series-only or shunt-only, as shown in Figure 9–9. Alternatively, it may have both types of field coils and be called a compound motor. All DC motors have electrical connections to the rotor through the brushes.

An important concept in understanding DC motor operation is *counter electromotive force (CEMF).* The CEMF concept also applies to AC motors and to generators. It will be discussed here as it applies to the DC motor. When power is initially applied to the motor, the motor is at standstill. The current is $I = V/R$, where I is starting current, V is terminal voltage applied, and R is effective motor resistance. For example, with 200 V applied to a 5 Ω motor, the current will be 200/5 or 40 A. As the motor picks up speed and load, the current falls off considerably. This falloff is due to a CEMF back-acting voltage. The CEMF voltage is created by the interaction of the current in the armature with the field coil's magnetic field. The running Ohm's-law formula becomes $I = (V - \text{CEMF})/R$. For

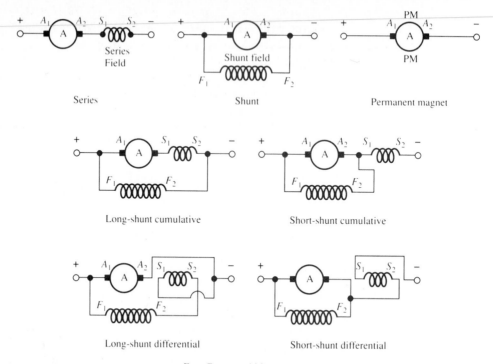

Four Compound Motors

FIGURE 9–9
DC Motor Connection Schemes

example, if the CEMF at a given point is 150 V, the current will be $(200 - 150)/5 =$ 10 A.

As the motor continues to speed up, the current will be reduced further. The CEMF is approximately directly proportional to motor speed, assuming that pole magnetic field is constant. Assume that speed is measured at 2200 RPM for 150 V CEMF. At 100 V CEMF, speed will be $2200 \times (100/150) = 1467$ RPM. The back CEMF will never reach 200 V, as there must be some net voltage. Net voltage is needed to furnish current to overcome friction and other losses and keep the motor running.

The CEMF concept leads to two other fundamental motor parameters: *motor power consumption* and *motor torque* (turning capability with load). The motor power is approximately equal to CEMF $\times$ I in the armature. That is, $P = E \times I$. With current at 10 A in the example given earlier, $P = 150 \times 10 = 1500$ W. At 746 W/HP, the horsepower is 1500/746, or 2.01 HP.

Motor torque is directly proportional to armature current, again with a constant magnetic field. In the example motor, assume 18.5 ft-lb starting torque. This

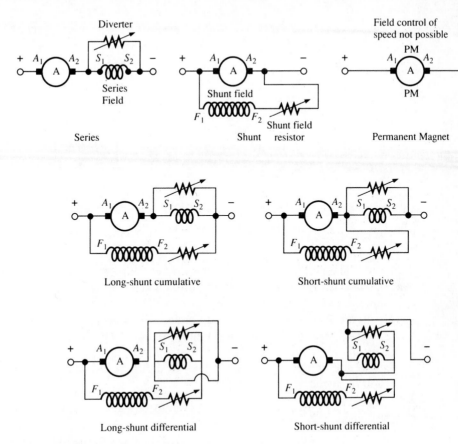

FIGURE 9–10
DC Motor Connections for Field Control of Speed

would be with 40 A armature current. As the motor speeds up, the torque produced falls off with the current drop. At 150 V CEMF and 10 A, the torque would be $18.5 \times 10/40 = 4.625$ ft-lb.

The speed of any DC motor varies as its terminal voltage is varied. Normally, higher voltage means greater speed. Varying the voltage to field only, to armature only, or to both will result in variations of motor speed. In many control systems the field is controlled electrically by an electronic system as shown in Figure 9–11. Very precise speed and torque control is possible when using this electronic control. Chapter 12 will have examples of the use of this control.

Note in Figure 9–10 that shunt field control uses a series resistor for speed control. Varying the resistor value varies field current and, consequently, speed. For series fields a parallel resistor is used. This parallel resistor is called the

FIGURE 9–11
Electronic DC Motor Control

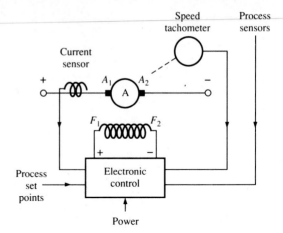

diverter. The parallel resistor is used for field current control. If a series resistor were used, it would also affect armature current. Series resistors are not used in conjunction with series fields.

Figure 9–12 shows the speed-versus-torque curves for the various types of DC motors that have been discussed. An ideal DC motor has a straight-line characteristic; that is, the torque output is constant at all possible motor speeds. Some of the important characteristics of different DC motors are illustrated by this curve, including the following:

- A DC series motor has very high torque at low speeds. However, speed varies greatly as load varies.
- A DC shunt motor has fairly constant speed throughout the speed range. However, it has a lower starting torque.
- A DC compound motor tends to combine speed and torque characteristics of series and shunt motors.

FIGURE 9–12
DC Motor Speed-Torque Curves

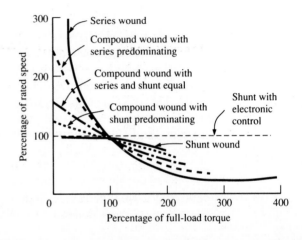

□ The electronically controlled DC shunt motor has a straight-line speed-torque curve. As load varies, the speed can be maintained at a preset, precise, constant value.

□ A DC motor can "run away." With no mechanical load on a series motor, the speed of the motor becomes very high. On large DC machines, the loss of field power can cause the same effect. The DC motor will accelerate rapidly, overspeed, and sometimes even fly apart.

A summary of the operating characteristics of the major types of DC motors is shown in Figure 9–13.

Detailed motor control is covered in Chapters 11 and 12. Basic DC motor control—starting, reversing, and stopping—will be discussed briefly here.

Starting of small DC motors is accomplished by simply connecting the motor directly across the line. Larger DC motors require reduced voltage for starting. Reduced voltage may be accomplished by lowering line voltage and then raising it as the motor accelerates. Where the line DC voltage is fixed, series resistors are used during starting to reduce inrush current. The starting series resistor is then bypassed during full-speed running.

To reverse the rotation of the DC motor, it is necessary to reverse the connections of the field windings with respect to the armature windings. Reversing the polarity of total-motor external connections will result in both armature and field reversal. Reversing both armature and field will cause the motor to rotate in the same direction as it did originally.

Stopping can be accomplished by simply letting the motor decelerate to a standstill. If faster deceleration is needed, mechanical brakes may be used. Applying reverse polarity for abrupt stopping is possible but seldom used. A more common method of rapid DC-motor deceleration is *dynamic braking*. This consists of turning the motor into a generator and feeding the resulting output power into a small-value resistor. Later chapters will cover dynamic braking in detail.

Electronic means of controlling starting and stopping are often used and will be described in Chapter 12.

9–6   THREE-PHASE AC INDUCTION MOTORS

The three-phase induction motor can be connected directly to the power company lines, through transformers if necessary. The transformers have up to 99% efficiency and introduce negligible power losses. In contrast, for a DC motor, the power requires electrical rectification from AC to DC. The rectification requires expensive equipment. Furthermore, the efficiency of the supplying rectifier is only 60%–80%.

The three-phase motor has the simplest construction of any type of motor. Unlike a DC motor, it has no brushes, no commutator, and no wires wound in its rotor. In contrast to the single-phase AC motor (discussed in the next section), the three-phase motor has no starting windings or starting centrifugal switch. A major

*Speed Regulation	Speed Control	Starting Torque	Breakdown Torque	Applications
SERIES				
Very high. Varies inversely with load. Theoretically, infinite no-load speed.	From zero to maximum speed, depending upon control and load.	Very high (300 to 375%). Varies as square of voltage applied to armature.	High (approximately 200%). Limited by commutation and heating.	Where very high starting torque is required and speed can vary with load changes. Hoists, gates, movable bridges, street cars, locomotives, and transfer cars.
SHUNT				
3 to 10% maximum.	Speed range up to 6 to 1 by field control.	Medium (125 to 200%). With constant field, varies directly with voltage applied to armature.	High (150 to 250%). Limited by commutation and heating.	Where constant or adjustable speed is required and starting conditions are not severe. Paper making machines, printing press drives, conveyors, wood-working and metalworking machines.
COMPOUND				
7 to 25% depending upon the amount of the series field.	Speed range up to 4 to 1 by field control.	High (180 to 260%). Limited by starting resistor.	High (175 to 250%). Limited by commutation and heating.	Where high starting torque and fairly constant speed are needed. Punch presses, shears, crushers, geared elevator plunger pumps, reciprocating compressors.

*Speed Regulation $= \dfrac{\text{No load speed} - \text{Full load speed}}{\text{No load speed}}$

FIGURE 9-13
Summary of DC Motor Operational Characteristics

advantage of the three-phase motor is that it is easily connected and reconnected for dual-voltage operation or for reverse rotation.

One major characteristic of the three-phase AC induction motor is that it runs at essentially constant speed. It does not run at a perfectly constant speed, but very nearly so. Controlled, wide speed variations are not possible without extensive control equipment. The variable-speed control system will be discussed in a later chapter. Three-phase motors are relatively easy to construct. They are normally built in ratings of 2 HP and up.

Figure 9–14 shows the construction of a three-phase motor in a cutaway view. Note the simplicity of construction. As in the DC motor, coils are wound in slots in the stator. The electrical power is fed into these stator coils by externally connected wiring. In contrast to DC motors, the rotor has no coil wires wound in it. As in the DC motor, both the rotor and the stator are made up of stacks of heat-treated laminations.

For the three-phase motor rotor, the die-casting process creates aluminum bars in the slots and also aluminum end rings, which hold the laminations together. There is therefore a "squirrel cage" of aluminum bars and end sections. The aluminum bars and end sections are relatively good electrical conductors, in contrast to the laminations, which are relative insulators.

When an input AC voltage is applied to the stator windings, the resulting current produces a rising and falling magnetic field. The changing stator magnetic field links the aluminum rotor bar conductors. This magnetic field induces current

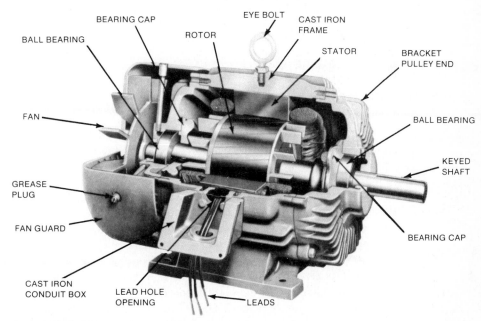

FIGURE 9–14
Cutaway View of an AC Three-Phase Motor (Courtesy of Marathon Electric)

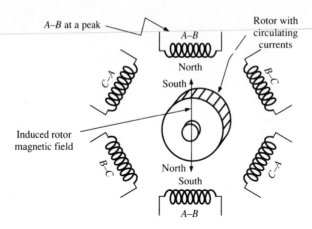

FIGURE 9–15
Magnetic Field Creation in the Rotor

in the aluminum rotor bars. The rotor current, in turn, produces a rotor magnetic field, as shown in Figure 9–15. The induced rotor field is opposite in magnetic polarity to the stator field. The interaction of these two fields is used to produce rotary motion.

It should be noted that in some designs of very large three-phase motors, copper bars are put into the rotor slots instead of the die-cast aluminum. The copper bars are then welded to copper end rings. The induction effect described for die-cast rotors is the same in these large, fabricated copper-bar rotor motors. In large motors the copper gives improved performance over aluminum bars, at added cost.

In a three-phase motor, the differing phase angles of the three alternating voltages create the rotor rotation. The three power voltages peak at different times during each AC cycle, as shown in Figure 9–16. The three input voltages are used to create a rotating magnetic field. If three-phase-power voltage values are plotted against time, the result is as shown in the figure. There are three different voltage patterns, one from each line-to-line combination. These combinations are A to B, B to C, and C back to A, as shown. The three-phase motor has three sets of two windings. As in the DC motor, the coils in each winding set are directly across from each other. Each incoming phase is attached electrically to one pair of windings.

An analysis of rotation will be begun at the point in the waveform where phase A–B is zero and increasing, point 1. At point 1 the field coils for A–B have no voltage, current, or magnetic field. Phases B–C and C–A do have voltage applied at the instant of point 1. Therefore, B–C and C–A do have magnetic fields. These two fields have opposite polarity and are wound so that they have an additive magnetic effect on the rotor. Since the B–C and C–A fields have equal

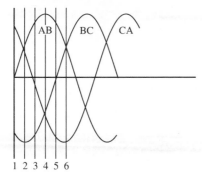

A. Three Phase Waveforms

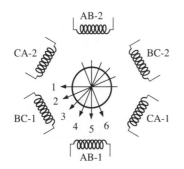

B. Rotor Orientation

Point	Predominate	Add	Oppose/Cancel
1	None	BC/CA	None
2	BC-1	None	AB/CA
3	None	AB/BC	None
4	AB-1	None	BC/CA
5	None	AB/CA	None
6	CA-1	None	AB/BC
7	None	BC/CA	None
8	BC-2	None	AB/CA
9	None	AB/BC	None
10	AB-2	None	BC/CA
11	None	AB/CA	None
12	CA-2	None	AB/BC

Not Shown { (points 7–12)

C. Magnetic Field Status

FIGURE 9–16
Three-Phase Motor Rotation Principle

magnitudes, the rotor field will tend to line up midway between them. This position is indicated by the arrow in the "Position 1" diagram in the rotor-orientation section of Figure 9–16. The field status of position 1 is indicated in the magnetic field status table of the figure.

Next, point 2 in the waveform is chosen, 30°, or 1/12 cycle, later than point 1. At this point all three fields have voltage, current, and magnetic fields. Field *B–C* has a negative maximum value. Fields *A–B* and *C–A* have equal values. The coils are wound with such polarity that the *A–B* and *C–A* fields cancel each other. Therefore, the *B–C* field predominates. The rotor then tends to line up with the *B–C* coils, 30° from point 1. This position is shown as position 2 in the figure for rotor orientation. The magnetic field status is shown in the table for point 2.

This process continues every 30°, and the rotor rotates 30° each time. Four more points, 30° apart, are shown in Figure 9–16. The process would repeat for 11 points and start over at the 12th. As a result of this process, the rotor will rotate one revolution (ideally) each electrical cycle.

FIGURE 9–17

Synchronous Speed and Slip Calculations

Synchronous speed is determined by the frequency (cycles per second) of the input AC power:

$$\frac{60 \text{ cycles}}{1 \text{ s}} \times \frac{60 \text{ s}}{1 \text{ min}} = 3600 \text{ RPM}$$

or

$$\frac{50 \text{ cycles}}{1 \text{ s}} \times \frac{60 \text{ s}}{1 \text{ min}} = 3000 \text{ RPM}$$

Assume 2 poles per phase for 60 cycles:

If slip is 5%, speed is 95% of 3600, or 3420.

If speed is 3425 RPM, slip is

$$\frac{3600 - 3425}{3600} \times 100 = 4.86\%$$

The ideal AC motor would make one revolution per electrical cycle, but a real rotor does not follow the rotating stator pattern exactly. It actually runs at a speed slightly less than the field rotation speed. The difference between running speed and theoretical speed is called *slip*. Slip can be expressed as a decimal but is normally expressed as a percentage of *synchronous speed*. Synchronous speed is defined as the speed at which the stator field rotates. It is expressed in revolutions per minute, or RPM. For no load or for very low-torque loads applied to the motor shaft, the slip approaches zero, but does not reach zero. Normal-motor full-load slip runs in the neighborhood of 5%. Some typical calculations for motor slip are shown in Figure 9–17.

For motor synchronous-speed calculations, the number of poles the motor contains is used as a base. The number of poles in a three-phase motor is determined by counting the number of poles for only one of the three phases. In the example of Figure 9–16, the motor, which has a total of 6 poles for three phases, would be designated a two-pole machine. As in DC motors, the number of poles can increase in multiples. A four-pole three-phase motor would have a total of 12 poles, a six-pole would have 18 poles, and so on. The speed varies inversely with the number of poles, as shown in Figure 9–18.

Three-phase motor control will be discussed more completely in Chapters 11 and 12. Rotation control will be discussed briefly here. Three-phase-motor rotation control involves reversing, stopping, and starting.

Reversing the rotation of a three-phase motor is accomplished by interchanging any two power leads to the motor. This interchange is illustrated in Figure 9–19.

A three-phase motor can be stopped by use of a mechanical brake. Or applying reverse power to a three-phase motor while it is rotating will stop it very quickly. The second method is used more often. Reversing two leads and quickly

FIGURE 9–18
Synchronous Speed versus Number of
Poles

$$\text{Synchronous speed in RPM} = \frac{(2 \times \text{cycles/s}) \times 60}{\text{Poles per phase}}$$

For 60 Hz (cycles/s) and 2 poles:

$$\text{Synchronous speed} = \frac{2 \times 60 \times 60}{2} = 3600 \text{ RPM}$$

For 60 Hz (cycles/s) and 4 poles:

$$\text{Synchronous speed} = \frac{2 \times 60 \times 60}{4} = 1800 \text{ RPM}$$

Poles per Phase	Synchronous Speed
2	3600
4	1800
6	1200
8	900
10	720
12	600

reapplying power will cause the motor to slow down to zero speed and then reverse direction. To stop at zero, the reverse power must be disconnected when the motor reaches zero speed. Circuits to accomplish stopping at zero will be covered in Chapter 12.

Applying reverse power is usually called "plugging." Note that plugging must be applied prudently. For example, a large inertia load, such as a conveyor,

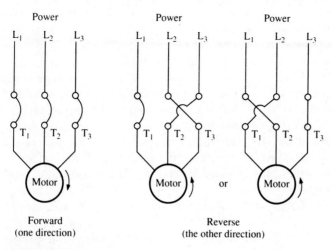

Forward
(one direction)

Reverse
(the other direction)

FIGURE 9–19
Three-Phase Motor Reversal

may be connected to the motor shaft. Plugging the motor may overload it electrically as the inertia load is decelerated. In such cases an antiplugging circuit is used to prevent the application of reverse power until the motor coasts to a stop.

Starting large three-phase motors requires special starting systems. Large three-phase motors, like DC motors, require reduced voltage when starting from a standstill.

All motors, including three-phase, require five to ten times their running current when starting if full voltage is applied. This inrush current must be considered even though it is of short duration. The inrush surge has two bad effects. First, the inrush can overload and overheat the motor and cause shortened motor life. Second, the surge will overload the power system, causing low voltage elsewhere. To reduce the large inrush current surge, a number of methods are used. These methods include the use of series resistors, transformers, a delta-wye system, part-winding start, and electronic systems. These will be discussed in Chapters 11 and 12.

After reduced-voltage starting of large motors, a switching method is used to connect the motor directly to the line during normal running at rated speed. To start small motors, power is applied initially across the line. The small input surges do not affect the power system enough to cause any performance problems.

Three-phase motors with 3 leads can be used only at the specified single voltage. Other motors, with 9 leads brought out, can be used at two voltages. The leads are connected to the input power leads in different configurations for the two voltages. The dual voltages are normally one given value and its multiple of two—for example, 220 and 440 or 120 and 240. For dual-voltage motors, each winding in the motor must have two identical subsections. The two identical sections are connected in series for the low voltage or in parallel for the high voltage. Dual-voltage connection schemes are shown in Figure 9–20. Some three-phase motors have 12 leads, not 9, for use in special motor-control applications. The three extra wires can, as an option, be brought out from the center point.

Not all three-phase designs of motors of a given horsepower and speed are identical. There are different design and application parameters. Three-phase motors may be designed for high or low slip. They may also be designed for high or low starting torque, compared with running torque. The different designs are standardized by NEMA. The design categories are identified by code letters. Figure 9–21 illustrates the commonly used code designs and their characteristics. One design might have higher starting torque but a lower running efficiency, for example. It is important to match the motor design to the application.

One important time to consider code-letter designation is during motor replacement in case of motor failure. Design A is normally furnished if a code letter is not specified. If a design A motor replaces a design C motor, the new motor may not work properly. If it turns the load—and it may not—it may burn out and fail.

Another factor to consider when applying motors is the *service factor*. This factor is listed on the nameplate. It is the factor by which the motor can be

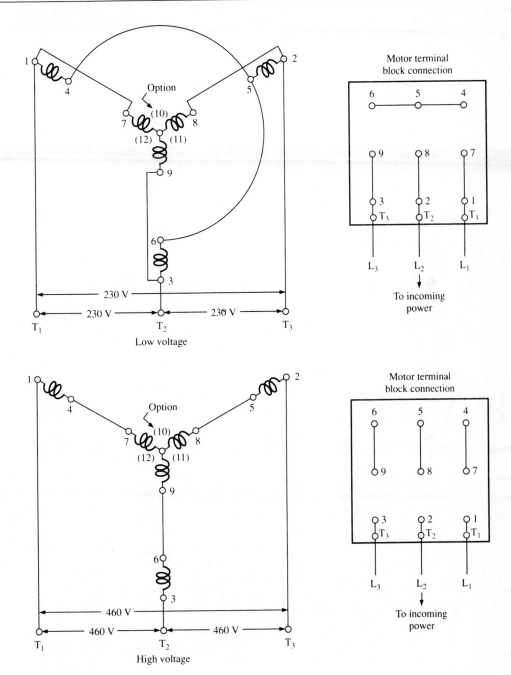

FIGURE 9–20
Dual-Voltage Motor Connection Schemes

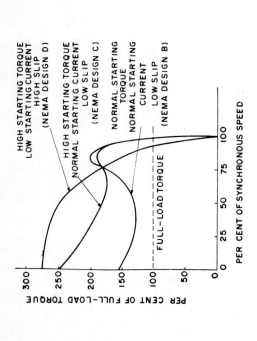

CHARACTERISTIC	NEMA B	NEMA C	NEMA D
Slip or Speed Regulation	*2 to 5%	*2 to 5%	5 to 13%
Starting Torques (% of Full Load)			
Full Voltage Starting	*100 to 275%	*200 to 250%	275%
Breakdown	*200 to 300%	*190 to 200%	275%
Full-Voltage Starting Current (% of Running Amperes)	*600 to 725%	*600 to 650%	550 to 650%
Efficiency (approximate)			
Full-Load	85 to 93%	84 to 92%	80 to 88%
¾-Load	86 to 93%	85 to 92%	81 to 88%
½-Load	83 to 89%	82 to 88%	78 to 84%
Initial Cost (Dripproof Designs)			
10-50 hp, 1800 r/min	100%	105%	125 to 160%
60-200hp, 188 r/min	100%	115%	140 to 170%
Typical Applications Characteristics	Constant speed, fairly constant load, infrequent starting and reversing	Hard-to-start loads, constant speed, fairly constant load, infrequent starting and reversing	Cyclical loads, very hard to start loads, frequent starting and reversing
Examples	Fans, blowers, and centrifugal pumps and motor-generator sets	Conveyors, crushers, reciprocating pumps, and compressors	Punch presses, elevators, small cranes and hoists

*Higher values are for lower hp rated motors

FIGURE 9–21
Three-Phase Motor Design Listing

overloaded, provided voltage and environmental factors are ideal. Consult the motor manufacturer before using this factor in your application.

9–7   SINGLE-PHASE AC MOTORS

The single-phase motor can be run where only single-phase power is available. Furthermore, it can be run on any one phase of a three-phase system, provided the voltage is correct.

In contrast to the three-phase motor, the single-phase motor has to have some internal system for starting. The starting system normally consists of a centrifugal device on the rotor, a stationary switch on one end frame, and a start winding. The start winding pair is wound 90° (electrically) from the main winding pair in the stator.

Like the three-phase induction motor, the single-phase induction motor has no brushes or commutator and no windings in the rotor. There are, however, some single-phase noninduction motors, which do have brushes, commutators, and wound rotors. These will be discussed later in this chapter.

Single-phase motors can be built for dual rotation, dual voltage, and multiple speed. They are, as are three-phase motors, an essentially constant speed device. If a single-phase motor is overloaded so that its speed decreases to approximately 90% of running speed, the motor stalls. When speed then decreases to 80% of full speed, the start winding comes back ON. The motor will cycle back and forth in this situation. Start windings are designed to be ON for a short time during periodic starting. Single-phase motors will burn out if the start winding is constantly cycled ON and OFF.

Figure 9–22 shows a cutaway view of a single-phase motor. It is constructed the same way as a three-phase motor, except that a rotary centrifugal device with a stationary switch is included. The construction of the induction rotor and stator is the same as for three-phase motors. Windings are constructed similarly except that there are two types of windings: main and start, or auxiliary.

Most single-phase motors are of fractional ratings, under 1 HP. A few go up to 10 HP, 220 V.

The rotary centrifugal switch is mounted on the shaft. It is adjusted to actuate when the motor reaches approximately 80% of its rated speed. At this 80% point, weights rotate outward, causing a bobbin or cone to move inward toward the rotor core. The face of the rotary device rides on a stationary switch, which is attached to one of the end brackets. The stationary part of the switch is held closed from zero speed up to 80% of full-load speed. When the 80% speed is reached during starting run-up, the displacement of the bobbin or cone allows the stationary switch to spring open. This switching action is used to open the circuit to the start winding above 80% speed.

Figure 9–23 schematically illustrates the single-phase motor with only two winding coils. It will not rotate when the coils shown are energized. Adding more coils will not help. All that this motor will do is sit still and hum. If the rotor of the motor is given a spin, the motor will then slowly accelerate to running speed.

CAPACITOR · CAPACITOR COVER

BRACKET, OPPOSITE PULLEY END

STATOR

ROTOR · STEEL FAN WITH BACK

STATIONARY SWITCH

BALL BEARING

CENTRIFUGAL SWITCH

KEYED SHAFT

CENTRIFUGAL SPRING SWITCH

STEEL FRAME

C-FACE BRACKET, PULLEY END

FIGURE 9–22
Cutaway View of an AC Single-Phase Motor (Courtesy of Marathon Electric)

The next parts of this section will cover motor types and connections that do actually work. All of these motors have start windings in them. The first to be covered is the *capacitor* (or *capacitor-start*) *motor*. The capacitor motor has a large electrolytic capacitor mounted on top and is used in a variety of applications. To start the rotation of the single-phase induction motor, start windings are added as shown in Figure 9–24. The start winding is wound as shown, 90° from the main winding. The diagram shown is for a two-pole winding. For a four-pole motor, the start windings would be placed between the mains at 45°. Also shown in the diagram are the required starting switch and the capacitor for phase shifting for starting torque for the capacitor motor.

FIGURE 9–23
Single-Phase Motor with No Start Winding

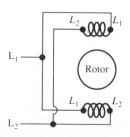

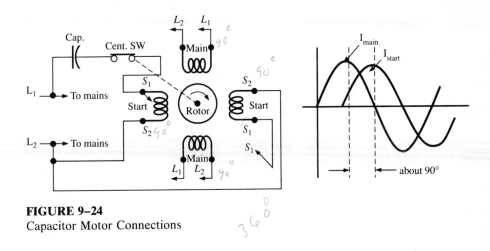

FIGURE 9–24
Capacitor Motor Connections

When power is applied with the motor at rest, both windings are energized. The starting effect is similar to that of the three-phase motor. The same voltage source is connected to both main and start windings. As shown in the diagram, the current and the magnetic field of the main winding peak first. The rotor produces a field of opposite polarity by induction. The capacitor in the start circuit delays the current to the start windings with respect to the main windings. Thus, the magnetic field buildup for the start windings is delayed. The capacitor value is chosen so that the delay is close to 90°. Because of the delay effect (with phase reversal) of the capacitor, the start winding peaks 90° later in the cycle. The rotor, with an induced field, then rotates to line up with the start winding when it peaks. Next, the rotor rotates to line up with the next main pole when it peaks 90 electrical degrees later. The resulting continuous turning effect on the rotor is the same as for the three-phase winding.

When the motor reaches about 80% of full-load speed, the switch opens. The start winding is disconnected from power and is no longer used. The motor then runs on the main windings alone. Rotor speed is close enough to synchronous speed that the rotor rotates almost to the opposite main pole when line voltage polarity changes. The line AC voltage changes polarity twice each cycle. Therefore, the time required to go from one main pole to the other is 1/2 cycle. Rotation is therefore maintained by the main windings only.

The single-phase motor slips slightly behind synchronous speed in the same manner as the three-phase motor. Calculations of slip and synchronous speed are the same as for the three-phase motor.

The capacitor motor is used where a good deal of starting torque is needed—for example, in saws and in other applications with large starting loads.

Note that the switch may not open, or may not reclose, because of a malfunction of some kind. The switch also may not open if motor overloading prevents the motor from attaining the 80% speed. In either event the motor could

FIGURE 9–25
Split-Phase Motor Connections

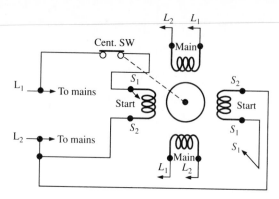

quickly burn out. The start windings are designed to operate for only a few seconds during acceleration. If they are energized longer, the motor can quickly overheat. Conversely, if the switch (or capacitor) is open-circuited, the start has no power and no magnetic field. The motor is then the same inoperative humming motor shown in Figure 9–23. Motor burnout will occur quickly unless the motor overload relay opens in a short time.

The *split-phase motor,* shown in Figure 9–25, operates similarly to the capacitor motor. It has a centrifugal switch but no capacitor. The switch is again set to open at about 80% of full-load speed. All slip and synchronous-speed considerations are the same as for the previously discussed motors.

The split-phase motor has lower starting torque but costs less than a capacitor motor. It is used on such loads as fans and blowers, where required starting torque is less than torque required at full speed.

In the capacitor motor the phase shift for the start winding field with respect to the main winding field was produced by capacitor action. The phase shift for the split-phase motor is accomplished by winding the start windings so that they have a different inductance from the main windings. As a result of the differing inductances, the phase shifts for the main and the start are different. Phase-shift differences of 45° are typical—compared to the nearly 90° of the capacitor motor.

Another type of capacitor single-phase induction motor, the two-capacitor type, is shown in Figure 9–26. It has both a start capacitor and a run capacitor. During starting, both capacitors are connected to the start winding for phase shifting. Since the capacitors are in parallel, the total capacitance is the sum of the two. The switch opens at 80% of full-load speed as in the motors described previously. At switching speed only one capacitor—the start capacitor, which has a larger capacitance—is switched out of the circuit. The run capacitor, which has a smaller value, remains connected to the start winding during running. The start winding is therefore also used as auxiliary running winding. The start/auxiliary winding is designed to run full time at a fraction of the current it carries during starting. Running torque is therefore increased by having the start/auxiliary winding in the circuit during running. Additionally, the motor power factor is improved.

FIGURE 9–26
Two-Capacitor Motor

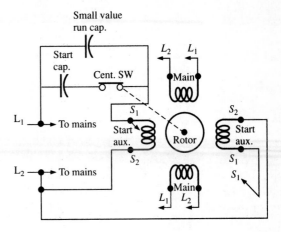

This motor costs more than other capacitor motors, but it does have more running torque and better running characteristics. Typical applications are compressors and other devices with heavy running loads.

Another type of capacitor motor, the permanent-split capacitor (PSC) motor, is shown in Figure 9–27. This type of motor is sometimes called the single-value capacitor motor. This motor has no centrifugal switch but does have a capacitor. The capacitor is in series with the auxiliary windings. The auxiliary windings are energized all the time. The turning torque, both starting and running, is produced by the phase shift of the capacitor, as before. This motor is less expensive than other capacitor motors but has less torque. It is electrically less efficient, so it is normally produced in smaller horsepower ratings only. However, it does have the advantage that there is no switch to cause malfunctions or maintenance problems.

FIGURE 9–27
PSC—Permanent-Split Capacitor Motor

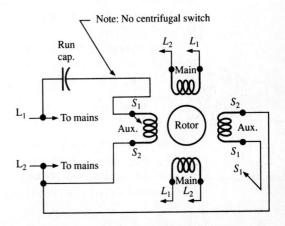

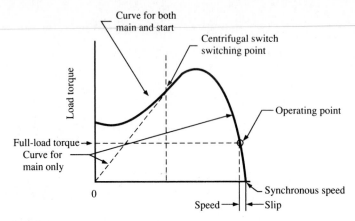

FIGURE 9–28
Typical Speed-Torque Curve

How does the speed-versus-torque curve look for all four types of motors discussed in this chapter section? It is shown in Figure 9–28. All motors start from rest at zero speed. At this starting point the main windings have no torque capabilities by themselves. However, combined with the start, or auxiliary, windings, they produce starting torque as shown. Starting torque must exceed the required turning torque of the load—full-load torque—for the motor to accelerate. Once started, the motor proceeds to the right along the solid line shown. When switching speed is reached, the motor's overall characteristics change to the values on the main-only curve. Ultimately, the motor reaches the operating point, on the far right, where the motor torque equals the load torque. Note that the speed at the operating point is slightly less than synchronous speed. The difference is the motor slip, as previously defined.

Starting and stopping of single-phase motors is straightforward. Since they are always low-horsepower, they draw limited line current. No reduced-voltage starting-control system is required. Stopping is accomplished normally by coast-down. For faster stopping, a mechanical brake is often used. Plugging by reversing polarity does not work for single-phase motors and is not used.

The variable operations possible for single-phase AC motors are as follows:

1. Dual rotation: clockwise (CW) or counterclockwise (CCW)
2. Dual voltage: 110 and 220 or 220 and 440
3. Multiple speed: Full speed, one-half speed, one-third speed, and so on

The advantage of dual rotation or voltage is product versatility. One motor model can be stocked or purchased for varying applications. The dual feature does, however, increase the cost of the motor. Dual operation also introduces a greater opportunity for error in the connection and reconnection of wires in the motor's

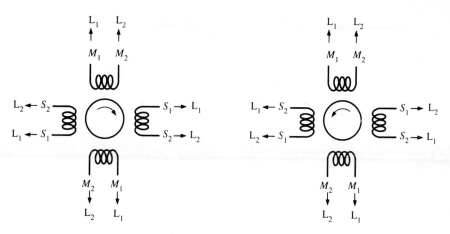

FIGURE 9–29
Reversing a Single-Phase Motor

terminal box. Errors in reconnection can cause motor malfunction and even burnout.

Dual-rotation connection is shown in Figure 9–29. To reverse direction of rotation, the start winding must be reversed in polarity with respect to the main. The reversal is accomplished by reconnecting wires to terminals found in the motor's terminal box. Connection instructions are found on the motor nameplate or on a decal in the terminal box. The reversing wires can also be brought out to an appropriate external switch, if the motor is to be reversed often. To be effective, the connection reversal must take place while the motor is stopped. The start winding is not in the circuit during running. Reversing the start during running has no effect.

Dual-voltage connections are shown in Figure 9–30. The general principles of dual rotation apply to dual voltage. The main winding halves are connected in series for high voltage and in parallel for low voltage. The start, which needs low voltage to prevent overheating, is connected directly for low voltage. For high-voltage connection, the high voltage is connected to the main winding. The start winding is connected across only one main winding. The start winding then has low voltage by a 2-to-1 transformer effect. There are some motors available for both dual rotation and dual voltage. Their reconnection schemes are quite complicated.

A typical multispeed scheme is shown in Figure 9–31. For four-pole operation, about 1800 RPM, the motor is connected normally with alternating poles (*A*). For two-pole operation, about 3600 RPM, poles IA and IB could be used alone (*B*). This use of only two poles would result in a reduced rating for the motor. To maintain nearly the same rating, IA and IIA are combined into one pole (*C*). Also, IB and IIB are combined into the other pole. For other speeds, a motor with more poles and similar multiple combinations can be used.

Low Voltage (Parallel)

M_{1-1} to M_{1-3} to L_1
M_{1-2} to M_{1-4} to L_2
M_{2-1} to M_{2-3} to L_1
M_{2-2} to M_{2-4} to L_2
S_{1-1} to S_{2-1} to L_1
S_{1-2} to S_{2-2} to L_2

High Voltage (Series)

M_{1-1} to M_{2-1} to L_1
M_{1-4} to M_{2-4} to L_2
M_{1-2} to M_{1-3}
M_{2-2} to M_{2-3}
S_{1-1} to $M_{1-2}M_{1-3}$ junction*
S_{1-2} to $M_{2-2}M_{2-3}$ junction*
S_{2-1} to S_{1-1}
S_{2-2} to S_{1-2}

* Connect for $1/2$ voltage (120 V).

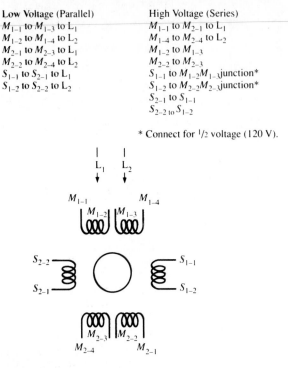

FIGURE 9–30
Single-Phase Dual-Voltage Motor Connections

Dual- or multiple-speed motors are used mainly for fans and similar applications. Dual-speed motors are not very efficient, so they are used only in smaller-horsepower applications.

9–8   OTHER COMMON AC MOTORS

The layout of the shaded-pole motor is shown in Figure 9–32. It is constructed with a squirrel-cage induction rotor. It has only main poles and windings. The figure shown is for a two-pole motor. More of the same poles would be added for a four- or six-pole motor. The turning torque for starting and running is produced by adding a shorted turn at one side of each pole. A slot is cut in the face of the pole, and a copper ring is inserted as shown.

During operation on AC voltage, a current is induced in the copper coil as the field flux builds up during pole voltage buildup. The induced current in the copper coil produces its own field of opposite polarity. The net effect is that the field between shaded poles builds up and recedes slightly later than the main field. When the main field reaches a peak N to S, an opposite field, S to N, is produced

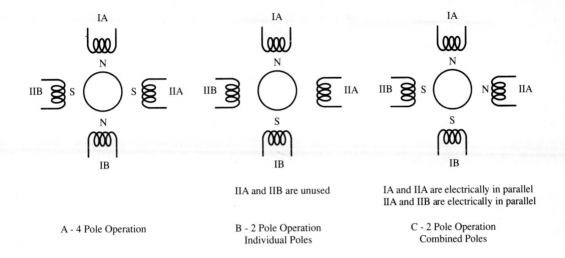

IIA and IIB are unused

IA and IIA are electrically in parallel
IIA and IIB are electrically in parallel

A - 4 Pole Operation

B - 2 Pole Operation
Individual Poles

C - 2 Pole Operation
Combined Poles

Note: Main Windings only shown

FIGURE 9–31
Multispeed Single-Phase AC Motor Connections

in the rotor. As the main field decreases after peak, the delayed field reaches a peak. The rotor tends to rotate toward the delayed peak. This process is repeated continually, and continuous rotation results.

The shaded-pole motor has very low torque. It is simple to construct. If it is stalled, it will not burn out. Since it is very inefficient, it is built only in very small ratings. A typical application is for a small blower motor for equipment.

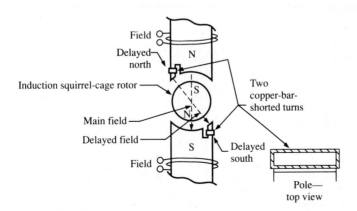

FIGURE 9–32
Shaded-Pole Motor

The *universal motor* is built exactly like a series DC motor. The universal motor does not follow the general rule that AC motors are essentially constant-speed devices. *Universal* means that it can be run on either AC or DC of the proper voltage. It is specially designed for variable-speed operation on AC or DC. A series DC motor cannot be run as a universal motor, even though both motors look the same internally and externally. They are built and wound differently. Figure 9–33 shows how a universal motor is connected. There are one or two

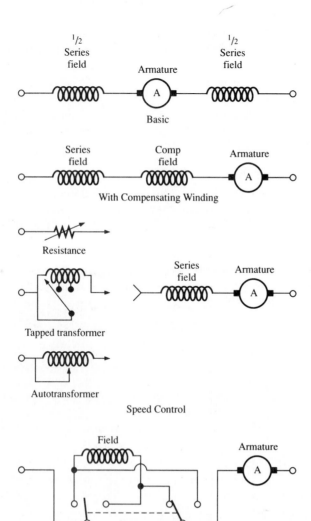

FIGURE 9–33
Universal Motor Connections

parts to the series field. When the series field is divided in two, one part is connected on each side of the armature.

Some universal motors have compensating windings wound 90 electrical degrees away from the main series windings. The compensating windings help the motor operate more efficiently on AC by reducing armature reactance voltage.

Speed is controlled by varying the voltage to the motor terminals. Three standard methods are shown in the diagram: series-resistor voltage reduction, a tapped transformer, and a continuous-voltage autotransformer. Additionally, a speed-governor system is used for speed control on some universal motors. The speed of the motors goes from just above 0 RPM to a multiple of synchronous speed. A regular two-pole motor runs just below 3600 RPM. A two-pole universal motor can run at 7500 RPM, and faster, on 60 cycle voltage. Reversing the relative polarity of field and armature, as shown, reverses the rotation of the universal motor.

Why not use the universal motor in industrial applications? First, it is only 25%–35% efficient. For a given horsepower rating, three times as much power is needed as with other AC induction motors. Second, the motor has poor speed regulation. As it is mechanically loaded, it slows down considerably. It is applicable where small amounts of high-speed torque are needed, as in drills and mixers.

Repulsion motors are used where a very large amount of starting torque and a good amount of running torque are needed. They are, however, more costly and can require more maintenance than capacitor or split-phase motors. Their rotation is easily reversed. Reversal is accomplished by shifting the brush mechanism about 15° to either side of a neutral position.

Repulsion motors are built with the usual windings in the stator. The rotor is a wound rotor similar to that of a DC or universal motor. The rotor windings are connected to a commutator, as in the DC and universal motors. The repulsion motor commutator may be radial, as in DC motors, or axial. The axial commutator's surface faces the end of the motor. Axial brushes ride inward, toward the rotor core.

There are three subtypes of repulsion motors. Their operating schemes are shown in Figure 9–34. The first type of repulsion motor is the straight repulsion motor, as shown in Figure 9–34A. The brushes that ride on the commutator are not connected to the power source. They are connected to each other by shorting wires. Torque is produced in the motor by the high short-circuit currents produced in the rotor by this shorting arrangement. The straight repulsion motor is the simplest type but is subject to a lot of brush wear.

The second type, Figure 9–34B, is the repulsion-start, induction-run motor. It has an additional feature not found in the straight repulsion motor. It has a centrifugally operated short-circuiter. The short-circuiter is in the form of a continuous coiled spring. It flips inward and makes contact with all commutator segments at about 80% of full speed. The short-circuiting action essentially converts the rotor to a squirrel-cage induction rotor. Another feature of some of these motors is a brush-lifting device. When the motor reaches full speed, the brushes are lifted to reduce brush wear.

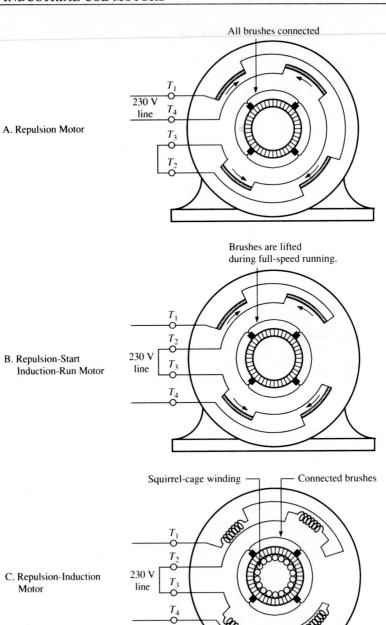

All brushes connected

T_1
230 V
line T_4
T_3
T_2

A. Repulsion Motor

Brushes are lifted
during full-speed running.

T_1
T_2
230 V
line T_3
T_4

B. Repulsion-Start
   Induction-Run Motor

Squirrel-cage winding ——   —— Connected brushes

T_1
T_2
230 V
line T_3
T_4

C. Repulsion-Induction
   Motor

FIGURE 9–34
Repulsion Motors

The third type, Figure 9–34C, is the repulsion-induction motor. It is similar to the previous two types. Additionally, it has cast rotor bars in the rotor. These added rotor bars make the motor run like a split-phase motor. The repulsion part of the motor makes it capable of high starting and running torque.

With all repulsion motors the trade-off for high torque is higher cost and higher maintenance.

The *wound-rotor motor* is shown in Figure 9–35. Its major attribute is that it runs on AC, with no rectification necessary. It runs without stalling from 0 RPM

A. Cutaway View

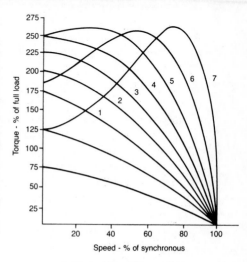

B. Typical Speed-Torque Curve

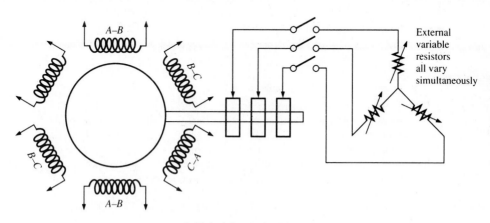

C. Typical Connection Diagram

FIGURE 9–35
Wound-Rotor Motor (Parts A and B from Humphries, J. T. *Motors and Controls*. Merrill/Macmillan: 1988. Figs. 5.21 and 5.23)

to full speed. It also has extremely high torque at very slow speeds. A typical application is for a stone crusher. The speed is varied by varying the settings of external resistors. The external resistors are electrically connected to three rotor windings through brushes and slip rings, as shown. The slip ring differs from the commutator in that it is continuous, with a single connection to rotor coils.

The wound-rotor motor is very costly and also is very inefficient because of the power loss in the control resistors. A typical curve of torque versus speed for various resistor values is shown for reference.

Synchronous motors run at exactly synchronous speed. A two-pole synchronous motor runs at exactly 3600 RPM when supplied from a 60 cycle source. A four-pole motor runs at exactly 1800 RPM. Small synchronous motors were used in clocks and timing devices until the advent of the quartz types.

Large industrial synchronous motors are used where two or more machines must have precisely identical speeds. For example, in a paper mill, input and output rollers for a process must run at the same speed. Otherwise, the paper will tear or bunch up. Electronically controlled drive systems are replacing large synchronous motors as solid-state technology advances.

Synchronous motors require some type of auxiliary device to run them up to speed. These starting systems can be very large and expensive. When synchronous speed is reached, the motor is connected directly across the power source. Run-up devices include an auxiliary motor and special extra windings in the motor itself. When synchronous speed is reached, switching from start to run must be carried out. In addition to proper speed, proper phase angle must be attained before switching. The run switch must be closed at the proper time in the cycle for proper pole-to-rotor lineup.

One use of the synchronous motor is for power-factor correction. It can be used in place of expensive correcting capacitors. Power-factor correction will be discussed in detail in Section 16–4.

SUMMARY

☐ Motors function according to the principle of magnetic attraction.

☐ All industrial motors are constructed similarly. Their major differences are in rotor construction.

☐ Direct-current (DC) motors have continuous high torque from start through low speed to high speed. However, they require special electrical sources, they are expensive, and they require maintenance on brushes and commutators.

☐ Counter electromotive force (CEMF) is created by the rotation of motors. It creates a back-voltage, which decreases line current. Motor power and torque are related to CEMF.

☐ Practically all alternating-current (AC) motors, single-phase and three-phase, are essentially constant-speed motors. Their speed does vary a little as their loading changes.

☐ Three-phase AC motors are very efficient, low-maintenance motors. Their cost per horsepower is the lowest of any motor type.

☐ Single-phase AC motors are used for low horsepower ratings and in places where only single-phase power is available. They require internal starting devices composed of switches and start windings.

☐ Shaded-pole motors are simple, low-torque, inefficient, stallable motors.

☐ The universal motor is built like a DC motor but operates on AC or DC. It is a variable-speed device but has poor efficiency and poor speed regulation.

☐ Repulsion motors are of three types. All have very high starting and running torques. They run on AC. They are expensive and require more maintenance than other motors.

☐ Wound-rotor AC motors have extremely high torque but are inefficient.

☐ Synchronous motors run at exactly constant speeds. Large synchronous motors require expensive and complicated starting schemes.

EXERCISES

1. What type of DC motor would you choose for the following applications, and why?
   a. a hoist
   b. a small fan to blow cooling air over a circuit board
   c. a paper-mill roller that requires precise speed control
   d. a general-use machine in which speed must be maintained with load within 10%

Refer to Figures 9–9 and 9–10 for Exercises 2–5.

2. In what type or types of DC motor does the armature current equal the field current?

3. In what type or types of DC motor does the armature current plus the field current equal the line current?

4. All else being the same, which type of compound DC motor would have the largest total stator magnetic field, cumulative or differential?

5. What happens to motor system efficiency as resistors are added for field control in Figure 9–10? Which efficiency is more greatly affected, series or shunt? Use Ohm's law to answer this question.

6. The example used to illustrate CEMF in this book had the following given: $R = 5\ \Omega$, V applied = 200 V, 2200 RPM at 150 V CEMF, and 18.5 ft-lb of torque at start (CEMF = 0). At 85 V CEMF, find the current, the power in watts and in horsepower, the speed, and the torque.

7. What would be the maximum possible CEMF in Exercise 6?

8. A DC motor has the following electrical characteristics: 185 V applied, 3.5 Ω resistance, 1050 RPM at 100 V CEMF, and starting torque (CEMF = 0) of 7.25 ft-lb. Determine starting current. Find the following at 172 V: I, P in watts and horsepower, speed, and torque.

9. List the main parts of a three-phase motor.

10. Find the percentage of slip for the following numbers of poles and speeds (at 60 Hz):

Poles	Speed (in RPM)
2	3250
4	1680
6	1050
2	3682

11. Find motor speed for the following numbers of poles and percentages of slip (at 60 Hz):

Poles	% Slip
2	6.3
4	3.7
6	12.5
12	8.6

12. What would happen to the values in the table of Figure 9–18 in Europe, where the power frequency is 50 Hz?

13. In Figure 9–16, what would happen to rotation if one of the three input lines were open or disconnected?

14. In Figure 9–16, what would happen if the phase sequence of A–B and B–C were suddenly reversed?

15. In Figure 9–19, is there another possible combination for reversing the motor? If so, what is it?

16. In Figure 9–19, in which direction would the motor go if L_1 were connected to T_2, L_2 to T_3, and L_3 to T_1?

17. Specify by design-code letter the three-phase motor you would choose for each of the following qualities:
   a. lowest cost to run
   b. maximum torque
   c. most constant speed

18. What type of AC single-phase induction motor would you choose for each of the following applications, and why?
   a. furnace blower motor
   b. stone crusher (for small stones)
   c. fan
   d. silo unloader
   e. machine tool with heavy running load

19. Describe the motor reversing system for each of the following:
   a. split-phase motor
   b. capacitor-start motor
   c. permanent-split capacitor motor
   d. two-capacitor motor

20. Two motors have identically wound main windings. One is a split-phase motor, and the other is a capacitor-start motor. What is the major difference between their running characteristics?

21. The speed of a motor as measured by a tachometer is 1225 RPM. What does this tell you about the number of motor poles and the power frequency (or the tachometer accuracy)?

22. What would a three-voltage (110, 220, and 440) split-phase motor's winding scheme look like?

23. A single-phase motor has only the main windings in it. At a 5% overvoltage (for example, 241.5 V for a 230 V motor), which way will it start to rotate? If the motor is given a start CW, which way will it rotate? What if it started CCW?

24. A motor has a total of 8 main poles. For a 50 cycle power system, what are the possible approximate speeds of the motor, depending on connections?

25. A motor has two leads. Can it be run in reverse? At dual speeds? On more than one multiple of voltage?

26. In Figure 9–31, how would the motor be connected for eight-pole operation? What would the resulting speed be?

For the following questions (except Question 36) refer to Section 9–8:

27. Explain how rotation can be reversed for each of the five motors discussed in this section.

28. Which of the five motors has slip? How much?

29. How can the speed of each of the five motors be controlled?

30. List two possible applications for each of the five motors, in addition to the applications listed in the chapter.

31. What would be the running speed of an eight-pole synchronous motor on 50 cycle voltage?

32. Which of the five motors would run well on DC? Which might run somewhat well?

33. List the starting and running torques of which each of the five motors is capable.

34. Which of the five motors have wound rotors? Which have commutators?

35. Which of the five motors is relatively energy efficient?

36. Define and describe the following terms:
  a. motor frame size
  b. motor centrifugal switch—internal
  c. motor centrifugal switch—external
  d. commutator
  e. slip ring
  f. short circuiter
  g. interpole

APPLICATION ASSIGNMENT ANSWER

For the small blowers, shaded-pole motors would be appropriate.

The grinder motor could be DC, preferably a series type. A universal motor would not be appropriate, because of the high horsepower needed. A repulsion motor would not work, as repulsion motors run at a maximum of about 3450 RPM for a two-pole machine under load. A speed of 4000 RPM could not be attained with any AC motor using 60 Hz without an expensive gearing or belt system.

Because of the high horsepower needed for the pump, a three-phase motor would be needed—a NEMA B. Single-phase motors are not made above 10 HP.

Single-phase 110 V would be needed for the blower motors. The AC voltage requirement for the DC motor would depend on the input power required for the AC-to-DC rectification system chosen. Three-phase 440 V would be appropriate for the three-phase motor—or perhaps 220 V or even 208 V, depending on the power voltage available.

10

Special-Purpose Motors

CHAPTER OUTLINE

10–1 Introduction □ **10–2** Stepper Motors □ **10–3** Synchros and Resolvers □
10–4 Servomotors—DC and AC □ **10–5** Other Special Motors

LEARNING OBJECTIVES

- □ Describe the principles and the construction of the stepper motor.
- □ Outline the various control systems for stepper motors.
- □ Using schematics, describe the operation of various types of synchros.
- □ Using schematics, describe the operation of the resolver.
- □ Describe the construction and the operational principles of the DC servomotor.
- □ Describe the construction and the operational principles of the AC servomotor.
- □ Describe the operation, the construction, and the application of the disk and Hall motors.
- □ Describe the operation, the construction, and the application of the hysteresis and reluctance motors.
- □ Describe the operation of linear and torque motors.

APPLICATION ASSIGNMENT

An *XY* table has been delivered to your department with no drive motor for either axis. You have been assigned to develop specifications for the drive motors. From the specifications, you are then to procure motors and controls. What steps would you follow?

10–1   INTRODUCTION

In the previous chapter, Chapter 9, industrial motors were covered. Industrial motors are used to convert electrical energy to mechanical energy. Industrial motors are neither precision-speed nor precision-positioning devices. For many automated devices and systems, precise speed control is needed. In other automation systems, precise positioning is required of electrical motors. In many instances, both precise speed and precise positioning are required.

This chapter will cover special electric motors, most of which have characteristics that give precise speed and positioning control. There are many different types of special motors. Some of the major types, representative of the many types, will be discussed.

The first type to be covered is the *stepper motor*. The stepper motor is relatively inexpensive and easy to control. Special IC chips developed for stepper motor control will be described.

Next, two common types of rotary position-indicating devices will be covered. These positioners are the *synchro* and the *resolver*.

Control of the positioning of large automated devices requires large amounts of torque and the ability to hold a specified static position. Such automated devices use DC or AC *servomotors* for positioning control. Both DC servos and AC servos will be discussed in this chapter.

A discussion of some of the other types of special motors will be included in the final chapter section. Such motors as the brushless DC motor and small AC synchronous motors will be covered.

10–2   STEPPER MOTORS

EXM: chart recorder

Stepper motors are used in a variety of automation applications in which a relatively small amount of torque is needed. Typical applications include rotary table control, wire-harness assembly, laser or pen positioning, and office peripheral-equipment control. Stepping motors can be used for precise positioning, without the need for a complicated position-indicating feedback system.

When a pulse is fed to the stepper motor, the motor rotates a given angle—for example, $1.8°$. Each input pulse results in rotation through the stepper's rated angle. For example, in 41 pulses a $1.8°$ stepper would rotate precisely $73.8°$. The radial tolerance for positioning is typically $3\%–5\%$ of one step. For the $1.8°$ example, the tolerance would be, at 4%, 0.04×1.8, or about $0.07°$.

Steppers normally do not need feedback position indicators, as they step reliably and precisely to a given point when programmed properly. In critical applications, however, feedback indication must be used for precise control.

Figure 10–1 shows the appearance of a typical product line of stepper motors. Figure 10–2 gives the typical standard step angles available. The table lists some standard step angles available, along with the typical maximum step rate. The table also includes the number of steps per revolution. The relationship between steps per revolution and step angle is given by the following formula:

FIGURE 10–1
Stepper Motors (Courtesy of Superior Electric Co.)

$$\text{Step angle} = \frac{360°}{\text{number of steps per revolution}}$$

The torques of various sizes of stepper motors run from 0.5 oz-in to 5000 oz-in. The rotors of the stepper motor are either permanent-magnet or variable-reluctance. In this discussion the type with a permanent-magnet rotor will be

FIGURE 10–2
Standard Stepper-Motor Ratings

Steps (Full) per Revolution	Step Angle (degrees)	Typ. Max. Run Rate (steps per second)
400	0.72	1000
200	1.8	2000
96	3.75	1000
48	7.5	1000
24	15	600
20	18	500

considered. The stepper motor typically has five or six wires for connection. A typical connection scheme is shown in Figure 10–3. In some units, the black wire and the white wire are combined in one common connection. The connection scheme is the same for all models with the same number of steps per revolution. The connection is shown for a two-phase stepper. A two-phase stepper has two control voltages fed to it, two wires per voltage. A four-phase stepper would have more connections because it would have four voltages fed to it as inputs.

The stepper can operate in the full-step mode or in the half-step mode. To illustrate full- and half-stepping, a 30°, full-step stepper, as shown in Figure 10–4, will be used. In the figure, the stepper has a four-pole stator and a six-pole rotor. There are four control switches. Each switch turns ON the coil of a corresponding pole, as shown. When ON, a pole is a north pole. When OFF, the pole is magnetically neutral. The rotor has six magnetic poles of the polarity shown. When switch 1 is ON, the rotor-stator alignment is initialized as shown. Turning switches ON and OFF in sequence enables the motor to step. For full steps, the rotor steps 30°, and for half steps, 15°.

The switch ON-OFF patterns for full and half steps are illustrated in Tables A and B in Figure 10–4. For full steps, as shown in Table A, P_1 lines up with S_1 initially. When S_2 is ON, the rotor rotates 30°, until P_3 lines up with S_2. The ensuing sequence of switch positions for each step angle is illustrated in Table A. For half-stepping, the sequence of switching is shown in Table B. The initial positioning is the same as in Table A. For 15° of rotation, S_3 and S_4 are both turned ON, as shown in the table. Rotor pole P_3 lines up halfway between S_2 and S_3, as S_2 and S_3 have equal magnetic strength. The sequence for 15° steps continues as shown in Table B.

If a motor requires switching a number of times per second, it cannot use switches for control. An IC chip that produces appropriate pulses for operation must be used. The required pulsing sequence is shown in Figure 10–5 for a step rate of 10 steps per second. The sequences for both CW and CCW are shown.

FIGURE 10–3
Stepper-Motor Connections

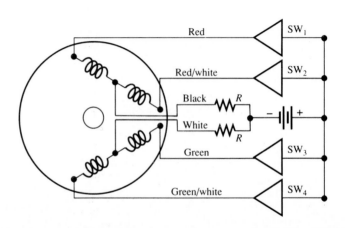

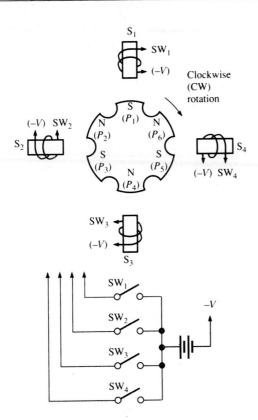

Table A. Full Steps (CW)

Degrees	Switch ON (x)			
	SW$_1$	SW$_2$	SW$_3$	SW$_4$
0	x			
30		x		
60			x	
90				x
120	x			
150		x		
180			x	

↓

Table B. Half Steps (CW)

Degrees	Switch ON (x)			
	SW$_1$	SW$_2$	SW$_3$	SW$_4$
0	x			
15	x	x	x	
30		x		
45	x	x		x
60			x	
75	x		x	x
90				x
105	x	x		x

↓

FIGURE 10-4
Operational Principle of the Stepper Motor

A typical IC stepper-motor controller to accomplish stepper motor solid-state control is shown in Figure 10–6. The figure shows the terminal identification of the IC along with basic specifications and the step patterns available. This particular IC is for two-phase steppers. Other ICs are available for four phases and for other numbers of phases.

The function of each terminal of the IC is as follows:

1. *Out$_B$*   Output voltage to line *B* of the stepper (phase *B–D*)
2. *VE$_2$*   Current sensing of *B–D* current for overload protection and control
3. *Out$_D$*   Output voltage to line *D* of the stepper (phase *B–D*)
4. *RC$_2$*   OFF time control for phase *B–D*
5. *Ground*   Electrical ground; also area for heat sinking for heat dissipation
6. *Ground*   Same as 5
7. *V$_{refl}$*   Voltage reference for overload trip—phase *B–D*

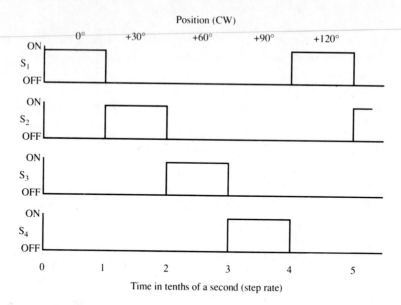

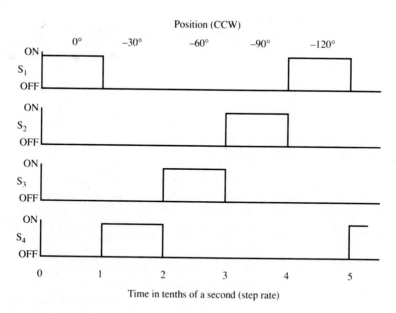

FIGURE 10–5
Stepper Pulse Control

UCN5871B

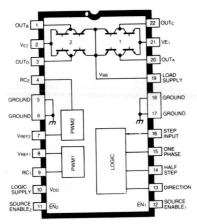

Dwg. PP-002

WAVE-DRIVE SEQUENCE

Half-Step = L, One Phase = H				
Step	Out$_A$	Out$_B$	Out$_C$	Out$_D$
POR	H	Z	L	Z
1	H	Z	L	Z
2	Z	H	Z	L
3	L	Z	H	Z
4	Z	L	Z	H

TWO-PHASE DRIVE SEQUENCE

Half-Step = L, One Phase = L				
Step	Out$_A$	Out$_B$	Out$_C$	Out$_D$
POR	H	L	L	H
1	H	L	L	H
2	H	H	L	L
3	L	H	H	L
4	L	L	H	H

DIRECTION = L   ↓   ↑ DIRECTION = H

HALF-STEP DRIVE SEQUENCE

Half-Step = H, One Phase = L				
Step	Out$_A$	Out$_B$	Out$_C$	Out$_D$
POR	H	Z	L	Z
1	H	Z	L	Z
2	H	H	L	L
3	Z	H	Z	L
4	L	H	H	L
5	L	Z	H	Z
6	L	L	H	H
7	Z	L	Z	H
8	H	L	L	H

Z = High Impedance
POR = Power on Reset

ABSOLUTE MAXIMUM RATINGS
at T$_A$ = 25°C

Motor Supply Voltage, V$_{BB}$ . . . . **45 V**
Output Current, I$_{OUT}$ (Continous) . . . . **1.0 A**
   (Peak) . . . . **1.5 A**
Logic Supply Voltage, V$_{DD}$ . . . . **7.0 V**
Logic Input Voltage
   Range, V$_{IN}$ . . . **−0.3 V to V$_{DD}$ + 0.3 V**
Sense Voltage, V$_E$ . . . . **1.0 V**
Package Power Dissipation, P$_D$. **See Graph**
Operating Temperature Range,
   T$_A$ . . . . **−20°C to +85°C**
Storage Temperature Range,
   T$_S$ . . . . **−55°C to +150°C**

FIGURE 10–6
IC Stepper Control System (Courtesy of Sprague Electric/Semiconductor Group)

8.	V_{ref2}	Voltage reference for overload trip—phase A–C
9.	RC_1	OFF time control for phase A–C
10.	LS	Voltage supply (+5 V) for logic circuits
11.	SE_2	HIGH enables phase B–D; LOW disables B–D
12.	SE_1	Same as 11 for A–C
13.	Dir	HIGH—stepper runs CW; LOW—stepper runs CCW
14.	$Half\ step$	HIGH—stepper runs half-step; LOW—stepper runs full-step
15.	$One\ phase$	HIGH—only B–D or A–C can go ON; LOW—one or both can go ON

16. *Step input*   Stepper steps when this pin goes from LOW to HIGH.
17. *Ground*   Same as 5
18. *Ground*   Same as 5
19. *Ld Sup*   Motor operational supply voltage—typically 45 V DC
20. *OUT$_A$*   Output voltage to line *A* (phase *A–C*)
21. *VE$_I$*   Same as 2 for phase *A–C*
22. *OUT$_C$*   Output voltage to line *C* (phase *A–C*)

When this IC is used in conjunction with other appropriate digital, pulsing ICs, it can precisely control stepper-motor action.

Two factors that must be considered in a stepper application are illustrated in Figure 10–7. The first, shown in Figure 10–7*A,* is for factor Ld Sup in the IC pin function listing (Figure 10–6). The stepper, in the half-step mode in this case, automatically starts at a lower step rate than the running rate. It starts at 150 half steps per second and gradually speeds up until it reaches the running rate of 400. The acceleration rate is adjustable by applying the appropriate voltage to the S pin (Figure 10–6). The acceleration and deceleration needed in the program depend on the application. If the stepper started at full rate, it would probably miss a step or two before it got up to speed. The acceleration function prevents this possible skipping. The same principle applies to deceleration, as shown. If the stepper stopped abruptly, it could overshoot one or more steps. Deceleration prevents possible overshooting of the final required position.

Another application consideration is shown in Figure 10–7*B.* As with most motor types, as the motor runs faster, the ability to carry load decreases. A typical load curve is shown. As the speed increases, the load capability decreases. If this curve's limitations are not considered, the stepper can start skipping steps under heavier torque loading.

How can the number of steps per revolution of a stepper motor be increased? For more steps, there will be a smaller step angle. How about a gear train? The disadvantage of a gear train is added bulk and cost. Also, inaccuracies would be introduced by the gears' inertia and backlash. A better system is the *microstepping system.* The microstepping system involves the application of varying voltages to the rotor poles. For regular stepping the voltage to each pole is either ON or OFF. Varying ratios of voltage are applied to adjacent poles for microstepping. As these relative voltages vary, the rotor positions itself at various intermediate positions between stator poles. An example of microstepping is shown in Figure 10–8. For illustration, only one rotor pole and two stator poles are shown.

The stepper motor itself is the same as for regular stepping. Only the control system is changed. Obviously, the controller must be more sophisticated than for straight stepping, and more costly. In the example shown in Figure 10–8, a step is divided into five parts. For a stepper with 200 steps per revolution, there are now 1000 possible steps. The resolution increases from 1.8° to 0.36°. Microstepping comes in standard ratings of 1/5 (our example), 1/10, 1/32, and 1/125. Accuracy and smoothness of stepping are achieved by using microstepping. In many applications the extra cost and complexity of control are worth it.

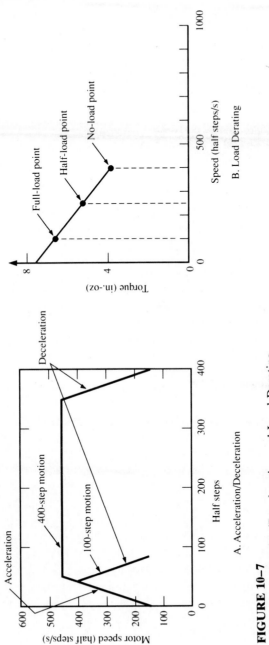

FIGURE 10–7
Stepper Acceleration/Deceleration and Load Derating

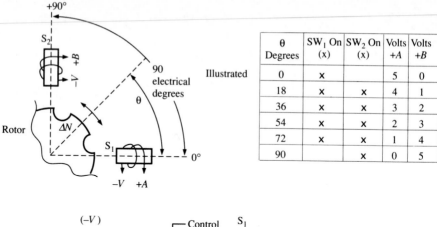

θ Degrees	SW₁ On (x)	SW₂ On (x)	Volts +A	Volts +B
0	x		5	0
18	x	x	4	1
36	x	x	3	2
54	x	x	2	3
72	x	x	1	4
90		x	0	5

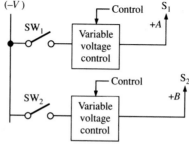

FIGURE 10–8
The Microstepping Principle

The table in Figure 10–8 lists four intermediate positions between the two positions directly opposite the stator poles. An angle of 90° is used for illustration. In actual applications a smaller angle is divided into five parts. A table of voltages applied to the end and intermediate points is shown. The voltages applied would be in a ratio to the values shown. The variable pole voltages are supplied as shown by variable voltage supplies. The relative voltages are determined by IC chip control for each position. The voltages may be straight analog levels. More often, they are varied by pulse-width modulation (PWM) techniques, which give smoother and more effective control. PWM is covered in detail elsewhere in the text. Here it is discussed only as it applies to microstepping.

Figure 10–9 shows the varying voltages used for microstepping in Figure 10–8 versus time. Amplitude-modulation (AM) control (Figure 10–9A) is shown for the two end steps and the four intermediate steps of Figure 10–8. At 0° the rotor is aligned with S_1, as S_1 is maximum, 5 V, and S_2 is 0 V. For 18° the S_1 voltage is decreased to 4 V, and S_2 is increased to 1 V. Therefore, the rotor is aligned 18° from S_1 toward S_2. For the other step angles, the voltages are changed to the values shown.

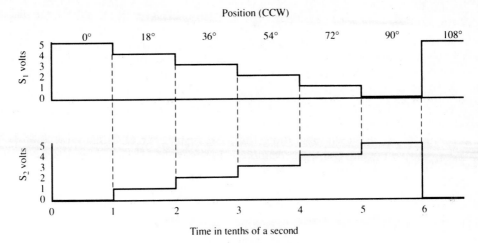

Position (CCW)

A. Amplitude Modulation

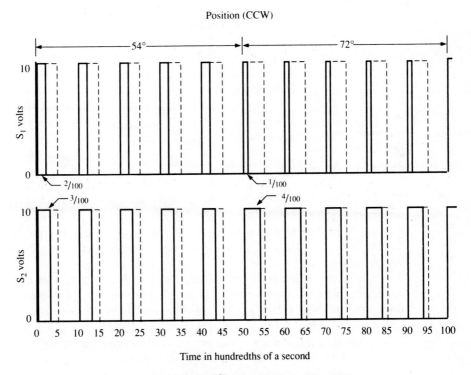

Position (CCW)

B. Pulse Width Modulation (PWM) for 54° and 72°

FIGURE 10–9
Microstepping Control Voltages

For PWM control (Figure 10–9B) pulse widths of 0.05 s, a pulse ON ratio of 50%, and a peak pulse voltage of 10 V are assumed. When the pulse is ON for the full 0.05 s, the average voltage is 5 V with the 50% ON ratio. For 54°, S_1 is to average 2 V. To accomplish this requirement, the pulse is only ON for 0.02 s. Over one cycle, the pulse will average 2 V. Stator pole S_2 has a pulse width that will cause it to average 3 V, as shown. The rotor will align itself according to these average voltages, 2 V and 3 V, at the required 54°. For 72°, S_1 is 0.01 s, and S_2 is 0.04 s, as shown. Other positions and voltages have similar ratios.

In actual operation, the pulse height, the ON time ratio, and the frequency will be different. The sample values have been chosen arbitrarily to illustrate the PWM operation for steppers.

10–3   SYNCHROS AND RESOLVERS   *LOW TORQUE*

Synchro is a general name for a series of motors that indicate or duplicate rotary-motion positions. The *resolver* is sometimes considered a subtype of the synchro. In this book the resolver will be considered a different type of device. Both synchros and resolvers are low-torque devices. If a large device, like a robot arm, is to be positioned, synchros and resolvers will not work. A servo device is needed for higher torque. If a synchro is used in a high-torque system, it must have its signal converted and amplified for control of a servo device. Servomotors will be covered in Section 10–4.

Synchros are small cylindrical motors varying in diameter from 1/2 in. to 4 in., depending on their power output. The synchro functions by converting a single-phase input to three unbalanced voltages that are 120 electrical degrees apart. These three voltages may then be converted back to a single-phase voltage. The seven major types of synchros are listed in Figure 10–10.

The synchro has three windings on the stator, as illustrated in Figure 10–11. A rotor with one winding is fed a fixed supply voltage through two brushes via two slip rings. As the rotor changes angle, the rotor magnetic field rotates accordingly. The resulting three voltages induced in the three stator windings vary accordingly. The schematic symbol and the connections for the synchro are also shown in Figure 10–11.

Three applications of the synchro will be illustrated. The first is a torque transmitter (TX) driving a torque receiver (TR). As the TX is rotated to a given angle, the TR will rotate to the same angle. This follower system is shown in Figure 10–12. The stators of both units are energized by the same single-phase 115 V, as shown. Transformer action induces different voltages in the three TX rotor windings. The three voltages vary in amplitude and in phase angle, depending on the angle of rotor rotation. The three resulting voltages are then fed to the three TR windings as shown. The three stator voltages in TR have the same amplitude and the same phase angle as in TX. The rotor of TR will therefore rotate to the same angle as TX. The mathematics of this transformation is quite involved and can be found in textbooks listed in the bibliography. Note that this system is

FIGURE 10–10

Synchro Types

Functional Classification	Military Abbreviation	Input	Output
Torque transmitter	TX	Rotor positioned mechanically or manually by information to be transmitted	Electric output from stator identifying rotor position supplied to torque receiver, torque differential transmitter, or torque differential receiver
Control transmitter	CX	Same as TX	Electric output same as TX but supplied only to control transformer or control differential transmitter
Torque differential transmitter	TDX	TX output applied to stator; rotor positioned according to amount; data from TX must be modified	Electric output from rotor (representing angle equal to algebraic sum or difference or rotor position angle and angular data from TX) supplied to torque receivers, another TDX, or a torque differential receiver
Control differential transmitter	CDX	Same as TDX but data usually supplied by CX	Same as TDX but supplied only to control transformer or another CDX
Torque receiver	TR	Electrical angular position data from TX or TDX supplied to stator	Rotor assumes position determined by electric input supplied
Torque differential receiver	TDR	Electrical data supplied from two TDXs, two TXs, or one TX and one TDX (one connected to rotor, one to stator)	Rotor assumes position equal to algebraic sum or difference of two angular inputs
Control transformer	CT	Electrical data from CX or CDX applied to stator; rotor positioned mechanically or manually	Electric output from rotor (proportional to sine of the diffference between rotor angular position and electric input angle)

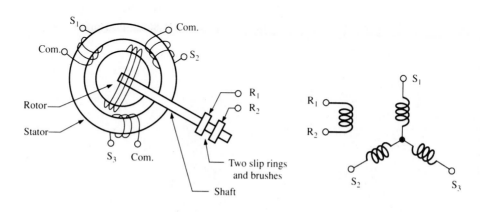

Connections

Symbol

FIGURE 10–11

Synchro Connections and Symbol

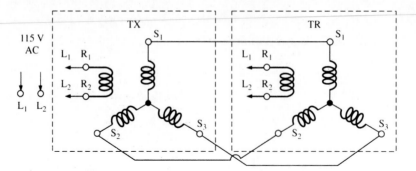

FIGURE 10–12
Synchro TX-to-TR Rotary System

reversible. If TR is forcibly rotated to a given position, TX will assume the same position, if it is free to rotate.

Another use of synchros is shown in Figure 10–13. The TX is connected to the control transformer (CT) as shown. The output from the stator of the CT is used as a controlling signal voltage. The TX and CT shaft angles are both usually set initially at 0°. The TX is then rotated to vary the CT output. The CT output in this case is very nearly a full-wave rectified sine wave when plotted linearly against the angle of TX. If the CT shaft angle does not start at 0°, the output graph will be displaced accordingly by the offset angle.

Besides being positioning devices, synchros can be used to add or subtract rotary angles. The way this operation is accomplished is shown in Figure 10–14. For the addition mode shown, the TR position is the sum of the TX angle and the torque-differential-transmitter (TDX) angle. For example, if TX is set at 43°, and TDX at 28°, TR will rotate to a position of 71°. The connections shown in Figure 10–14 are for addition. For subtraction the same equipment is used. To use this

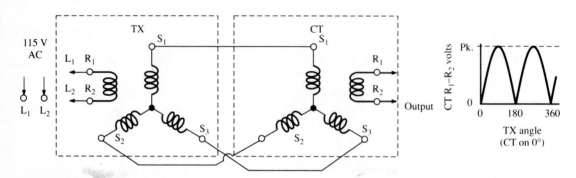

FIGURE 10–13
Synchro TX-to-CT Control System

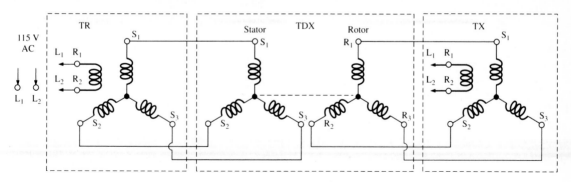

FIGURE 10–14
Synchro Add or Subtract System

system for subtraction, two or more appropriate wires are interchanged on the stator or the rotor or both.

The resolver is a device similar to the synchro. It differs from the synchro in the following ways:

☐ The electrical displacement is 90°, not 120°.
☐ There are two stator windings and two rotor windings.
☐ Input can be either to the rotor or the stator.
☐ Resolvers are usually not used as followers. Voltage V_{out} is utilized.

The construction and the symbol for the resolver are shown in Figure 10–15. There are two stator windings, wound 90° apart. In most applications, only one

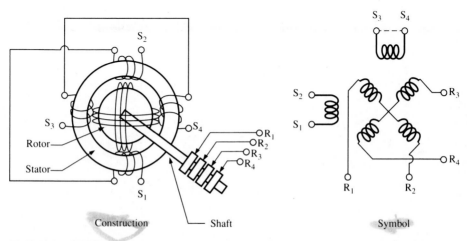

FIGURE 10–15
Resolver Connections and Symbol

stator winding is used, and the other is shorted out. The two rotor windings' connections are brought out through slip rings and brushes.

As has been stated, either the rotor or the stator may be the input. Typical application connections are shown in Figure 10–16. In each case illustrated, only

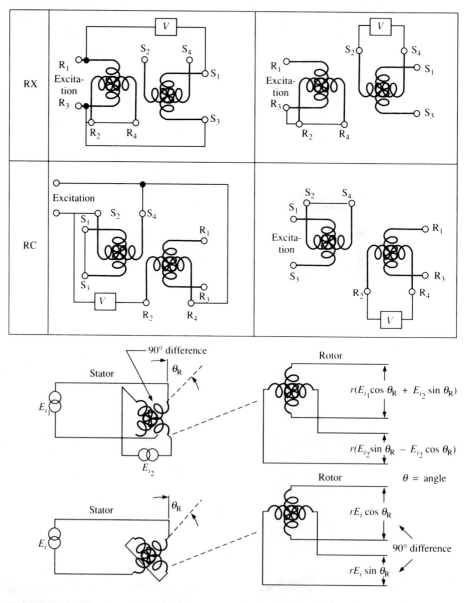

FIGURE 10–16
Typical Resolver Application Connections

one stator winding is used. Also, in all cases shown, the two rotor voltages are 90° out of phase with each other.

A typical resolver application might be in navigation or height determination. These two applications are illustrated in Figure 10–17. In the first example, a resolver is to be used to determine the distance to the destination. The range to a base station is determined by a radar range device. The angle θ from base station to destination is also determined directly. For this example, the amplifier scale is 6.5 V per 100 mi. The distance is 237 mi, which is represented by 15.4 V. The angle is determined by the indicating device to be 51.5°. Set the resolver at 51.5°, and apply the 15.4 V to R_3R_4. The voltage at S_1S_2 represents the destination distance. Assume a 1:1 rotor-stator ratio. The voltage at S_1S_2 will be 15.4 V/cos θ = 15.4/0.623 = 24.7 V. According to the scale used (6.5 V per 100 mi), this represents 411 mi for D.

Another application is height determination. Suppose the distance, D, to the top of a building is found by a range finder to be 645 ft. The scale of the amplifier to the resolver stator is 3 V per 100 ft. The representative voltage is then 3 × 6.45, or 19.35 V. This voltage is applied to S_1S_2 of the resolver. The angle of declination, read from the resolver scale, is 62.3°. The height of the building is found at the R_1R_2 terminals in voltage form. Assume a 1:1 rotor-stator ratio and the same amplifier ratio for the rotor output (3 V per 100 ft). The voltage will be D voltage × sine θ = 19.35 × 0.885 = 17.1 V. The voltage scales out to 17.1/3 = 5.7 hundred feet. The building is 570 ft high.

Note that in using the resolver, there is no need to go through trigonometric calculations for either example. The answer comes out directly in both cases.

There is a conversion system between resolvers and synchros. The system uses the Scott tee transformer connections shown in Figure 10–18. The Scott tee transformer combination shown in Figure 10–18A changes three-phase power to two-phase. To convert two-phase to three-phase, the system is reversed, as shown in Figure 10–18B. The synchro is a type of three-phase system, and the

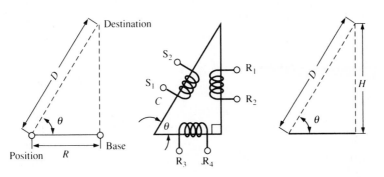

Navigation Application   Key   Height Determination Application

FIGURE 10–17
Resolver Applications

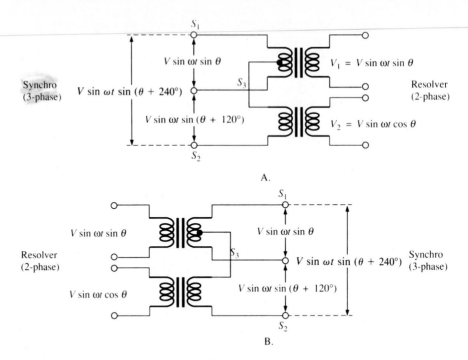

FIGURE 10–18
Synchro-Resolver Signal Conversion

resolver is a two-phase system. In Figure 10–18*A,* the three synchro signals are sent to S_1, S_2, and S_3. When the output of this circuit is sent to an appropriately connected resolver, the resolver will assume the same position angle as the synchro. To go from a resolver to a synchro, the circuit in Figure 10–18*B* would be used for direct angle correlation.

10–4   SERVOMOTORS—DC AND AC

Many control applications require high torque capabilities. These applications require both precise speed and precise position control at high torque. The devices discussed so far in this chapter do not have high torque capabilities. For example, as has been stated, a large robot could not be driven by steppers, synchros, or resolvers. High-torque applications usually use AC or DC servomotors. Some actual control systems for servomotors will be covered in Chapters 13 and 14. In this section, only the drive motors themselves will be covered.

Servomotors differ in application capabilities from industrial motors in the following ways:

☐ The servomotor must produce high torque at all speeds.

☐ Servomotors must be capable of holding a static (no-motion) position.

☐ At lower speeds or at standstill, servomotors must not overheat.

☐ Servomotors should be able to reverse directions quickly.

☐ To reach a position or rate of speed quickly, the servomotor must be able to accelerate and decelerate quickly.

☐ Servomotors must return to a given position time after time and not drift.

Both DC and AC servomotors look like AC and DC industrial motors. The main difference from industrial motors is that more wires may come out of the motor, for power and also for control. The servomotor wires go to a controller, not to the electrical line through contactors. Also, a speed-indicating device (often a tachometer) is mechanically connected to the motor shaft. Sometimes special blower or fan units may be attached to assure motor cooling at low speeds.

The DC servomotor is similar to a regular DC motor. As a matter of fact, any standard DC commutator motor may be used as a servomotor with proper control. In actual practice, any DC commutator motor has a modified design to cope with slow speeds and static conditions. However, DC servo systems have the disadvantage of system drift in their controls, so they are not used as often as AC.

An AC servomotor looks like a regular industrial induction motor externally. The most common AC servomotor is constructed internally in a manner similar to that of the AC induction motor. Some major differences between AC servomotors and standard motors are as follows:

☐ The start/auxiliary winding is the control winding in the servomotor.

☐ The servomotor rotor bars are smaller and have higher resistance.

☐ The servomotor can stay energized at zero speed (that is, when it is stalled) without overheating.

☐ The servomotor control winding is controlled by an electronic controller.

Figure 10–19 illustrates the electrical layout of a typical two-phase, two-pole servomotor. As with industrial motors, the AC servomotor can have four, six, or more poles and can be constructed to use three-phase power. The rotational speed and torque of the rotor are controlled by the phase difference between main and control windings. Reversing the phase difference from leading to lagging, or vice versa, reverses the motor direction.

Included in Figure 10–19 are some waveforms that will produce certain motions. Normally, V_S and V_C are equal in amplitude. For maximum torque, waveforms are 90° apart. For lesser speeds, the waveforms are less than 90° apart. For no motion, the waveforms are perfectly in phase.

10–5   OTHER SPECIAL MOTORS

In this last section of Chapter 10, seven types of motors will be discussed. The seven motors are typical of special motors not previously covered. They are given in the following list:

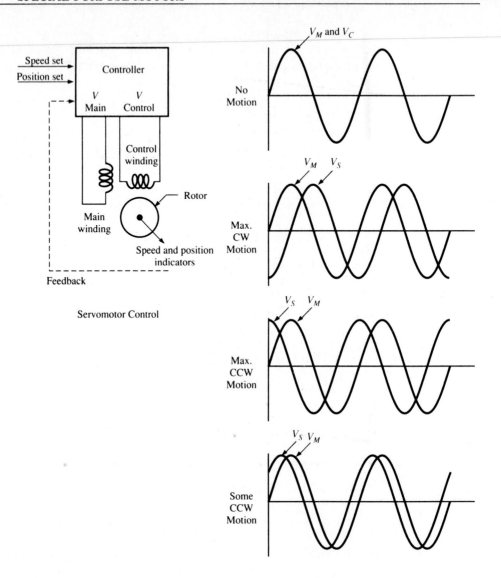

Servomotor Control

Servo Control Waveforms

FIGURE 10-19
Two-Phase, Two-Pole AC Servomotor Layout

FIGURE 10–20
Disk Rotor Motor (Courtesy of PMI Motion Technologies, division of Kollmorgen Corp.)

1. disk rotor motor
2. Hall-effect motor
3. hysteresis motor
4. reluctance motor
5. linear motor
6. torque motor
7. inside-out motor

The *disk rotor motor* is also called a *moving coil motor*. An exploded view of this type of motor is shown in Figure 10–20. The round, short cylinders in the end bells are permanent magnets. The permanent magnets are of alternating north–south polarity. The disk rotor is fed electrical current through a commutator. The rotor has current paths much like those on a double-sided PC board. The rotor therefore functions in the disk motor the same way as in a DC motor or a universal AC motor. The disk motor differs from others in that the rotor has a very low inertia. The ratio of motor torque to inertia is very high. This type of motor is used where rapid acceleration, deceleration, and reversing are essential.

The *Hall-effect motor* is a low-power, reliable, efficient DC motor. Its operation is based on the Hall effect. The Hall effect takes place when a magnetic field and a current are applied to a special semiconductor material or a crystal. The magnetic field and the applied current produce a voltage across the semiconductor material. This resulting DC voltage is proportional to the product of the field and the current. In AC operation, voltage is proportional to field times current times the cosine of the angle between field and current.

For the DC Hall motor, the Hall effect is used as shown in Figure 10–21. Two Hall devices, HG_{11} and HG_{12}, are supplied with control current, I_{C11} and I_{C12}.

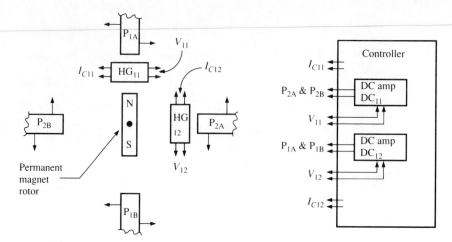

FIGURE 10–21
Hall Motor Principle

The control current is constant. The Hall devices sense magnetic field changes due to changes in the rotor's rotary position. The Hall device does not sense the field due to the pole magnetic strengths of P_{1A}, P_{1B}, P_{2A}, and P_{2B}. In the rotor position shown in the figure, HG_{11} is subject to maximum rotor magnetic effect. The large magnetic field causes a maximum V_{11}. That V_{11} is fed to DC amplifier DC_{11}. The output of DC_{11} goes to maximum. The output of DC_{11} is fed to motor field poles P_{2A} and P_{2B}. The poles are connected so that they have a north pole for this situation. The rotor therefore turns clockwise to align with P_{2A} and P_{2B}.

Note that, because of minimum reaction of HG_{12} with the rotor, P_{1A} and P_{1B} are minimum when the rotor is aligned as shown. When the rotor rotates 90°, it affects HG_{12}. Field poles P_{1A} and P_{1B} are then maximum strength, while P_{1A} and P_{1B} are minimum. The process continues 90° later, with the rotor south pole near HG_{11}. Hall device HG_{11} puts out a voltage of the opposite polarity from before. The DC amplifier, DC_{11}, puts out maximum reverse polarity to P_{2A} and P_{2B}. The south pole then continues clockwise. The process continues, and the rotor keeps turning. Hall motors run at synchronous speed, as the poles must line up at voltage peaks.

The *hysteresis motor* layout is shown in Figure 10–22. The hysteresis motor is essentially an induction motor that runs at synchronous speed and has a special rotor. Its stator may be of any type that produces a rotating magnetic field. It is a low-power, low-efficiency motor. It is used for smooth-start, synchronous applications, such as timers.

The rotor consists of a core of solid, nonmagnetic material. An outer layer of magnetic material is added as shown. On small motors the rotor is solid. On larger motors the outer layer may be laminated to reduce eddy current and heat loss. The principle of operation of this motor is the delay of buildup of rotor magnetic field

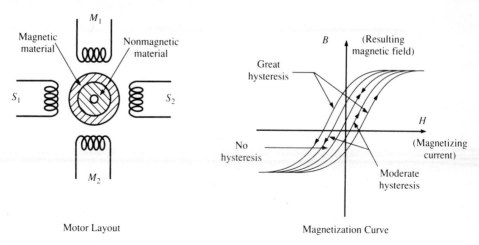

Motor Layout   Magnetization Curve

FIGURE 10–22
Hysteresis Motor Principle

caused by the hysteresis effect. The material chosen for the rotor surface must
have a high hysteresis characteristic. Alloys containing cobalt or vanadium have
this high hysteresis characteristic and are usually used.

Refer to the magnetization curve in Figure 10–22. The outer rotor material
for the hysteresis motor has the great-hysteresis curve. If the material had no
hysteresis, it would follow the inner, no-hysteresis curve. The field buildup would
be in phase with the stator field and current. The motor would not work properly.
The great-hysteresis material works better for this motor.

The hysteresis motor starts as a normal induction motor. The rotor-induced
flux lags the stator flux by an angle corresponding to the hysteresis curve. This lag
produces the starting and running torque. The motor will accelerate to a full speed
equal to synchronous speed.

The *reluctance motor* is another motor that runs at synchronous speed. Its
principle of operation is shown in Figure 10–23. Its advantage over a conventional
synchronous motor is that it has no windings in the rotor. The lack of windings
eliminates the need for slip rings, brushes, and a DC supply for the rotor. Eliminating those elements reduces maintenance. However, the reluctance motor is a
lower-torque, low-power-factor, and low-efficiency motor.

The reluctance motor rotor has indentations between poles, as illustrated.
The rotor has copper bars placed axially in the poles as shown. Each group of bars
is connected at each end. The rotor has different magnetic reluctances at different
rotor angles. At 0° the path has least reluctance, allowing a maximum flux pathway. At 45° there is less magnetic pathway, the reluctance is highest, and the flux
path is minimum. This relationship is shown by the graph in the figure. The stator
can be any type that produces a rotating magnetic field. The reluctance motor
starts as an induction motor does. When it gets almost to full speed, it locks in at

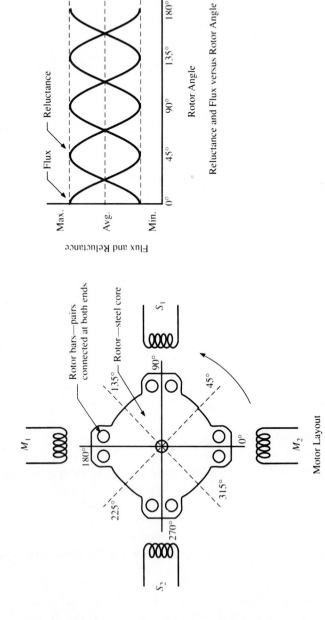

FIGURE 10–23
Reluctance Motor Principle

synchronous speed. This lock-in effect is the result of the rotor's lining up sequentially during rotation with the lowest-reluctance path.

Linear motors are essentially variable-position solenoids. Figure 10–24 illustrates four types of linear motors. Figure 10–24A is a *one-way linear motor* used as part of an indicating meter. As current flows in the coil, the magnetic force attracts or repels the permanent-magnet cup. As the magnetic cup moves, the dial pointer moves accordingly. The dial can be calibrated for the possible range of the input electrical value.

A *two-way linear motor* is shown in Figure 10–24B. A moving member moves back and forth according to the magnetic strengths and polarities of the two coils. The moving member may alternatively have a coil-induced field, instead of a permanent magnet. A typical application for this type of linear motor is to move the pen in a chart recorder. There is no mechanical linkage, so the pen has a fast response to changes of input.

The *continuous linear motor* is, in effect, an uncoiled induction motor. Two arrangements are illustrated, in Figure 10–24C and D. The advantage of these

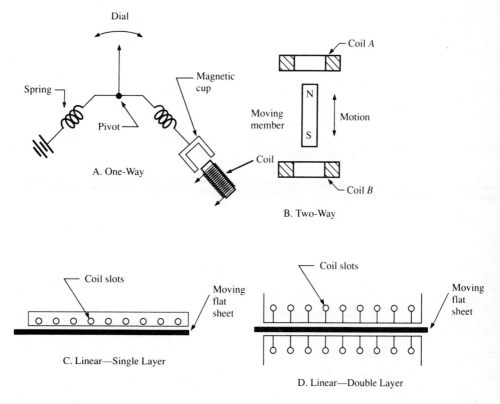

FIGURE 10–24
Linear Motor Principle

FIGURE 10–25
Torque Motor Principle

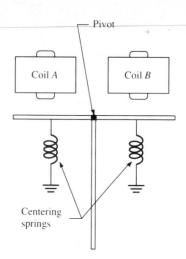

motors is simplicity. They do have a limited range of movement. One problem to watch for is that they can be nonlinear at the ends of travel. Also, a special control system is needed, since 60 cycle current has too fast a change for good operation. Continuous linear motors work on the principle of magnetizing a flat sheet of magnetic material by energizing one coil. When the next coil is appropriately energized, the sheet moves to line up with the new magnetic field. Coils are energized alternately or sequentially to produce the motion.

The *torque motor* is a special type of linear motor. Its principle is illustrated in Figure 10–25. The torque motor is typically the size of a fist. Torque motors are quite expensive. The movement is in thousandths of an inch. The advantage of these motors is extreme linearity between input signal and mechanical displacement. A typical application is in a hydraulic servovalve. The torque motor motion is used to vary the flow of a pilot hydraulic valve. The hydraulic valve, in turn, drives a larger, control hydraulic valve. In the servovalve operation, motion is extremely linear with respect to input signal—within the operating range. Also, the torque motor has a relatively high frequency response and can follow rapidly varying input signals.

The *inside-out motor* is not illustrated but does deserve a mention in this chapter. The part normally designated "rotor" is stationary. Coils are wound on this stationary member, and power leads are brought out from it. The part normally designated "stator" is the part that rotates. The outer, rotating member is made of permanent-magnet material. The inner, stationary member is fed AC or DC in pulses to create a revolving magnetic field. The inside-out motor is found in some types of office machines in special applications.

SUMMARY

☐ Stepper motors are used for reliable device positioning. For absolute reliability of positioning, a feedback system is needed.

☐ Stepper motors are available in many torque ratings and step angles.

☐ Stepper motors are usually controlled by IC chips.

☐ Microstepping of stepper motors (the creation of more increments of stepping) can be accomplished by the use of special, more expensive control systems.

☐ Synchros are low-torque, angular following devices.

☐ Synchros can be used for adding, subtracting, and other positional functions.

☐ Resolvers are also low-torque position indicators.

☐ Resolvers produce an output voltage that is related to angular position.

☐ Synchros and resolvers can be used in tandem.

☐ A DC servomotor is a modified DC motor that is capable of holding a static position.

☐ The operation of an AC servomotor depends on the phase difference between two or more signals sent to different sets of motor windings.

☐ Disk motors are low-inertia, high-torque motors.

☐ Hall motors are small synchronous motors requiring special controls.

☐ Hysteresis motors are small AC motors that run at synchronous speed.

☐ Reluctance motors are synchronous motors that require no power to the rotor.

☐ Linear motors are, in effect, continuous positional solenoids.

☐ Linear motors can be one-way, two-way, or continuous.

☐ Torque motors are high-cost, precise linear motors in which mechanical output is almost perfectly proportional to input signal.

☐ Inside-out motors have a stationary inner member that is wound and a permanent-magnet outer shell that rotates.

EXERCISES

1. A 3.75°/step stepper motor starts at 78.75°. Full-step pulses are applied for 102 steps CW and then for 27 steps CCW. At what angle does the motor end up?

2. A 400 steps per revolution stepper motor starts at −45.36°. Full-step pulses are applied for 67 steps CW (the positive direction) to get to angle A. Then, 42 half steps are applied CW to get to angle B. Finally, 206 half steps are applied CCW to arrive at angle C. What are the values of angles A, B, and C?

3. For Exercises 1 and 2, the tolerance of the steppers is 4%. How far off could they be at their final position?

4. Figure 10–5 is a plot of the step pattern for Figure 10–4A. Plot a step pattern like Figure 10–5 for Figure 10–4B, half steps.

5. Refer to Figure 10–4. Draw a similar layout for an 8-pole stator/12-pole rotor, stepper motor.

6. Chart the full-step pattern like Figure 10–4 for the stepper of Exercise 5, CW.

7. Same as Exercise 6, CCW.

8. Extra credit problem: Plot the CW, half-step pattern for the stepper of Exercise 5, CW.

9. Explain the reason for not starting and stopping a stepper motor at full stepping rate. If the full stepping rate were one pulse per second, would slower starting and stopping be needed?

10. Explain mechanical rotary-load derating for stepper motors.

11. Refer to Figure 10–8. The 90° span is divided into 10 steps, for 3.5 V maximum. Make up a chart similar to the one in the figure for this pattern.

12. Refer to Figure 10–9. Sketch the amplitude modulation pattern for the stepper motor of Exercise 11.

13. Refer to Figure 10–9. Sketch the PWM pattern for the stepper motor of Exercise 11 at 18° and at 63°. (*Note:* Again use 3.5 V.)

14. Refer to Figure 10–10. Briefly explain the function of each class of synchro.

15. Refer to Figure 10–11. Could you use four windings 90° apart in each unit instead of the three normally used? Why or why not?

16. In Figure 10–12, what would happen in each of the following situations:
    a. L_1 and L_2 reversed to TR only
    b. L_1 and L_2 reversed for both TX and TR
    c. S_1 and S_2 connections reversed on TX only

17. Refer to Figure 10–13. If TX is set at 45°, not 0°, how will the output graph change?

18. Refer to Figure 10–14. Can these units be made to multiply and divide? Why or why not?

19. Refer to the navigation example in Figure 10–17. The scale is 0.125 V per mile. Angle θ is found to be 41.3°. Voltage D is found to be 62.4 V. What is R in volts and miles?

20. Refer to the height example in Figure 10–17. The scale is 3.25 V per 100 ft. Angle θ is determined to be 62.1°. Voltage D is found to be 5.45 V. What is the height in volts and feet?

21. In Exercise 20, what is the distance to the base of the structure in volts and feet?

22. In your own words, what would be the difference between a DC servomotor and a conventional industrial DC motor?

23. Repeat Exercise 22 for AC.

24. Refer to Figure 10–19. What would the servomotor control diagram look like for a three-phase AC servomotor?

25. Refer to the listing of the seven types of motors discussed in Section 10–5. Tell whether and how well each would work on DC, single-phase AC, and three-phase AC.

26. Refer to Figure 10–21. How would an 8-pole Hall motor be laid out functionally? Would it work?

27. Refer to Figure 10–22. How would a three-phase hysteresis motor be laid out functionally? Would it work?

28. Refer to Figure 10–23. How would a three-phase reluctance motor be laid out functionally? Would it work?

29. Explain the linear motor principles. Give two possible applications of linear motors.

30. Explain how a torque motor works. Give two possible applications for a torque motor.

APPLICATION ASSIGNMENT ANSWER

Questions to ask in determining the specifications for the XY drive motors include the following:

1. How much does the driven device weigh?
2. What speeds do you need?
3. What precision do you need for the positioning of the driven device?
4. What is the voltage type of the input interface?

You would probably choose two stepper-motor drives with appropriate IC controls. The small load is adaptable to stepper motors. The specific stepper chosen would match the specifications. Servos would be too big and expensive. You do not need the larger torques developed by servos. Synchros and resolvers are not appropriate for this application, since the XY table requires a large-scale linear output.

11

Electrical Power-Control Devices

CHAPTER OUTLINE

11–1 Introduction □ **11–2** Enclosures □ **11–3** The Electrical Magnetic Contactor □ **11–4** Tandem Contactor Applications □ **11–5** DC Contactors □ **11–6** Low-Voltage Starting Equipment for Motors □ **11–7** Drum and Centrifugal Devices □ **11–8** Controllers for Large Synchronous and Wound-Rotor Motors

LEARNING OBJECTIVES

□ List the types of electrical controller enclosures available, and tell where they are used.
□ Describe the AC across-the-line starter.
□ Show how contactors are used in tandem for such applications as forward-reverse control.
□ Describe the DC contactor.
□ Describe the motor-control-system components for each of the following types of AC reduced-voltage motor control: primary-resistor, transformer, wye-delta, and part-winding start.
□ Describe the connections for drum switch control.
□ Describe motor-speed-indicating centrifugal devices.
□ Describe the motor-control-system components for large synchronous and wound-rotor motors.

APPLICATION ASSIGNMENT

Your manufacturing plant has received a piece of preowned process equipment. The equipment is driven by a motor whose control requirements you are to analyze. You have checked the motor nameplate for horsepower, voltage, and other characteristics. The nameplate says the motor is a 25 HP, three-phase, 1725 RPM, 440 V motor. You have disassembled the motor to verify that its characteristics match the nameplate. You are assigned the task of determining the size and type of motor starting equipment for the motor. What are some of the factors in this equipment determination?

11–1   INTRODUCTION

This chapter deals with the hardware used in electrical power and motor control. It also deals with the basic internal wiring of such devices. Chapter 12 will deal with the detailed control wiring of these devices.

The types of enclosures available will be discussed first. Enclosures must be chosen that will protect the electrical equipment from its environment.

The current to higher-power devices is not controllable directly by a switch. A special form of relay called the *contactor* is used. Contactors are used for the following reasons:

1. The current to be controlled is too high for a low-current switch to turn ON and OFF.
2. The high-load current needs to be controlled remotely from one or more switching locations.
3. The power voltage is high—440 V or 4160 V, for example. High voltage is not safe for a small switch in proximity to people.
4. Switches and relay contacts do not handle arcing well. Many large industrial devices are highly inductive. When the circuit is opened for switching for those inductive devices, arcing occurs at the switch contacts. Similar arcing takes place with large capacitive electrical loads. To handle arcing, contactors have enclosed contacts. Some large contactors also have special attachments to extinguish arcing.

The basic *across-the-line contactor* will be discussed. The across-the-line AC contactor is used for all kinds of electrical devices. The use of two contactors together will then be covered. Forward-reverse control uses a combination of two interlocked contactors. Direct-current contactors, which are quite similar to AC contactors, will be discussed. Special types of DC contactors will be illustrated.

Low-voltage AC and DC motor starters will be shown and their operation explained. The types of reduced-voltage starters to be covered include *primary-resistor, transformer, wye-delta*, and *part-winding*.

Drum and centrifugal power-control devices will also be covered. Finally, the power-control devices for wound-rotor and synchronous motors will be shown.

11–2   ENCLOSURES

Figure 11–1 shows some types of electrical equipment enclosures. Enclosed electrical equipment operates in different environments. More severe environments require more substantial enclosures. Electrical codes mandate the type of enclosure to use. Enclosures for more severe environments are more costly. Therefore, the enclosure is matched to its required performance.

There are two general types of enclosures: nonhazardous-location and hazardous-location. Nonhazardous-location enclosures are further subdivided into

NEMA 12 Enclosure

NEMA 7 – Class I Groups C and D
Hazardous Location Enclosure

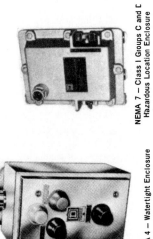

NEMA 4 – Watertight Enclosure

NEMA 1 – General Purpose Enclosure

NEMA 13 – Oiltight, Dusttight

NEMA 9 – Class II Groups E, F, and G
Hazardous Location Enclosure

NEMA 4X – Watertight, Corrosion-Resistant Enclosure

NEMA 3 – Dusttight-Raintight Enclosure

FIGURE 11-1
Electrical Equipment Enclosures (Courtesy of Square D Company)

FIGURE 11–2

Classification of Hazardous Atmospheres (based on NEC and UL)

Class	Division	Group	Typical Atmosphere/Ignition Temps.	Devices Covered	Temperature Measured	Limiting Value
I Gases, vapors	1 Normally hazardous	A	Acetylene (300C, 571F)	All electrical devices and wiring	Maximum exterior temperature in 40C ambient	By group: A–280C (536F) B–280C (536F) C–180C (356F) D–280C (536F)
		B	Hydrogen (585C, 1085F), manufactured gas, and equivalent			
		C	Ethyl-ether vapors (180C, 356F), ethylene (450C, 842F), cyclopropane (498C, 928F)			
		D	Gasoline (280C, 536F), hexane (261C, 502F), naptha (232C, 450F), benzine (288C, 550F), butane (405C, 761F), propane (466C, 871F), alcohol (343-428C, 650-900F), acetone (538C, 1000F), benzol (526C, 1044F), lacquer solvents, natural gas			
	2 Not normally hazardous	A B C D	Same as Division 1 Same as Division 1 Same as Division 1 Same as Division 1 (Not normally hazardous means that the gases aren't normally present.)	Lamps, resistors, coils, etc., other than arcing devices (see Div. 1)	Max. inside temp. not to exceed 80% of ignition, T_i	80% T_i = A–240C (464F) B–468C (874F) C–144C (291F) D–224C (435F) (or depending on gas)
II Combustible dusts	1 Normally hazardous	E	Metal dust, including aluminum, magnesium, and their commercial alloys, and other metals of similarly hazardous characteristics	Devices not subject to overloads (switches, meters)	Max. exterior temp. in 40C ambient with a dust blanket	No overload: E–200C (392F) F–200C (392F) G–165C(392F)
		F	Carbon black, coal, coke dust	Devices subject to overload (motors, transformers)		Possible overload: E, F, G– 120C (248F) but not to exceed no-overload values at overload
		G	Flour, starch, grain dusts			
	2 Not normally hazardous	G	Same as Division 1	Lighting fixtures	Max. exterior temp. under conditions of use	165C (329F)
III Easily ignitible fibers and flyings	1, 2	G	Same as Class II, Division 1	Lighting fixtures	Same as Class II, Div. 2	165C (329F)

the following categories: general-purpose, watertight, oiltight, and dusttight. The general-purpose nonhazardous-location enclosure is least costly. Hazardous-location enclosures are very costly, but necessary in some applications. Hazardous-location, explosion-proof enclosures involve forged or cast material and special seals with precision-fit tolerances. The explosion-proof enclosures are constructed so that an explosion inside will not escape the enclosure. If an internal explosion were to blow open the enclosure, a general-area explosion and fire could ensue. Figure 11–2 is a table describing the environments in which the various hazardous-location enclosures are required, by code, for operational safety.

11–3   THE ELECTRICAL MAGNETIC CONTACTOR

Electrical magnetic contactors are used to turn a large variety of electrical loads ON and OFF. These contactors are a major component in the across-the-line magnetic starter. The contactor's coil is energized electrically to close two or more contacts. Closure is by solenoid axial action. Power is applied to the load or motor when the contacts are closed. By definition, a contactor differs from a motor starter in that the motor starter has added overload current relays in each power line. A motor starter may also have an added disconnect switch in the same enclosure. The combined disconnect switches are placed electrically ahead of the contactor in the starter. The disconnect switches may also include fuses in each line. Figure 11–3 shows a typical magnetic contactor.

FIGURE 11–3
Magnetic Contactor (Courtesy of Allen-Bradley Company)

WIRING DIAGRAM   **ELEMENTARY DIAGRAM**

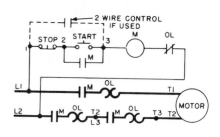

A. Single-Phase

WIRING DIAGRAM   **ELEMENTARY DIAGRAM**

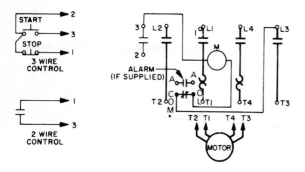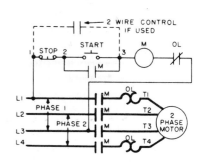

B. Two-Phase

WIRING DIAGRAM   **ELEMENTARY DIAGRAM**

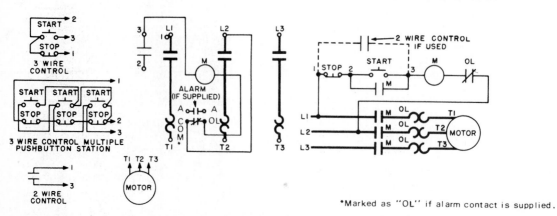

*Marked as "OL" if alarm contact is supplied.

C. Three-Phase

FIGURE 11–4

Across-the-Line Magnetic Starter Diagrams (Courtesy of Square D Company)

Figure 11-4 shows the wiring and elementary (line) diagrams for three types of across-the-line motor starters. Each starter includes a contactor, overload relays, and internal wiring. Figure 11-4A is for a two-wire, single-phase AC motor control. Figure 11-4A would also apply to small DC motors. Figure 11-4B is for a two-phase, four-wire AC motor. Note that the two-phase motor is rare in the industrial world. Figure 11-4C is for a three-phase, unidirectional AC motor starter.

Each of these three starter circuits uses a standard start/stop-seal control circuit. Pushing the start button starts the motor. A contactor auxiliary contact in parallel with the start button closes when the main contacts close. The parallel path seals the circuit ON. When the start button is released, the motor continues to run. Pushing the stop button opens the motor starter coil, unsealing the circuit and stopping the motor. If the motor is overloaded, the overload contact will open, turning the motor OFF. The motor cannot be restarted until the overload condition has been cleared and the overload has been manually reset. This start/stop-seal control circuit will be discussed in more detail in Chapter 12.

Note that in Figure 11-4 the line voltage is used as the control voltage for all three examples. In some starters the control voltage is supplied separately, through a transformer and an individually fused circuit. The control voltage can be different from the line voltage in those cases.

11-4   TANDEM CONTACTOR APPLICATIONS

Tandem applications of contactors use two adjacent contactors. Typical applications are forward-reverse starters and two-speed starters.

Forward-reverse control of electric motors is normally applied to three-phase induction motors. A basic system for such control is shown in Figure 11-5. There are two separate contactors for the controller. One set of contacts is closed for forward and another set is closed for reverse. To reverse the direction of rotation of the three-phase motor, two incoming power lines to the motor are interchanged. This reversal principle was explained in Chapter 9. In the figure, T_1 and T_3 are interchanged to accomplish reversing.

This forward-reverse control is applicable to devices other than motors—for example, a bidirectional hydraulic cylinder.

Note that it is imperative that there be no way for both the forward contacts and the reversing contacts to be closed at the same time—even for an instant. Closing both, in the system of Figure 11-5, would short L_1 and L_3, causing a direct short in the incoming power. The forward contacts must open before the reversing contacts are closed, and vice versa. This circuit is called break-before-make (BBM) in relay-logic terms, as explained in Chapter 2. For many larger controllers, the F and R coils are also mechanically interlocked by a pivot bar. The pivot bar is like a miniature teeter-totter. Each end of the pivot bar is attached to one of the two contactor plungers. Only one end of the pivot bar can be down (ON) at a time.

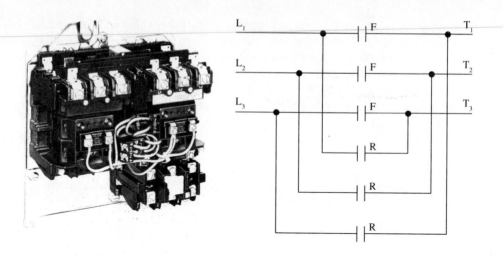

FIGURE 11–5
Forward-Reverse Magnetic Starter (Courtesy of Allen-Bradley Company)

The operation of the two-speed-motor starter is similar to that of the forward-reverse starter. The appearance of the two-speed starter is the same as for the forward-reverse starter of Figure 11–5. A two-speed-motor starter circuit for a single-phase (two-pole) motor and a three-phase (three-pole) motor is shown in Figure 11–6. One or the other contactor is closed, depending on the speed chosen for operation. The two contactors must be interlocked to prevent motor damage. The single-phase two-speed motor shown has one common lead and one other lead for each speed, as shown. The three-phase two-speed motor has two sets of three leads. One set of leads is for one speed, and the other set is for the other speed. There is a start button for each speed, and there is a master stop button. The high-speed and low-speed coils are electrically interlocked by NC contacts from the opposite coil. The CR relay and its circuit are included to make sure that the motor is started initially at low speed for this motor application.

FIGURE 11–6
Two-Speed Magnetic Starter

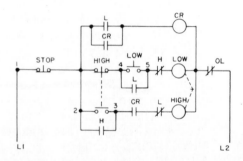

A few motors are capable of running at three or more speeds. For these motors, three or more interlocked contactors are used for motor control.

11–5 DC CONTACTORS

Direct-current contactors are similar in construction to AC contactors. A DC contactor is shown in Figure 11–7. The contacts are enclosed to confine the arc, as with AC contactors. The coil can be chosen to operate on DC or AC at a series of available voltages. A DC contactor can be single-pole or double-pole. Control systems of a DC motor often include forward-reverse tandem contactors, just as AC controllers do. Often another contactor is used for dynamic braking, as will be explained in Section 12–5. In addition, many DC motors have separate field power-control contactors. Further, DC motors with stepped voltage values for starting will have one or more additional contactors on the control panel.

As previously stated, an AC contactor and a DC contactor of the same rating are not interchangeable. Their design is different. As with AC contactors, the environment must be considered in choosing a DC contactor and enclosure.

FIGURE 11–7
DC Contactor (Courtesy of Square D Company)

11–6   LOW-VOLTAGE STARTING EQUIPMENT FOR MOTORS

Some typical three-phase motor starters are shown in Figure 11–8. In this section equipment components used for reducing motor voltage at start will be described. The components used depend on the type of reduced-voltage start employed.

The five major types of reduced-voltage start are as follows:

electronic

primary-resistor

FIGURE 11–8
Three-Phase Motor Starters (Courtesy of Eaton Corp./Cutler Hammer® Products)

transformer

wye-delta

part-winding

Electronic-control starting equipment will be covered in Chapter 12. For the other four types of reduced-voltage control, one contactor is needed for each direction of rotation. As stated previously, for forward and reverse two contactors are needed. One or more additional contactors are required for low-voltage starters. Also, one timer is normally required for each step of starting. The timers are set at appropriate times depending on the acceleration characteristics of the motor.

Primary-resistor control requires three resistors for three-phase AC or one resistor for DC. The resistors are in the circuit during starting but are bypassed during running. The resistors are low-value, high-wattage types used for a short duty cycle. Duty cycle is the percentage of time a device is on, which is small for these restrictors. The resistors can therefore be smaller in size than if they were ON continuously. For more steps of starting, more resistors are needed for AC, or tapped resistors for DC. The resistors are required in multiples of the number of steps. For example, three steps of low-voltage starting require nine resistors for three-phase AC. Three steps for DC require a full resistor with two intermediate taps. Each step of starting requires an additional contactor to short out the resistors or to change taps as starting proceeds.

Motor starters, whether across-the-line or reduced-voltage, come in various sizes. For larger motors with greater current requirements, physically larger contactors and associated equipment are required for control. Of course, starters become more costly as their size increases. Figure 11–9 illustrates the minimum standard starter size required for motors of various sizes and voltages. For example, according to the table, for a 50 HP three-phase AC motor rated at 230 V, at least a size 4 starter, which is rated at 135 continuous amps, is required. The table in Figure 11–9 covers only basic AC starters. For DC, multiple-function, and special motors, there are other tables. More detailed information on starters is given in various manufacturer and industry standard publications. As previously stated, as the starter size goes up, the starter cost also rises. Therefore, it is not normal practice to use a starter that is larger than necessary.

Figure 11–10 shows the circuitry for control of AC and DC motors for reduced-voltage primary-resistor starting. Figure 11–10A shows the control for AC. When start is depressed, contactor M contacts close, and M is sealed ON. Motor power current goes through the three resistors, lowering motor terminal voltage. Timer TR starts at the same time. After it times out, its contact closes contactor A. The resistors are then bypassed, and full voltage is applied to the motor.

The DC primary-resistor control system, shown in Figure 11–10B, is similar to the AC system. At start, contactor M closes and starts TR_1. Power to the motor is reduced by a voltage drop across the entire resistor. After TR_1 times out, it closes coil $1A$. This reduces the amount of resistance and therefore increases motor terminal voltage. When TR_1 times out, it also starts TR_2. When TR_2 times

FIGURE 11–9

Motor Starter Size for Horsepower and
Voltage Rating

3 Phase, 3 Pole

Max Hp

200 Volts	230 Volts	460/575 Volts	Continuous Amp Rating	Starter Size
1½	1½	2	9	00
3	3	5	18	0
7½	7½	10	27	1
10	10	15	40	1¾
10	15	25	45	2
15	20	30	60	2½
25	30	50	90	3
40	40	75	115	3½
40	50	100	135	4
50	75	150	210	4½
75	100	200	270	5
150	200	400	540	6

Single Phase, 2 Pole

Max Hp

115 Volts	230 Volts	Continuous Amp Rating	Starter Size
⅓	1	9	00
1	2	18	0
2	3	27	1
3	5	35	1P
3	7½	45	2
5	10	60	2½

out, its contact closes contactor 2*A*. The 2*A* contacts completely bypass the start resistor. Full voltage is then applied to the motor. This process is two-step starting, in contrast to the single-step AC starting previously shown.

Transformer low-voltage starting requires two transformers (not three) for three-phase. Obviously, DC motors cannot use transformer starting. The starting-voltage-reducing transformers need a number of taps equal to the number of steps of starting. Autotransformers are used, as they function properly and cost less than the tapped primary/secondary type. As with resistor starting, an additional contactor is required for each step of starting. Figure 11–11 shows a transformer reduced-voltage three-phase AC starter.

The wye-delta starter (Figure 11–12) has the advantage of needing no resistors or transformers. This reduces cost and maintenance. Wye-delta uses the three-phase principle for starting. The three-phase principle will be discussed in detail in Chapter 16. The output voltage in delta is 1.732 (the square root of 3) times the wye output voltage. Starting in wye provides reduced voltage. For example, a 440 V motor has 254 V at starting in wye. After time for starting, the full 440 V is applied in delta. Since current is also reduced by the same factor (1/1.732 = 0.578), power at start is one third (0.578 × 0.578) of running power. This

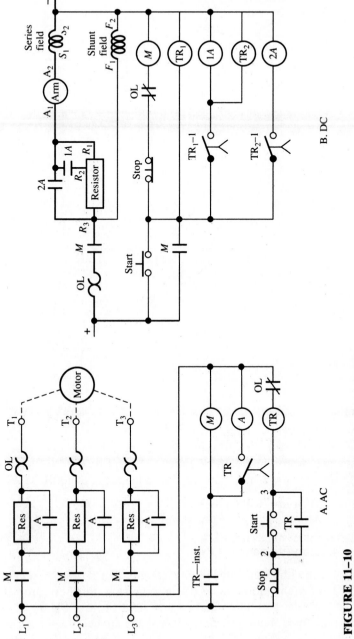

FIGURE 11–10
AC and DC Low-Voltage Primary-Resistor Starting

357

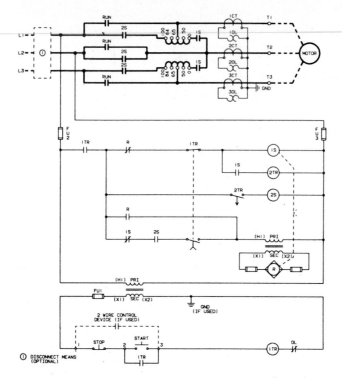

FIGURE 11–11
Three-Phase AC Transformer Motor Starter

type of starting does require one extra contactor for operation in switching the power from wye to delta.

In Figure 11–12 two types of starting systems, open-transition and closed-transition, were illustrated. Figure 11–13 will be used to help explain the difference. In open-transition starting the sequence is as follows:

1. Turn low voltage ON.
2. Allow a period of time to elapse.
3. Turn low voltage OFF.
4. Turn full voltage ON.

A problem with open-transition starting is that during the elapsed time between steps 3 and 4, the motor slows down somewhat. Furthermore, when full voltage is applied, a relatively large current transient occurs. Figure 11–13A shows some open-transition characteristics: the voltage applied versus time and the resulting current values versus time. Note the high current when full voltage is applied. The high current is due to the intermediate slowing down of the motor.

Closed transition, on the other hand, gives smoother starting (at extra cost). During the intermediate switching period, voltage continues to be applied. The

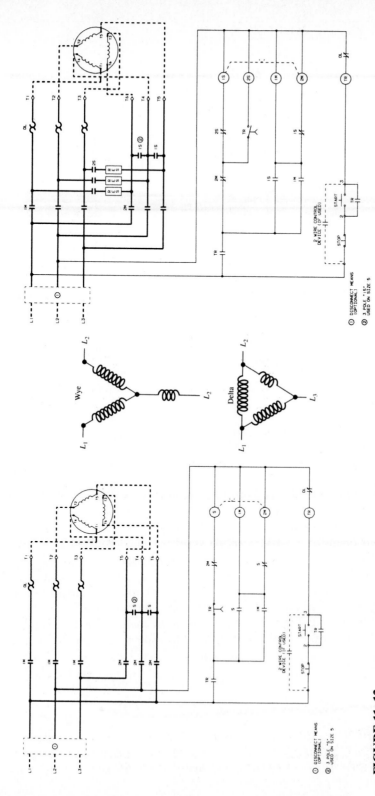

FIGURE 11-12
Wye-Delta Reduced-Voltage Starting

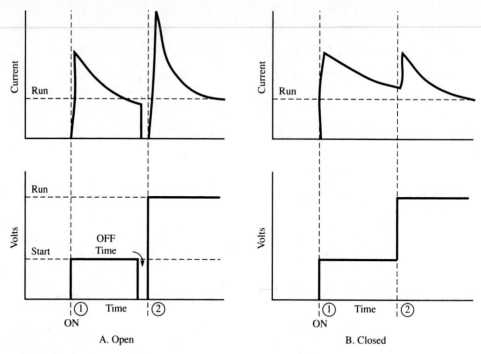

A. Open   B. Closed

FIGURE 11–13
Open- Versus Closed-Transition Starting

current transient value is reduced considerably, as shown in Figure 11–13*B*. Furthermore, the motor does not slow during switching for closed transition.

The motor application and performance requirements will determine whether open- or closed-transition starting should be used.

The part-winding starter (Figure 11–14) requires two contactors for unidirectional motors and four for forward-and-reverse motors. One contactor is for starting and the other is for running. The start contactor energizes leads from the motor, which use only part of the motor windings. When the motor nears full speed, the main contactor comes in also. The main contactor energizes additional motor leads for full-speed operation. At full speed all the motor windings are then utilized.

11–7   DRUM AND CENTRIFUGAL DEVICES

Drum switches are used in a variety of control applications. Drum switches are multiple-pole, multiple-position switches. They may have up to 16 poles, with 5 or even 7 throws. Drum switches are high-current devices.

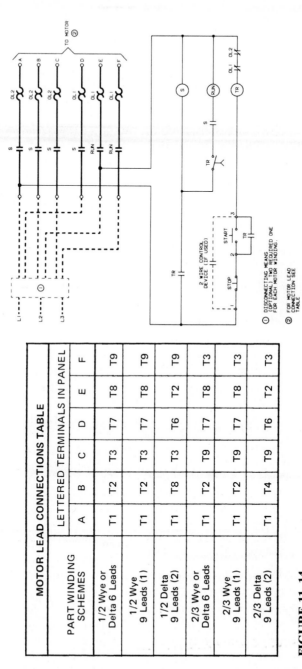

MOTOR LEAD CONNECTIONS TABLE

PART WINDING SCHEMES	LETTERED TERMINALS IN PANEL					
	A	B	C	D	E	F
1/2 Wye or Delta 6 Leads	T1	T2	T3	T7	T8	T9
1/2 Wye 9 Leads (1)	T1	T2	T3	T7	T8	T9
1/2 Delta 9 Leads (2)	T1	T8	T3	T6	T2	T9
2/3 Wye or Delta 6 Leads	T1	T2	T9	T7	T8	T3
2/3 Wye 9 Leads (1)	T1	T2	T9	T7	T8	T3
2/3 Delta 9 Leads (2)	T1	T4	T9	T6	T2	T3

FIGURE 11–14
Part-Winding Reduced-Voltage Starter

A typical drum switch is shown in Figure 11–15. This drum switch is a 3-pole, 3-position switch. Its switching pattern is shown in "Internal Switching." The pattern can take many connection forms. Note that the switching pattern is not like that of a 3-pole triple-throw toggle switch.

To illustrate the switch function, five examples for motor reversing are shown in Figure 11–15. To reverse electric motors, one or more motor wire pairs from the motor must be reversed in polarity. The drum switch accomplishes this reversal in all five examples shown.

To reverse a three-phase motor, two of the three leads must be interchanged (as explained in Chapter 9). For the three-phase motor of Figure 11–14, when the switch is in the center position (OFF), no power is fed to the motor. With the switch in the forward position, the connections are L_1 to T_2, L_2 to T_1, and L_3 to T_3. The motor will rotate in one direction. When the drum switch is switched to the reverse position, the connections are L_1 to T_1, L_2 to T_2, and L_3 to T_3. The motor will run in the opposite direction because of the interchanging of the T_1 and T_2 connections to the three-phase line.

For the single-phase motor shown, reversal is accomplished by reversing the polarity of the run winding with respect to the start winding (as explained in Chapter 9). For the two DC motors in the figure, reversal is the result of changing the polarities of the field connections with respect to the armature connection (again, as explained in Chapter 9).

In some motor applications, it is necessary to know that the motor is at or very near zero speed. In other applications, it is necessary to know when the motor reaches a certain speed. Motor speed can be indicated by a centrifugal switch attached to the motor shaft, usually externally. The switch terminals are brought out to slip rings and brushes. Of course, other, tachometer-type, noncontact devices can be used. However, the centrifugal switch indicating zero speed is often a mechanical switch. A typical mechanical centrifugal motor switch is shown in Figure 11–16. For indicating direction of rotation, these switches can be any one of three types: CW, CCW, or both. The last type operates ON for either direction.

11–8   CONTROLLERS FOR LARGE SYNCHRONOUS AND WOUND-ROTOR MOTORS

Figure 11–17 shows the elementary power and control diagram for a reduced-voltage starter for a synchronous motor. This starting system has the same two-step transformers that induction-motor starters have. Additionally, it has other control components. The synchronous motor starts as an induction motor. When near full speed, it locks in to become synchronous and does not slip. The additional control components include two field relays, timers, and a speed-sensing synchronizing system.

The starter-system elementary diagram for a wound-rotor motor is shown in Figure 11–18. The wound rotor and the resistors take the place of the other

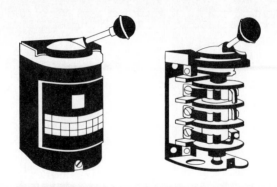

Left   Up   Right

Handle End		
Forward	Off	Reverse
1 o———o 2	1 o o 2	1 o̶ o̶ 2
3 o———o 4	3 o o 4	3 o̶ o̶ 4
5 o———o 6	5 o o 6	5 o———o 6
Internal Switching		

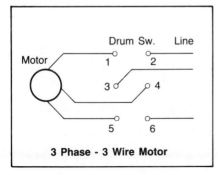

3 Phase - 3 Wire Motor

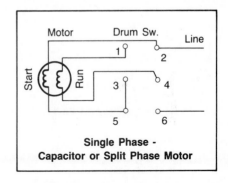

**Single Phase -
Capacitor or Split Phase Motor**

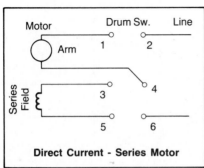

Direct Current - Series Motor

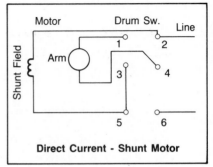

Direct Current - Shunt Motor

FIGURE 11–15
Drum Switch Connections and Applications

363

FIGURE 11–16
External Centrifugal Motor Switch

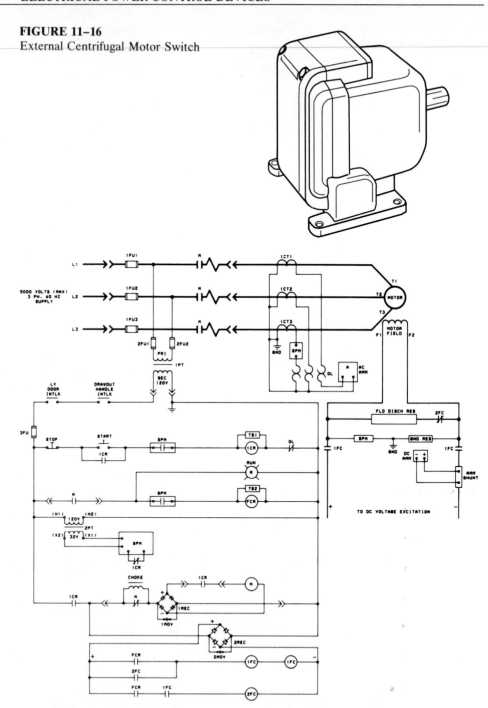

FIGURE 11–17
Synchronous-Motor Starting System

FIGURE 11–18
Wound-Rotor-Motor Starting System

ELEMENTARY DIAGRAM

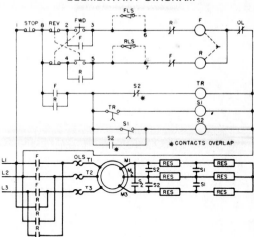

reduced-voltage starting systems. This starting equipment includes three, six, or nine resistors and one, two, or three contactors to connect the resistors sequentially to the rotor through the rotor slip rings. Low-resistance, high-wattage resistors are used. The control system uses zero, one, or two timers for operation.

There are wound-rotor motor applications in which the resistors are used for motor speed control. Such control systems have one set of three resistors in the positions shown. The resistors are variable, not fixed. Varying the resistors' values varies the speed at a given torque load.

SUMMARY

- ☐ Electrical equipment enclosures are available in many classes. The classes are rated according to the operational environments to which they are suited.
- ☐ Hazardous-location enclosures are the most costly. They are grouped in various subclasses according to the explosive materials in their environments.
- ☐ Electrical contactors are similar to relays but have enclosed contact areas and large contact areas for inductive or capacitive surges.
- ☐ Contactors are used in pairs to control forward and reverse and other dual functions of motors.
- ☐ A DC contactor differs from an AC contactor of the same electrical rating. The AC and DC contactors of a given rating are not interchangeable.
- ☐ Large AC and DC motors can be started by electronic means or by primary-resistor starting systems.
- ☐ Large AC motors can be started by other control systems. These control systems include transformer, wye-delta, and part-winding.
- ☐ Drum switches are often used for direct motor reversing control and for other electrical functions.

- External centrifugal motor switches are available to indicate whether motor speed is zero or the motor is running CW or CCW.
- Synchronous-motor starters require extra components for control to switch from starting status to synchronized running status.
- Wound-rotor-motor controllers utilize resistors for starting or speed control. These resistors are connected to the rotor by slip rings through contactors.

EXERCISES

1. List the major classes of electrical equipment enclosures. Describe the application of each.

2. To play it safe, why not use explosion-proof enclosures in all applications?

3. What type of enclosures would you use for each of the following environments?
   a. paint booth
   b. outside in the weather
   c. feed mill
   d. inside a plant, for lathe controls
   e. inside a cheese factory

4. List the characteristics of a contactor. Which characteristics are the same and which are different from those of an electromechanical relay?

5. What is the function of the auxiliary contact in Figure 11–4?

6. Refer to Figure 11–4. Explain how each circuit functions.

7. Refer to Figure 11–5. What happens if F and R contactors are both energized at the same time?

8. Refer to Figure 11–6. What happens if the High, Low, and stop push buttons are pressed in the following order with the motor at rest? Running at low speed? Running at high speed?
   a. stop   Low
   b. stop   High
   c. Low   High
   d. High   Low

9. How does a DC contactor differ from an AC contactor?

10. List the four major types of nonelectronic low-voltage motor starting. Explain in general terms how each works. Use Ohm's law in the explanation.

11. Refer to Figure 11–10. Construct a control and power diagram for the AC motor for two steps of resistance starting.

12. Refer to Figure 11–10. Construct a control and power diagram for the DC motor for three steps of reduced-voltage starting.

13. Refer to Figure 11–11. Construct a step-by-step sequence of operation.

14. Refer to Figure 11–12. Construct a step-by-step sequence for open-transition starting. Extra credit: Same for closed-transition starting.

15. What are two advantages of closed-transition starting over open-transition starting? Are there any disadvantages?

16. Refer to Figure 11–14. Construct a step-by-step sequence of operation.

17. Refer to Figure 11–15. Explain how each of the four circuits shown performs reversing for the given type of motor.

18. Refer to Figure 11–16. How does this external switch differ from a motor-internal centrifugal switch?

19. Refer to Figure 11–17. Explain in general terms how the circuit works.

20. Refer to Figure 11–18. The circuit shown is for a three-step starter. How would two- and four-step starter circuits be different?

APPLICATION ASSIGNMENT ANSWER

First, you verify that the motor is a three-phase induction motor. A motor with such a large horsepower rating usually is. A look at the rotor verifies that it is three-phase, since it has no rotary switch. Next, you consider the voltage and speed of the motor. From the motor horsepower and the voltage rating, you find the required starter size from the table of Figure 11–9. Next you determine whether reduced-voltage starting is needed. For reduced-voltage starting, if needed, you have a choice of primary-resistor, transformer, or wye-delta. If the motor is a part-winding-start motor, a special part-winding-start (PWS) starting system is required.

12

Control of Motors

CHAPTER OUTLINE

LEARNING OBJECTIVES

□ Define *seal, jog, plugging, dynamic braking,* and *reduced-voltage* start.
□ Draw ladder diagrams for and explain the operation of the following AC and DC motor control systems:
   starting, stopping, sealing, and jogging
   hand/automatic operation
   interlocks, forward/reverse, and sequencing
   pilot-light indication of status
□ Create multiple-station diagrams for the above ladder diagrams in the preceding objective.
□ Draw circuits for and explain motor-reversing systems.
□ Draw circuits for and explain sequencing, timing, plugging, antiplugging, and dynamic braking.
□ Explain the operation of the major types of reduced-voltage starting.
□ Describe the variable-frequency control systems for AC motor speed control.
□ Explain electronic reduced-voltage AC motor starting.
□ List other electronic motor-control functions, including the energy-saving function.
□ Describe the electronic variable-speed control and other controls for DC Motors.

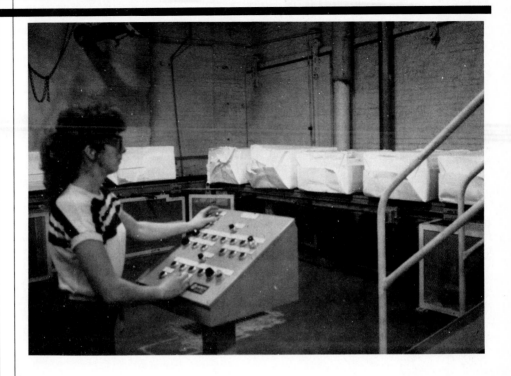

APPLICATION ASSIGNMENT

At the paper mill where you are a technician, a motor that drives a large conveyor system has been burning out. The conveyor carries large stacks of paper from one department to the next. Failure usually occurs when the conveyor is rapidly reversed a few times. The operator reverses the conveyor to return and recheck material previously put on the conveyor. What changes in circuitry or equipment might be made to prevent these motor burnouts?

12–1 INTRODUCTION

The basic control of motors is the same as for such other devices as solenoids, lights, and valves. However, electric motors have special characteristics, which necessitate special control systems. These characteristics are as follows:

- ☐ Motors require a momentary starting current surge of five to ten times running current.
- ☐ Going from forward to reverse requires special considerations due to the current surges and the inertia of the motor and its attached load.
- ☐ The time to reach operating speed and the time to slow to a stop must be considered.
- ☐ To vary motor speed in many cases, the control system must vary input voltage for DC or frequency for AC.

This chapter will first cover the starting and stopping principles for unidirectional motor operation. Next, it will cover the controls for bidirectional operation, forward and reverse. The controls for multiple-motor sequencing and motor stopping will also be discussed.

The need for, and the methods of, reduced-voltage motor starting will be explained. Electronic control for variable frequency of AC motors will next be covered. Special controls for electronic starting and for energy saving for AC motors will be explained. Finally, the electronic control of variable-speed DC motors will be discussed.

12–2 BASIC START, STOP, SEAL, AND JOG CONTROL AND PILOT LIGHTS

The simplest control of an AC or DC motor is straight ON/OFF control, as shown in Figure 12–1. This control is often called two-wire control, since only two wires are required between the ON/OFF switch and the motor controller coil. To the right of the coil is the overload contact, as discussed previously. The advantage of this circuit is its simplicity. However, it does have a drawback. Suppose a toggle switch is used as a pilot device. The toggle switch is closed, and the motor is running. Then, for some reason, the overload trips, opening the overload contact. The motor stops. A person then goes near the motor and the drive to see what is wrong. The overload cools off and the power goes back ON. The motor restarts if the person forgot to turn OFF the switch before approaching the machine. The person could be injured as a result. The same thing can happen if the overload is successfully reset while the person is near the machine.

A more common control is the start/stop seal circuit shown in Figure 12–2. This circuit is often called three-wire control, as three wires are required between the start/stop station and the AC or DC motor control coil. The circuit requires two push buttons, one for start and one for stop. When the start push button is depressed, M is energized. When M is energized, its auxiliary contact, located

FIGURE 12–1
Straight ON/OFF Control

Pilot device, such as limit
switch, pressure switch, etc.

under the start button in the diagram, closes. When the start button is released and reopens, M stays energized. A complete circuit is made through the M–1 contact. Whenever the power goes OFF or the overload opens, the motor coil is de-energized. When power is restored, the motor does not restart. The start button must be depressed to restart the motor. Whenever the stop button is depressed, the motor stops. Sometimes this seal is incorrectly called a latch. Latching, as previously discussed, is a different type of control.

The diagrams used here are ladder diagrams, not the connection diagrams discussed in Chapter 3. Also, only the control is explained here in ladder diagram form. The ladder or connection diagram for power is not included, as power diagrams are very straightforward. For reference, a connection diagram for a start/stop seal is shown in Figure 12–3. The power wiring is shown by heavier lines. The connections shown match the ladder diagram in Figure 12–2. Throughout the remainder of this chapter, only control ladder diagrams will illustrate circuits.

Jogging is a momentary-ON control capability with one push button. Jogging is used for a final adjustment of a machine's position. For this reason, jogging is sometimes called "inching." Jogging can be accomplished by the use of a selector switch that disables the seal circuit. A typical jog using a selector is shown in Figure 12–4. For a normal start/stop seal, the selector closes a path from A_2 to 2. Opening A_2-to-2 causes the start button to be a momentary-ON, or jog, button. The disadvantage of this circuit is that a selector has to be changed to choose between jog and seal-ON.

In some control circuits, it is desirable to choose either jog or seal-ON directly. Figure 12–5 shows a circuit that will accomplish this transition. Push jog and momentary operation continues as long as the jog push button is depressed.

MUST KNOW !
DRAW

FIGURE 12–2
Start/Stop Seal Control

THReE
WIRE →

most
COMMONly USED

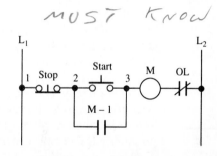

FIGURE 12–3
Wiring Diagram for Figure 12–2 with
Power Wiring

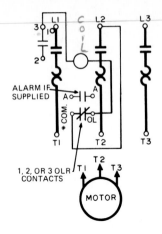

FIGURE 12–4
Start/Stop/Jog Control with Selector Switch

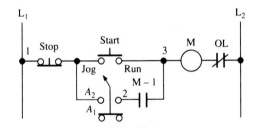

FIGURE 12–5
Questionable Start/Stop/Jog Circuit

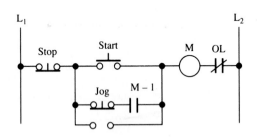

Release JOG and the motor goes OFF. Push start and the circuit seals ON. If the motor is running sealed, pushing jog turns it OFF, as the button leaves the top contacts. A very short time later, when the push button reaches the bottom contact, the circuit is in the jog mode.

This circuit is a questionable one to use. Suppose the jog push button is down. When it is let up, the motor is expected to go OFF. If the jog button returns up quickly, before M–1 opens, the motor will seal ON. This result is a particular possibility for a large contactor that drops out slowly. Therefore, this circuit is not generally used.

FIGURE 12–6
Positive-Unseal Start/Stop/Jog Circuit

BEST

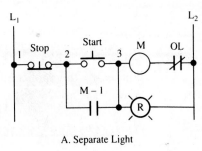

A better circuit, to prevent this unwanted seal-ON with the jog button re-turn, is shown in Figure 12–6. It is necessary to push stop to go from run/seal to jog, which was not required in the previous circuit. This circuit requires an added logic relay for the control circuit. Pressing start energizes CR, which energizes M through CR-2. Contacts CR-1 and M-1 complete the seal circuit. Depressing jog during running has no effect. If the motor is stopped, pushing jog causes M to go ON momentarily, with no possibility of sealing taking place.

Many motor control systems require a pilot light to indicate when the motor is ON. This is accomplished by wiring a light in parallel with the motor starting coil as shown in Figure 12–7. The light must have the same voltage rating as the motor starter coil and the supply voltage. Figure 12–7A is for a separate pilot light. Figure 12–7B is for a pilot light that is part of the start button.

FIGURE 12–7
Pilot Lights for Run Indication

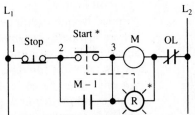

A. Separate Light

* Pushing on pilot light operates start contacts.

B. Light Integral to Start Button

FIGURE 12–8
Pilot Light Variations

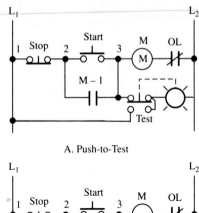

A. Push-to-Test

B. Pilot Light ON When Motor OFF

There are many possible pilot light configurations. Two are illustrated in Figure 12–8. Figure 12–8A illustrates a push-to-test system. Suppose that you depend on a pilot light to determine whether a motor is ON when it should be. You cannot observe or hear the motor. If the light is not ON, one of two things is possible: the motor is not running, or the pilot light is burned out. To eliminate the latter possibility, a push button is supplied to check the light. First test to see that the light is not burned out by pushing push-to-test.

Another pilot light variation is shown in Figure 12–8B. In this case the pilot light is ON only when the motor is *not* ON. Other variations are possible as required by the application.

12–3 MULTIPLE-STATION AND ALTERNATE CONTROL

Very often, for a given process, an AC or DC motor will need to be started and stopped from more than one location. Multiple-location starting and stopping are accomplished by wiring the stops in series and the starts in parallel. A circuit for accomplishing multiple start/stop control is shown in Figure 12–9. The figure shows a circuit for three locations. For more or fewer locations, start and stop buttons would be added or deleted as required. Note that only one seal contact is needed, no matter how many starts there are. Pressing any one start button will energize the M contactor. Pressing any one stop button will deenergize M and unseal the M contactor at M-1.

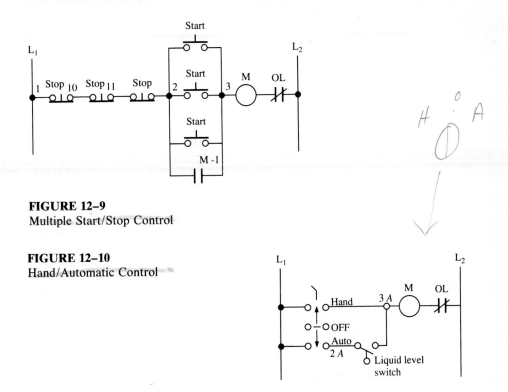

FIGURE 12-9
Multiple Start/Stop Control

FIGURE 12-10
Hand/Automatic Control

One type of alternate control is hand/automatic control. A typical circuit for this is shown in Figure 12–10. A typical application is a sump pump. For normal daily operation, the pump should go ON when the liquid-level switch closes, indicating water is to be pumped out. For servicing or checking, the pump should stay OFF. For periodic checks to make sure the pump works, switching to the hand position will make the pump run. If it does not run in the hand position, it will not run when it rains. A three-position selector switch is used to select hand, OFF, or automatic as shown. During normal operation, the switch is in the automatic position.

12-4 DUAL OPERATION: ROTATION AND SPEED

To reverse a three-phase motor, two contactors are needed, one for forward operation and one for reverse. Power is connected through the contactors and to the motors as described in Chapter 11. A basic, mutually interlocked forward-reverse circuit is shown in Figure 12–11. This circuit can be used for three-phase AC or DC motors. It is not usable for regular single-phase motors, as explained in the motor chapter. The circuit can also be used for other control applications, such as in-and-out operation of a pneumatic or hydraulic cylinder.

FIGURE 12–11
Forward-Reverse Control

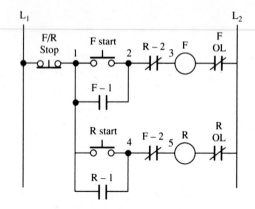

The circuit consists of two separate start/stop seal controls with a common stop button. It also includes two NC interlock contacts. If forward is started by pushing the F start button, reverse cannot be started, as the F-2 contact is open when F is running. A complete circuit to the reverse coil is not possible. The inverse is true for the forward contact when reverse is running. Closing both F and R at the same time will short out the incoming power. The mutual interlocking prevents F and R from being closed at the same time.

The stop button must first be depressed before the circuit can be switched from F to R or from R to F in Figure 12–11. Sometimes, it is desired to go directly from F to R or from R to F. Figure 12–12 is a circuit that allows this. Using this scheme is not always a good idea. For example, when there is a big inertia load on a motor, direct instant reversal can be harmful to the motor. The circuit shown is triple-interlocked to prevent both F and R from being on at the same time. It has

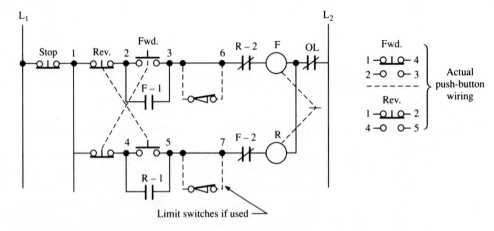

FIGURE 12–12
Direct Forward-Reverse Control

the interlock NC contacts as before. In addition, the dotted lines to the right of F and R indicate a mechanical interlock, usually a pivot lever. Also shown is the interlock use of optional limit switches, which are used on very large contactors. These limit switches are installed to contact the plunger of the coil when the coil is energized.

Direct reversal is accomplished by using both the NO and the NC contacts of the F and R buttons. Actual wiring of the buttons is shown on the right, for clarity. In the control ladder diagram, the two parts of both buttons are shown apart for circuit operational clarity. Suppose the motor is running in the forward direction. Pushing the R button deenergizes F by opening 1-to-2. When the button reaches the lower contacts, 4 and 5, the R coil is energized. The inverse is true when going from R to F.

Two examples of control circuits for dual-speed motor operation will be covered next. Both have two contactors, one for high speed and one for low speed. The first circuit, Figure 12–13, can be started up in either low or high speed. Also, it can go directly from low to high speed. However, before the motor can go from high to low speed, stop must be pushed. This circuit is used where going directly from high to low speed is harmful to the motor or to its driven load. If the motor is going at high speed, pressing the low-speed button will have no effect. As already said, the stop button must be depressed first.

Another two-speed control scheme is shown in Figure 12–14. This is used when the motor must start initially at low speed. An example of an application is a motor driving a big flywheel. Starting in high speed would overload the motor and possibly the motor electrical supply system. This circuit requires an extra electromechanical relay for control (CR). The relay used as CR is sometimes designated the *compelling relay*. Pressing the high button will not start the motor initially, as contact CR-2 is open. However, if the motor has been running at low speed, the CR relay is energized and sealed. Contact CR-2 is closed. In that situation, press-

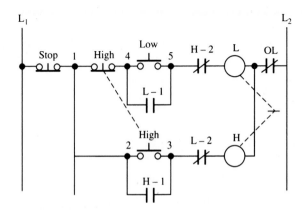

FIGURE 12–13
Two-Speed Control with Stop before High-to-Low

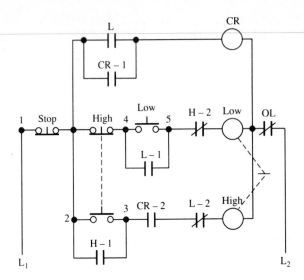

FIGURE 12–14
Two-Speed Control That Requires Low-Speed Start

ing the high button turns off low (the upper section of the push button) and energizes the high contactor. Note that the motor can be started in low when it is at rest.

12–5   SEQUENCING, TIMING, PLUGGING, AND DYNAMIC BRAKING

Sequencing control systems are of various types. The most common type involves making sure one output is ON before another can be turned ON. An example of this situation is shown in Figure 12–15. Suppose a milling machine must have lubrication flowing before it is turned ON. Starter M_1, the lube pump starter, is controlled by a start/stop seal system, as shown. Starter M_2, the milling machine starter, is controlled by a separate start/stop seal system. Note that M_2 cannot be started unless M_1 is started first, because of the M_1-2 contact. Furthermore, if M_1 goes OFF, M_2 is turned OFF also, protecting the milling machine from lack of lubrication. There are numerous variations and combinations of this type of control, all of which operate on the same principle.

Often a timed interval is needed between the energizing of one starter and the energizing of one or more others. The schemes shown in Figure 12–16 will accomplish this timing. Both schemes use timers. In the first case (Figure 12–16A), the output is to go ON a set time after the input switch is closed. When SW_1 is closed, the timer is energized. After a period of time set on the timer, a timer contact closes the output's starter.

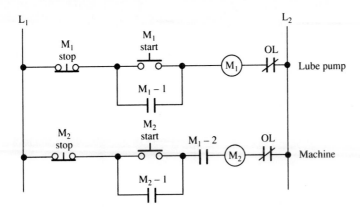

FIGURE 12–15
Interlock Sequence Control

In Figure 12–16*B*, two outputs are involved. Contactor 1 comes ON when the start button is depressed, and it stays sealed ON. The timer coil, which is in parallel with the contactor 1 coil, comes ON at the same time. The timer runs for the set period of time. At the end of the time interval, a timer contact closure causes contactor 2 to energize.

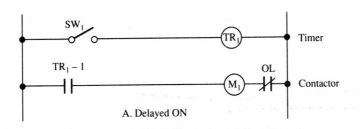

A. Delayed ON

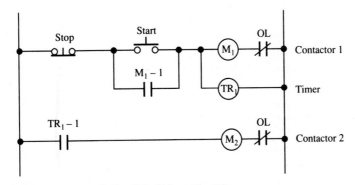

B. One ON—Delay—Two ON

FIGURE 12–16
Timed Sequence Control

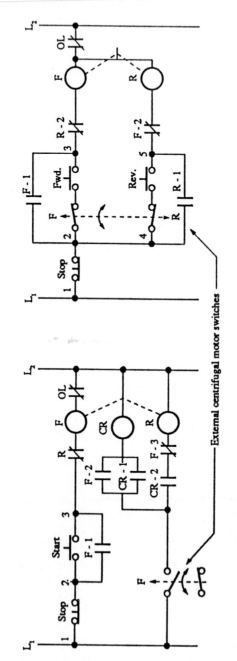

B. Two-way antiplugging

A. One-way automatic

External centrifugal motor switches

FIGURE 12–17
AC-Motor Plugging

Plugging is defined as the immediate application of reverse polarity to an AC three-phase motor. The concept of plugging was previously discussed for DC motors in Chapter 9. A complete control circuit for plugging will now be shown. When the motor is running forward, for example, stop is not depressed before reverse power is applied. Reverse power is applied at the same time (or immediately after) the forward power is disconnected. If reverse power were permanently applied, as in the forward-reverse control circuit, the motor would simply reverse direction. Plugging to a stop is more complicated. The idea is to disconnect reverse power at the moment the motor decelerates to a stop, and not go beyond. This disconnection is accomplished automatically with the typical plugging-to-a-stop circuit shown in Figure 12–17A.

The circuit in Figure 12–17A requires the motor to have an external centrifugal switch, which is mounted on one end of the motor shaft. This centrifugal switch is open when the motor is stopped. When start is depressed, the motor starts and seals ON. As the motor starts in the forward direction, the switch closes, and it remains closed during forward rotation. Also, the CR relay is turned ON, and it seals itself through CR-1. When stop is pushed, F goes OFF and R comes ON through three contacts. The three are the closed F centrifugal switch; CR-2, which is closed; and F-2, which has reclosed. The motor then slows down quickly with reverse three-phase AC power applied. When the motor reaches zero speed, the F centrifugal switch opens, R opens, and CR is turned OFF. The motor is then at rest and ready for restarting as required.

A variation of AC plugging is antiplugging, shown in Figure 12–17B. This circuit is designed to prevent reversal until the motor has slowed to a stop. The circuit is used where large inertia loads on the motor make instant reversal impractical. It contains a double centrifugal switch, which must be mounted on the motor shaft. One switch opens when the motor is rotating forward, and the other when it is rotating in reverse. Assume motor control is sealed in the F direction. Pushing R will not do anything, because of normal interlocking. After stop is pushed, R is immediately pushed again. Nothing happens, as the forward centrifugal switch is still open. Until the motor comes to a stop, it cannot be reversed. The same is true for going from R to F.

Basic DC dynamic braking is illustrated in Figure 12–18. The circuit shown is a simplified version of those normally used. For this circuit, both the power diagram and the control diagram are shown, to illustrate its operation better. When the power contactor, F, is turned ON and sealed, power is applied to the DC motor armature, and the motor runs. The field power is already turned ON. When the F contactor is turned OFF by pressing stop, the motor stops quickly. The quick stop is provided by the dynamic-braking resistor. When F is turned OFF, F-3 recloses, placing the resistor electrically across the armature. The field remains ON, and the rotating motor armature is thus temporarily turned into an output generator armature. The rotating energy of the armature is quickly dispersed when current and power are applied to the resistor. The armature consequently slows at a very rapid rate. Low-resistance, high-power resistors are used. This system is faster and more durable than mechanical brakes.

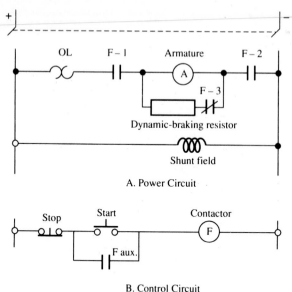

A. Power Circuit

B. Control Circuit

FIGURE 12–18
Basic DC Dynamic Braking Control

12–6   REDUCED-VOLTAGE STARTING

All the previous examples in this chapter have been for across-the-line starting. In other words, the motor is started by direct connection to the power leads. Motors of larger horsepower cannot be connected directly across the line when starting for two reasons. First, the power drain on the electrical supply system is too great. A motor that runs at 440 V AC, 85 A, will draw an initial surge of up to 425 A, five times the running current. It is very expensive to build a distribution system to successfully deliver the 425 A. Furthermore, when the motor starts, the lights in the vicinity will dim because of the line voltage drop due to the 425 A. The voltage dip can also cause equipment malfunctions, particularly in computers.

Second, a motor designed for full-voltage starting will be much larger and more expensive than the same rating of motor for reduced-voltage starting. All large motors are designed for reduced-voltage starting. Some method of reducing the starting current is needed.

There are many ways to reduce the motor current inrush:

primary-resistor control

transformer control

wye-delta control

part-winding-start control

electronic control

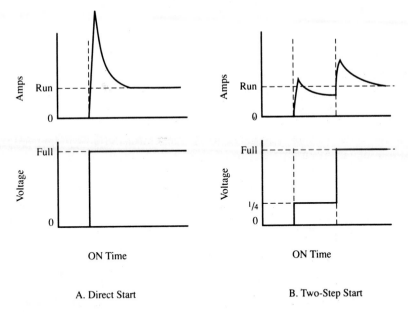

FIGURE 12–19
Comparison of Direct and Two-Step Starting

All the methods except electronic control were explained from a control stand-point in Chapter 11. Electronic control will be discussed in the remaining sections of this chapter.

Reduced-voltage starting applies to large DC and large three-phase AC motors, since single-phase AC motors are of small ratings. Reduced-voltage starting does not apply to small-rating motors, because they have relatively low start current. Only the resistor and electronic methods can be used to start DC motors. Three-phase AC can use any of the listed methods.

What is the effect of reduced-voltage starting? Figure 12–19 compares the start current over time for direct start with that for two-step reduced-voltage start. The reduced-voltage system keeps the current to a much lower peak value during the start cycle. The value is typically one fourth the value for across-the-line starting. Any reduced-voltage starting system, of course, costs more than a direct-start, contactor-only system.

12–7 ELECTRONIC VARIABLE-SPEED CONTROL OF AC MOTORS

In previous chapters AC motors have been described as essentially constant-speed devices. This description is true for the electrical supply and control systems discussed thus far. A relatively new electronic control has been developed for varying the speed of three-phase AC motors. This electronic control is not applicable to single-phase AC motors—only three-phase.

Until the late 1970s the only system available for high-horsepower variable-speed operation of AC motors was motor generator sets. The variable-speed system consisted of four rotating machines as shown in Figure 12–20. An AC motor drives a DC generator. The DC generator output is then used to drive a DC motor. The DC motor's speed can be varied by varying its input voltage. Its input voltage is the voltage output from the DC generator. This DC voltage is varied by varying field-current control to the DC generator. The DC motor, in turn, drives an AC generator. The output of the AC generator is a variable frequency, directly proportional to speed. Finally, the AC generator output can be used to drive any three-phase AC motor at varying speeds. The speed of the driven AC motor is controlled by the frequency of the AC supplied to it.

This system was expensive, required maintenance of four machines, and was about 50% efficient. Another problem was the arcing and the maintenance of the DC brushes. This machine system is shown for reference, to compare it with the advantages of the newer, electronic systems now in use.

In the early 1980s electronic variable-frequency systems were developed for use in mines, grain mills, and other hazardous areas. The variable-speed AC motors and controls replaced DC motors, which had brush sparking. The brush sparking could ignite gases or grain dust in the area. Originally the cost of these electronic control systems was thousands of dollars. Each year the electronics, especially the power electronic systems, were improved. By the late 1980s, a 3/4 HP, variable-frequency drive could be bought for less than $500. A general block diagram of a variable-frequency drive is in Figure 12–21.

The control system shown in Figure 12–21 will only vary the speed of the electronically driven motor. Electronic variable-speed drive units perform many other control functions. Some added control functions are standard, and some are optional, extra-cost features. The features vary from manufacturer to manufacturer. A more complete block diagram, indicating some of the standard features, is in Figure 12–22.

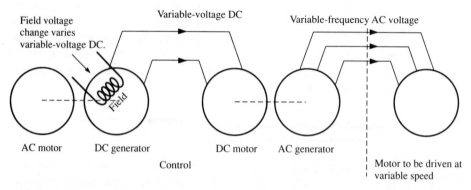

FIGURE 12–20
Multimachine Variable-Frequency Control

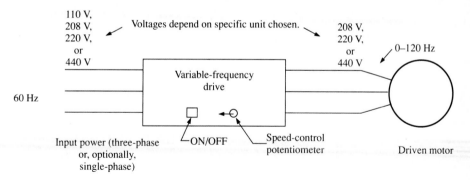

FIGURE 12–21
Basic Block Diagram for a Variable-Frequency Drive

A variable-speed electronic control can produce output frequencies from 2 to 90 Hz, or more, from a 60 Hz line. Therefore, a four-pole motor rated at 1725 RPM can be made to run at speeds from about 60 RPM (2 Hz) to about 2700 RPM (90 Hz). Some other controllers produce up to 120 Hz and even 400 Hz. Other basic control capabilities of a typical standard unit include reverse rotation by changing a jumper, control of acceleration (spin-up) time, and control of deceleration (spin-down) time. Other typical standard features include input undervoltage protection, input overvoltage protection, and protection from input line surges, shorts, and grounds. A listing of some typical features is given in Figure 12–23. The categories include basic control functions, device and motor protective features, and adjustments that may be made to the control. Also included are some typical extra-cost modifications, which may be obtained as required by each particular application.

Variable-frequency electronic-drive units do not put out a sine wave. To run an AC motor, a perfect sine wave is not necessary. The electronic-drive outputs approximate a sine wave. Many output waveforms look like the one shown in Figure 12–24. The output waveform is close enough to a sine wave that the motor runs within a reasonable range of sine-wave efficiency and performance. This multiple-step wave is much easier to produce electronically than a sine wave. Other variable-frequency systems produce other output waveforms, some on the same principle as the switching power supply.

The stepped waveform also produces one other effect. The motor runs at a slightly higher noise level, especially when accelerating or decelerating.

The motor being run by a variable-speed drive must be derated somewhat from its rating when run straight from a 60 Hz line. There are two major reasons for the derating. One reason is the change in waveform from a normal sine wave, as previously described. Another reason is that there is less motor cooling at lower speeds—when the fan inside (or on) the motor is moving less cooling air. Two derating charts are shown in Figure 12–25.

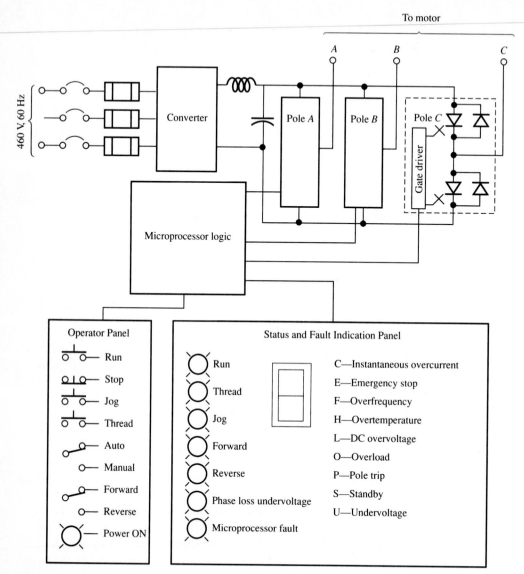

FIGURE 12–22
Block Diagram for a Typical Variable-Speed Unit

Standard Basic Control Functions
- Controller start, stop, and power-ON indicating lights
- Speed-control potentiometer terminations for door- or remote-operator-station-mounted devices
- Linear, independent timed acceleration and deceleration adjustable from 1 to 30 Hz/s (4–120 s)
- Unidirectional operation with controlled decelerated stopping
- A constant-torque range of 2–60 Hz and a constant-horsepower range of 60–120
- Frequency stability of ±0.5% and voltage regulation of ±2% of maximum rated torque
- Insensitive-to-line phase rotation
- Fixed dwell time on start to increase motor starting torque

Standard Protective Functions
- Input AC circuit breaker with an interlocked, padlockable handle mechanism
- AC input-line current-limiting fuses for fault-current protection of AC-to-DC converter section
- Electronic overcurrent trip for instantaneous and inverse-time overload protection
- Electric ground-fault protection
- AC input-line undervoltage and phase-loss protection
- Overfrequency protection
- DC overvoltage protection
- Overtemperature protection
- Surge protection from input AC line transients

Optional Modifications (at Added Cost)
- Improved speed regulation of ±0.5%
- Electronic reversing
- Sensitivity to line phase rotation
- Higher output frequency, 480 Hz, for greater motor speed
- Capability for 50 Hz, 380 V input line
- Space heaters for low-temperature environments
- Up-to-speed relay
- Dusttight, air-conditioned enclosure for force-ventilated enclosure
- Motor overcurrent relay
- Remote emergency-stop capability
- Jog at a presettable speed different from running speed
- Dynamic braking for quick slowdown or stop
- Energy saving logic
- Automatic reset on return of power following a utility outage
- Output load ammeter, voltmeter, and frequency-indication device

Standard Adjustments
- Minimum speed
- Maximum speed
- Torque (current) limit
- Acceleration time
- Deceleration time
- Low-frequency voltage boost
- Volts per hertz

Status and Fault Panel	
Run	E—Emergency stop
Jog	F—Overfrequency
Forward	H—Overtemperature
Phase loss	O—Overload
Phase undervoltage	P—Polarity
Computer fault	S—Standby

FIGURE 12–23
Variable-Speed Drive Specifications

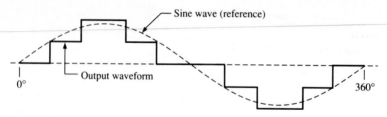

FIGURE 12–24
Output Waveform for a Typical Variable-Speed Electronic Drive

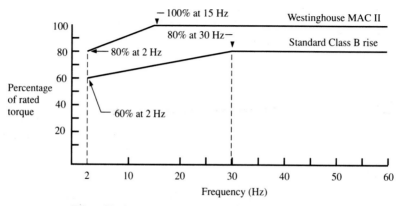

Effect of Reduced Cooling on Motor-Torque Capability Versus Speed

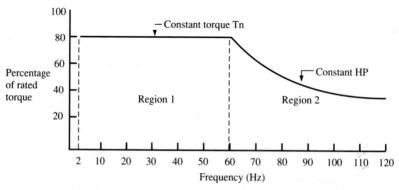

Induction-Motor Torque/Speed Characteristic for 2–120 Hz Class B Rise

FIGURE 12–25
Motor Derating Charts for Variable-Speed Operation

A typical diagram showing the major external connections for a variable-frequency drive is illustrated in Figure 12–26. For this particular unit, four connections need to be made for basic operation. First, incoming power is connected to lines L_1, L_2, and L_3. For this unit, 230 V three-phase power is connected. However, single-phase 230 V will operate the unit also, L_1 to L_3. If an internal jumper is changed (not shown), other input voltages can be used instead.

The second connections to be made are the three connections to the motor being driven. These are at U, V, and W. Third, connections for control power are made from L_1 and L_3 to R_{23} and T_1. The control transformer ratio is adjustable internally for different supply voltages. Finally, a jumper must be connected from ST to COM if an external remote start button is not connected. All of the other external wiring shown is for optional remote control of various functions as labeled in the figure.

The control units themselves are not very big. Up to 3 HP, a typical enclosure is $21 \times 11 \times 18$ in. An enclosure for a 10 HP unit is typically $27 \times 13 \times 26$ in. A great space saver is that there is no contactor in the unit. Power connection and disconnection are performed electronically by variation of the triggering angle of the SCR.

12–8   SOLID-STATE AC MOTOR CONTROL

An electronic means is available for starting large three-phase motors. Electronic starters, which use SCRs as power drive devices, are often called "soft-start" controllers. They have no moving parts and often no contactors.

Limitation of starting torque and prevention of excess line-voltage drop were accomplished by the reduced-voltage starters of Chapter 11. Those starters applied voltage in two or more steps. Solid-state, electronic starters also carry out this reduced-voltage starting. In addition, the solid-state starter/controller is adjustable to produce a smooth, continuously increasing voltage during start. The maximum allowable current is also adjustable. Torque, spin-up time, and other parameters are adjustable as well. A typical connection diagram for a solid-state starter/controller is shown in Figure 12–27.

Other functions available in the solid-state starter/controllers include the following:

- □ *Phase-loss trip.* When one phase is lost, starter trips OFF after a 0.5 s delay to prevent nuisance trips.
- □ *Undervoltage protection.* Starter trips at 75% voltage, after a 2–3 s delay.
- □ *Current unbalance.* Starter trips if unbalance exceeds 20%.
- □ *No-line-power trip.* If the power is OFF for more than 5–6 cycles, the starter trips OFF.
- □ *Current limit.* The limit can be set at two to six times running current.

Also available, at extra cost, are overtemperature protection, improper-line-phase rotation, and other special protective features. A block control diagram for a

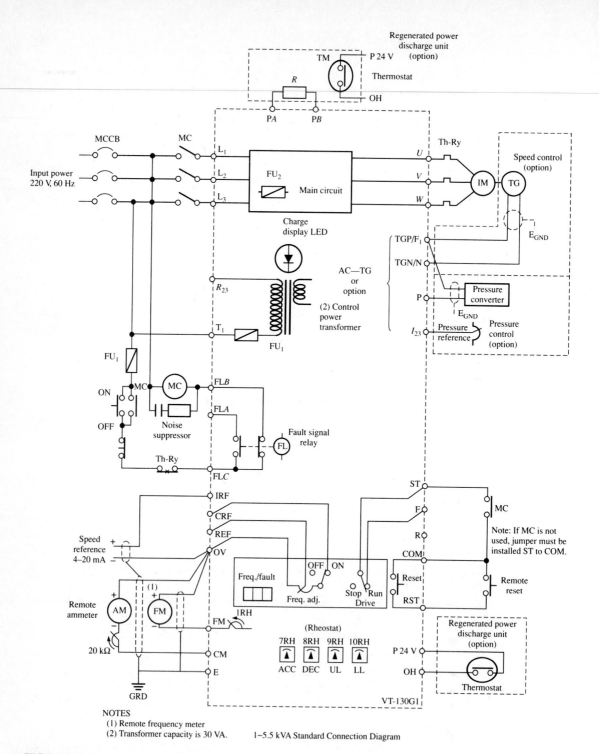

FIGURE 12–26
Variable-Speed AC Drive Connection Diagram

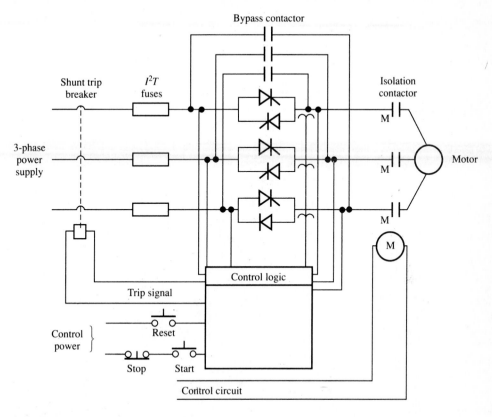

FIGURE 12–27
Solid-State, Reduced-Voltage AC Start and Control

typical solid-state starter/controller is shown in Figure 12–28. The figure includes a typical control-panel status display. Status lights go ON for correct operation or as problems arise.

Another feature available at additional cost as part of the solid-state starter/ controller is a feature to reduce motor energy consumption during light motor loading. The illustrations in this book use information from Westinghouse, whose trademark for this feature is Energy Saver. The unit functions by sensing motor current and torque loading. The load-sensing signal is compared with a fixed reference voltage. The reference voltage is individually set for the motor being controlled. The difference in the two signals causes logic circuits to appropriately adjust the voltage supplied to the motor. Solid-state power SCRs accomplish this voltage control through automatic triggering-angle adjustments.

Under light-load conditions the Energy Saver reduces the voltage to the motor. Cooler motor operation and higher electrical efficiency result from the reduced voltage. The motor does, however, operate at the proper speed and with

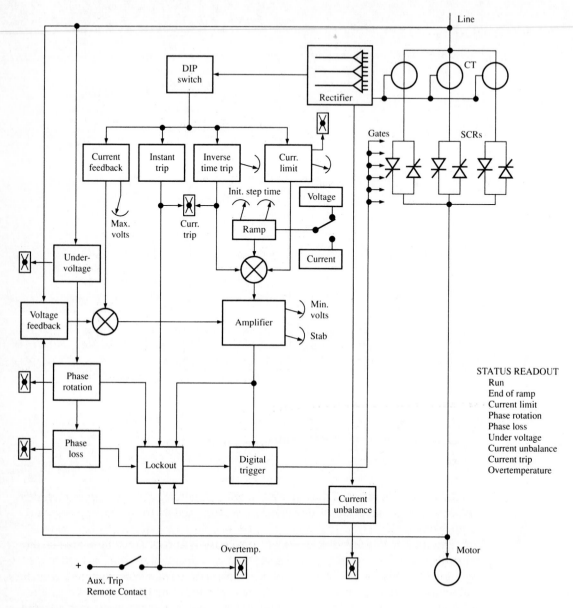

FIGURE 12–28
Solid-State AC Motor Control

proper torque at the lower voltage. Additionally, the power factor at lighter loads is improved by the Energy Saver. Motors running lightly loaded have poor power factor when connected directly to the power line. Improved power factor results in less voltage-line loss and helps reduce electrical power bills (as explained in

FIGURE 12–29
Percentage of Power Saved by Using Energy
Saver™

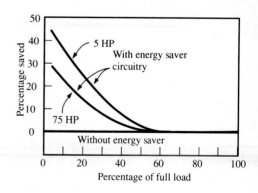

Chapter 16). Figure 12–29 is a graph that shows as a percentage the possible power saving when the Energy Saver is used. Note that if the motor controlled by the Energy Saver is loaded above 60% of full-load at all times, no power is saved.

An energy-saving system is also available for single-phase AC motors. It is called the Induction Motor Energy Saver, or IMES. The system consists of a monolithic control-IC package and a triac. The IC, the control PC board, and the triac are physically very small. The package can be installed in the end shield of almost any AC single-phase motor. The economics of buying a motor with the IMES depends on the application and the loading of the motor.

12–9   DC MOTOR-CONTROL DRIVES

Chapter 9 included a discussion of speed control of DC motors using only resistors, in series or in parallel with the motor armature or the field. Chapter 11 covered DC motor starting control with series resistors. In this section electronic control for starting and controlling DC motors will be discussed. Electronic controllers for DC motors do more than provide reduced-voltage start, dynamic braking, and general speed control. They precisely control speed under starting, running, stopping, and reversing conditions. Additionally, DC electronic controllers give added motor protection from overcurrent and overloading. Figure 12–30 shows the appearance of a DC electronic drive powered by a three-phase AC input line. The figure also includes a block diagram of the drive system's operation.

Major application features of a DC drive include the following:

- *Linear acceleration control.* Acceleration and deceleration rates are smooth and adjustable.
- *Speed regulation.* As the load changes, speed stays within 1.0%. Better regulations, with variations as low as 0.1%, are available at added cost.
- *Jog capabilities.* The motor can be jogged at separately adjustable speeds for forward and reverse.

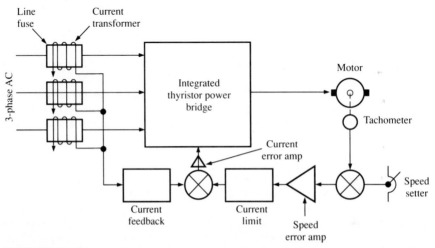

FIGURE 12–30
Solid-State DC Motor Control (Courtesy of Magnetek Drives and Systems)

☐ *Follower.* Operation is controllable from external tachometers or transducers.

☐ *Cooling ventilation.* Controlled-speed motor-blower ventilation is available. The DC motor under electronic control is cooled by a separate motor and blower at low speeds and under high horsepower (and high motor-heating) operation.

☐ *Extended range.* The drive provides operational capabilities above nominal motor speed by limiting horsepower to a constant value.

☐ *Metering.* Indicators are available for voltages, amperages, and speed. The meters and indicators are useful for monitoring, installation, and system setup.

A DC drive also provides electrical protection for the motor and for the electronic drive itself, including the following protective features:

☐ Thermal protection of the SCR power bridge.

☐ Short circuit protection through the use of special fast-acting semiconductor fuses.

☐ Various types of overcurrent protection. These include 300% instantaneous overcurrent protection, maximum steady-state current (adjustable), and sustained overload current.

☐ Incoming AC line protection against single-phasing and improper phase rotation.

☐ Undervoltage of the control-logic voltage.

☐ Incoming line transient and DV/DT protection.

☐ Simultaneous forward-and-reverse firing protection. This prevents firing of the SCRs in one bridge until current is zero in the other bridge.

Other, additional control and protective features are available from the electronic control manufacturer, as needed for a particular application.

SUMMARY

☐ Start/stop seal control of motors is preferred over straight ON/OFF control. It is safer and lends itself to multiple-station control.

☐ Multiple-station control of motors is accomplished by connecting stop push buttons in series and start push buttons in parallel. Also, NC control contacts are connected in series, and NO in parallel.

☐ Motors can be controlled for different speeds and dual rotation by proper controls with appropriate interlocking contacts.

☐ For more than one motor, various interlocking schemes are available to make sure one motor runs only when another (or others) is running or only when the other (or others) is not.

☐ Sequencing the starting and running of more than one motor can be accomplished by the proper use of timers.

☐ Plugging, which is three-phase sequence reversal, can be used to stop a motor quickly. Plugging can also be used for instant rotation reversal.

☐ Dynamic braking is used to stop a DC motor quickly by temporarily turning it into a generator.

☐ Reduced-voltage starting is required for large DC motors and large AC three-phase motors. There are numerous control methods to accomplish the reduced voltage.

☐ Variable speed of AC three-phase motors can be accomplished electronically, with some efficiency lost. Speeds from 1/30 normal full-load speed to 2 times normal full-load speed are possible.

☐ Special electronic controls can be used to start motors by slowly increasing the applied voltage to the motor. These special electronic controls can also be augmented to reduce power consumption when motors are periodically lightly loaded.

☐ A DC motor drive can accurately control motor speed under varying mechanical-load conditions.

EXERCISES

1. Refer to Figure 12–1. Assume that a motor is to be started and stopped from any of three locations in a room. How would you connect the three toggle switches—in series, in parallel, or neither? Why or why not?

2. Refer to Figure 12–2. What would happen if the NO M–1 seal contact were changed to a NC contact?

3. Refer to Figure 12–4. Design a circuit for two-station operation. Each station has a stop push button, a start push button, and a selector switch.

4. Refer to Figure 12–6. Design a two-station start/stop/jog circuit.

5. Refer to Figure 12–9. Assume that there are two (not three) control stations. Each has one stop button and two start buttons 3 ft apart. It is necessary to push both start buttons to energize the start coil. This arrangement is to ensure that both hands are out of the work area during starting. Design an appropriate circuit.

6. Refer to Figure 12–10. Assume that a start/stop seal system is to run the motor in the hand position. Also, a pilot light is to indicate when M is ON. Another pilot light is to indicate when the system is OFF. Design a circuit to meet these requirements.

7. Refer to Figure 12–11. Construct a circuit with two control stations. Each is to be able to stop the motor and to control forward and reverse operation. Each station is to have two pilot lights, one for forward and one for reverse.

8. Applying the principles of Figure 12–12, construct another circuit to meet the requirement of Exercise 7.

9. Refer to Figure 12–13. Construct a circuit with two control stations. Each is to be able to perform the same high-low functions as the single-station circuit.

10. Refer to Figures 12–11 and 12–15. There are three motors, each with its own start/stop seal control. The motors are *A*, *B*, and *C*. Design circuits for the following situations:
    a. Only one of the motors can run at a time.
    b. Any two of the motors can run at a time.
    c. Motor *A* must run before *C* can run. Also, *C* must run before *B* can run.
    d. If both *A* and *B* are running, *C* cannot run. Otherwise, *C* can run.

11. Refer to Figure 12–16. Design a circuit with four motors, *A*, *B*, *C*, and *D*. Motor *A* starts 5 s after start is pushed. Motor *B* starts 6 s after *A*. Motor *C* starts 8 s after *B*.

Motor D starts 7 s after A. Additionally, if any motor overload relay opens, all motors stop running.

12. Refer to Figure 12–17. Design two circuits—one with one-way automatic plugging, the other with two-way antiplugging. Each system is to have two control stations. Each station is to have two pilot lights indicating forward or reverse operation.

13. What is the difference between plugging and dynamic braking?

14. Can plugging be applied to DC motors? Explain.

15. Can dynamic braking be applied to AC induction motors? Explain.

16. List some good applications for a variable-speed AC drive.

17. Refer to Figure 12–25. You have a standard Class B motor that normally produces 75 lb-ft of torque. When used with a variable speed drive, what will be the motor torque capability at 15 Hz and at 55 Hz? Per the lower curve, what is the torque capability of a motor with 8.5 lb-ft rated torque at 15, 70, and 110 Hz?

18. Refer to Figure 12–29. What would be the energy savings for a 5 HP motor at 15%, 30%, 50%, and 80% of full load? Repeat for a 75 HP motor.

19. What are some of the advantages of the DC drive shown in Figure 12–30 over the DC motor control discussed in Chapter 11 and shown in Figure 11–10?

20. What are some of the advantages of a DC drive over the control of a DC motor shown in Figures 11–10 and 12–18?

APPLICATION ASSIGNMENT ANSWER

The problem is caused by rapidly reversing the motor. Applying reverse power while the motor is running forward (or forward power while the motor is running in reverse) causes an extremely large motor-current surge, which heats up the motor rapidly. The inertia of the conveyor and its contents compounds the problem by increasing surge current further.

The solution might be to add a timer to the motor-control system. The timer would prevent rapid reversing of the motor. When the stop button was depressed, the timer would start timing. Timer contacts in the control circuit would then prevent reversal from taking place for a set period of time. The timing interval would be set for the time it takes the motor speed to go from full speed to zero speed (stop). Allowing the motor to slow before reversing would eliminate the reversal surge current.

An even more positive solution would be to use the antiplugging circuit explained in the chapter. This would, of course, require the addition of a centrifugal device on the motor itself. This system would ensure a complete stop before power could be applied for the opposite direction of rotation.

IV

CLOSED-LOOP CONTROL

13

Analog Controllers

CHAPTER OUTLINE

LEARNING OBJECTIVES

□ Given the description of a particular control system, classify it as open-loop or closed-loop.
□ Describe the characteristics of ON/OFF, proportional, integral, proportional-integral, and PID controllers.
□ Given the characteristics of a particular process, decide which type of control would be most appropriate.
□ Given a particular ON/OFF controller, calculate its deadband.
□ Given a particular proportional controller, calculate its proportional band.
□ Given a particular controller, along with its characteristics and error input, graphically describe how the output should respond.

APPLICATION ASSIGNMENT

Suppose that a proportional controller similar to that of Figure 13–6 is being used to control the temperature of the combustion air for a furnace. For some reason, however, wide variations in the air temperature are detected. Inspection of the controller reveals that the output voltage varies greatly at regular intervals.

As production technician, you have been asked to inspect the circuit and report your findings. Since the controller has never been in operation before, you immediately suspect a calibration or adjustment problem. Before inspecting the circuit, you study the schematic and try to determine any likely problems. Suggest one maladjustment that could cause the controller oscillation that has been described.

Upon completion of this chapter, you will be able to diagnose this problem. One possible diagnosis is given at the end of the chapter.

13–1   INTRODUCTION

A control system can be described as a group of components arranged in such a way that the operational characteristics of a particular process are maintained at the desired level. Although control can be achieved mechanically, electrically, or through fluid power, this book is concerned with those systems that control the process by using electrical and electronic components.

To understand the need for control in even the most basic systems, consider the problem of regulating the temperature in an average household during the winter months. Technically, cycling the furnace ON and OFF at given intervals would keep the house at a relatively uniform temperature. Determining the proper duty cycle might require some experimentation, but after an appropriate cycle time had been determined, the temperature should remain constant. However, there are many variables, or system disturbances, that can change the duty cycle requirements. The outside temperature variations will change the rate at which heat is transferred to the outside. People entering and leaving the house will allow heat to escape. The number of people present in the house, along with their level of activity, will change the inside temperature. Obviously, constant monitoring and adjustment will be required if, in fact, the inside temperature is to remain constant. Rather than making these adjustments manually, it would be much more efficient (and less tiresome) to have the adjustments made automatically through the intervention of control devices.

In industrial processes the goal of control is usually much more critical than maintaining a constant household temperature. Automatic control systems are implemented in areas that would be hazardous or impossible for human operators to work in. They perform functions with greater speed, precision, and repeatability than are otherwise obtainable. The quality of the control systems in use can determine, for example, how well two machined parts will fit together or whether or not the welds on an auto chassis will break. That is, good or bad, identical items rolling out of a manufacturing facility should be just that—identical. Good electronic control allows products to be manufactured cheaper and faster and with a higher level of quality control.

Generally, a control system is classified according to the process that is being controlled. That is, temperature control systems control temperature, flow control systems control fluid flow, and level control systems control the height of material in holding bins or reservoirs. The nature of the controlling components also plays a part in system classification. In this respect, systems are usually classified as either analog or digital, depending on the electronic device technology employed. Control systems are also classified according to the presence or absence of feedback. Open-loop systems employ no feedback; closed-loop systems constantly monitor the process through a feedback loop.

Since all control systems share common structural characteristics, block diagrams are often used to describe them concisely. Depending on the particular textbook, application specification, or sales representative you are dealing with, the nomenclature associated with these block diagrams will vary. Each of the

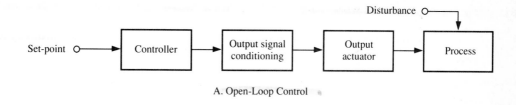

A. Open-Loop Control

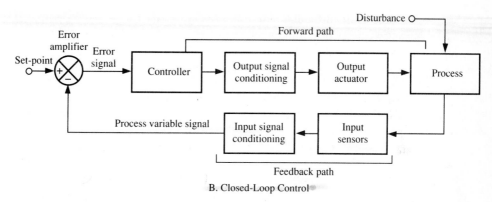

B. Closed-Loop Control

FIGURE 13–1
Typical System Control Diagrams

terms used in Figure 13–1*B* will now be explained, and some of the more common alternative names for them will be given. *Set-point, command,* and *reference* are all names used to describe the input that determines the desired or ideal operating point for the process. This is usually provided by a human operator, although it may also be supplied by another electronic circuit. The *process variable,* or *measured value,* or *controlled variable,* is the signal that contains information about the current process status. In other words, this is the feedback signal. Ideally, the process variable and the set-point should match (indicating that the process is operating exactly as desired). *Error amplifier, error detector, comparator,* and *summing point* are names given to the electronic components that determine whether the process operation matches the set-point. Usually, the device that makes the determination is nothing more than a differential amplifier. The output of the error amplifier is referred to as the *error signal* or the *system deviation signal.* The magnitude and the polarity of this signal will determine how the process will be brought back under control. The *controller* is the group of electronic components that, depending on the error signal, will produce the appropriate corrective signal. The *output actuator,* or *final correcting device,* is the component that directly affects a process change. Motors, heaters, fans, and solenoids are all examples of output actuators. For the controller output to drive the output actuator, the signal may have to be modified. For example, for an op amp to drive a 120 V AC motor, there must be some sort of power interface, as well as an isolation circuit. Likewise, in the sensing of the process status to produce the process variable signal, some modification or conversion may be required. The components or circuits that provide this function are called *signal conditioners*.

As shown in Figure 13–1A, an open-loop control system consists of a controller, signal conditioning circuitry, an output actuator, and a process. Since the only input to the controller is the set-point, it is apparent that an open-loop system controls the process blindly. That is, the controller receives no information concerning the present status of the process, the need for any correction due to process disturbances, or the responses of the process to previous controller action. The previously described duty-cycle-controlled furnace controller operated in an open-loop mode. The set-point was established by a human operator, and the furnace responded to that set-point alone, ignoring completely any variation in room temperature due to external influences. Open-loop control systems are considerably cheaper and less complex than their closed-loop counterparts. The inevitable result of their use, however, is poor process control.

The presence of the feedback loop in Figure 13–1B classifies this system as closed-loop. Since the controllers in such systems receive information about the process status, they can compensate for external disturbances. In other words, they can automatically adjust process conditions to ensure that external events do not cause the process to stray from the set-point. To convert the duty-cycle-controlled furnace to a closed-loop system, only temperature-measuring devices and an error amplifier must be added. The error signal can then be used to control the length of time that the furnace stays on. Or the furnace can be allowed to run continuously, and the error signal can be used to control the amount of fuel burned. Closed-loop systems, although more expensive and complex than open-loop control, allow more flexibility and afford the operator tighter control over the process.

13–2 THE ROLE OF THE CONTROLLER

At the heart of any electronic control system is the controller. This is the circuitry that accepts information from the input transducers and the signal conditioners and provides the signals that will ultimately be used to correct the error condition. The purpose of the controller is to maintain the process variable within an acceptable distance of the set-point. Under ideal circumstances, the process variable would never stray from the set-point. If conditions within the system attempted to change, the controller would sense this tendency and instantly produce an output that would prevent deviation from the set-point. The quality of a controller, therefore, depends on how closely the process variable tracks the set-point (how small an error can be maintained) and how quickly the controller can respond when the two do not match. The type of controller that is chosen for any particular application depends on the desired speed of response, the allowable system error, and the process dynamics.

13–3　ON/OFF CONTROLLERS

The *ON/OFF*, or *two-position, controller* attempts to control the process variable with an output that is either fully ON or fully OFF. If the sensed magnitude of the process variable is below a given threshold (the set-point), the controller output turns completely OFF. This action turns OFF the output actuator, and the process variable comes back into the vicinity of the set-point. If, on the other hand, the sensed magnitude of the process variable is above the set-point, the controller output turns completely ON, providing full power to the output actuator.

The preceding description is of a *direct-acting controller*—one in which the controller output and the process variable move in the same direction. An *inverse-acting controller* would produce a full-ON output if the process variable were below the set-point. Likewise, if the process variable were above the set-point, the controller output would be OFF. Changing a controller from direct- to inverse-acting (or vice versa) simply involves an inversion of the controller output.

For an ON/OFF controller to be practical, it must exhibit some degree of hysteresis. In other words, the magnitude of the process variable that forces the controller output ON must be different from the value that forces the output OFF. If no hysteresis exists, the output will oscillate, causing a loss of process control and possible damage to system components. Figure 13–2, the transfer curve for an ON/OFF controller, demonstrates this principle. For the controller output to be driven to full-ON, the error signal must decrease to the magnitude indicated by −Error. Once the output is ON, however, it will not turn OFF again until the error signal is above the magnitude indicated by +Error. A significant difference may exist between these two points.

In an inverse-acting temperature controller, for example, if the temperature rises above the set-point (creating a positive error) by a sufficient amount, the output of the controller will shut OFF the heater. This will cause the temperature to begin to decrease. A temperature decrease sufficient to cause the error signal to fall to zero will not turn the heater back ON. Instead, the temperature must decrease below the set-point (creating a negative error) by a certain amount before the controller output will again turn the heater ON.

FIGURE 13–2
Hysteresis in an ON/OFF Controller

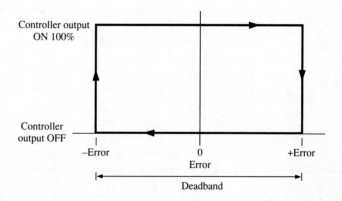

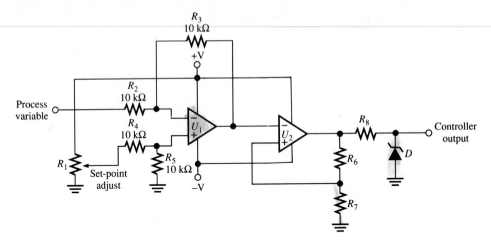

FIGURE 13–3
Typical Op-Amp ON/OFF Controller

This error "window" is called the *deadband* for the controller. Basically, a deadband means that the error signal can change without affecting the output of the controller. Formally, the deadband is the difference between the error signal that turns the controller output fully ON and the error signal that turns the controller output fully OFF. That is,

$$\text{Deadband} = \text{Error}_{\text{ON}} - \text{Error}_{\text{OFF}} \tag{13.1}$$

The deadband may be expressed as a voltage or in terms of the process variable.

In some cases, the deadband is an inherent part of the system, caused by mechanical hysteresis, thermal lag, or some other characteristic. In many cases, however, a deadband must be implemented electronically. This implementation allows for adjustment and calibration of the system for individual applications. Remember that an ideal controller would reduce the error signal to zero. Because a deadband is required, an ON/OFF controller will always have a finite error. That is, the process variable will never be maintained at the desired set-point. The system will continually "hunt" for the proper value.

Figure 13–3 shows a typical ON/OFF controller. Op-amp U_1, a differential amplifier, serves as the error amplifier. Here it is set for unity gain, so its output (the error signal) is simply the difference between the set-point voltage and the voltage at the process variable input. This error voltage is then compared with a fraction of the output from U_2. The amount of feedback to the inverting input of U_2 establishes the controller deadband. To drive the output of U_2 to negative saturation, the error signal must increase to a magnitude greater than the voltage across R_7 (the voltage across R_7 is positive at this time, since the output of U_2 is at $+V_{\text{sat}}$). As soon as the output of U_2 changes, however, the voltage across R_7 will be negative with respect to ground. Therefore, to drive the output back to positive saturation, the error signal will have to decrease to a magnitude that is more negative than the voltage across R_7. The span of error voltage (the deadband) that is required to drive the output of U_2 from $+V_{sat}$ to $-V_{sat}$ (or from $-V_{sat}$ to $+V_{sat}$) is twice the voltage across R_7. The equation for this relationship is as follows:

$$\text{Deadband} = 2V_{sat} \left(\frac{R_7}{R_6 + R_7} \right) \qquad (13.2)$$

In many cases, control will not be maintained if the output of the controller is allowed to swing from $+V_{sat}$ to $-V_{sat}$. For example, if the controller is driving the coil of a DC relay, the contacts will be pulled in whether the controller provides an output of $+V_{sat}$ or $-V_{sat}$. Therefore, the output must often be limited to values between 0 V and $+V_{sat}$ (or between 0 V and $-V_{sat}$). In Figure 13–3 resistor R_8 and the zener diode serve to clamp the output of U_2 to positive values. Note that when U_2 is at $-V_{sat}$, the zener diode is forward-biased, and the resultant output is clamped to -0.7 V. When U_2 is at $+V_{sat}$, the output cannot increase beyond the reverse-avalanche zener voltage. If the zener chosen for this circuit avalanched at 4.5 V, the controller output would be limited to values between -0.7 V and $+4.5$ V, even though the supply voltage to U_2 might be considerably higher. The value of R_8 should be chosen so as to limit the forward and reverse current through the zener to a reasonable value.

EXAMPLE 13–1

Given the circuit of Figure 13–3, suppose that the op amps are powered with a supply of ± 12 V, $R_6 = 100$ kΩ, $R_7 = 22$ kΩ, and D has an avalanche voltage of 6 V.

Calculate the Circuit Deadband
The deadband may be calculated from equation 13.2. Since U_2 is powered with a supply of ± 12 V, it will saturate at about ± 10.5 V. Inserting this value and the values of R_6 and R_7 into the equation yields the following result:

$$\text{Deadband} = 2(10.5 \text{ V}) \left(\frac{22 \text{ k}\Omega}{122 \text{ k}\Omega} \right) = 3.8 \text{ V}$$

That is, if the error signal increases above zero by $+1.9$ V, the output of U_2 will be driven to negative saturation. To drive the output to positive saturation, the error signal will have to decrease to a value of -1.9 V (a span of 3.8 V).

Draw the Circuit Transfer Curve
To draw the transfer curve, four distinct points are required: the deadband end-points ($+$Error and $-$Error), the minimum controller output, and the maximum controller output. Since the deadband has already been calculated, only the output conditions must be determined. Note that even though the output of U_2 may switch between $+10.5$ V and -10.5 V, the actual controller output is limited by R_8 and the zener to either -0.7 V or $+6$ V.

With the error signal along the x-axis, and the controller output along the y-axis, the following points define the hysteresis curve (Figure 13–4) that is characteristic of this circuit. It is important to include an arrow along the curve to indicate the direction of the process.

Error Signal (V)	Controller Output (V)
−1.9	−0.7
−1.9	+6
+1.9	−0.7
+1.9	+6

FIGURE 13–4

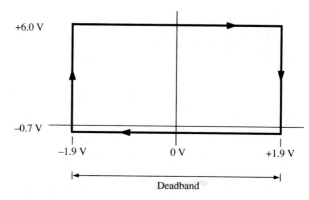

Deadband

13–4   PROPORTIONAL CONTROLLERS

In many instances, the error window of an ON/OFF controller cannot be tolerated. A *proportional controller* allows tighter control of the process variable, because its output can take on any value between fully ON and fully OFF, depending on the magnitude of the error signal. As indicated by the name of this controller, the output changes proportionally with small changes in the error input. A proportional controller usually has a linear response (see Figure 13–5). That is, if the error input doubles, the controller output also doubles. Linearity, however, is not a requirement for this type of controller. For example, the output may be proportional to the square or the logarithm of the error signal. Such outputs, however, are the exception and not the rule.

Large error inputs can still drive the controller to positive or negative saturation. Proportional controllers, however, are operated in a region called the *proportional band*. Like the deadband of an ON/OFF controller, the proportional band is described as the change in error signal that will cause the controller output to swing from full-OFF to full-ON. Likewise, it may be expressed as an absolute voltage or in terms of the process variable. The proportional band for a linear controller can be calculated in terms of the controller output, as follows:

$$\text{Proportional band} = \frac{V_{\text{out}_{max}} - V_{\text{out}_{min}}}{A_V} \tag{13.3}$$

where A_V is the controller gain.

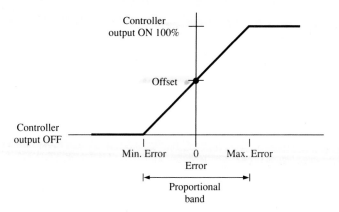

FIGURE 13–5
Typical Proportional Controller Response Curve

There are three points of interest on the proportional-controller transfer curve in Figure 13–5. The first is the magnitude of error signal that drives the controller to full-ON. The second is the magnitude of error signal that drives the controller output to full-OFF. These two points need not be equally spaced on either side of the zero-error point. The factor that determines where the two points fall is called the *offset*. This is the point where the curve crosses the y-axis. More precisely, it is the output that the controller produces when the error signal is zero. Note that when the offset is exactly halfway between $+V_{sat}$ and $-V_{sat}$ (as it is in Figure 13–5), $+$Error and $-$Error are spaced equally on either side of the zero error point. If, however, the offset is decreased, the effect is to move the entire curve to the right, without disturbing the slope of the line. Thus, the magnitude of the offset does not affect the magnitude of the proportional band.

Since the proportional band says nothing about the actual magnitude of the error signals, it may be necessary to calculate them. Naturally, the minimum error signal will produce the minimum controller output, and the maximum error will produce the maximum controller output. Both of these values, however, are modified by the offset voltage. The minimum and maximum error signals can be calculated by the use of the following equations:

$$Error_{min} = \frac{V_{out_{min}}}{A_V} - V_{offset} \qquad\qquad (13.4)$$

$$Error_{max} = \frac{V_{out_{max}}}{A_V} - V_{offset} \qquad\qquad (13.5)$$

Shown in Figure 13–6 is a typical proportional controller. Differential amplifier U_1 provides the error signal. As in the ON/OFF controller, it is set for unity gain, so the error voltage is the difference between the set-point voltage and the sensed process-variable voltage. Op-Amp U_2 is an inverting summing amplifier. Thus, it can add an offset voltage to the error voltage before amplification. The

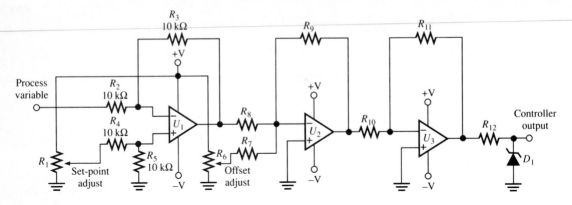

FIGURE 13–6
Typical Op-Amp Proportional Controller

significance of the offset is that it allows the controller to maintain an output even when the error signal is near zero. If the controller output decreased to zero when there was no error (when the process variable and the set-point matched), the output actuator would turn OFF. This action would, in turn, cause the process variable to stray from the desired set-point. For this reason, proportional controllers, like ON/OFF controllers, are unable to maintain an error signal of zero. If the controller output is offset by a value that is exactly halfway between $+V_{sat}$ and $-V_{sat}$ for U_2, it is easier (and faster) for the controller to correct for changes in the process variable. When the error signal is zero, the controller output may be, for example, +5 V. This may not be the correct output required to maintain the error at zero. Perhaps an output of +7 V is required. With an offset of +5 V, the controller output needs to change by only 2 V (as opposed to the entire 7 V). The offset will not allow the controller to maintain an error of zero; it simply facilitates correction of the process variable. The error can be reduced, however, by increasing the gain of U_2. This increase results in a narrower proportional band (a smaller error change will cause the output to swing full-scale). This action would be similar to reducing the deadband of an ON/OFF controller. There is a practical limit to the extent to which the error signal can be reduced by increasing the gain. The system will most likely oscillate long before the error can be totally reduced. Note that U_2 controls two aspects of the transfer curve. The gain of U_2 determines the slope of the line, and the magnitude of the offset positions the entire curve about the zero-error point.

EXAMPLE 13–2 Given the circuit of Figure 13–6, suppose that the offset input is adjusted to +1 V. Assume that R_7 and R_8 are each 10 kΩ and that R_9 is 20 kΩ. In addition, assume that U_3 is operating with unity gain.

Calculate the Circuit Proportional Band

Equation 13.3 can be used to calculate the proportional band. However, keep in mind that the output of the controller is clamped to -0.7 V and $+6$ V. Therefore, the minimum and maximum outputs are not $\pm V_{\text{sat}}$ for U_2.

$$\text{Proportional band} = \frac{+6 \text{ V} - (-0.7 \text{ V})}{2} = 3.35 \text{ V}$$

Draw the Circuit Transfer Curve

The transfer curve is most easily drawn by calculating the minimum and maximum error inputs and simply plotting and connecting them. Since the gain of U_2 is 2 and the offset is adjusted to $+1$ V, U_2 should produce an output of $+2$ V in response to an error of zero. Therefore, it is only necessary, as a quick check, to make sure that the curve crosses the y-axis at $+2$ V. Equations 13.4 and 13.5 give the following results:

$$\text{Error}_{\text{min}} = \frac{-0.7 \text{ V}}{2} - 1 \text{ V} = -1.35 \text{ V}$$

$$\text{Error}_{\text{max}} = \frac{+6 \text{ V}}{2} - 1 \text{ V} = 2 \text{ V}$$

With the error signal along the x-axis and the controller output along the y-axis, the following coordinates define the endpoints of the curve (Figure 13–7):

Error Signal (V)	*Controller Output (V)*
-1.35	-0.7
$+2$	$+6$

FIGURE 13–7

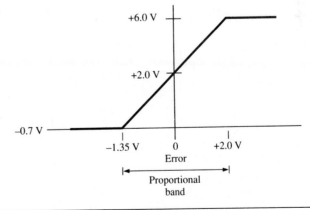

13–5 **INTEGRAL CONTROLLERS**

When implementing ON/OFF or proportional controllers in an electronic control system, the user must be able to tolerate a certain amount of error. In an ON/OFF temperature-control system, for example, one must be content with the fact that the system will never quite stabilize at the desired temperature—it will fluctuate around the set-point. An *integral controller,* however, is capable of driving the error to zero and keeping it there. The main element of an integral controller is an integrator.

Remember from Chapter 6 that the magnitude of the output of an integrator is not proportional to the input. Rather, it is the rate of change of the output that is proportional to the input. The greater the input, the faster the output changes. If the input is increasing, the output changes at an increasing rate. One of the most important characteristics of an integrator, however, is its ability to maintain an output even after the input has decreased to zero. That is, when the input has fallen to zero, the output no longer changes; it simply maintains the output that was present when the input fell to zero.

Formally, the output of an integral controller may be expressed in terms of its error input as follows:

$$V_{out} = - \frac{1}{R_i C_f} \int_0^t V_{error} \, dt + V_0 \qquad \textbf{(13.6)}$$

where V_0 is the initial capacitor voltage, R_i is the op-amp input resistor, and C_f is the op-amp feedback capacitor. For the special case in which the error input is a steady DC, equation 13.6 simplifies to the following equation:

$$V_{out} = - \left(\frac{V_{error}}{R_i C_f} \right) t + V_0 \qquad \textbf{(13.7)}$$

That is, when the error input is constant, the controller output changes linearly (*ramps*) as a function of time. Note that the actual output from an op-amp integrator is inverted (as indicated by the minus sign in equations 13.6 and 13.7). This inversion is of little consequence, however, since an extra gain or inversion stage can be easily added at the output.

Three important assumptions are made for equation 13.6. First, it is assumed that the output of the integrator is ideal. Very few electronic circuits provide ideal performance. Second, it is assumed that the error signal (the input to the integrator) can be accurately represented by a mathematical equation. Depending on the nature of the controlled variable, it may be very difficult to describe the error signal mathematically. The function will be, at best, an approximation. Finally, it is assumed that the person calculating the output signal will have more than a cursory familiarity with the process of integration. The overall result is that unless the error input is a simple step function, the magnitude of the output of an integral controller may be very difficult to calculate. By analyzing some typical inputs, however, it is possible to predict the response of the output. Shown in Figure 13–8 are some common input functions, along with the ideal integrator outputs they would produce.

The integral controller of Figure 13–9 consists of three main blocks. As usual, the unity-gain differential amplifier, U_1, provides the error signal. If, in

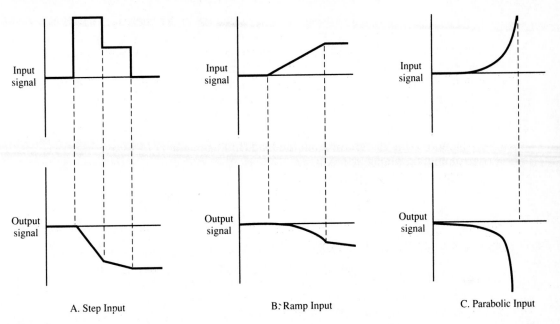

FIGURE 13–8
Typical Integrator Output Responses

fact, a higher gain is required, it is a simple matter to adjust the values of R_2 through R_5. Op-amp U_2 performs the actual integration. Remember from equation 13.6 that the values of R_6 and C will determine the integration constant. As the error signal fluctuates, C will charge and discharge through R_6. Keeping in mind

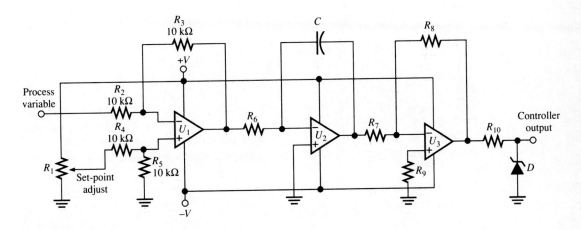

FIGURE 13–9
Op-Amp Integral Controller

that the integrator is to operate with its capacitor in the linear region of its charge curve, it is important to prevent the capacitor from fully charging. Whether the capacitor fully charges or not depends on three factors: the magnitude of the error signal, the length of time that the error signal is present, and the values chosen for R_6 and C. Knowledge of the system characteristics is required to properly choose these values. That is, an integral controller must be precisely "tuned" to respond properly for a given application. Op-amp U_3 has two main functions. It provides an inversion/gain stage for the output of the integrator. Moreover, it provides isolation of the integrator from the load. In this manner, U_2 is able more easily to maintain the capacitor voltage. That is, the capacitor is not constantly being discharged directly through the load.

EXAMPLE 13–3

Given the integral controller of Figure 13–9, along with the error signal shown in Figure 13–10, determine the controller output. Assume that the values of R_6 and C are 10 kΩ and 0.05 μF, respectively, and that the initial charge on the capacitor is zero. Assume also that U_3 is operating with unity gain.

Since the error signal is a step function, equation 13.7 can be used to calculate the controller output. The problem is best approached by dividing the input into five separate intervals and analyzing each interval individually.

Interval 1 (0–0.5 ms)
From 0 to 0.5 ms, the input is zero. Since the initial capacitor charge is zero, the output will remain at zero.

Interval 2 (0.5–1.5 ms)
From 0.5 ms to 1.5 ms, the error signal is a constant +1 V. At 0.5 ms, the output will begin to ramp (increase linearly), and 1 ms later (after a total elapsed time of 1.5 ms), the output will have attained a value of +2 V, found by using equation 13.7:

$$V_{out} = \left[\frac{+1 \text{ V}}{(0.05 \text{ } \mu\text{F})(10 \text{ k}\Omega)} \right] 1 \text{ ms} = +2 \text{ V}$$

Interval 3 (1.5–2 ms)
At 1.5 ms the output will have attained a value of +2 V. At this same instant, however, the error input steps up from +1 V to +2 V. The higher input voltage will cause the output voltage to ramp with a steeper slope. Throughout this 0.5 ms interval, the capacitor will accumulate enough charge to increase the output voltage by +2 V:

$$V_{out} = \left[\frac{+2 \text{ V}}{(0.05 \text{ } \mu\text{F})(10 \text{ k}\Omega)} \right] 0.5 \text{ ms} + (+2 \text{ V}) = +4 \text{ V}$$

Since the initial charge was already +2 V, the resultant output at the end of 2 ms will be +4 V.

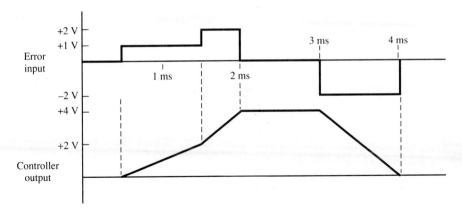

FIGURE 13–10

Interval 4 (2–3 ms)
Since the error input falls to zero during this interval, the rate of change of the output of the controller is zero. Therefore, the output will remain at +4 V.

Interval 5 (3–4 ms)
At the beginning of this interval, the error signal steps down from zero to −2 V, and it remains at that level for the remainder of the interval. Equation 13.7 is used to find that at the end of this 1 ms interval, the output has changed by −4 V:

$$V_{out} = \left[\frac{-2 \text{ V}}{(0.05 \text{ } \mu F)(10 \text{ k}\Omega)} \right] 1 \text{ ms} + (+4 \text{ V}) = 0 \text{ V}$$

Since the initial capacitor voltage was +4 V, the resultant output will be zero.

13–6 PROPORTIONAL-INTEGRAL CONTROLLERS

Integral controllers almost never exist as isolated controllers. They are usually used in conjunction with some other type of controller. The reason is that although they are capable of driving the error signal to zero, they have very poor transient response. That is, the output responds slowly to rapid changes in error signal. Examining Figure 13–10 should make this quite obvious. Notice that as the error signal steps up to a higher value, the output of the controller only begins to ramp. Depending upon the value of R and C, it may be a considerable amount of time before the output reaches the proper value. So, even though the error signal will be reduced to zero in a system that incorporates integral control, the system will react sluggishly to dynamic system disturbances.

By combining proportional and integral principles in the same controller, the advantages of each can be realized. The proportional controller reacts quickly but cannot reduce the error to zero. The integral controller reacts slowly but, over a

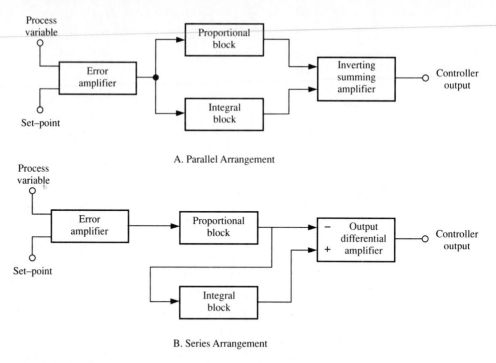

A. Parallel Arrangement

B. Series Arrangement

FIGURE 13–11
Typical Proportional-Integral Controller Arrangements

period of time, can eliminate the error. The proportional-integral controller exploits the fact that these two controllers naturally complement each other.

Generally, the proportional block and the integral block may be combined in one of two ways. As shown in Figure 13–11, the two blocks may be arranged in series or in parallel. In either case, the input to the controller is the familiar differential error amplifier, and the output is a summed combination of the outputs of the integral and proportional controller sections. The main difference between the two arrangements is the speed of response. Since the integral block in a series arrangement can receive an amplified error signal, the output in a series arrangement will change at a greater rate than the corresponding output in a parallel arrangement. This action will tend more rapidly to force the error to zero.

To understand the operation of a typical proportional-integral controller, consider the circuit of Figure 13–12. Since the characteristics of proportional and integral controllers have been discussed, the controllers are, for simplicity, represented as blocks. For this analysis, assume that the error amplifier, the proportional block, and the output summing amplifier are all operating with unity gain. In addition, assume that the product RC for the integral block is 1 s (the product of resistance and capacitance is in units of time). Since the proportional block has a gain of 1, its output will simply be an inverted image of the error signal. The

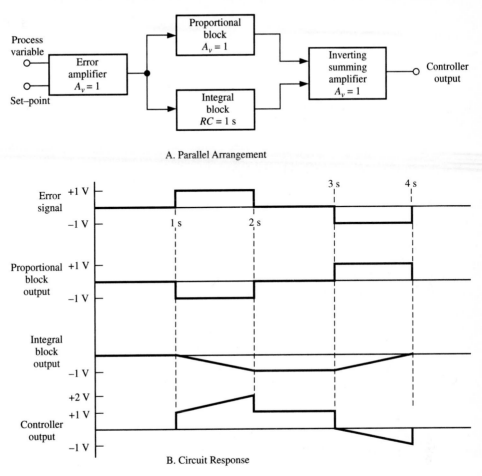

A. Parallel Arrangement

B. Circuit Response

FIGURE 13–12
Response of Parallel Proportional-Integral Controller to a Step Input

integral block receives the same error input, and after 1 s (when the error signal steps up to +1 V), the output begins to ramp. According to equation 13.7, after an interval of 1 s (a total elapsed time of 2 s) with a constant input of +1 V, the integral block output will be −1 V (remember that the integrator is still an inverting amplifier). At this time, the error signal falls to zero. This input causes the output of the integral block to remain constant at the −1 V level. After a total elapsed time of 3 s, the error signal steps down to a value of −1 V for a duration of 1 s. Again the integral block output begins to ramp. After 1 s the output due to this error step will be +1 V, but since the initial capacitor voltage was −1 V, the net output will be 0 V. Since the inverting summing amplifier has a gain of 1, the outputs from the integral and proportional blocks are algebraically added (and inverted) to obtain the overall controller output. The result is shown at the bottom of Figure 13–12.

EXAMPLE 13–4 Given the series proportional-integral controller of Figure 13–11B, along with the error signal shown in Figure 13–13, accurately sketch the controller output. Assume that the error amplifier and the differential amplifier are operating with unity gain. Assume also that the proportional block is operating with a gain of 2, and that the RC product for the integral block is 1 s.

As with the integral controller example, it is best to segment the error signal into separate intervals and analyze each individually. It can be seen that four

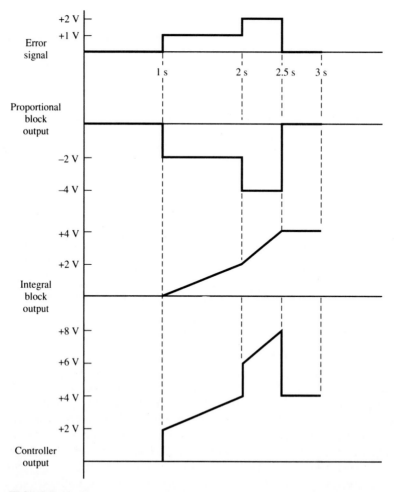

FIGURE 13–13

separate intervals exist. Remember that it is important to analyze the intervals where the error signal is zero, because the integral block will maintain an output during those times.

Interval 1 (0–1 s)
If the error input to the proportional block is zero, the output must also be zero. Since the input to the integral block is provided by the proportional block, the integral block output will not change. We assume that the initial capacitor voltage in the integral block is zero, so the output of this block will remain at zero. Since the final controller output is the algebraic difference between the integral and proportional block outputs, the controller yields 0 V for this entire interval.

Interval 2 (1–2 s)
After 1 s the error signal steps up to 1 V, and it remains at this level for a duration of 1 s. In response, the proportional block output steps down to a value of -2 V for the same duration (remember that the proportional block has a gain of 2). This -2 V is applied to the input of the integral block. Equation 13.7 can be used to determine the resultant output at the end of the interval:

$$V_{out} = - \left(\frac{-2 \text{ V}}{1 \text{ s}} \right) 1 \text{ s} + 0 \text{ V} = +2 \text{ V}$$

The output of the integral block, therefore, will begin to ramp linearly in the positive direction as soon as the error signal steps. At the end of a 1 s interval, the output will have achieved a value of $+2$ V. The controller output can be described by the following equation:

$$\text{Controller output} = V_{out_{integral}} - V_{out_{proportional}}$$

Since both inputs to the differential amplifier are linear, it is necessary to calculate the final output only at the beginning and the end of the interval and to connect those points with a straight line. For this interval the endpoint calculations are as follows:

$$\text{Endpoint 1} = 0 \text{ V} - (-2 \text{ V}) = +2 \text{ V}$$
$$\text{Endpoint 2} = +2 \text{ V} - (-2 \text{ V}) = +4 \text{ V}$$

Plotting and connecting these points will yield the controller's overall response for this interval.

Interval 3 (2–2.5 s)
For this interval the error input steps up an additional 1 V. The input to the proportional block is now $+2$ V. The output of this block, therefore, will be driven to a potential of -4 V for a duration of 0.5 s. This potential is applied to the input of the integral block for the same duration. The output of the integral block will begin to ramp linearly in the positive direction. At the end of the 0.5 s interval, the output will be

$$V_{out} = - \left(\frac{-4 \text{ V}}{1 \text{ s}} \right) 0.5 \text{ s} + 2 \text{ V} = +4 \text{ V}$$

At the beginning of this interval, the integral output starts at +2 V, and it linearly ramps positive to a final value of +4 V.

The controller output throughout this interval can again be determined by calculating the endpoint voltages:

$$\text{Endpoint 1} = +2 \text{ V} - (-4 \text{ V}) = +6 \text{ V}$$
$$\text{Endpoint 2} = +4 \text{ V} - (-4 \text{ V}) = +8 \text{ V}$$

Plotting and connecting these points will yield the overall controller response throughout this interval.

Interval 4 (2.5–3 s)
For the duration of this interval, the error signal has fallen to zero. This will cause the proportional block output to fall to zero, and the integral block output to hold its current output constant at +4 V. Since neither output is changing, there is only one calculation to be made to determine the overall controller response:

$$\text{Controller output} = +4 \text{ V} - 0 \text{ V} = +4 \text{ V}$$

13–7   DERIVATIVE CONTROLLERS

In many cases a process may possess an inherent inertia or hysteresis. For example, in the heating of water, it takes a significant addition of energy to raise the temperature. Therefore, there will be a lag from the time heat is applied to the time the temperature actually rises. The significance of this phenomenon is that a system disturbance will not immediately result in a deviation from the set-point. More important, however, once the error has been detected, the system responds just as slowly to the corrective action. To overcome this sluggish response, an exaggerated correction is required. In other words, if the controller produces a large corrective signal in response to a minute error, the system will be brought back under control more quickly (even though it possesses a large inertia). There is a flaw with this theory, however, for if the corrective signal remains large, the controller will overcompensate for the error and possibly break into oscillation. What is really required is a corrective action that is initially large but tapers off as time goes on. Such an output is characteristic of a *derivative controller*.

The derivative controller has as its basic element a differentiator. Remember from Chapter 6 that the output of a differentiator is proportional to the rate of change of its input. Mathematically, the output may be expressed in terms of the input as follows:

$$V_{\text{out}} = -RC \left(\frac{dV_{\text{in}}}{dt} \right) \tag{13.8}$$

In this equation dV/dt (the derivative of V with respect to t) is simply the time rate of change of the input voltage. If the input is zero or a constant DC, the output of the differentiator will be zero. If the input is a step function, the steep slope will surely send the differentiator into saturation. If, however, the input changes at a

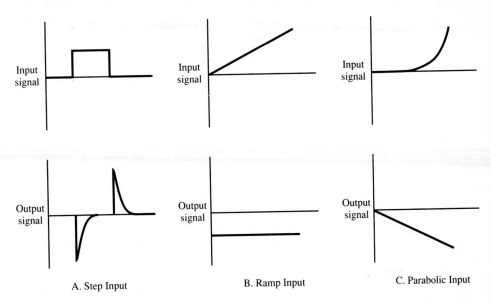

FIGURE 13–14
Ideal Differentiator Output Responses

linear rate (a ramp), the output will maintain a constant value that is equal to the slope of the function (expressed in volts/second) multiplied by the product RC. More complex input functions naturally produce more complex outputs. Calculation of these requires a familiarity with calculus.

Equation 13.8 assumes ideal circuit operation. It also assumes that the input function is known and can be accurately described by a mathematical equation. Remember that no circuit will produce ideal results. In addition, very often the input function is nothing better than an approximation. By analyzing some typical inputs, however, the response of the output can be predicted. Shown in Figure 13–14 are some common input functions, along with the ideal differentiator outputs that would be produced.

Examination of equation 13.8 shows that as the rate of change of the input signal increases, the output of the differentiator increases proportionally. High-frequency transients (noise) that are induced in the circuit will produce substantial outputs, at times even saturating the amplifier. This tendency can be reduced by inserting a resistor in series with the input capacitor. The high-frequency gain is then limited to the ratio of the feedback resistor to the input resistor.

A more significant drawback is that a derivative controller responds only to *changes* in the error signal. That is, if the system has a steady-state error, a derivative controller will take no corrective action. Unlike ON/OFF, proportional, and integral controllers, a derivative controller is never used alone. Instead, it is implemented in the proportional-integral-derivative (PID) controllers that have become the industry standard.

13–8   PROPORTIONAL-INTEGRAL-DERIVATIVE (PID) CONTROLLERS

As may be inferred from the name, proportional-integral-derivative (PID) controllers make use of the attractive attributes of all three controllers. The proportional block provides fast response to system disturbances, the derivative portion ensures that sudden disturbances will be met with an aggressive attempt to correct the error, and the integral section provides a means of eventually eliminating the error altogether.

Although many PID variations are possible, a common parallel configuration, shown in Figure 13–15, will be examined here. Each element receives the same error signal, and the outputs of all the elements are added through a summing amplifier. Since the response characteristics of all three blocks have been discussed, it should be a simple task to predict the output response to a change in error signal, as long as the error is represented by a relatively simple function.

The process of adjusting each of the three blocks in a PID controller is called *tuning*. The manner in which a PID controller must be tuned depends on the configuration of the controller, the characteristics of the process being controlled, and the desired controller performance. That is, if the same controller configuration was applied to two different processes, each would require a different tuning procedure. The procedure of tuning is by no means a simple task. Literature published by the manufacturer of the controller is often used as a guideline, although computer simulation programs have become popular because results can be observed quickly, without the necessity of starting up the process. Like any software simulation, however, the accuracy of the results depends on how well the system response can be modeled.

Two precautions must be observed when implementing PID control. Both stem from the fact that the action of the integral or derivative block can mask the effects of the other blocks in the controller. For example, if there is a sudden change (step) in error, the derivative block will most likely saturate, causing a corresponding saturation to occur in the summing amplifier. This sudden error change could be caused by a disturbance in the process or by a change in the set-

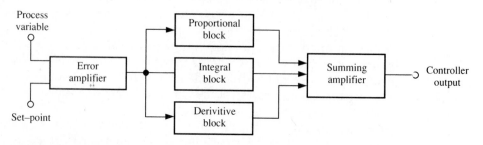

FIGURE 13–15
Block Diagram of a Typical Parallel PID Controller

point. The result may be an overcompensation, causing the process to oscillate. As another example, if a large error is present in the system for a substantial period of time, the output of the integral block may be forced into saturation. Even if the error is driven back to zero, the integral output will remain at saturation. This output, too, will cause the process to overshoot, until the resultant negative error brings the integral block out of saturation.

EXAMPLE 13–5 Given the circuit of Figure 13–15, along with the error signal shown in Figure 13–16, sketch the resultant controller response. Assume that the error amplifier, the proportional block, and the summing amplifier are all operating with unity gain. In addition, assume that the *RC* products for the integral and derivative blocks are 1 s and 0.2 s, respectively.

Examination of the given error signal reveals that there are three distinct time intervals to be analyzed. For each interval, the output signal from each of the three controller blocks must be calculated and sketched. When all three responses have been obtained, it will be a simple matter to add them algebraically to obtain the overall controller response.

Interval 1 (0–1 s)
For this first interval, the error remains at zero. Here it will be assumed that each of the outputs was previously at zero. Since the error is not changing, the outputs of the integral and derivative blocks will remain at zero. Because the error is at a level of 0 V, the proportional block will likewise output 0 V for this interval.

Interval 2 (1–3 s)
The error signal steps down to a level of −1 V at the beginning of this interval, and it remains at that level for a duration of 2 s. In response, the proportional block, having a gain of 1, will step up to +1 V for the same duration. The integral block output will begin to ramp linearly, and at the end of the interval, it will have attained a value calculated as follows:

$$V_{out} = -\left(\frac{-1 \text{ V}}{1 \text{ s}}\right) 2 \text{ s} + 0 \text{ V} = +2 \text{ V}$$

Therefore, the integral block output will ramp linearly from 0 V at 1 s to +2 V at 3 s. The differentiator, on the other hand, will be driven to positive saturation immediately in response to the negative step of the error signal at 1 s. As the differentiator capacitor charges, the derivative block output will fall. Due to the 0.2 s time constant of the resistor and the capacitor, the capacitor should be fully charged after 1 s (5 time constants). The output, therefore, should follow the characteristic exponential decay curve from $+V_{sat}$ at 1 s to 0 V at 2 s.

Interval 3 (3–4 s)
At the end of the preceding interval, when the error stepped back up to 0 V, the proportional block output likewise fell to 0 V. The integral block will maintain its +2 V output, since its input has fallen to zero. The derivative block, however,

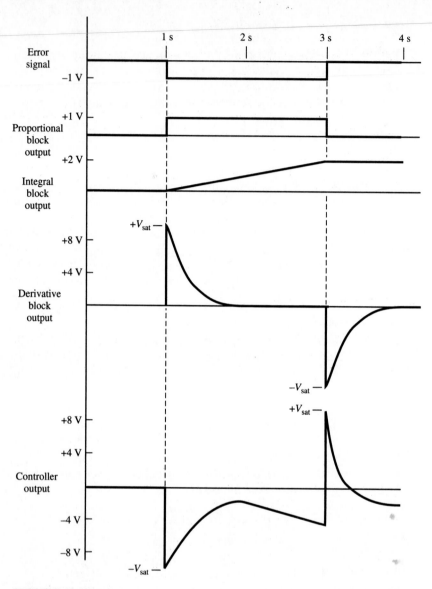

FIGURE 13–16

exhibits much more dynamic behavior. The tremendous positive slope of the error signal step will drive the output of the derivative block into negative saturation. Because of the same capacitor change characteristic described for interval 2, the output will return to zero by the exponential decay curve. After a period of 1 s, the output of the derivative block will again be zero.

13–9   DIGITAL CONTROLLERS

If a controller employs digital circuit technology, it is classified as a digital controller. Although digital techniques are often employed in signal conditioning and data transmission, it is uncommon to encounter discrete digital control circuits made up of individual small-scale-integration (SSI) or medium-scale-integration (MSI) components. If, in fact, a stand-alone, dedicated controller is implemented, it will most surely be constructed with microprocessors or microcontrollers. An average 8-bit microprocessor can be obtained today for less than $2. An 8-bit single-chip microcontroller with on-chip input/output ports and erasable programmable read-only memory (EPROM) sells for about $8. It simply is not economical to produce discrete digital control circuits. The rule is that if the circuit is too simple to warrant the use of a microprocessor, it probably will not be implemented digitally.

Moreover, there is a growing class of general-purpose digital controllers that have become permanent residents in industry. These are the programmable logic controllers (PLCs). In 1988 more than 65 companies manufactured and marketed their own versions of this versatile piece of control equipment. Today's PLCs can input and output analog or digital data. They can compute complex PID algorithms, make decisions based on user or electronic input, and control a wide variety of output actuators. In addition, the controller can be reconfigured simply by changing the control program. What has made the PLC so popular lately is the price. Small controllers can be had for as little as $100. It is not hard to understand why dedicated discrete digital control has all but disappeared. For more information on PLCs, see Chapter 15.

SUMMARY

- Electronic control systems are used to compensate automatically for external disturbances to processes. Control systems are used to carry out operations in hazardous locations, to reduce variations in product quality, and to reduce production costs.
- The electronic "brain" behind the control system is a group of components that are collectively called the controller. The controller accepts the system error signal and produces an output that is used to correct the error condition.
- An ON/OFF controller attempts to maintain control of the process by switching the output actuator ON or OFF according to the magnitude of the error signal. An ON/OFF controller cannot reduce the system error to zero. The range of system error that will cause the controller output to cycle from ON to OFF is referred to as the deadband.
- A proportional controller maintains control by generating an output proportional to the system error signal. Although a proportional controller usually results in a residual system error lower than that of an ON/OFF controller, it still cannot reduce the error to zero. The range of system error for which the controller will produce a proportional output is referred to as the proportional band.
- An integral controller produces an output whose rate of change is proportional to the magnitude of the system error signal. Since the controller can maintain an output even when its input is zero, this controller can eliminate system error altogether.

☐ Proportional-integral controllers combine the swift response of proportional controllers with the error elimination capability of integral controllers. Since integration is a function of time, these controllers must be tuned to ensure that the output can respond to variations in the process being controlled.

☐ Proportional-integral-derivative (PID) controllers produce outputs that depend on the magnitude, duration, and rate of change of the system error signal. Sudden system disturbances are met with an aggressive attempt to correct the condition. A PID controller can reduce the system error to zero faster than any other controller. Because it has an integrator and a differentiator, however, this controller must be custom-tuned to each process being controlled.

EXERCISES

1. Give two examples of open-loop control systems. Identify the controller, the output actuator, and the process for each.

2. Give two examples of closed-loop control systems. Identify the error amplifier, the controller, the output actuator, and the process for each.

3. State one common disadvantage that ON/OFF and proportional controllers share.

4. For the circuit of Figure 13–3, what circuit change or modification would increase the circuit deadband?

5. How would an ON/OFF controller respond if the deadband were too narrow?

6. List two ways to decrease the proportional band of the controller in Figure 13–6.

7. If a home heating system is controlled by an ON/OFF controller, what will be the effect of widening the deadband?

8. If more precise control is desired from a proportional controller, should the proportional band be increased or decreased?

9. In Example 13–2, if the full-scale error is 8.5 V, what is the proportional band, expressed as a percentage?

10. Which of the following output actuators is likely to be driven by an ON/OFF controller?
   a. DC motor
   b. power amplifier
   c. solid-state relay
   d. solenoid valve

11. Which of the following output actuators is likely to be driven by a proportional controller?
   a. DC motor
   b. power amplifier
   c. solid-state relay
   d. solenoid valve

12. Give an example of a process for which it would be advantageous to use PID control instead of proportional-integral control.

13. When a proportional-integral or PID controller is tuned, what controller characteristic is actually being adjusted?

14. Given the circuit of Figure 13–3, assume that $R_6 = R_7$. Draw the response curve for the circuit.

15. For Exercise 14, if the full-scale error is 10 V, what is the circuit deadband, expressed as a percentage?

16. In Figure 13–6, suppose that $R_7 = R_8 = 1$ kΩ. Suppose also that $R_9 = 3.3$ kΩ. If U_3 is operating with unity gain and the offset is adjusted to 0.5 V, draw the circuit response curve, and calculate the controller proportional band.

17. Given the information in Exercise 16, rework the problem for the situation in which R_{10} and D have been removed. (Assume that U_2 and U_3 saturate at ± 10 V.)

18. For Example 13–3, redraw the output response for $R_6 = 1$ kΩ. Use the given error signal.

19. For Figure 13–12, redraw the output for the condition where the proportional block has a gain of 2. Use the given error signal.

20. For Example 13–4, redraw the output response for the condition where the proportional block has a gain of 1 and the RC product for the integral block is 0.5 s. Use the given error signal.

21. For Example 13–5, show how the output response would change if the RC time constants for both the integral and derivative blocks were changed to 0.5 s. Use the given error signal.

APPLICATION ASSIGNMENT ANSWER

There are actually two controller adjustments that can cause the output to oscillate: the gain and the offset. If the gain is set too high, the slightest change at the input may force the output to saturate. When the air temperature changes and the sensor responds, the temperature will overshoot (or undershoot) the set-point. The period of the oscillation will depend on the time it takes to change the air temperature and on the response time of the sensor.

If, on the other hand, the offset is set too low, every time the air temperature is in compliance with the set-point, the controller output will fall to zero. The zero output will decrease the temperature of the air. When the sensor detects the discrepancy between the set-point and the actual temperature, the controller output will be forced to increase, only to fall to zero again when the temperature has risen to the proper level. The oscillator period for this problem is likewise dependent on the sensor response time and the time it takes to change the air temperature.

14

Closed-Loop Systems

CHAPTER OUTLINE

LEARNING OBJECTIVES

- □ Develop a systematic approach to analyzing an electronic control system.
- □ Discuss the implementation of ON/OFF control in practical applications.
- □ Discuss the implementation of proportional control in practical applications.
- □ Discuss the concept and the operational characteristics of pulse-width-modulated (PWM) proportional controllers.
- □ Discuss the concept and the operational characteristics of phase-controlled proportional controllers.

APPLICATION ASSIGNMENT

In an industrial research facility, the PWM controller shown in Figure 14–4 is being used to regulate the temperature of a steam pressure vessel. Although the controller has been used successfully in other applications around the facility, the temperature of the vessel cannot be brought under control. Regardless of variations of the temperature adjust, offset adjust, and gain adjust potentiometers, the controller output seems always to be fully ON.

Since you suspect a circuit failure, you decide to send one of your technicians to investigate the problem. Suggest at least three faults likely to result in the operation observed. In addition, describe how each can be verified.

14–1   INTRODUCTION

Previous chapters have examined the various components that make up an electronic control system. Most of these components, however, were presented as isolated subsystems. Although it is important to develop a strong understanding of individual control-system components, it is just as important (if not more so) to become familiar with the operational characteristics of an entire system.

It is not the intent of this chapter to present new concepts or ideas. Rather, a series of examples will show how the various control system components interact to maintain control of a single process variable. Commercially available components will be used; actual part numbers will be given, and any applicable characteristics will be discussed.

14–2   ON/OFF TEMPERATURE CONTROL

Overview

For certain chemical reactions to progress properly, the reactants must be combined in a controlled atmosphere. In many cases the temperature in the reaction chamber must be maintained within a given range. Figure 14–1 shows just such an arrangement.

In this particular case, three gas mixtures are to be combined in the reaction chamber. For the proper products to be obtained, the temperature within the chamber must be maintained between 300 °C and 500 °C throughout the reaction.

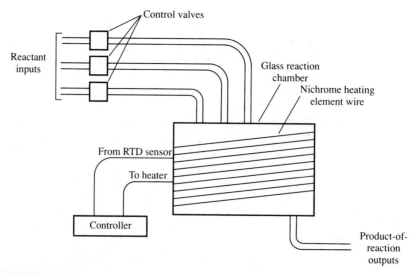

FIGURE 14–1
System for Combining Gases at Constant Temperature

As the reactants enter the chamber, the internal temperature will drop. In addition, depending on the nature of the gases, the mixtures may actually generate heat during the course of the reaction, raising the internal temperature. Therefore, some method must be employed to regulate the temperature.

System Description

Gas mixtures have a very low thermal mass. That is, their temperature can be changed relatively quickly. In addition, since the required temperature range is so wide (200 degrees), the system will tolerate a certain amount of controller overshoot. Therefore, the ON/OFF temperature controller shown in Figure 14–2 should provide satisfactory control for this application.

    The system consists of a 240 V AC, 2000 W heater that is switched ON and OFF by a solid-state relay. The relay, in turn, is driven by the output of an op-amp ON/OFF controller. Process-variable feedback is provided by an RTD sensor assembly suspended inside the reaction chamber.

    The RTD sensor is made up of a fixed resistor and an Omega Engineering Co. 1Pt100K2028 RTD. This particular RTD has a resistance of 100 Ω at 0 °C and a positive temperature coefficient of 0.00385 $\Omega/\Omega/°C$. Basically, this means that the RTD resistance will change by 0.385 $\Omega/°C$. In addition, the usable temperature range of the RTD extends from -200 °C to $+750$ °C. The series resistor, R_2, creates a voltage divider that yields a temperature-dependent voltage. Examination of Figure 14–3 shows that this voltage decreases as the temperature of the RTD increases. Moreover, it shows the voltages that will be produced within the required temperature range for this system. At 300° the sensor yields an output of 9.5 V, and at 500° the output is 7.5 V.

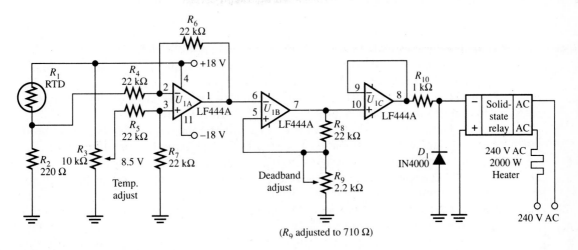

FIGURE 14–2
ON/OFF Temperature Controller for Reaction Chamber in Figure 14–1

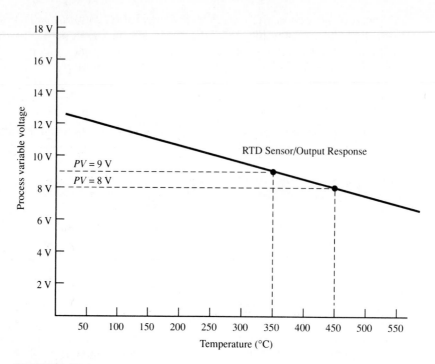

FIGURE 14–3
RTD Sensor Temperature versus Output Response

Since ON/OFF controllers characteristically overshoot their thresholds, it would not be wise to use 300° and 500° as the thresholds for this controller. If the heater is shut off just as the temperature reaches 500°, the temperature will continue to rise for a short period of time. Likewise, applying power to the heater as the temperature falls to 300° will allow the temperature to fall below 300° until the heater is hot enough to raise it to the proper level again. Instead, the thresholds will be set at 350° and 450°, allowing a 50° overshoot to occur while still maintaining the required temperature range. Figure 14–3 shows that 350° and 450° correspond to sensor voltages of 9 V and 8 V, respectively.

The controller consists of a unity-gain differential amplifier and a buffered comparator with adjustable positive feedback. The output of the differential amplifier (the error voltage) reflects the difference between the "temperature-adjust" and RTD sensor voltages. Since U_{1A} amplifies only the difference between its input terminals, its output can swing either positive or negative. It is, in fact, necessary for U_{1A} to produce both positive and negative outputs, because the polarity of the comparator threshold shifts each time the comparator output changes. With the temperature adjust set at 8.5 V (as it is in Figure 14–2), the error voltage varies from -0.5 V at 350°, to $+0.5$ V at 450°. To maintain this temperature range, the comparator must switch when one of these two thresholds is crossed. That is, R_9 must be adjusted so that the feedback voltage is +0.5 V when U_{1B} saturates positive, and −0.5 V when it saturates negative. With saturation potentials of +16 V and −16 V, R_9 must be adjusted to 710 Ω to generate these feedback voltages.

The output is isolated from the solid-state relay by a voltage follower. The purpose of U_{1C} is to ensure that U_{1B} is not loaded by the solid-state relay, resulting in asymmetrical saturation voltages. Any current demanded by the solid-state relay is provided by the voltage follower rather than by the comparator directly.

The solid-state relay requires a DC input anywhere between 3 V and 32 V to reliably trigger the output. The input current requirement is only about 5 mA. Since the input is optically coupled to the output internally, there is virtually no danger that a circuit fault will apply 240 V AC directly to the controller. The output will operate properly with load potentials anywhere between 24 V AC and 280 V AC and will withstand continuous currents of 10 A (or a 90 A surge current for 20 ms).

In addition, this relay has a zero-voltage-switching (ZVS) characteristic. That is, even if the input is provided with the necessary trigger voltage and current, the output will not switch into conduction unless the load potential is less than about 10 V AC (technically, relays of this type should be called near-zero-voltage-switching). Naturally, after the output has switched ON, it will remain ON for the remainder of the cycle. This action prevents the possibility of applying a high voltage to a cold heating element. When a heating element is cold, its resistance may be just a few ohms. Application of a high voltage would result in excessive current. As the temperature of the element increased, so would its resistance. Eventually (within several cycles of the AC line), the resistance of the element would have risen to its normal operating value. For an instant, however, the solid-state relay might have been subjected to several hundred amperes of surge current. Besides the detrimental effects high current surges would have on the relay, they would produce an undue amount of electromagnetic interference (EMI). Use of a ZVS relay ensures that the initial current surge through the load is held to a very low value.

System Operation

Suppose that the system has been shut down for a period of time. The heater will have returned to ambient temperature. If the system is turned ON, the following sequence of events will occur:

According to Figure 14–3, at room temperature (about 25 °C) the process-variable voltage will be nearly +13 V. With the temperature-adjust voltage set at +8.5 V, the output of U_{1A} in Figure 14–2 will be approximately −4.5 V. Since the heater is presently OFF, the output of U_{1B} must be at negative saturation. The feedback voltage present at pin 5 of U_{1B} will therefore be −0.5 V (remember that R_9 is adjusted to 710 Ω). The output of U_{1A}, being even more negative than −0.5 V, will cause U_{1B} to saturate positive (as it should, since the temperature is far too low). The feedback voltage will now be +0.5 V. As the heater increases the temperature in the reaction chamber, the process-variable voltage will fall. This voltage drop will in turn cause the output of U_{1A} to increase. To cause the output of U_{1B} to switch to negative saturation again, however, the output of U_{1A} must increase above +0.5 V. It does so when the process-variable voltage is at +8 V (corresponding to a temperature of 450°). Even after the heater shuts OFF, the chamber temperature may continue to rise slightly. Eventually, however, the temperature will begin to decrease. Since the feedback voltage at pin 5 of U_{1B} is now −0.5 V, the temperature will have to fall to a value low enough to cause the output of U_{1A} to fall below −0.5 V. At that temperature, 350°, the heater will be turned back ON.

14–3   PWM PROPORTIONAL TEMPERATURE CONTROL

Overview

Suppose that the required temperature range for the system shown in Figure 14–1 is much more critical. In particular, suppose that the range must be much more narrow. The previously discussed ON/OFF controller will no longer provide satisfactory control.

The system shown in Figure 14–4 is a proportional pulse-width-modulated (PWM) controller. It will maintain the temperature much closer to a set-point by varying the ratio of ON time to total cycle time (ON time + OFF time) for the solid-state relay. In this manner, the effective amount of energy imparted to the load (and so also the temperature of the load) can be precisely controlled.

System Description

The system consists of an op amp proportional controller, a triangle-wave oscillator, and a pulse-width modulator. Process-variable feedback is provided by a thermocouple whose signal has been conditioned by a thermocouple transmitter. The 240 V AC, 2000 W heater is switched ON and OFF by a solid-state relay driven by the output of the pulse-width modulator.

The thermocouple sensor is made up of a type E thermocouple, and an Omega TX30E thermocouple transmitter. The thermocouple has a useful temperature range much wider than the range to which it will be exposed in this application. The TX30E transmitter performs several functions. First and foremost, it produces a linear output over the selected temperature range. It also provides cold-junction compensation for the thermocouple (therefore, the TX30 must be matched to the thermocouple type being used). In addition, the TX30E provides DIP switches and trim pots that permit the selection of output type (current loop or voltage), output range (0–5 V, 0–10 V, 4–20 mA, etc.), and temperature range. For example, in this particular application, the TX30E has been configured to produce an output of 0–10 V over a temperature range of 0°–600° (0 V at 0° and 10 V at 600°). With this information, output can be plotted against temperature to give an accurate indication of the actual chamber temperature (see Figure 14–5).

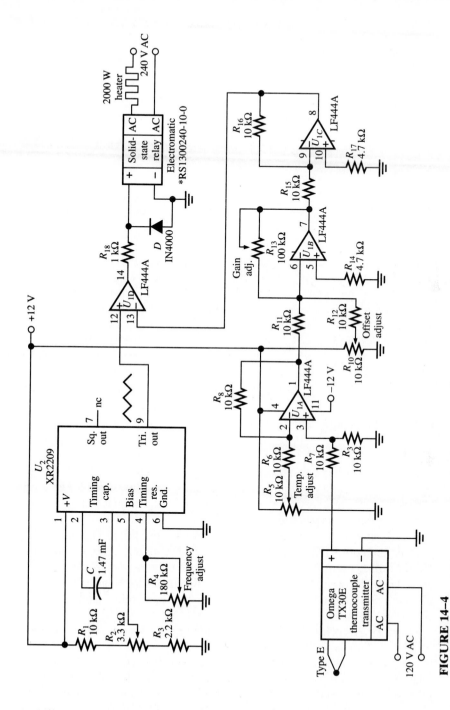

FIGURE 14–4
PWM Temperature Controller for Reaction Chamber in Figure 14–1

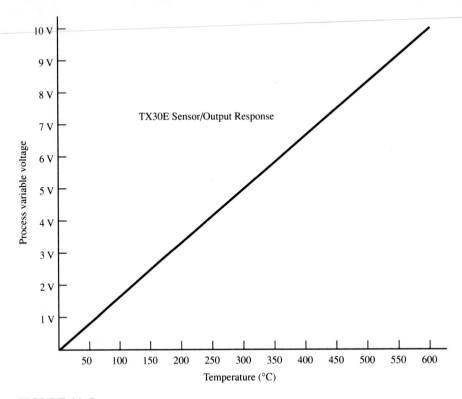

FIGURE 14–5
Output of the TX30E Temperature Transmitter

The controller consists of a unity gain error amplifier, an inverting summing amplifier, and a unity gain inverter. The error signal produced at the output of U_{1A} represents the difference between the temperature-adjust voltage (the set-point) and the process-variable feedback voltage (the TX30E output). In essence, U_{1B} and U_{1C} together form a noninverting summing amplifier. The main function is to provide the controller with an output offset adjustment. This adjustment is important because, when the chamber temperature coincides with the set-point temperature (zero error), it prevents the controller from turning OFF the heater completely. The potentiometer R_{13} adjusts the controller gain, and so also its sensitivity to temperature changes.

An EXAR XR2209 integrated function generator serves as the triangle-wave oscillator. The oscillator output frequency is given by the reciprocal of the product $(R_4 \times C)$. In this application, R_4 is adjusted to yield an output frequency of 6 Hz. The signal is centered at a voltage 0.6 V above the bias voltage (the voltage present at pin 5), and its amplitude is approximately one half of the supply voltage. The values of R_1, R_2, and R_3 have been chosen to allow adjustment of the bias voltage so that the output signal varies between 0 V and about +6 V.

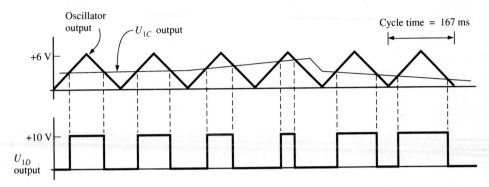

FIGURE 14–6
Output of U_{1D} in Figure 14–4

   The pulse-width modulator is nothing more than an op-amp comparator. The noninverting input is driven by the triangle-wave oscillator, and the inverting input is connected to the controller output. The comparator output will be at positive saturation (approximately 10.5 V) as long as the oscillator output is at a higher potential than the controller output. Naturally, when the controller output voltage exceeds the oscillator output voltage, U_{1D} will be driven to negative saturation. The output of U_{1D} is clamped to -0.7 V by R_{18} and D so that the solid-state relay is never subjected to an excessive negative voltage. Figure 14–6 shows how the output of U_{1D} is affected by its input signals. Note that with the triangle-wave oscillator frequency set at 6 Hz, the pulse width is variable between zero and 167 ms.
   The solid-state relay used in this system is identical to that used in the ON/OFF temperature controller described previously. It will reliably switch power to the load with an input voltage anywhere between $+3$ V DC and $+32$ V DC. In addition, it will handle load currents of up to 10 A at 240 V AC. The most important aspect of its operation, however, is its zero-voltage switching. Remember that this solid-state relay will switch power to the load only if the line voltage is within about 10 V of zero at the time the relay receives an input signal. Therefore, in this application, power will be delivered to the load in integral numbers of half cycles of the line. That is, the duty cycle is variable between 0% and 100% in increments of 5%. Figure 14–7 shows the relationship between the output of U_{1D} and the load voltage.

System Operation

As with the last system discussed, it will be assumed that this system has been shut down long enough for the heater to have returned to ambient temperature. It will also be assumed that the temperature-adjust voltage is set at 6 V, the offset-adjust voltage is set at 1.5 V, and R_{13} is adjusted to yield a gain of 2.

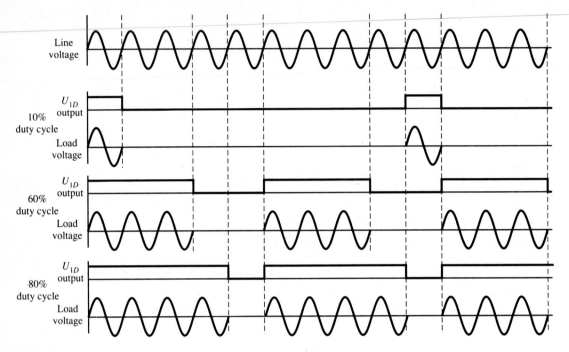

FIGURE 14–7
Duty Cycle of Output of U_{1D} in Figure 14–4, Related to Load Voltage

Ambient temperature (about 25 °C) corresponds to an output of 0.5 V from the temperature transmitter (see Figure 14–5). With the temperature-adjust voltage set at 6 V, the output of U_{1A} will be −5.5 V. The output of the summing amplifier is simply the gain multiplied by the sum of the offset-adjust and U_{1A} output voltages. At 25° this is 2(1.5 V − 5.5 V) = −8 V. Therefore, at all times, the oscillator output will be at a higher potential than the output of U_{1C}. Therefore, the output of U_{1D} will be saturated positive, and the solid-state relay will be providing continuous power to the heater (as it should, since the temperature is too low). Mathematically, the output of U_{1C} can be described as follows:

$$V_{U_{1C}} = 2[1.5 \text{ V} + (V_{TX30E} - 6 \text{ V})] \tag{14.1}$$

As the temperature begins to rise, the output of U_{1C} changes. From equation 14.1 and Figure 14–5, the expected output can be calculated for various temperatures.

Temperature (°)	U_{1C} Output (V)
100	-5.5
150	-4.0
200	-2.2
250	-0.6
300	$+1.0$
350	$+2.8$
400	$+4.4$
450	$+6.0$
500	$+7.8$

Note that the output of U_{1C} does not increase above zero until the temperature has risen to a point between 250° and 300°. Since the oscillator output varies between 0 V and 6 V, the solid-state relay will be continuously ON until this temperature is reached. When the temperature has reached 300°, the solid-state relay will not be turned ON until the oscillator output increases above $+1$ V. This corresponds to a duty cycle of about 80%. The temperature will still rise, although at a slower rate. By the time the temperature reaches 350°, the output of U_{1C} will be nearly $+3.0$ V, and the duty cycle will have decreased to about 50%. Note that when the error voltage (U_{1A} output) has decreased to zero, the output of U_{1C} will be $+3.0$ V. That is, 350° corresponds to zero error. This is the theoretical temperature setting of the controller.

It is difficult to say whether a 50% duty cycle will cause the chamber temperature to rise, to fall, or to stay the same. What is certain, however, is that if the temperature does continue to rise, the duty cycle will further decrease. If the temperature begins to drop, the duty cycle will increase. In any case, the controller will vary the duty cycle proportionally over a temperature range of approximately 250°–450°. The actual chamber temperature will not vary nearly this much, however. Instead, it will vary slightly around the set-point. The amount by which the actual temperature fluctuates depends on the thermal characteristics of the chamber and, to a lesser extent, on the controller gain. Increasing the gain will cause the controller to vary the duty cycle more drastically for smaller temperature variations. In other words, changing the controller gain changes its sensitivity. If, instead, the temperature-adjust voltage is increased, the chamber will attain a higher temperature before the duty cycle decreases enough to begin cutting back on power to the heater.

14–4   ON/OFF SUMP PUMP CONTROL

Overview

Most buildings with foundations that extend more than 4 ft underground require a system that will channel water away from the walls. This prevents hydraulic pressure from cracking the foundation and undermining the building substructure.

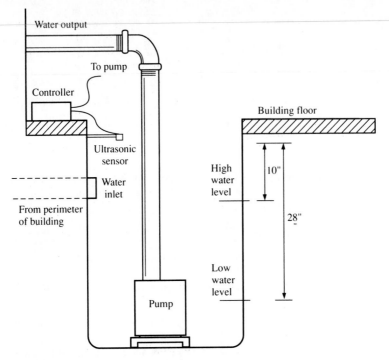

FIGURE 14–8
Typical Sump Pump Arrangement

As water filters down along the foundation, it is collected in semiperforated pipe that surrounds the structure, and it is channeled into a central pit inside the building (see Figure 14–8). As the pit fills, the water is pumped out at a controlled rate and is deposited somewhere distant from the building. To ensure that the pump is not run continuously, it should be switched ON only after the water in the pit has risen above some high-level threshold, and it should not be switched OFF until the water has fallen below some low-level threshold. The purpose of the electronic control system that controls the pump is not to maintain a given water level in the pit, but rather to maintain (as nearly as possible) an empty pit.

Although many systems have been devised to handle the problem of controlling such a pump, most are rather simple, employing electromechanical float-switches of some sort. Regardless of the quality of the switch, there is always the possibility that the switch may jam, become clogged with debris, or become corroded to the point of failure. The system shown in Figure 14–9 eliminates this possibility by utilizing a noncontact ultrasonic level sensor.

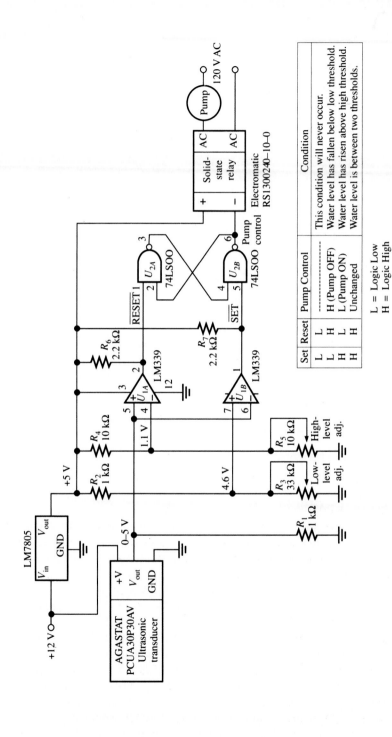

Set	Reset	Pump Control	Condition
L	L	-----------	This condition will never occur.
L	H	H (Pump OFF)	Water level has fallen below low threshold.
H	L	L (Pump ON)	Water level has risen above high threshold.
H	H	Unchanged	Water level is between two thresholds.

L = Logic Low
H = Logic High

FIGURE 14–9
ON/OFF Controller for Sump Pump

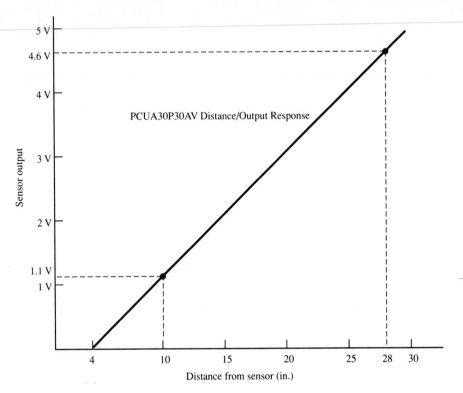

FIGURE 14-10
Output of PCUA30P30AV Ultrasonic Sensor

System Description

The sensor chosen for this system is an Agastat PCUA30P30AV. It is a self-contained, threaded, cylindrical device, requiring only three electrical connections—two for power (+12 V) and one for the output. This device bounces ultrasonic acoustic energy off the water surface and, from the echo time, calculates the distance that the surface is from the sensor. As is evident from Figure 14-10, the sensor output varies linearly between 0 V and 5 V, depending on the target distance. As the target distance increases, so does the sensor output. The minimum sensing distance is 4 in. (producing an output of 0 V), and the maximum sensing distance is 30 in. (producing an output of 5 V).

The controller in Figure 14-9 is made up of a window comparator and a cross-coupled NAND flip-flop. The comparator provides the high- and low-level threshold adjustment, and the flip-flop introduces the necessary hysteresis. The LM339 is a quad single-supply comparator. In this application it is supplied with +5 V to ensure compatibility with the input requirements of the 74LS00. Resistors

R_6 and R_7 function only as pull-up resistors, as the LM339 has open-collector outputs. Resistor pairs R_2 and R_3, and R_4 and R_5 form voltage dividers that establish the low- and high-level thresholds for the controller. The output of the ultrasonic sensor is compared with each of these.

Current through the pump motor is controlled by a solid-state relay. This particular device will trigger with input voltages anywhere between +3 V DC and +32 V DC and will control 120 V AC or 240 V AC loads at currents up to 25 A. With +5 V DC connected to the positive input of the relay, it is necessary only to provide a current path from the negative input to ground to cause the relay to trigger. The output of U_{2B} provides this path. Whenever the output of U_{2B} is at a logic LOW, there is a low-resistance path from the output to ground.

System Operation

To fully understand the operation of this system, it is necessary to follow it through one complete cycle. Start by assuming that the water has just been pumped down to the low-level threshold, the pump has been shut OFF, and water continues to enter the sump pit. Suppose that at the instant the pump shuts off, the output of the ultrasonic sensor was +4.65 V. According to Figure 14–10, this corresponds to a distance of about 28 in. from the sensor. Given the reference voltages set by R_3 and R_5 in Figure 14–9, the outputs of U_{1A} and U_{1B} will be +5 V (logic HIGH) and 0 V (logic LOW), respectively. With the $\overline{\text{SET}}$ input to the flip-flop at a logic LOW, the pump-control output will be forced to a logic HIGH. This will inhibit current through the input of the solid-state relay and will shut OFF the pump motor. Since water continues to enter the sump pit, the water level will begin to rise, causing the sensor output to decrease. When the sensor output falls just below +4.6 V, the output of U_{1B} will switch to a logic HIGH (note that the output of U_{1A} will remain unchanged). Because of the latching effect of the cross-coupled NAND gates, the pump-control output will be unaffected by the change of state in U_{1B}, and the pump motor will remain OFF. Not until the water rises to a level about 10 in. from the sensor (corresponding to a sensor output just below the +1.1 V reference voltage for U_{1A}) will the pump motor be switched ON. When the water rises to this height, the output of U_{1A} will be driven to a logic LOW (note that the output of U_{1B} will be at a logic HIGH at this time). This action will force the pump-control output to a logic LOW, allowing current through the input of the solid-state relay. Power will therefore be applied to the pump motor. As the water level begins to fall, the outputs of U_{1A} and U_{1B} will again be forced to logic HIGHs (having no effect on the state of the motor). Only after the water has been pumped down to a level that corresponds to a sensor voltage just above the +4.6 V reference for U_{1B} will the pump motor again be shut OFF.

Since this system does not attempt to maintain a particular water level, the term *set-point* has little meaning. Therefore, there is no set-point adjustment. It is, however, possible to adjust the controller deadband by adjusting the values of R_3 and R_5. Moreover, it is possible to adjust the high and low water levels independently of each other. For example, if the voltage across R_3 is decreased to +3.1 V, the pump will shut OFF when the water level has fallen to a level that is only 20 in. (as opposed to 28 in.) away from the sensor. If the voltage across R_5 has not been changed, the pump will still be switched on when the water level rises within 10 in. of the sensor.

With R_3 and R_5 set as they are in Figure 14–9, the pump will be forced to pump out approximately 70 gal of water (if the pit diameter is 2.5 ft) every time it is turned ON. If water continues to enter the pit while the pump is ON, the pump will have to be able to pump at an even faster rate. If the pump cannot empty the pit as fast as it is being filled, a decrease in the high-level threshold (a decrease in the value of R_5) may be desired. This will force the pump to pump out smaller volumes of water more often. Then water from the building perimeter will no longer get a 70 gal head start on the pump.

14–5   PROPORTIONAL MOTOR-SPEED CONTROL

Overview

As certain plastic films are manufactured, small thermoplastic pellets are fed into a hopper and melted. The molten plastic is forced out through a narrow slit and forms a continuous sheet. This sheet is drawn across chilled rollers and wound onto a spool (see Figure 14–11). To maintain the proper sheet thickness, it is imperative that the take-up spool draw the film away at a constant speed. Turning

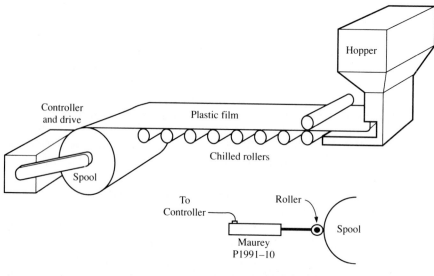

FIGURE 14–11
Process for Manufacturing Plastic Film

the take-up spool at a constant speed is not the solution, for as the diameter of the roll increases, so does its circumference. As a matter of fact, for every inch that the diameter increases, the circumference of the roll increases by more than 3 in. If the take-up spool is turned at a constant speed, the film will be drawn away at a speed that is ever increasing. The film will become increasingly thin and will eventually break.

The velocity of the sheet is related to the radius of the roll by the following equation:

$$v_{\text{sheet}} = 2\pi rf \qquad\qquad (14.2)$$

where f is the rotational frequency of the roll. If the take-up spool were turned at a constant frequency of 20 RPM, the sheet velocity would increase from 8.4 in./s to 16.8 in./s as the roll radius increased from 4 in. to 8 in. It can be seen from equation 14.2 that to keep the sheet velocity constant, the rotational frequency of the take-up spool must be decreased as the radius of the roll increases. In addition, the turning frequency and the roll radius are linearly related. That is, if the radius doubles, the rotational frequency must be reduced by one half if the sheet velocity is to remain constant.

System Description

Figure 14–12 shows a system that will control the speed of the take-up spool in accordance with the radius of the roll. The spool is driven by a gear-reduced universal motor. The speed of the motor is controlled by varying the angle at which the triac (Q_2) fires during each half cycle of the AC line.

To gain control over each half cycle of the AC line, a line-synchronized UJT trigger circuit is employed. Note that T_1, D_1, and D_2 provides full-wave unfiltered DC to the trigger circuit. Even though the rectifier output peaks at approximately 34 V, the voltage across the zener diode, D_3, will never rise above 20 V. Therefore, the trigger circuit will be provided with 20 V synchronization pulses that coincide with each phase of the 120 V AC line (see Figure 14–13). *Synchronization* means that the trigger circuit is provided with a 20 V DC power supply that is turned ON at the beginning of each half cycle of the 120 V AC line and turned OFF at the end of each half cycle.

At the beginning of each synchronization pulse, C is charged to an initial voltage determined by the voltage divider R_2, R_3 and R_4. The capacitor, C, charges to the voltage drop across R_3 almost instantly, because of the low values of R_2 and R_3. The purpose of D_4 is simply to prevent C from discharging through R_4. The capacitor, C, then continues to charge toward 20 V through R_5 and R_6. This method of triac triggering is called *ramp and pedestal control*, for the capacitor begins charging (ramping) from a variable-voltage pedestal.

The value of R_3 (in conjunction with R_4) determines the initial charge on C. Since this will determine how far into each half cycle of the 120 V AC line the triac fires, R_3 controls the initial motor speed (set-point). The speed may then be modulated by variations in R_4, the roll-radius sensor. Resistor R_5 controls the

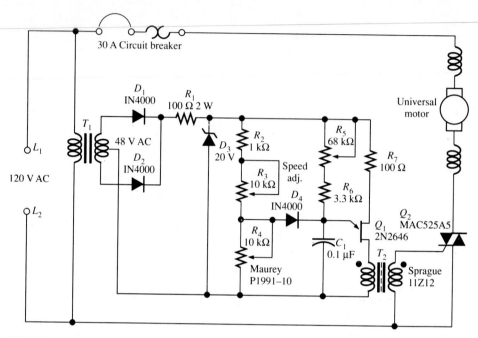

FIGURE 14–12
System for Proportional Motor-Speed Control

charge rate of C after it has been initially charged by the voltage across R_4. This control effectively changes the sensitivity of the circuit to variations in the value of R_4, so R_5 is called the *circuit gain adjust*. Resistors R_2 and R_6 function as protection devices. If the value of R_3 or R_5 is adjusted to zero, R_2 and R_6 will prevent excessive UJT emitter current.

The transducer employed in this system is a Maurey Instrument Corporation type P1991-10 linearly actuated potentiometer (R_4 in Figure 14–12). The wiper of the device is connected to a shaft that has a maximum mechanical travel of 10.25 in. Each inch of travel produces a resistance change, from wiper to either terminal, of 1 kΩ. It should be recognized that R_4 must be connected in such a way that as the roll radius increases, the transducer resistance decreases.

The unijunction transistor, Q_1, has a typical intrinsic standoff ratio of 0.7 and a typical interbase resistance of about 7 kΩ. For the UJT to break over into conduction, the emitter must be biased 0.5 V higher than the voltage drop from emitter to base 1. That is, the emitter peak-point (trigger) voltage is

$$V_{EB_1} = 0.7(20 \text{ V}) + 0.5 \text{ V} = 14.5 \text{ V}$$

(Here it is assumed that the voltage drops across R_7 and the primary of T_2 are both negligible.) Thus, the UJT will conduct after the voltage across C increases to about 14.5 V. When the UJT does break over into conduction, the voltage drop

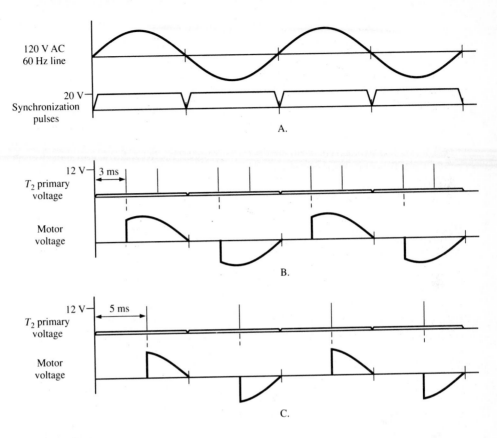

FIGURE 14–13
Synchronization Pulse and Delayed Trigger Pulses

from emitter to base 1 (V_{EB_1}) will fall to perhaps 2 V. If the capacitor is currently charged to 14.5 V, and V_{EB_1} falls to 2 V, 12.5 V will be imposed across the primary of T_2.

Transformer T_2 is a high-frequency transformer with a 1:1 turn ratio between primary and secondary. Transformers of this type, called *pulse transformers,* are designed specifically for triggering thyristors such as triacs and SCRs. Each time the UJT breaks over into conduction, C discharges through the primary of T_2. Because of the low resistance in the discharge path, C provides a very short-duration current pulse to this winding. This pulse is coupled through to the secondary and to the gate of the triac, Q_2. As the triac latches into conduction, power is provided to the motor. The speed of the motor depends on the conduction angle of the triac in much the same way as the temperature of the heater depended on the pulse width of the controller discussed in Section 14–3.

System Operation

Technically, this system operates in an open-loop mode. Although the UJT trigger circuit can control how far into each half cycle the triac fires, it receives no feedback with which to determine whether the motor speed has been correctly adjusted. Furthermore, the circuit is not sensitive to motor load changes.

To determine the conduction angle of the triac, it is necessary first to determine the time it takes C to charge from the pedestal voltage to the UJT trigger voltage. In general, for a constant charging voltage, V_S, the time it takes a capacitor to charge from an initial voltage, V_I, to a final voltage, V_F, is given by the following equation:

$$\Delta t = RC \ln \left(\frac{V_S - V_I}{V_S - V_F} \right) \tag{14.3}$$

For this circuit, R represents the combined values of R_5 and R_6, and C represents the value of C. To illustrate how variations in the value of R_4 will change the motor speed, it will be assumed that R_5 is adjusted so that the RC product in equation 14.3 is 5 ms (R_5 would have to be adjusted to 46.7 kΩ). It will also be assumed that, currently, the value of R_4 is such that the pedestal voltage is 10 V. Substituting these values into equation 14.3 gives the following result:

$$\Delta t = 5 \text{ ms } \ln \left(\frac{20 \text{ V} - 10 \text{ V}}{20 \text{ V} - 14.5 \text{ V}} \right) = 2.99 \text{ ms}$$

That is, 2.99 ms after the synchronization pulse is applied to the UJT trigger circuit, the UJT will break over into conduction. The resultant voltage pulse from T_2 (see Figure 14–13B) will fire the triac and allow motor current to flow. Although C will recharge in another 2.99 ms and trigger the UJT into conduction again, this event will have no effect on the triac, as the triac will continue to conduct until the line voltage crosses zero. Within each 8 ms half cycle of the 120 V AC line, the UJT may trigger the triac once, twice, or even 20 times (depending on the values of R_4 and R_5). However, it is always the first gate pulse that latches the triac into conduction.

Suppose that the value of R_4 decreases because of the increasing radius of the take-up spool. The pedestal voltage will likewise decrease. If, in fact, the pedestal voltage decreased to 5 V, the triac would not receive a gate pulse until 5 ms after the start of the synchronization pulse. This fact can be verified by again substituting the appropriate values into equation 14.3. It can be seen from Figure 14–13C that as the value of R_4 decreases (as the roll radius increases), the conduction angle of the triac likewise decreases. The speed of the motor, therefore, will vary proportionally with the radius of the roll.

Examination of equation 14.3 reveals why R_5 functions as a gain adjust. Note that the RC product acts as a multiplier for the right-hand portion of the equation. Although the value of R_4 directly affects only the initial capacitor voltage, the value of R_5 determines the overall range of control.

14-6   PROPORTIONAL MOTOR-SPEED CONTROL WITH FEEDBACK

Overview

In the previous section the speed of a universal motor was phase-controlled by the position of a linear potentiometer shaft. That system, though accurate and stable, cannot compensate for speed variations due to motor load changes. The arrangement shown in Figure 14–14 adds the necessary circuitry to sense the motor speed indirectly and to generate a feedback signal that may be used to correct any system error. Although this system may be used as a general-purpose motor-speed controller, it will be applied here in the same manner as the system in Section 14–5.

System Description

Close inspection of the circuitry of Figure 14–14 will reveal that it is very similar to that of Figure 14–12. The line synchronization and the ramp and pedestal section are virtually identical. The main difference is the substitution of a 2N6027 programmable unijunction transistor (PUT) for the 2N2646 unijunction transistor (UJT) of Figure 14–12. In addition, a current transformer, a rectifier, and a transistor voltage divider have been added to sense the motor current.

Any changes in motor speed due to load variations will be reflected by a proportional change in motor current. The purpose of the current transformer is to indirectly detect motor speed variations by sensing the motor current variations. The current transformer used here is an Electromatic MI5. It is a through-primary variety that produces 0.8 V of secondary voltage for every ampere of primary current. Since the secondary voltage is alternating, it must be rectified and filtered to produce a useful DC bias signal for the base of Q_2. As the wiper of R_9 is moved, the bias on Q_2 changes. This varies the voltage present at the gate of the PUT. Since the voltage at the base of Q_2 is dependent on the motor current, the motor current varies the gate bias on the PUT. Specifically, as the motor current increases, the PUT gate voltage will decrease.

A PUT such as Q_1 in Figure 14–14 has a trigger voltage that varies with the voltage present at its gate. For the device to break over into conduction, the anode voltage must exceed the gate voltage by approximately 0.7 V. Note that the gate voltage in Figure 14–14 is an adjustable portion of the rectified and filtered output of the current transformer, T_3. The position of the wiper of potentiometer R_9 is one of the variables that determines the motor set-point speed. Suppose that when the motor is running at the desired speed, under normal load, the voltage at the wiper of R_9 is adjusted to provide a PUT gate voltage of 8 V. The PUT will not fire until the voltage across C_1 increases to approximately 8.7 V. This means that, depending on the settings of R_3, R_4, and R_5, the required 8.7 V trigger voltage may be achieved very early or very late in each half cycle of the 120 V AC line.

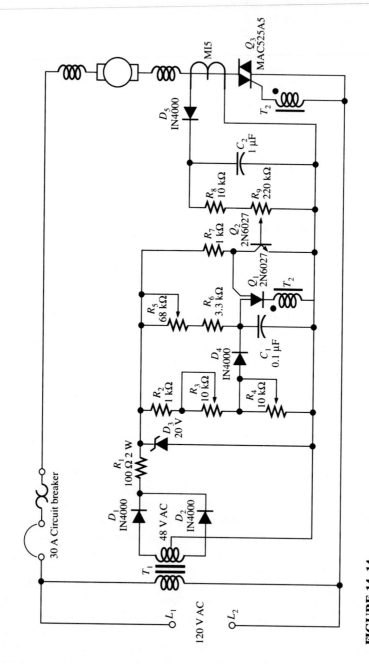

FIGURE 14-14
Proportional Motor-Speed Controller with Feedback

It quickly becomes apparent that a variation in the value of R_3, R_4, R_5, or R_9 will change the motor speed. Remember, however, that R_3 is used to compensate for the initial position of the linear-potentiometer radius-sensor R_4. In most cases the value of R_3 will be set once and changed only occasionally to compensate for mechanical variations in roll size. In fact, if R_4 is replaced with a user-variable potentiometer (a manual motor-speed adjust), R_3 may be eliminated altogether. The value of R_5 will likewise be changed only occasionally, as it sets the gain, or sensitivity, of the controller. Under normal operation, the only variables will be the voltage at the wiper of R_9 and the value of R_4. Note that this situation is typical of a closed-loop proportional controller. This arrangement is somewhat unusual, however, since there are actually two process-variable feedback mechanisms (the roll-radius sensor and the motor-current sensor) that may vary the motor speed. Both of these will be changing under normal operation.

System Operation

Suppose that R_3 is adjusted so that, with R_4 at its initial position, the firing delay angle for the triac is about 90°. The motor will begin to turn the take-up spool. The motor current at this time will determine the voltage available at the wiper of R_9. If the wiper of R_9 is adjusted to provide a low PUT gate voltage, C_1 will be able to accumulate sooner the charge necessary to trigger the PUT (the firing delay angle will decrease). This situation will effectively increase the motor speed. As the speed increases, however, the current through the primary of T_3 will increase, thus tending to oppose the decrease in PUT gate voltage. This tendency will result in an unexpected sluggish response to speed adjustment with R_9. Adjusting R_3 will also change the motor speed. However, R_3 also determines the minimum pedestal voltage that may be achieved by the voltage divider composed of R_3 and R_4. In other words, the value of R_3 may be decreased to increase the initial motor speed, but doing so may limit the extent to which R_4 may lower the spool speed as the roll radius increases.

Suppose that R_3 is adjusted to provide a 5 V pedestal for C_1 at the beginning of each synchronization pulse. Assume that R_5 is adjusted so that the RC product of R_5, R_6, and C_1 is 5 ms. If the motor is lightly loaded, the current will be low, as will be the available bias voltage for Q_2. This situation will result in a low collector current for Q_2 and a correspondingly high voltage (near 20 V) at the gate of the PUT. With the gate voltage so high, it will be far into each half cycle when C_1 accumulates enough charge to trigger the PUT. For example, if the PUT gate voltage is currently +16 V, the triac will not fire until 6.6 ms into each half cycle. Therefore, power will be applied to the motor for only 1.7 ms out of each 8.3 ms half cycle. The motor will therefore regulate at a low speed.

In Section 14–5 it was shown that if the value of R_4 decreases (because of an increase in the roll diameter), the pedestal voltage will also decrease. Capacitor C_1 will then take even longer into each half cycle to accumulate the charge necessary to trigger the PUT. That is, the motor will slow down even more. If, however, the motor becomes more heavily loaded, the current through the primary of T_3 will increase. A higher voltage will be applied across C_2, and Q_2 will be driven further into conduction. This condition will lower the bias voltage at the gate of the PUT, allowing it to trigger sooner (even though R_3, R_4, and R_5 have not changed). Since Q_3 will be triggered sooner in each half cycle, power will be applied to the motor over a proportionally greater period of time. This longer application of power will have a tendency to speed the motor up, or—more precisely—it will tend to oppose the decrease in motor speed due to the increased load.

14–7   PROPORTIONAL HUMIDITY CONTROL

Overview

The strength of concrete building materials (blocks, bricks, roof tiles, prefabricated wall panels, etc.) depends, to a large extent, on the curing time. Generally, the slower the material cures, the stronger it will be. Many manufacturers employ large curing rooms, where their products are kept at a constant temperature and humidity for several weeks.

Figure 14–15 shows an overhead ventilation system used to control the humidity in one such curing room. A blower at one end of the vent pushes air through a heat exchanger. The warm air picks up water vapor from the mist nozzle as it passes by, and then it distributes the humidity throughout the room through slots in the bottom of the vent. To maintain a controlled humidity, a linear humidity transmitter is located inside the vent, about midway between the heat exchanger and the end.

System Description

The humidity-control system shown in Figure 14–16 consists of an op-amp proportional controller, a humidity transmitter, and a motor-driven valve actuator. As the humidity of the room increases above the set-point, the valve actuator stem causes the valve to close a proportional distance. This closing decreases the

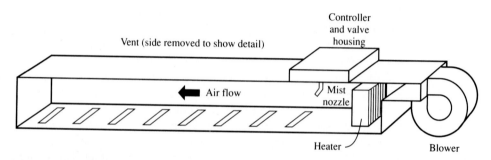

FIGURE 14–15
Detail of Ductwork to Control Humidity in a Concrete-Curing Room

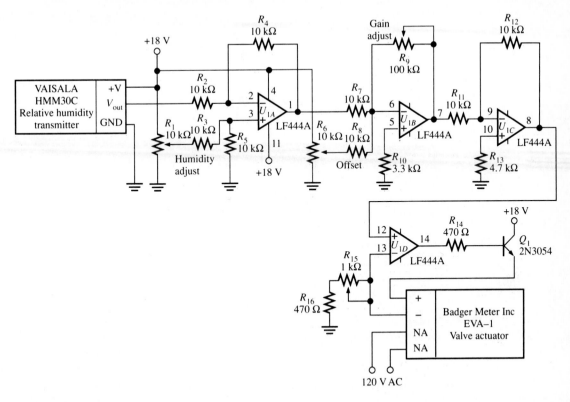

FIGURE 14–16
Humidity Controller

amount of water delivered to the mist nozzle. Likewise, if the humidity falls below the set-point, the valve opens, delivering more water to the nozzle.

The controller consists of a unity-gain differential amplifier, a noninverting summing amplifier (made up of U_{1B} and U_{1C}), and a voltage-to-current converter. Operation is typical of a proportional controller. The output of the differential amplifier (the error signal) is the difference between the humidity adjust (the set-point) and the output of the HMM30C humidity transmitter. The sum of this error voltage and the offset voltage (provided at the wiper of R_6) is amplified by the gain of U_{1B}. A unity-gain inverter (U_{1C}) has been added at the output of U_{1B} to ensure that signals of the proper polarity are provided at the input of U_{1D}.

The one unusual feature of this controller is its output. Rather than providing a variable voltage output, it provides a variable current output. This output is accomplished through the use of an op amp configured as a voltage-to-current converter, U_{1D}. Close inspection of U_{1D} will reveal that it is nothing more than a noninverting amplifier. The feedback resistor consists of R_{14}, the base-emitter junction of Q_1, and the input of the EVA-1 valve actuator. Remember that the

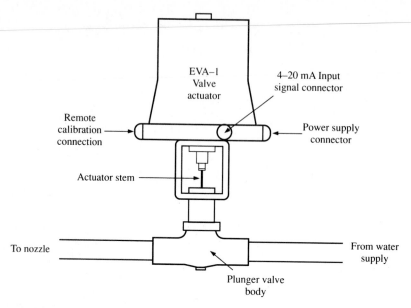

FIGURE 14–17
Valve Actuator and Plunger Valve

output of a noninverting op-amp configuration will continue to change until the potential present at the inverting input is equal to the potential present at the noninverting input. Suppose that the output of U_{1C} is $+1$ V. The output of U_{1D} will increase in the positive direction until the voltage across the R_{15}/R_{16} combination is equal to $+1$ V. If R_{15} is adjusted so that the value of $R_{15} + R_{16}$ is 1 kΩ, then 1 mA will be required to provide this voltage. The purpose of Q_1 in this application is to provide current amplification. That is, to force 20 mA through the input of the valve actuator, U_{1D} will have to provide only a fraction of this current through the base-emitter junction of Q_1. With the value of $R_{15} + R_{16}$ set at 1 kΩ, this converter will provide a 0–20 mA output in response to a 0–20 V input.

Process variable feedback is provided by a Vaisala HMM30C humidity transmitter. The main component of this sensor is a thin-film polymer capacitor. As the polymer film absorbs water vapor, the capacitance changes. This change is reflected by a change in sensor output voltage or current. This particular sensor (the HMM30C) produces a 0–10 V signal that changes linearly with humidity. That is, at 0% relative humidity, the sensor output will be 0 V; at 50% relative humidity, the output will be 5 V; and at 100% relative humidity, the output will be 10 V. This sensor unit is made up of two parts: a probe, which contains the humidity transducer, and the electronic circuit that produces the 0–10 V output in response to signals from the probe. This arrangement allows the electronic circuit to be located remotely from the probe. This capability is important because the probe will withstand temperatures up to 160 °C, whereas the electronic transmitter is reliable only at temperatures up to 55 °C.

Because of the nature of the transducer, the response time of the HMM30C is slow. It will take a full 15 s for the sensor to indicate a full-scale change in humidity. This delay is quite acceptable, however, since the moisture content of the air will not change instantaneously.

The valve actuator (see Figure 14–17) is a motor-driven linear device. It will accept an input current between 4 mA and 20 mA; in response, it will provide up to 0.56 in. of linear stem travel at up to 40 lb of thrust. This actuator is meant to be attached to a plunger valve, which restricts flow until the stem is pushed down (against spring pressure), forcing a rubber plunger out of the flow path.

The response of the actuator is linear with respect to the input signal. The stem will not move from its initial position until the input current exceeds 4 mA. With an input of 8 mA, the actuator will open the valve one quarter; with an input current of 12 mA, the valve will open halfway; with a 20 mA input, the actuator will open the valve fully. The following table shows how the valve will respond to varying levels of actuator input current:

Actuator Current (mA)	Valve Response (% Open)
4	0
6	12.5
8	25
10	37.5
12	50
14	62.5
16	75
18	87.5
20	100

Being motor driven, the actuator operates relatively slowly. The speed depends quite heavily on the required stem thrust. At a stem thrust of 10 lb, the full-scale travel time of the actuator stem is about 3.5 s, but at a required stem thrust of 40 lb, it will take the actuator a full 8.5 s to complete the same travel. Although this limitation may seem to be a drawback, it is actually a requirement. If the valve actuator operated much more quickly than the humidity sensor, the system would oscillate. Suppose that the humidity sensor called for more water vapor to be added to the air. If the valve actuator could respond quickly, the humidity would rise quickly. If the humidity sensor was slow in responding, it would essentially be unaware of the increase in humidity. The valve actuator would therefore continue to add water vapor to the air. By the time the humidity sensor responded, the air would be much more humid than desired. The valve actuator would then receive a signal to close, and by the time the humidity sensor responded to the decrease in humidity, the air would be too dry.

System Operation

There are four user-adjustable variables in Figure 14–16. Values must be assigned to all of them before a meaningful evaluation of the system's operation can be made. It will be assumed that the humidity adjust is set at $+7$ V, the offset is set at $+3$ V, the gain is 2, and the value of $R_{15} + R_{16}$ is 500 Ω. Given this information, it can be seen that $+10$ V at the output of U_{1C} will result in 20 mA through the valve actuator input (20 mA through 500 Ω will result in a 10 V drop). Likewise, an output of $+2$ V at the output of U_{1C} will result in a current of 4 mA into the valve-actuator input.

The output of U_{1C} can be expressed as follows:

$$V_{\text{out}} = A_V(V_{\text{Error}} + V_{\text{Offset}})$$

If the error signal is zero (that is, if the moisture content of the air is in agreement with the level set by R_1), the output of U_{1C} will be $+6$ V. This output will produce an output current of 12 mA from U_{1D}. Since this current level is exactly halfway between the minimum (4 mA) and maximum (20 mA) actuator inputs, the valve will be open halfway whenever the humidity is correct. As with all proportional controllers, this offset value will most probably not maintain the desired set-point. It will, however, facilitate correction of any system error. If the valve is open halfway, it will take an equal amount of time to force the valve fully open or fully closed. In other words, the response time of the system will be decreased by the offset.

Suppose that the system is powered up after a long period of being shut down. The relative humidity of the air will most probably be very low. In any case, it will surely be lower than the 70% called for by the humidity-adjust potentiometer (remember that the set-point chosen was 7 V). Suppose that, in fact, the relative humidity at the time of power-up is only 30%. This humidity will result in an output of $+3$ V from the HMM30C. This condition will produce an output of 4 V from U_{1A} (7 V $-$ 3 V). Adding this error to the offset and multiplying the sum by the gain of U_{1B} (2 for this example) shows that the controller will produce $+10$ V at the output of U_{1C}. This output will result in a current of 20 mA through the input of the valve actuator. The valve will therefore be fully open. The temperature of the air, the size of the valve, the water pressure, and the size of the nozzle orifice will collectively determine the actual amount of moisture added to the heated air. In any case, the humidity will increase. As it does, the output of the HMM30C will increase. The overall effect will be a decrease in valve actuator current. Even after the humidity level has reached the set-point (that is, after zero error has been achieved), the moisture content of the air may still rise. After all, the valve will still be open halfway. If the humidity does continue to increase, the output of the HMM30C will likewise increase. As the output of the HMM30C exceeds the humidity-adjust voltage, the output of U_{1A} will be driven negative. This negative output will decrease the output of U_{1C} to less than $+6$ V, moving the valve closer to fully closed. Suppose, for example, the output of the HMM30C is $+8.5$ V (corresponding to a relative humidity of 85%). The output of U_{1C} will be

$$V_{\text{out}} = 2(3 \text{ V} + (7 \text{ V} - 8.5 \text{ V})) = 3 \text{ V}$$

This output will produce only 6 mA of valve-actuator input current. The valve will therefore close from 50% open to only 12.5% open.

EXERCISES

1. Draw the transfer curve for the controller shown in Figure 14–2. Assume that R_9 is adjusted to 2 kΩ.

2. For the controller shown in Figure 14–2, explain how circuit operation would be affected if a negative-temperature-coefficient thermistor were substituted for R_1.

3. In Figure 14–2, what would be the effect of increasing the gain of U_{1A}?

4. In Figure 14–1, why is the controller located so close to the reaction chamber?

5. Draw the transfer curve for the controller of Figure 14–4. Assume that the gain of U_{1B} is set at 4.

6. Would the controller in Figure 14–4 operate equally well with an oscillator frequency of 1 kHz? Explain.

7. Would the controller in Figure 14–4 operate equally well if a random-trigger solid-state relay were substituted for the ZVS relay? Explain.

8. Could the need for U_{1C} in Figure 14–4 be eliminated by simply reversing the inputs of U_{1D}? Explain.

9. Draw the transfer curve for the controller of Figure 14–9. Assume that R_3 and R_5 are adjusted to 2.5 kΩ and 1 kΩ, respectively.

10. How would overall operation of the controller in Figure 14–9 be affected if the voltage across R_3 were less than the voltage across R_5?

11. How would operation of the controller in Figure 14–9 be affected if the ultrasonic sensor were raised further above the sump pit (see Figure 14–8)?

12. Why must the triac trigger pulses in Figure 14–13 be synchronized to the AC line?

13. How would the operation of the controller in Figure 14–12 be affected if D_1 or D_2 opened?

14. Suppose that a 1 μF capacitor were accidentally substituted for C in Figure 14–12. How would the system respond?

15. Suppose that Q_2 in Figure 14–14 shorted from collector to emitter. How would this affect circuit operation?

16. Suppose that D_5 in Figure 14–14 opened. How would this condition affect circuit operation?

17. Suppose that the motor in Figure 14–14 does not run at all. Give three reasons for this condition. (Assume that the condition is due to some circuit fault.)

18. Draw the transfer curve for the controller in Figure 14–16. Assume that R_6 is adjusted for an offset of 1 V and that R_9 is adjusted for a gain of 3.

19. Given the circuit of Figure 14–16, what is the overall effect of increasing the value of R_{15}?

20. Is there a practical reason for locating the humidity sensor in Figure 14–16 inside the vent? Why not locate it somewhere in the curing room?

APPLICATION ASSIGNMENT ANSWER

1. The solid-state relay output may be shorted. If the voltage across the load is a continuous 240 V AC, the output of the SSR may be either shorted or continuously triggered. To distinguish the difference, check the input (the intersection of R_{18} and D) of the SSR with an oscilloscope. If the input is pulsing, decrease the duty cycle by varying R_{10}. If the duty cycle changes but the load voltage does not, the SSR output is indeed shorted, and the SSR must be replaced. If the input is continuously held near +12 V, the SSR is most likely operational.

2. Pin 14 of U_{1D} may be held at a continuous positive potential. This condition would indicate that the potential at pin 12 of U_{1D} is higher at all times than that at pin 13. The signal at pin 12 should be a triangle wave that varies between 0 V and +6 V. If the signal never returns to zero, the problem may be that R_2 (the signal center adjust) has been tampered with. If the signal is held at a constant positive potential, U_2 must be replaced.

   If the signal at pin 12 of U_{1D} is oscillating properly and the output of U_{1D} is still held at a positive potential, the output of U_{1D} may be shorted to the positive supply. Since the output of U_{1D} should saturate at a potential somewhat lower than the positive supply, this fault should be easy to identify. Very simply, if pin 14 of U_{1D} is held at +12 V (exactly the same potential as the positive supply), a short is indicated.

3. A potential at pin 13 of U_{1D} that is always too low will also result in the observed symptoms. Variation of the offset potentiometer should change the potential at pin 13. If it does not, then U_{1C} is most likely defective, and U_1 should be replaced.

V

PROGRAMMABLE LOGIC CONTROLLERS AND POWER DISTRIBUTION EFFECTS

15

Programmable Logic Controllers

CHAPTER OUTLINE

LEARNING OBJECTIVES

- □ Construct a block diagram of a typical programmable logic controller (PLC)
- □ Describe the function of each section of a PLC.
- □ Describe the general operational procedure for a PLC.
- □ Construct programs for basic input/contact and output/coil circuits.
- □ Perform basic PLC programming in the digital logic format.
- □ Explain the use of registers and addresses in PLC operation.
- □ Program PLC timers and counters.
- □ Describe other major PLC functions and their use in control circuits.
- □ Construct a PLC sequencer program.
- □ Describe the use of PLC digital bits in process control.
- □ Describe PLC analog programming capabilities.
- □ Describe PLC feedback and PID systems and their application.

APPLICATION ASSIGNMENT

You are a technician assigned to convert an older relay-panel control system to a PLC control. What steps would you follow in the process?

15–1 INTRODUCTION

The *programmable logic controller* (PLC) is a user-friendly, microprocessor-based computer. A person relatively unskilled in computers can construct programs for a programmable logic controller. Only a knowledge of control circuits and ladder logic is needed. The programmable logic controller has replaced the relay logic systems used in past years.

This chapter will cover the basic principles of PLCs as they apply to control systems. First, the system layout of a PLC will be discussed. The various components of the PLC will be described, along with their operations and functions. Basic operational procedures will then be covered. Next, the construction and operation of PLC ladder diagrams for ON/OFF inputs and outputs will be illustrated. Both ladder and digital logic programming schemes will be covered.

The concept and usage of registers and addresses will be explained. Some of the other most often-used PLC functions, such as timers and counters, will be illustrated. The use of these other functions in control electronics systems will be shown. The operation of the PLC sequencer will next be shown, along with other schemes of using digital bits for control. Analog PLC operation will be illustrated as a process control system. Finally, PLC operations for feedback loops and PID control will be discussed.

15–2 GENERAL CHARACTERISTICS AND SYSTEM LAYOUT

The PLC has a number of important characteristics, including the following:

- □ user-friendly programming for persons who are not computer programmers
- □ relative low cost and compactness
- □ flexible programming capabilities, discrete and analog
- □ visible display of ladder circuitry and its operation
- □ high-speed solid-state operation and high reliability
- □ ability to enable a printout to be made of the circuit and of its operation
- □ security through key lock, preventing program alterations
- □ easy changes of circuitry
- □ easy pilot runs in a simulation mode
- □ drawbacks to its use with fixed programs
- □ susceptibility to solid-state environmental considerations

The layout of a programmable logic controller is shown in Figure 15–1. The parts of the system are as follows:

- □ central processing unit (CPU)
- □ program monitor (PM)
- □ input and output (I/O) module
- □ printer
- □ recorder-player—tape or disk

□ optional remote-location interconnection

□ optional connection to master computer or data highway

The CPU is the core of the system, as it is in all computers. The CPU contains fixed memory, set by the manufacturer. It also contains a programmable section and a memory for data and function storage. All the peripheral devices are connected directly to the CPU. The PLC process-operating program is entered into the CPU from the PM or a recorder-player. The CPU receives signals indicating the status of external switches and devices through an input module. It makes logic decisions based on the entered program and then sends status signals to the output devices through the output modules. The I/O signals may be discrete (ON/OFF) or analog (variable).

The program monitor for the PLC has many forms, depending on the price and complexity of the PLC. Programming is accomplished from a keyboard. A typical keyboard is shown in Figure 15–2. The keyboard is usually associated with

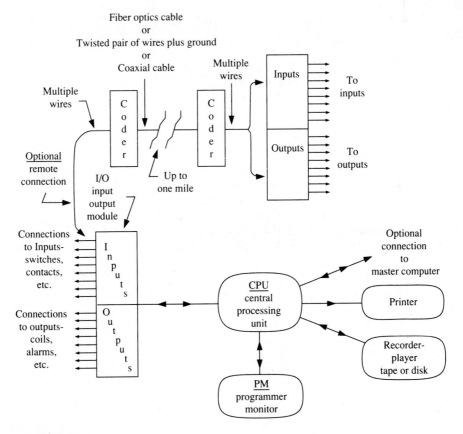

FIGURE 15–1
PLC System Layout and Connection

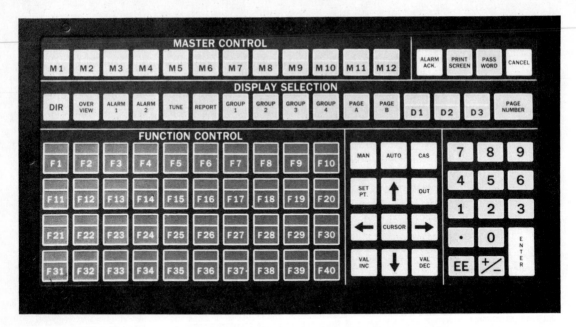

FIGURE 15-2
PLC Keyboard (Courtesy of Texas Instruments)

a monitor. Both are combined in one unit, as shown in Figure 15-3. Some keyboards are smaller and are combined with small monitor screens also as shown in Figure 15-3. Many PLCs can be programmed with a personal computer. A disk with a PLC program can be purchased from the PLC manufacturer. The program effectively converts the personal computer to a program monitor. The PLC program then appears on the personal computer screen. The personal computer keyboard is coded for the PLC functions by the disk program. The code is typically shown by a screen menu. A dedicated program monitor is normally much more costly than the disk system.

The I/O modules are used to connect devices to inputs, and outputs to the process. The modules are connected to the CPU by multiconductor ribbon cables. These ribbon cables carry low-voltage, digital electronic coded information. A typical large PLC may have 256 input ports for 256 input devices from the process. The ribbon cable does not have 256 wires. A typical ribbon cable has 18 wires, but because of the coding and decoding, those are enough. The output information is transmitted similarly. A typical I/O module is shown in Figure 15-4.

Input devices, such as switches, are connected between a common module terminal and another specified module terminal. Output devices are connected in the same manner. There are two major factors to consider with these modules: electrical rating and module switch settings.

FIGURE 15–3
PLC Program Monitors (Courtesy of General Electric, left, and Eaton Corp./Cutler-Hammer® Products, right)

Ratings of the modules are for voltage and current and for discrete or analog operation. A typical discrete-module rating is 115 V AC, 3 A, for both the input and the output modules. Other applications might require outputs of 24 V AC, or a DC voltage, depending on the electrical rating of the output device being controlled. Higher output currents may be required, also. Input module ratings must be determined similarly. Analog devices need a voltage-range rating that matches the application requirements. Manufacturers' catalogs list the standard module ratings available. Note that it is possible to mix and match module ratings for a given PLC in groups of 4 to 16.

The second I/O module consideration is switch setting. Each module has 4 to 16 terminals. The CPU recognizes numbers assigned to the terminals. For example, the first input module might contain inputs 1 through 16, and the second, 17 through 32. The output module of 8 terminals might be designated 81 through 88. How does each of the three modules know the numbers assigned to its terminals? Assignment of numbers is accomplished by the proper setting of individual

FIGURE 15–4
Typical Input/Output Module

SIP or DIP switches. The abbreviations SIP and DIP refer to small, multiple-toggle switch assemblies, the single in-line package and the dual in-line package. These small switches are located on the end of each module. The proper settings for the switches for each number series are listed in the operating manuals. Note that often internal switches must also be set for the required terminal identification.

In some cases the process to be controlled is at a location remote from the PLC. In such cases an extended remote I/O system may be used, as shown in Figure 15–1 at the top. Only two wires plus a ground/shield are required for distances up to 1 mi. Communication is by the telemetering technique. The remote I/O modules act the same as ones at the PLC with the same numbers assigned. Communication can be through wires, by fiber optics, or by RF telemetry signals.

Two possible added options for a PLC are a printer and a program recorder, as shown in Figure 15–1. The printer can be any regular computer printer that is properly interfaced to the CPU. The information and diagrams that can be obtained include the following:

□ complete ladder diagram including cross-references
□ forced-contact listing
□ status of registers or contacts
□ timing diagrams of circuit operation, registers, or contacts
□ other special programmed information

As with any computer, the data in the PLC can be recorded on tape or disk. When a program is lost or develops a malfunction, the original program can be quickly reinserted into the PLC CPU. Tape recorders must be of a special, high-speed type and are relatively costly. Any properly interfaced disk system will work with most PLCs.

One precaution in using printers and recorders is the proper setting of baud rates. For example, the CPU might normally operate at 9600 baud. The printer might work only at 2400, and the recorder at 1600. Before printing, recording, or playback, proper baud rates must be set. The baud rates are set or reset in the PLC CPU by following procedures listed in the appropriate PLC operating manual.

15–3   OPERATIONAL PROCEDURES

A simple program will indicate how to begin utilizing a PLC. Suppose that you wish to program and connect a PLC to accomplish the following discrete operational procedure: A relay coil is to actuate when two toggle switches and one limit switch are actuated.

The first step is to assign individual PLC identification numbers to the inputs and outputs. Inputs normally have the prefix IN. Outputs normally have the prefix CR (for Control Relay). The following numbers could be assigned:

Switch 1 for relay   IN 001

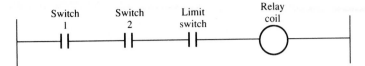

FIGURE 15–5
Ladder Logic Diagram for Relay Output

Switch 2 for relay	IN 002
Limit switch for relay	IN 003
Relay output	CR 001

Next, sketch a ladder logic diagram to represent the operational circuit. This is shown in Figure 15–5.

Next, figure out how the inputs and outputs will be connected to the input and output modules. Assume an eight-terminal input and an eight-terminal output. It is necessary to set the module switches so that the modules recognize signals as inputs 1 through 8 and outputs 1 through 8. The connections from the inputs and outputs then are made according to Figure 15–6. Note that each component is connected to one of the modules. No external interconnections are made.

Finally, the ladder program must be entered into the CPU by means of the keyboard. A general procedure for entering the program in ladder format follows:

1. Clear the PLC program memory with the CPU on Stop. The procedure will be outlined on a screen menu or in the operation manual for the PLC.
2. Insert the relay control line as follows, in the EDIT mode:
   a. Push the NO contact key.

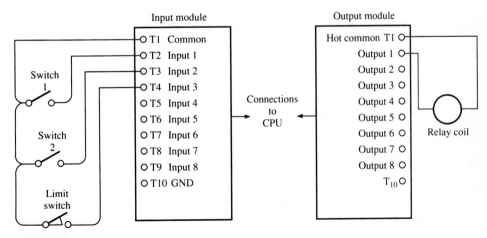

FIGURE 15–6
Connection Diagram for Figure 15–5 Circuit

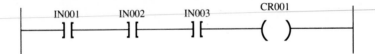

FIGURE 15-7
PLC Screen Ladder for Figure 15-5 Circuit

    b. Push the INPUT key.

    c. Push 001 numeric keys.

    d. Push the Enter key. The contact should appear on the monitor.

    e. Move the cursor one space to the right.

    f. Repeat steps a and b.

    g. Push 002 numeric keys.

    h. Push the Enter key. The second contact should appear on the monitor.

    i. Move the cursor one more space to the right, and repeat the process for 003.

    j. Continue the line to the right.

    k. Push the coil/output key. The coil should appear on the monitor.

    l. Push 001 numeric keys.

    m. Push Enter.

    n. If the line now looks correct (check it), push the ''insert ladder'' key and then Enter.

The resulting PLC diagram should look like the one shown in Figure 15-7. When the PLC switch is set to Run, the circuit will operate as outlined.

    What if the programmer were a small, hand-held one with no screen? The programming would be similar to steps a through k. Three differences in the sequence would be as follows:

    1. Instead of moving the cursor to the right in step e, you would press the AND key.

    2. Before pressing CR in step j, you would push a LOAD key.

    3. You might not be able to use the same number (001) for both an input and an output. Check how numbers are allocated to inputs and outputs by referring to the operating manual.

15-4 CONTROL OF COILS AND CONTACTS: DIRECT AND DIGITAL LOGIC

The basic function of the PLC is the use of discrete ON/OFF logic to turn outputs ON and OFF. The inputs are connected to the input modules as previously shown. Input devices include the following:

    toggle switches   push-button (momentary) switches

limit switches	selector switches
centrifugal switches	temperature switches
level switches	flow switches
pressure switches	proximity switches

As these input devices open and close, the input module sends their status to the CPU. A contact programmed by the CPU as normally open (sometimes called *examine open*) will change to closed when the input device contacts are closed. Conversely, a contact programmed as normally closed (sometimes called *examine closed*) will change to open when the input contacts are closed.

The CPU program is constantly scanned, typically in 5 ms per scan. As technology improves, this scan time is approaching microseconds. As the input devices change status, programmed logic functions are performed. The outputs are then turned ON or OFF according to the ladder program logic. The CPU performs the logic of the ladder diagram. It then puts out signals to the output modules, which turn the outputs ON or OFF. Some typical output devices are the following:

motors	coils
relays	pilot lights
meters	bells
annunciators	other alarms

Five examples will be used to illustrate this selection's principles:

1. a standard start/stop seal circuit
2. a forward-reverse control circuit with a stop before reversing
3. a forward-reverse control circuit with direct (plugging) reverse
4. a start/stop/jog (momentary-ON) circuit
5. a multiple alarm system

The first example, the start/stop seal circuit, is shown in Figure 15–8. The relay logic elementary and connection diagrams are shown at the top. Next, the PLC program and connections are shown. For the PLC program, stop is assigned IN001, and start, IN002. The output is assigned CR017 (control relay output 17). The PLC connection diagram is as shown. The seal contact is created internally in the PLC and is not wired from the contactor. Two typical formats are shown for the PLC. One is straight ON/OFF. The other uses a latch-and-unlatch system. Additionally, the latch/unlatch system uses an output number (out 7) different from the function identification number (608).

The programming of a hand-held digital PLC for Figure 15–8 would be as follows (slashes separate the steps):

IN / 02 / OR / CR / 17 / AND NOT / IN / 01 / LOAD / CR / 17

The second example, forward-reverse with stop before reversing, is shown in Figure 15–9. Stop, forward, reverse, and the output coils are assigned numbers

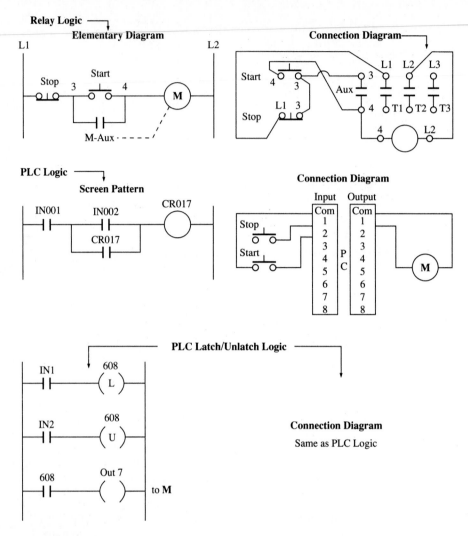

FIGURE 15–8
Start/Stop Seal Circuit

as shown in the screen pattern. Notice the simplicity of the connection diagram of the PLC, compared with the relay logic diagram. The advantage of the PLC for wiring begins to show in this example. Not only is the wiring simpler, but also any changes or corrections can be easily made by program alterations. For a relay system, physical wiring changes would have to be made. The hand-held-device digital programming for Figure 15–9 would be as follows:

IN / 02 / OR / CR / 17 / AND / IN / 01 / AND NOT / CR / 18 / LOAD / CR / 17

IN / 03 / OR / CR / 18 / AND / IN / 01 / AND NOT / CR / 17 / LOAD / CR / 18

Relay Logic

Elementary Diagram

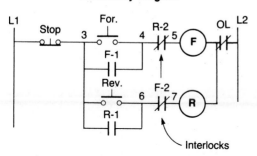

Note: Relay circuit above is "fail safe." It unseals when OL (overload) opens. The PLC circuit below is not "fail safe" since the OL does not unseal the circuit.

Note: IN001 is programmed open. Will close when control power is applied.

Connection Diagram

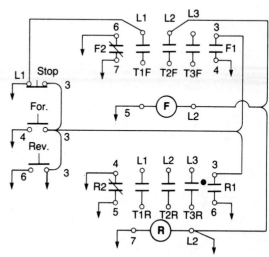

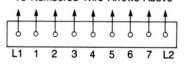

To Numbered Wire Arrows Above

Terminal Block

PLC Logic

Screen Pattern

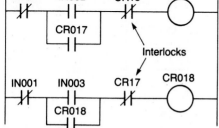

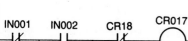

Connection Diagram

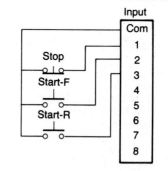

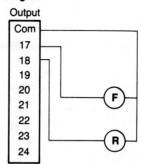

FIGURE 15–9
Forward/Reverse with Stop before Reversing

Relay Logic (Reference)

3 Wire Control - Reversing Starter
with Pilot Lights to Indicate
Direction Motor is Running

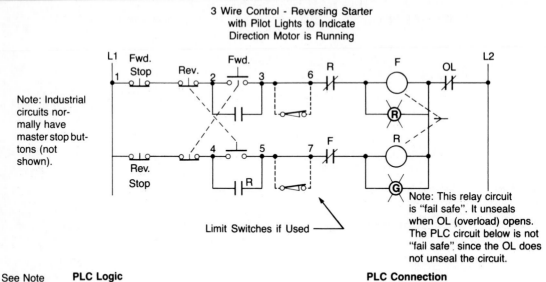

Note: Industrial circuits normally have master stop buttons (not shown).

Limit Switches if Used

Note: This relay circuit is "fail safe". It unseals when OL (overload) opens. The PLC circuit below is not "fail safe" since the OL does not unseal the circuit.

See Note   **PLC Logic**

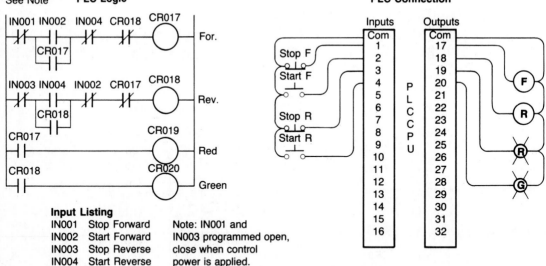

PLC Connection

Input Listing

IN001	Stop Forward
IN002	Start Forward
IN003	Stop Reverse
IN004	Start Reverse

Note: IN001 and IN003 programmed open, close when control power is applied.

FIGURE 15–10
Forward/Reverse with Direct (Plugging) Reverse

The third example, forward/reverse with direct (plugging) reverse, is shown in Figure 15–10. Inputs and outputs are assigned numbers as shown. The difference between this circuit and the previous one, in Figure 15–9, is that it is now possible to go directly from forward to reverse or directly from reverse to forward. There is no attempt here to show the relay logic diagram. It would be extremely complicated. For the PLC scheme, there are only four wires plus a common wire going in, and four wires plus a common wire going out.

For start/stop/jog (momentary-ON) control, the PLC scheme is shown in Figure 15–11. The connections are very simple. The logic is all generated internally in the CPU.

Programming of a hand-held digital PLC for the circuits of Figures 15–10 and 15–11 will be reserved as chapter exercises.

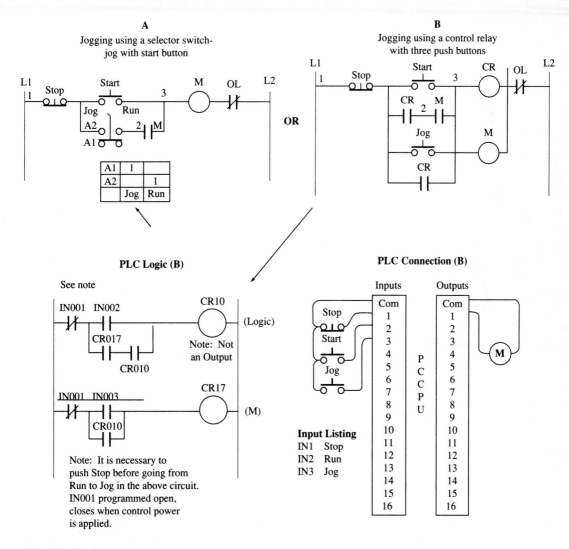

FIGURE 15–11
Start/Stop/Jog (Momentary-ON) Circuit

Inputs		Outputs	
A	IN 0001	Red pilot light	CR 0027
B	IN 0002	Alarm (siren)	CR 0018
C	IN 0003	Fire department notify	CR 0019
D	IN 0004		

PLC Logic

PLC Connection

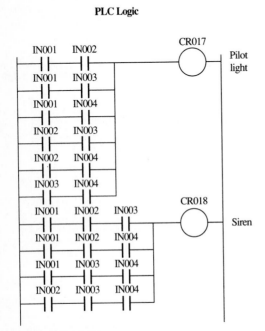

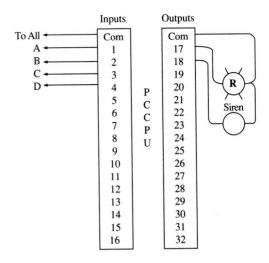

FIGURE 15–12
Alarm System

The final example, an alarm system, is shown in Figure 15–12. There are four inputs, as shown. Each represents a malfunction in a system. If any two malfunctions occur, a red pilot light is to come on. If any three malfunctions occur, a siren is to sound. The derived logic diagram is shown, along with its connection diagram. Again, note how simple the connection diagram is. For a relay logic system, the connections would be very complicated, and a number of logic relays would have to be wired.

The hand-held-PLC digital, or Boolean, program for the Figure 15–12 ladder diagram would be as follows:

IN / 01 / AND / IN / 02 / OR / IN / 01 / AND / IN / 03 / OR / . . . / LOAD / CR / 17

IN / 01 / AND / IN / 02 / AND / IN / 03 / OR / IN / 01 / . . . / LOAD / CR / 18

15–5   ADDRESSES AND REGISTERS

For the contacts-and-coil operation of the preceding section, the person doing the programming does not have to be concerned with the addresses or registers in the PLC CPU. However, most other PLC functions use the internal addresses or registers in their operation. A knowledge of the characteristics and operation of these addresses or registers is needed for an understanding of these other PLC functions.

Addresses and *registers* are different names for the same things, "slots" used to store information in computer memory. In computer operation these internal slots are called addresses. In this PLC discussion, they will be called registers. Registers can be 4, 8, 16, or 32 bits wide, depending on the PLC model you are using. For illustration, a 16-bit register will be used.

Figure 15–13 shows the relative location and number of registers available for three types of register. These are the internal *working,* or *holding, registers,* the *input registers,* and the *output registers.* The prefixes HR, IR, and OR will be used for these three types throughout this chapter.

Note that in small PLCs there are only HRs and they may not be accessible by the programmer. In most PLCs the registers can be keyboard-accessed. All PLC registers are numbered consecutively. Those PLCs which do not have register prefix designations have certain series of register numbers reserved for the different kinds of registers. The operation manuals for these PLCs tell which register numbers are for which type.

In a medium-sized PLC, there are typically 500 to 1000 holding registers. Normally, there are only around 10 input and 10 output registers. The input registers connect conveniently to certain input devices. The output registers connect conveniently to certain output devices.

In addition, PLCs have two other types of register configurations. These are input group (IG) registers and output group (OG) registers. The advantage of these group registers is that one single register can receive or control multiple inputs or outputs. For a 16-bit IG register, the status of 16 inputs can be recorded in one register. With regular programming techniques, 16 lines of program would be required to receive information from 16 inputs.

A typical IG register scheme is illustrated in Figure 15–14. The first 16 inputs, 1 through 16, have their ON/OFF status fed to IG register 001. The status of input 1 is recorded in the bit 1 position. An ON status is represented by a 1 and

FIGURE 15–13
PLC Registers

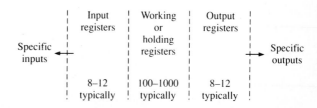

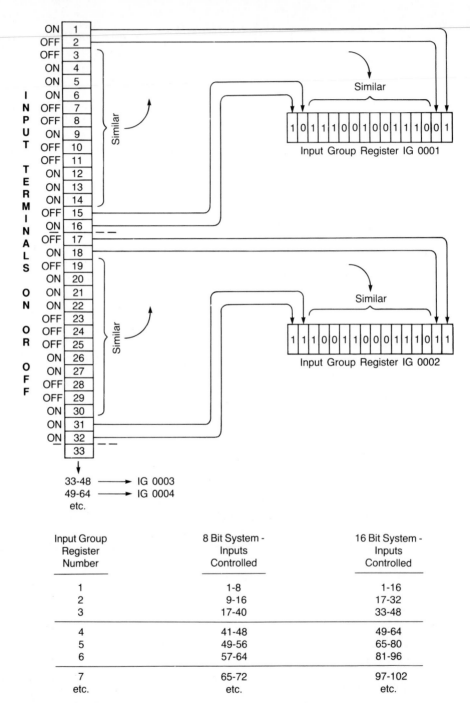

Input Group Register Number	8 Bit System - Inputs Controlled	16 Bit System - Inputs Controlled
1	1-8	1-16
2	9-16	17-32
3	17-40	33-48
4	41-48	49-64
5	49-56	65-80
6	57-64	81-96
7 etc.	65-72 etc.	97-102 etc.

FIGURE 15–14
Input Group Register Scheme

FIGURE 15–15
Group Register Numbering Scheme

Group Register Number	8-Bit System—Inputs or Outputs Controlled	16-Bit System—Inputs or Outputs Controlled
1	1–8	1–16
2	9–16	17–32
3	17–40	33–48
4	41–48	49–64
5	49–56	65–80
6	57–64	81–96
7	65–72	97–102
etc.	etc.	etc.

OFF by a 0. The status of the other 15 inputs is recorded consecutively as shown. Status conditions for inputs 17 through 32 are recorded in IG 002 as shown. Other groups of 16 would be recorded in other IG registers, consecutively in groups of 16.

Details of a system for numbering both IG and OG registers are shown in Figure 15–15.

The OG register operates in an inverse manner from the IG register. An OG register is loaded with a 0/1 bit pattern. Each OG register controls 16 output devices through 16 output module ports. The system of OG output control is illustrated in Figure 15–16. The bit pattern has been moved into the OG2 register, as shown, from register 276 (the data-moving process will be covered later in this chapter). The pattern of bits in OG2 is as shown. Output 17 is turned ON because its bit is a 1. Other corresponding outputs are turned ON or OFF similarly as shown.

15–6 TIMERS AND COUNTERS

In relay logic circuits, timers and counters are individual panel-mounted units that must be wired. There are many different types of these panel timers for different timing applications. For example, time-delay ON and time-delay OFF timers are constructed differently. For the PLC timing function, the PLC timer is generated in the program, by calling it up as a special function. There is only one type of timer function in the PLC, time-delay ON. Any other type of timing requirement can be carried out by using combinations of a number of timer functions in the PLC.

Figure 15–17 illustrates two typical timer formats for PLCs. When the timing function is called up on the keyboard, the timer block appears. The block is programmed for three parameters. The first is a timer number, TS017 or 31 in Figure 15–17. Next, the time value of the required timing interval is inserted, 14 s for each format. Finally, a register is specified in which the counting is to take place, HR101 or function 31. There are two inputs to the timing function. Input IN001 or IN7 is the input that starts the function's timing interval. Input IN002 or

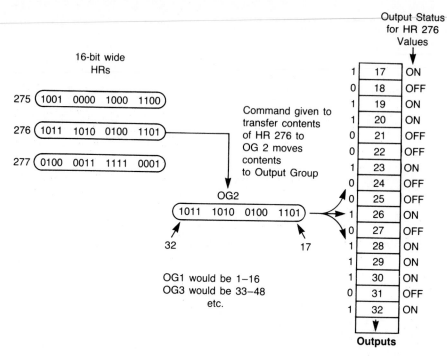

FIGURE 15–16
Output Group Register Scheme

IN8 is the Enable/Reset input. When Enable/Reset is OFF, the timer will not run even if the Run line is ON. When Enable/Reset is ON, the timer may run. When the Enable/Reset is turned OFF after the timer times out, the timer resets to zero. After an ON time of less than 7 s, the timer output will not go ON. However, the timer is cumulative. For example, there is a 7 s time setting. Close Run for 3 s and then turn it OFF. Do not reset. After Run has been closed another 4 s, the timer will go ON.

In the second format, note that the output is not direct to an output. A timer function contact, 31, is used to control the output, 78.

A simple application of a timing circuit is shown in Figure 15–18. In this circuit, light LT1 goes on immediately as a switch is closed. Eight seconds later a second light, LT2, is to go ON. The elementary and connection diagrams for a suitable relay system are shown for reference. The same diagrams are shown for a PLC, along with a timing diagram showing status versus time. As timing circuits become more complex, the advantage of a PLC over relay logic for connection increases greatly—as it does for contacts and coils.

The PLC counter function operates similarly to the timer function. Two typical formats for the PLC counter function are shown in Figure 15–19. When the Enable/Reset line is ON, the counter counts once each time the count line goes

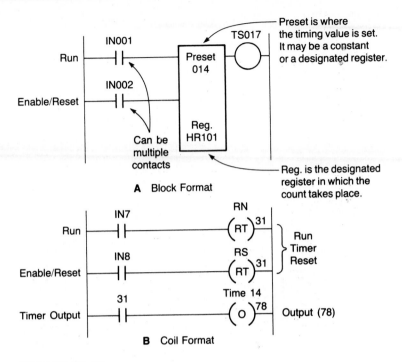

FIGURE 15–17
Basic PLC Timer Function

ON. When the preset count of 21 is reached, the outputs go ON. The preset count may be a constant, as shown, or may be contained in a register, as in the PLC timer function. When Enable/Reset is turned OFF, the counter resets to 0. Note that counters may be up counters, 0 to 21 in our example, or down counters, 21 to 0.

One application of counters is illustrated in Figure 15–20. The purpose is to keep a constant count of the parts on a conveyor. First, the count is set, in HR037, to the initial part count—247, for example. Sensor input IN004 pulses the number value in HR037 up one count every time a part is put on the conveyor at the front end. The count will go to 248, 249, and so on. At the end of the conveyor, parts are removed one at a time. Input IN005, from another sensor, pulses every time a part leaves the line. The count value in HR037 is decreased one count for every IN005 pulse—249, 248, and so on. Therefore, the net count of the parts on the line is always contained in HR037.

15–7 OTHER DISCRETE FUNCTIONS

There are a few dozen other PLC discrete functions. A complete listing of these functions is found in manufacturers' literature and in the reference books listed in the bibliography. The number of different functions in a PLC depends on the

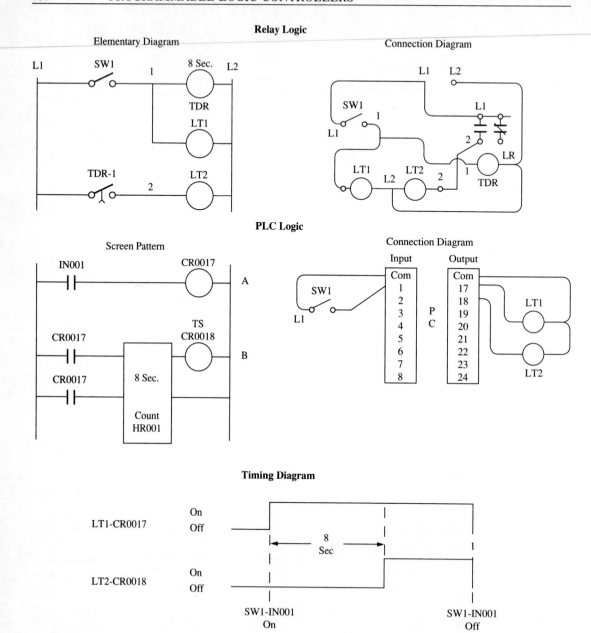

FIGURE 15–18
ON Time-Delay Circuit

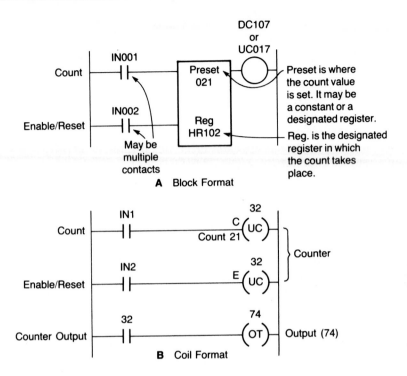

FIGURE 15–19
Basic PLC Counter Function

requirements of its application. The more complex the application, the more functions are needed, at greater cost.

This section will cover a few of the other commonly used discrete functions. Four mathematical functions and the number comparison functions will be described. Skip, the master control relay (MCR) function, and the basic Move functions will also be covered in this section.

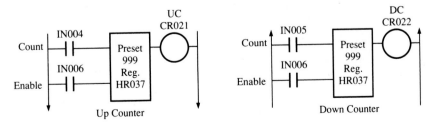

FIGURE 15–20
Two-Counter Net Count Circuit

The Add function is shown in Figure 15–21. When called up by programming, the block appears as shown. An output number is assigned to the function, AD0078. Operand 1 is the base number from a register. The number to be added to the base number is programmed as operand 2. In most PLCs operand 2 can be programmed as a constant or as the value from another register. A location must be provided for the sum. This is the destination—OR0013 in our example. The Enable line is programmed as is appropriate to the operation.

When IN0065 is turned ON for the ADD function, the function adds the operands, and the sum appears in the destination. If IR0062 changes after the addition, it is not automatically added to give a new sum. Input IN0065 must be turned OFF and back ON for the new addition to take place. There are normally four decimal digits, giving a maximum of 9999 for each operand. The destination is also four digits wide for a maximum number of 9999. When 345 and 5291 are added, for example, the sum, 5636, fits in the four-digit destination. However, if 8652 and 7693 are added, the sum, 16,345, does not fit in the destination. This problem is taken care of by having the coil go ON to represent the 1 of the 16,345. The destination would contain the rest of the figures—6345.

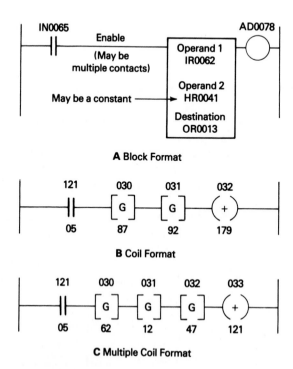

A Block Format

B Coil Format

C Multiple Coil Format

FIGURE 15–21
The PLC Add Function

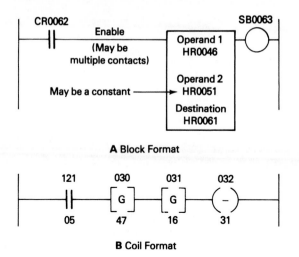

A Block Format

B Coil Format

FIGURE 15–22
The PLC Subtract Function

The Subtraction function, shown in Figure 15–22, operates in much the same way as the Add function. When CR0062 is turned ON, operand 2 is subtracted from operand 1. The result is in destination register HR0061. If the result is negative, the coil goes on to indicate the negative result.

The Multiply function is shown in Figure 15–23. It operates in a manner similar to that of the Add and Subtract functions. When enabled, operand 1 and operand 2 are multiplied together to give the product in the specified destination register. Note an important point: The destination register is normally automatically two registers wide. Register HR0068 has been specified as the destination. The PLC automatically uses two registers for the destination, HR0068 and HR0069, as shown. Small products, with only four or fewer digits, will appear in HR0069. Larger numbers will be contained in both registers, starting in HR0068.

The fourth arithmetic function is the Divide function, shown in Figure 15–24. It operates similarly to the other math functions. Operand 1 is normally two registers wide, as it can be a large number. Operand 2 is one register wide or can be a specified constant. The destination is two registers wide, as shown. The first destination register contains the dividend. The second register, OR0014, contains the remainder, not the decimal part as might be assumed. One common error in using this function is to have operand 1 and operand 2 in consecutive registers—for example, HR0045 and HR0046. Then, operand 2 and the second register for operand 1 are the same. This arrangement does not work. A cardinal rule with PLCs is not to use the same register for two purposes—it will not work.

The general Compare function is shown in Figure 15–25. The most common number-comparison function is the "Equal to" function. When the number in operand 1 exactly equals the number in operand 2, the output goes ON. Operand 1

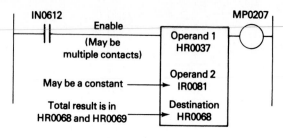

A Block Format

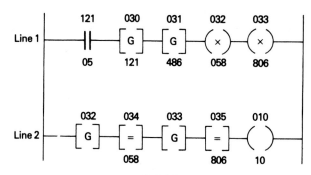

B Coil Format

Note: For result of less than 6 digits, use Line 1 only

FIGURE 15-23
The PLC Multiply Function

is the number in a specified register. Operand 2 is the number in another specified register or a specified constant.

In comparing two numbers, there are six possible mathematical comparisons, which are listed in Figure 15–26. Many PLCs contain only two comparison functions, "Equal to" (EQ) and "Greater than or equal to" (GE). From these two, the other four can be derived. For example, "Not equal" (comparison 2) would have a basic EQ function. In format 2, for "Not equal," an NC contact, 37, would be used to control the output. The other two comparison functions, GThan (comparison 5) and LE (comparison 6), involve using both basic functions with the same operands in each. Then, a contact from each function is connected as shown in Figure 15–26. The GThan function would have series contacts, and LE would have parallel contacts.

The Skip and master control relay (MCR) functions are two similar, often-used PLC functions. The Skip function is shown in Figure 15–27. It is programmed with a coil function that specifies two parameters. First, the function is assigned a coil output number—SK0216 in this example. Next, the number of consecutive ladder lines to be skipped is specified. The function is enabled by a contact or contacts in the normal manner.

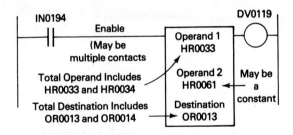

A Block Format

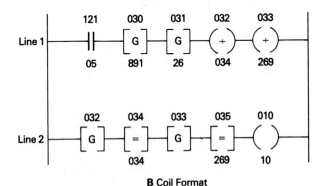

B Coil Format

Note: Value is in 032 and Decimal is in 033.

FIGURE 15–24
The PLC Divide Function

The Skip function is placed in the ladder sequence just ahead of the lines to be skipped. A typical skip sequence is illustrated in Figure 15–28. When Skip is not enabled, the ladder diagram functions in the normal manner, just as if the Skip were not present. When Skip is enabled, the specified number of lines are skipped and not scanned—three lines in this example. Lines 4, 5, and 6 in the illustration will remain in their previous state as long as Skip is on. Opening or closing the ladder inputs, IN0056 through IN0058, will have no effect. Coils CR0085 through CR0087 will remain ON or OFF as they were before Skip was energized.

The MCR function operates similarly to the Skip function with two differences. The MCR function is active when the function is not enabled or is OFF. This operation is for fail-safe purposes. The function is inactive when ON. The second difference is that when the MCR is OFF, the next specified number of lines are turned OFF. The coils are not left in their previous ON or OFF state as they are by the Skip function.

The MCR function is shown in Figure 15–29, and an MCR application is shown in Figure 15–30. Three lines are affected by the function in the application shown.

FIGURE 15–25
The General Comparison Function

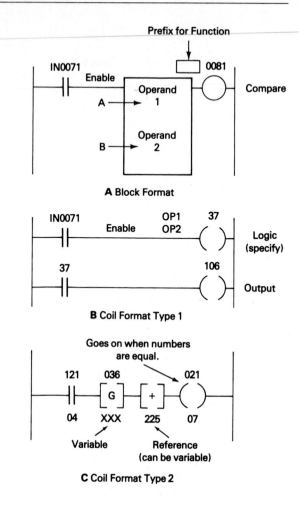

A Block Format

B Coil Format Type 1

C Coil Format Type 2

The Move function is a useful, often-included PLC function. It is shown in Figure 15–31. When enabled, it duplicates the value of the source register in the destination register. The data in the source register are unaffected. The previous data in the destination register are eliminated and lost. The register's previous value is replaced by the value from the source register. Other forms of the Move function move large blocks of data. Those other Move functions are covered in the textbooks listed in the bibliography.

An application of the Move function is shown in Figure 15–32. A timer is to have either a 7 s or a 15 s run time, depending on the product being processed. Rather than having the system use two timers, one of the times is moved into the one timer's preset register as needed. Each time is moved by enabling an appropriate Move function. This type of circuit is useful when the time is not fixed, but rather a variable. The time utilized is chosen according to the process requirements of the moment.

Comparison	Function	Equation	Circuit (conducts when equation is true)
*1	Equal (EQ)	A = B	EQ ─┤├─
2	Not equal	A ≠ B	EQ ─╫─
*3	Greater than or equal to (GE)	A ≥ B	GE ─┤├─
4	Less than	A < B	GE ─╫─
5	Greater than	A > B	GE ─┤├─ EQ ─╫─
6	Less than or equal to	A ≤ B	GE ─╫─ / EQ ─┤├─

* Basic Functions

FIGURE 15–26
The Six Comparison Functions

15–8 USING DIGITAL BITS: THE SEQUENCER

Section 15–4 included a brief look at the use of digital bits to control outputs in the ON or OFF state. This capability of the PLC will now be examined in more depth. When the PLC memory is cleared, all registers are set to 0. That is to say, all 16 bits are set at 0. To insert 1s in a register, the specific register is called up on the screen in the Monitor mode. The desired bit pattern of 1s and 0s is inserted in the register. In most PLCs the register call-up is accomplished by pushing Change

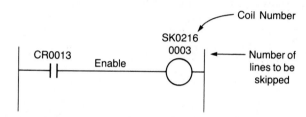

FIGURE 15–27
The PLC Skip Function

FIGURE 15–28
Skip Function Operation

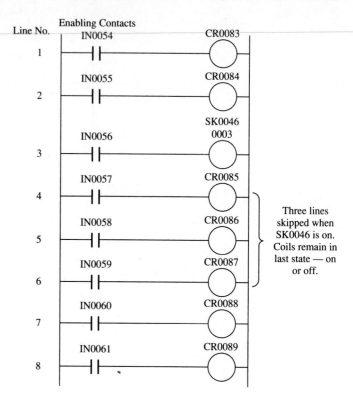

Three lines skipped when SK0046 is on. Coils remain in last state — on or off.

with the cursor over the register to be altered. Then the bit pattern is inserted in the register.

In Figure 15–33 an example is shown of an inserted bit pattern in HR0207. This HR0207 pattern will be used for illustration throughout this section. The binary value of this register is irrelevant for bit-pattern-control operations.

The register bits may be used for ON or OFF outputs, as shown in Figure 15–34. The first ten bits are used to control ten outputs, as illustrated. Any of ten outputs can be controlled by the use of Bit-pick function.

The system of output control by these bits is accomplished by the Bit-pick contact function as shown in Figure 15–35. When a contact is programmed on a

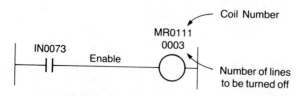

FIGURE 15–29
The PLC MCR Function

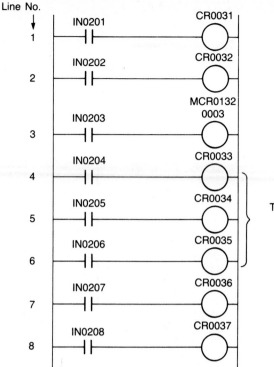

FIGURE 15–30
MCR Function Operation

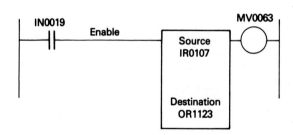

A Block Format

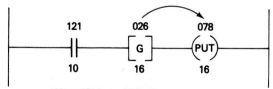

When 121 is on, Value in
location 026 (16) is duplicated
(effectively transferred) to location 078.

B Coil Format

FIGURE 15–31
The PLC Move Function

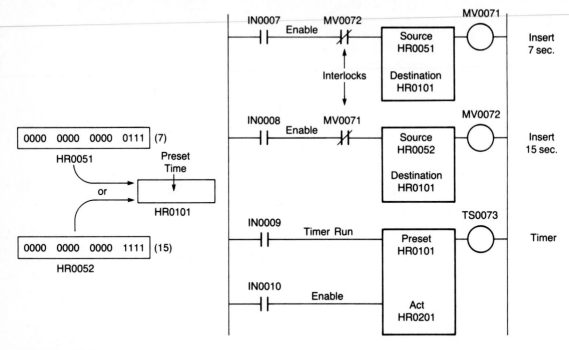

FIGURE 15–32
Move Application

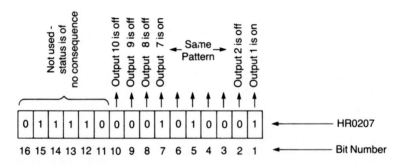

0111   1000   0101   0001   HR   0207

FIGURE 15–33
Register with a Bit Pattern

FIGURE 15–34
Ten Outputs Controlled by Ten Register Bits

FIGURE 15–35
Bit-Pick Contact Control

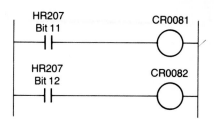

PLC, there are normally three choices for contact number designation. These are Input, CR (for logic from another output), or Bit-pick. These choices are shown on a screen menu, from which you select one. Bit-pick programming is shown in Figure 15–35. Register HR0207 has been specified as the source of information. The individual control bit number is also specified. Note that CR0081 is OFF, as bit 11 of HR0207 is a 0. Similarly, CR0082 is ON.

Suppose there is a 16-bit pattern for control, and an alternate control pattern is to be added, with only one bit changed. Another register could be created and used for alternate control. Instead, one of three bit-changing functions can be used on one given register. These three bit-changing functions are Bit-set (BS), Bit-clear (BC), and Bit-follow (BF).

The first function, BS, is shown in Figure 15–36. It is called up on the screen as one of the special functions available. It normally appears as a coil with two labels to be specified: the register number and the bit of that register to be altered. A switch, or other input control, is programmed to enable this BS function. Turning the BS function ON causes the specified bit to become a 1. Once the bit has been changed to a 1, it stays a 1 when the function enable is turned OFF. Note that if it was a 1 originally, it will have stayed a 1.

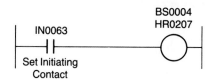

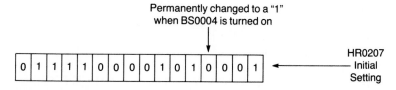

FIGURE 15–36
Bit-Set Function

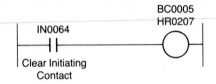

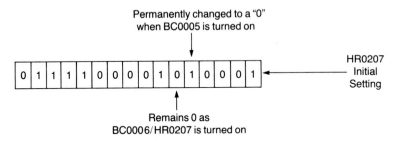

FIGURE 15–37
Bit-Clear Function

Figure 15–37 shows another bit-altering function, the BC function. It is programmed the same as the BS function. It performs the opposite of BS: it changes a 1 to a 0 when enabled, as shown in the figure.

The third bit-altering function is the BF function. Again, it is programmed in the same manner as BS and BC. The BF function changes the bit to a 1 and back to a 0 as it is turned ON and OFF. Turning it ON produces a 1 and turning it OFF produces a 0. The BF function cannot be used on the same bit as a BS or BC function; they are contradictory. Figure 15–38 illustrates the BF function.

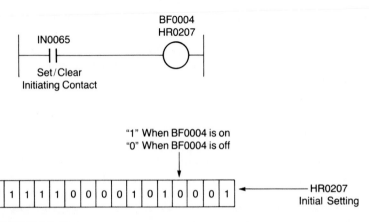

FIGURE 15–38
Bit-Follow Function

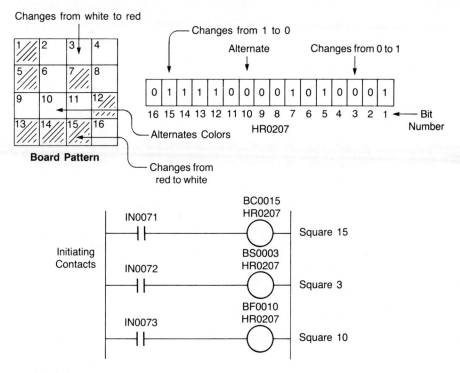

FIGURE 15–39
Bit-Control Painting Application

A simple application of bit patterns and bit-altering functions is shown in Figure 15–39. A board is to be painted with red paint as shown. There are 16 squares on a white board. There is a controlled spray gun above each square. The spray gun goes ON for a corresponding bit of 1. Using the pattern of HR0207 would produce the red-on-white pattern shown. Other models of boards have different patterns. For their painting, the HR0207 patterns needs modification of its bit pattern. One model needs section 3 to be red. Another needs 15 to be left white. In another model, section 10 must be capable of being either color. The three bit-altering functions are used to accomplish the pattern changes for these other spray patterns.

A common use of bit pattern control is the *sequencer*, or *drum-controller* (*DR*), function. It is used to control a number of outputs through a sequence of operations. A dishwasher cycle is used for illustration in Figure 15–40. The required dishwasher device functions are listed at the top. To accomplish wash, rinse, and dry, a planned series of step intervals is set up as shown, 1 through 11. The resulting ON/OFF patterns are put into a series of registers, starting with HR251.

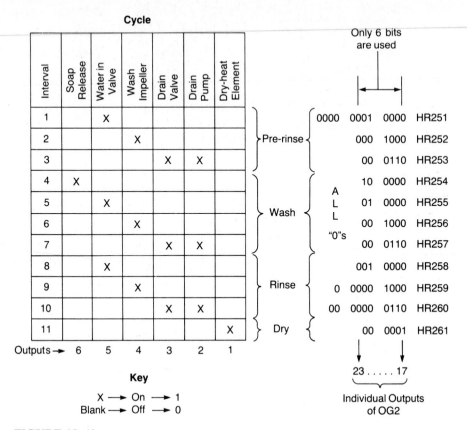

FIGURE 15–40
Dishwasher Function Matrix and Register Patterns

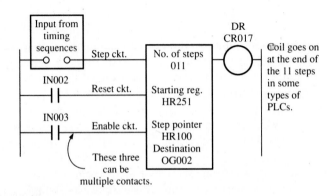

FIGURE 15–41
Sequencer Function for Figure 15–40 Matrix Control

The DR (sequencer) PLC function to run the dishwasher through its cycle is illustrated in Figure 15–41. The DR function is called up under Special function. There are five parameters to be specified for the DR function. First, the coil/ function number must be specified. Care must be taken to choose a number for DR that is not used anywhere else in the program. Next, the number of steps is specified—11 for our sequence. Then, the starting register is listed: HR251. The step pointer, an option, tells which of the 11 steps is being acted on at the moment. The step pointer is essentially a step counter. Finally, where does the output go? In this example, OG002 is specified; it contains outputs 17 through 32. Actually, only 17 through 23 are used for the dishwasher.

There are three input control lines for a DR function. The first is the step line. The first register, HR251, feeds the outputs initially. When the step line is turned ON, the function switches to the next register for the output pattern, HR252. Each time the step line is turned ON, the function moves to the next register. After step 11, the function starts over again. Special additional circuitry (not shown) is needed to stop the cycle after step 11. This input line must be run to a suitable timing circuit for the required operational time intervals. This timing circuit is not shown.

The second line, IN002, is a reset circuit. When it is ON, the unit may function. Any time it is turned OFF, the DR resets to the start state, at HR251. The third input line is the Enable line. It must also be ON for the function to operate. When it is turned OFF, the DR stops where it is at that time and does not reset to the beginning.

A final function using digital bit control is the shift register system. There are many different types of shift register functions available in PLCs. These are shift right, shift left, rotate, and multiple shifts. Only the single-shift, shift-right register is shown in Figure 15–42. Shift functions, including the SR shown, are capable of utilizing 1 to 16 registers for shifting. In Figure 15–42 only one-register and two-register shifts are shown.

The shift function has three parameters that must be specified. These are the function (SR) identifying number, SR0111; the register in which shifting starts, HR0207; and the number of registers involved, 1 or 2 (or more). There are three input lines for the SR function, as shown. The shift line, IN0051, when turned ON for the shift in Figure 15–42A, causes all bits to move one space to the right. The second control line determines whether a 1 or a 0 is reinserted in the empty bit space, 16. For "Data in" ON, a 1 is shifted in, and for "Data in" OFF, a 0. The third line is the function Enable. If the function Enable is OFF, the function does not step when "Line on" is turned ON. When the Enable is ON, the function can step. In some PLCs the coil, SR0011, is turned ON when a 1 is shifted out and is turned OFF when a 0 is shifted out. This OFF/ON operation is not carried out in all manufacturers' shift functions.

Figure 15–42A is for a single register. In Figure 15–42B two registers are used. There can be more registers involved, depending on the number of registers specified in the function. For two registers a shift through 32 steps is shown. The bit leaving HR0208 appears at the beginning of HR0207 as illustrated.

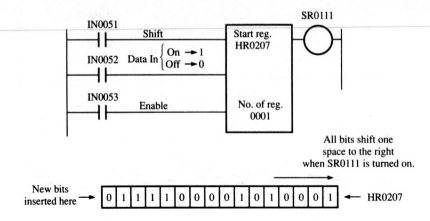

A. One Register

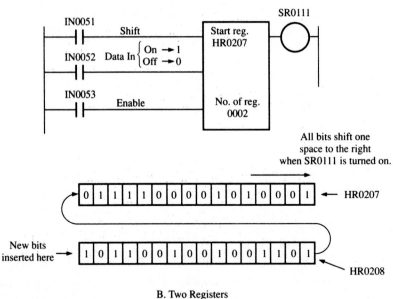

B. Two Registers

FIGURE 15–42
Shift Register Layout

Applications of this function include sequencing of a number of machines by appropriate bit-picking. This type of function can also be used for moving alphanumeric displays.

15-9   ANALOG OPERATION

So far, the discussion of PLCs has involved only discrete functions. Inputs are ON or OFF. After logic decision by the CPU, appropriate outputs are turned ON or OFF. This section will cover another capability of PLCs—the ability to analyze variable input values and produce appropriate variable output values. Most medium and large PLCs have this analog capability.

To carry out PLC analog operations, special input and output modules are needed. Some of the standard ratings of available modules are shown in Figure 15-43. These ratings are for either input or output modules. There are other special modules, not listed, for special devices, such as thermocouples and transducers. These special modules are designed to match a specific model of input or output device.

The input signals available have to be converted and scaled to match an available module. For example, you have a signal that varies from 0 V to 78 V AC, with 78 V representing 100% input voltage. You decide to use a 0–5 V DC input module. Therefore, you must convert 0–78 V AC to a linear 0–5 V DC, as shown in Figure 15-44. The DC voltage fed from the converter into the module is then in turn converted to a digital number. This digital number is sent from the analog module to an input register in the CPU, as shown in the figure.

How does the input conversion work? For illustration, trace 31 V AC. The converter analyzes the portion of 78 that 31 represents. This is 0.397. The converter, which you must design and supply, puts out a DC voltage that is this proportion of 5 V DC. This DC value, 1.987 V, is sent to the input module. Assume that the input module is an eight-bit base, which can hold a value up to 256 in decimal. The input module then takes this same proportion of 256, 102, and sends the value to a CPU input register. Which register receives the data, 102, depends on the setting of DIP switches on the module.

Note that the input is stepped, in 256 steps, and is not perfectly linear when the CPU receives data. The accuracy of this system is $1/256 = 0.0039$, or about

FIGURE 15-43
Typical Analog Module Ratings

2-10 mA
4-20 mA
10-50 mA
0 to + 5 Volts DC
0 to + 10 Volts DC
± 2.5 Volts DC
± 5 Volts DC
± 10 Volts DC

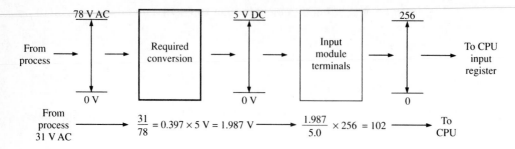

FIGURE 15–44
Analog Input Signal Path and Values

0.4% accurate. Other, more accurate input modules, of 10 and 12 bits, can be obtained, at greater cost, if needed in your application. These would have 1024 and 4096 steps, respectively.

How does the output signal get from the CPU to an output analog device? Figure 15–45 shows an output system. For illustration it is assumed that the signal in Figure 15–44 was multiplied by 2 in the CPU. The output ratio is then 0.794, as shown in Figure 15–45. This would be 203 on the 256-step scale. Assume that there is an output module feeding an op amp device with a range of −10 V to +10 V. The math shown indicates that the output would have a value of 5.9 V DC.

To illustrate how an analog system might be used, an Add application is shown in Figure 15–46. An output meter is to indicate the sum of two analog inputs. The two input values go through conversion and then through the input module. The digital values end up in IR01 and IR02, as shown. An Add function adds the two values when the Add function is enabled. The sum can be updated very quickly by rapid enabling of the Add function. The sum is put into OR01. The sum is then sent to an output module and then to the indicating meter. For illustration, 17 V and 41 V AC are added, to give an output of 7.8 V to the meter. The meter full scale could be set at 2 times 78, or 156 V AC, to match input and output scales.

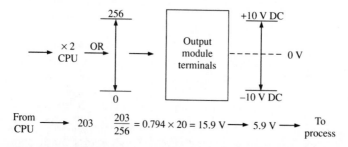

FIGURE 15–45
Analog Output Signal Path and Values

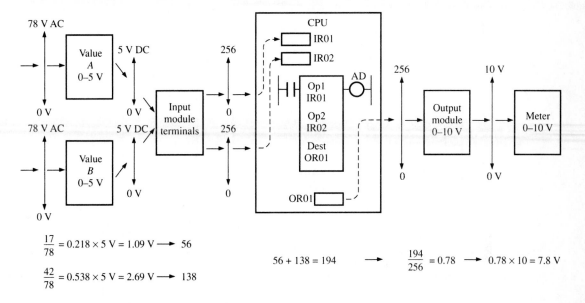

$$\frac{17}{78} = 0.218 \times 5 \text{ V} = 1.09 \text{ V} \longrightarrow 56$$

$$56 + 138 = 194 \qquad \longrightarrow \qquad \frac{194}{256} = 0.78 \longrightarrow 0.78 \times 10 = 7.8 \text{ V}$$

$$\frac{42}{78} = 0.538 \times 5 \text{ V} = 2.69 \text{ V} \longrightarrow 138$$

FIGURE 15–46
Analog Add Application

Many analog systems are operated, set, and observed by machine- and process-operating personnel. Therefore, for operator clarity the operating analog system needs to be in decimal form. This decimal form is called *multibit* or *binary coded decimal* (*BCD*). To handle the BCD system, special BCD modules are required for the PLC. A thumbwheel decimal input is shown feeding a BCD module in Figure 15–47. Also, a BCD output module is shown feeding a signal to a 0–9999-count LED output display.

In BCD each input decimal digit is represented by four bits, as illustrated in Figure 15–47. The BCD bit patterns are sent to an input register in the CPU for processing. After processing in the CPU, the resulting BCD number is placed in an output register. The output register contents are sent to the display.

All PLC CPUs calculate in the binary number system, not BCD. If the incoming BCD number or numbers are used for arithmetic processes, the mathematical results will be incorrect. The incoming BCD number must be converted to binary before effective use by the CPU. Appropriate number conversion functions are available in medium and large PLCs. These are the BCD-to-binary conversion function and the binary-to-BCD conversion function. An example of these conversions required to give a correct answer is shown in Figure 15–48. The output is to display a decimal figure that is half the input value. This is a simple example for illustration. In practice, the mathematical conversions are more involved. Also, multiple inputs and outputs are often involved.

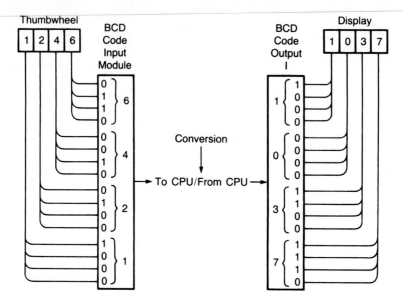

FIGURE 15–47
Analog BCD Input and Output System

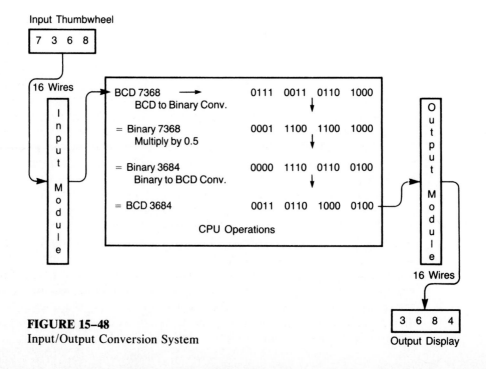

FIGURE 15–48
Input/Output Conversion System

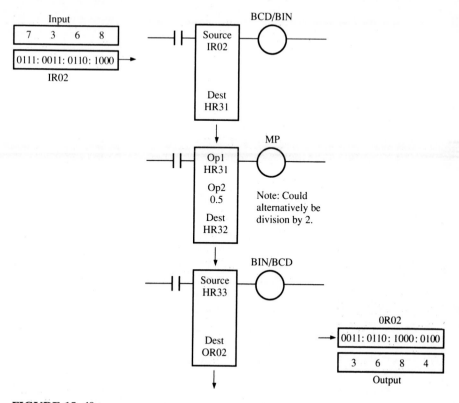

FIGURE 15–49
PLC Program for Figure 15–48 Conversion System

Figure 15–49 is a general format of the internal PLC programming to accomplish the conversion of Figure 15–48. The input number is brought in to IR02. The input value is converted from BCD to binary by the first programmed function. Then, multiplication (or division) is carried out by the second function. Finally, the resulting binary number is converted to BCD by the third function. The resulting output BCD value is sent out through OR02 and the output BCD module to the output display, as shown.

15–10 LOOP AND PID CONTROL

The last section covered the PLC's analog capabilities. A PLC has the further capability of expanding its analog capability to closed-loop control. This closed-loop control functional capability is found in loop-control and PID functions. These functions are available in basic form in medium-sized PLCs for a few hundred extra dollars. In larger PLCs these feedback-based functions are included in the higher-priced units.

FIGURE 15–50
Typical Loop-Control Function

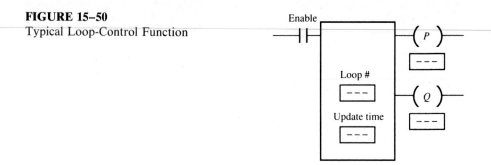

In Chapters 13 and 14 the principles of closed-loop control were discussed. The PLC can be programmed to control the processes outlined in those chapters. In this section the use of these PLC functions will be described only in general terms. Detailed programming information would require a whole chapter or even a number of chapters. One requirement for loop and PID operation is the use of special PID input and output modules.

Figure 15–50 is a representative loop-control PLC function. This loop-control function can contain typically one or two PID functional operations. When enabled, the function is operational, and its coil P goes ON. Assume that the update time has been set to 15 s. Every 15 s the PID function is sent through its operational sequence. The Q coil goes ON when the update time is reached and stays ON for one scan. A loop identifying number is entered in the functional block.

Figure 15–51 is representative of a PLC PID (proportional-integral-derivative) function. Some PLCs combine the previously described loop-control function with the PID function. This example is for the PID function alone.

This function can be described as follows:

Inputs:

EN—the usual function Enable line

MANUAL/AUTO—manual or automatic mode

HOLD—used for "clamping" and for logic transition control

Outputs:

P—the coil number assigned

Q—an output limit coil used in the logic

Functional Descriptions and Values:

PID number—the PID block identification number

INPUT—the register in which the process variable is stored

OUTPUT—the register in which the output algorithm is stored

FIGURE 15–51
Typical PID Function

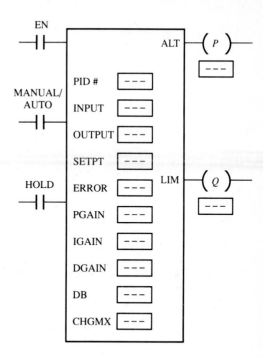

SETPT—the register in which the set-point is stored

ERROR—the register in which the value of ERROR = (SETPT − INPUT) is stored

DB—the register for the deadband value

CHGMX—the register in which the maximum allowable rate of change is stored

PGAIN—the register in which proportional gain is stored

IGAIN—the register in which the integral term is stored

DGAIN—the register in which the proportional term is stored

Most of these functional values are written as a percentage of the set-point. The values may be constant or may be moved in from a register in another part of the program.

SUMMARY

☐ The PLC is a microprocessor-based, user-friendly process controller.
☐ The PLC communicates with the process through special input and output modules.
☐ A small PLC is programmed in digital logic form from small, hand-held devices. For larger PLCs programming is carried out in ladder format on dedicated monitors or on specially programmed personal computers.

☐ Coil-and-contact programming is the most common PLC programming. Coil-and-contact circuit connection is very simple compared with relay logic connections.

☐ The use and control of internal registers is very important in PLC operation.

☐ Circuit logic flow, register values, and other parameters can be observed on a PLC in the Monitor mode.

☐ Timers and counters are easily programmed on the PLC. Combinations of more than one timer can be used to carry out any required timing configuration. Counters may be up or down counters and also may be used in combination.

☐ Dozens of other PLC discrete functions are available (at a cost), depending on the complexity of the application.

☐ The most common other functions are the mathematical functions, Compare, Skip, and the master control relay (MCR) function.

☐ The PLC uses digital bit patterns in registers to perform a number of functions. The most common of these functions is the sequencer. Other bit-based functions include various shift register functions. Bit patterns can be modified by applying appropriate PLC functions.

☐ A PLC can perform analog functions, both input and output. The usually used analog forms are the numerical analog and the BCD, or multibit, systems. Analog and discrete PLC inputs and outputs can be "mixed and matched" according to the application.

☐ Advance PLCs can perform closed-loop feedback control. More than one loop can be included.

EXERCISES

1. List the parts of a PLC controller. Which are essential, and which are options?

2. You are putting together a PLC system. There are 24 inputs and 16 outputs for discrete control. There are 8 inputs and 4 outputs for analog control. There is a disk drive for recording programs, and there is a large program monitor. Additionally, 8 of the 24 discrete inputs and 4 of the discrete outputs are also at a location 1000 ft from the PLC. Draw a block diagram of the system.

3. Draw figures similar to Figures 15–5 and 15–6 for the following operation:

   Switch 11 or switch 12 turns ON relay 1.
   Switch 13 and switch 14 turn ON relay 2.
   Switch 15 or limit switch 31 and relay 1 turn ON relay 3.

4. Refer to Figure 15–8. Suppose there are two start buttons and two stop buttons in the system, for two different control locations. Each location has a pilot light to indicate ON. Draw PLC logic and connection diagrams for this two-location system.

5. Refer to Figure 15–9. Redraw the PLC screen pattern and connection diagram for the case in which there are forward, reverse, and stop buttons at two locations.

6. Refer to Figure 15–10. Redraw the PLC logic and connection diagrams for a double-control-station system. Each of the two stations has forward, reverse, and stop buttons.

7. Refer to Figure 15–11B. Redraw the PLC logic and connection diagrams for two-station control. Each station has start, stop, and jog.

8. Refer to Figure 15–12. An alarm system has two characteristics in addition to the two functions given. First, if any one of the inputs comes ON, a yellow pilot light goes ON. Second, if all four inputs go on, an alarm bell sounds. Show a modified logic and connection diagram to accomplish these added functions. The other requirements still hold.

9. Look up register types in a number of manufacturers' manuals. List the types and the series of numbers reserved for each type.

10. Fill in the following table. Which outputs or inputs would the specified group registers control?

Group	For 8-Bit PLC	For 16-Bit PLC
OG3		
OG11		
IG4		
IG13		

11. What output group register would control the following? What would its bit pattern be for an 8-bit PLC?

CR065 ON	CR069 OFF
CR066 OFF	CR070 OFF
CR067 OFF	CR071 ON
CR068 ON	CR072 ON

12. Suppose the pattern in Exercise 11 were for CR05 through CR13. How could those outputs be controlled by an 8-bit machine output group register system? (*Hint:* Parts of two output group registers would be used.)

13. Refer to Figure 15–18. Make up a screen (logic) pattern and a PLC connection diagram for the time pattern in Figure 15–52, using two timers.

FIGURE 15–52

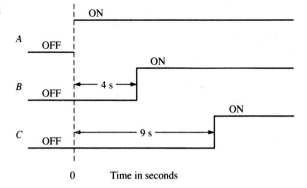

14. Refer to Exercise 13. Make up a screen (logic) pattern for the time pattern in Figure 15–53. (*Hint:* Starting ON and going OFF after a time interval is accomplished by using an NC contact for the timer output to another coil.)

FIGURE 15–53

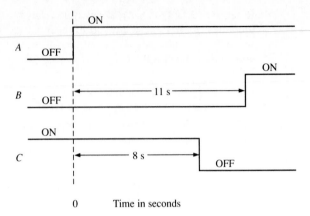

15. Refer to Figure 15–19. A light is to be ON if IN0003 has a count of 9 or more and also IN004 has a count of 16 or more. Make up PLC logic and connection diagrams to satisfy this requirement.

16. Suppose in Exercise 15 that the light is to be ON only if the respective counts for IN0003 and IN0004 are 9 and 16. Make up a PLC logic and connection diagram to satisfy this requirement.

17. Refer to Figure 15–20. Make up a counting system to give the number of parts on the conveyor shown in Figure 15–54. The conveyor has two input lines and one output line, as shown. The total count is to be in HR0045.

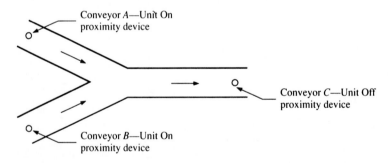

FIGURE 15–54

For Exercises 18–20, refer to Figures 15–21 through 15–24, and use the following registers:

$$
\begin{array}{ll}
T & \text{HR065} \\
A & \text{IR023} \\
B & \text{IR024} \\
C & \text{IR025}
\end{array}
$$

18. Develop a PLC math program for $T = A + B - C$.
19. Develop a PLC math program for $T = (A - B) \times C$.
20. Develop a PLC math program for $T = (A/C) - 8$.

For Exercises 21–24, assume that *A* is in HR0067 and *B* is in HR0068. Also refer to Figure 15–25. Develop a PLC comparison program for the requirements of each exercise.

21. Output CR0011 is ON if *A* equals either 7 or 15.

22. Output CR0018 is ON only if *A* equals 12 and *B* equals 22.

23. Output CR0045 is ON if *B* is not equal to 18.

24. Output CR0002 is ON if *A* is greater than *B*.

25. Refer to Figures 15–27 and 15–28. There are 12 ladder lines in a PLC process program, CR32 through CR43. Lines CR35 and CR36 are to be skipped when IN04 is ON, and lines CR38 through CR42 are to be skipped when IN05 is ON. Make up a ladder diagram for function and control of this process.

26. The problem is the same as in Exercise 25, except that the skipped lines are to be OFF. Refer to Figures 15–29 and 15–30.

27. Refer to Figures 15–31 and 15–32. A timer, TS0045, is to be able to time according to three different input values. The three times are 8 s (IR0004), 22 s (HR0062), and a time whose value varies according to the process action (IR0005). Make up a triple Move program to accomplish this timing requirement.

28. Refer to Figures 15–34 through 15–39. A 16-bit holding register has the following bit pattern: Register number is HR0062.

1001   0111   0100   1110

Bit-changing functions are to be used to accomplish the following:

Change bit 8 from OFF to ON.
Change bit 11 from ON to OFF.
Change bit 15 back and forth between ON and OFF.

Create a three-line PLC program to accomplish these changes.

29. How could you change bit 8 back to ON and bit 11 back to OFF in Exercise 28?

30. Refer to Figures 15–40 and 15–41. Create a similar PLC program for a clothes washer-dryer combination. Include register patterns. The washer-dryer (in one unit) contains the following functions:

Washer	*Dryer*
Soap release	High heat
Bleach release (optional)	Low heat
Water-in valve	Spin only
Tub spin	
Drain pump	
Wash impeller	

31. Refer to Figure 15–42. A 16-bit shift register's shift line is turned ON every 1 s. Create a PLC program, including register bit patterns, to give the following pattern for an output light. You may use the output coil, or you may bit-pick the last bit through which you shift.

2 s ON
3 s OFF
1 s ON
5 s OFF
8 s ON

1 s OFF

4 s ON

7 s OFF

3 s ON

6 s OFF

9 s ON

32. Refer to Figure 15–44. Trace 61.5 V through the system.

33. Refer to Figure 15–45. Suppose multiplication were by 0.5 instead of 2. The digital number would be 51. Trace 51 through the system.

34. Refer to Figure 15–46. Trace for inputs of 22 V AC for valve A and 31 V AC for valve B.

35. Refer to Figure 15–47. Suppose 8492 were an input and the CPU divided the input by four. What would the figure look like, including the register patterns?

36. Refer to Figure 15–48. Suppose that 2368 were an input and it were multiplied by 4 in the CPU. What would the figure look like, including the register patterns?

37. Refer to Figure 15–49. What would the PLC program for Exercise 36 look like?

APPLICATION ASSIGNMENT ANSWER

First, determine how many coils, contacts, timers, and other components are involved. Then, determine any additions that might be made to the control system. From this information you can determine the type and size of PLC to procure. Another consideration is the type of programming system, ladder or digital, that is required.

Determine the electrical ratings for the input and output modules. A normal rating is 110 V AC at about 2 A. Order the I/O modules accordingly. Then construct the general PLC ladder diagram. Program the resulting ladder diagram into the PLC.

16

Power Distribution Effects on Industrial Electronic Controls

CHAPTER OUTLINE

16–1 Introduction □ **16–2** Circuit Protection: Fuse, Circuit Breaker, and Overload Relay □ **16–3** Proper Wiring, Conduit, Connectors, and Other Hardware □ **16–4** Causes of Low Voltage: Line Loss and Poor Power Factor □ **16–5** Effects of Power Surges and Transients □ **16–6** Three-Phase versus Single-Phase □ **16–7** Electrical Grounding Systems and Protection □ **16–8** Power Cost Considerations □ **16–9** Standby Power Units—Generator and Electronic

LEARNING OBJECTIVES

- □ Describe how to apply fuses, circuit breakers, and overload relays to electrically protect power systems and to protect equipment.
- □ List the limitations of protective devices.
- □ Describe and calculate the low voltages to equipment caused by line loss and by poor power factor.
- □ Describe how to protect electrical and electronic equipment from power surges and power line transients.
- □ Describe the major differences between single-phase and three-phase power systems in the manner in which they are used to supply electrical and electronic equipment.
- □ List the characteristics and describe the construction of proper grounding systems for personnel safety and for proper equipment operation.
- □ List the ways to reduce monthly power costs for the three types of billing charges.
- □ Explain where standby power equipment should be used and how. Differentiate between the major types of standby power by application.

APPLICATION ASSIGNMENT

You are a technician working on installing control equipment in a manufacturing plant. New electrical equipment is being added weekly to increase production. There have been four problems in operation over the past few weeks:

1. Relays are dropping out periodically when they should stay energized.
2. Computers are periodically causing random erroneous equipment movements.
3. Motors on a machine are burning out in spite of the use of the lowest possible value of fuse in its feeder circuit.
4. In a general plant meeting, you find out production has risen 20% but power bills have risen 38%.

What actions for correction can you look into?

16–1   INTRODUCTION

Why is this chapter needed in an electronic control text? Why be concerned about anything beyond where you plug the equipment in? Engineers and technicians who have actually worked in the field find that a basic knowledge of electrical supply systems is essential. The characteristics and proper operation of the feeder power system are of great concern for trouble-free control system operation.

The authors' experience with control system problems has shown that the causes of equipment field malfunction are of three general types:

1. failure of the equipment
2. misuse, possibly due to poor communication between the equipment provider and the customer on equipment specifications
3. problems with the power system

The relative frequency of these three types of causes varies with the product or system. In general, each cause seems to be responsible for about one third of the malfunctions.

This chapter will deal with the third cause of equipment failure: power system problems. Consideration of power system effectiveness is needed for analyzing field malfunctions of equipment. Also, possible power variables must be taken into consideration in the original design and testing of a product (and for major changes in the product). What will your equipment see in the way of power variations, surges, and so on once it is in actual operation?

Proper circuit protection and proper wiring and connecting will be dealt with first. Causes of low voltage, which can cause electronic malfunctions, will be covered. Next will come consideration of line surges and line transients and advice on preventing them from getting to and damaging equipment. The uses of three-phase and single-phase power will then be compared.

Plant power costs and possible cost reduction will be illustrated. Finally there will be a discussion of standby power units, including identification of the correct type of unit to use for a given application.

16–2   CIRCUIT PROTECTION: FUSE, CIRCUIT BREAKER, AND OVERLOAD RELAY

Circuit electrical protection and equipment electrical protection are similar but not the same. Circuits are protected by fuses and circuit breakers so that the wiring and other distribution devices do not overheat and cause a fire. Equipment protection, on the other hand, is accomplished by some means to protect particular equipment from electrical overloading. For example, the fuse in your home fuse box protects the circuit. The circuit wiring, plugs, and so on are protected from being overloaded electrically by these fuses. Equipment protection is different. Some equipment, such as a home TV, is internally protected by an overload relay located on the back panel.

In this section some of the major types of distribution fuses and circuit breakers will be discussed, including ground-fault interrupters (GFI). The use of overload relays of various types that protect individual pieces of equipment will also be covered.

First, two general questions: Does a 25 A fuse blow instantly at 25.0 A or 25.1 A? Does it stay intact at 24.9 A? The answer to both questions is no. The actual blow time is best illustrated by the curve in Figure 16–1. This is a curve for a fuse series, but it also represents the action of breakers and other protection devices.

This curve, for a series of fuses 20–600 A, is typical. The 100 A fuse's curve will be analyzed. Various points on the curve give the following table of values:

Current (Amps)	Time (Seconds)
100	Minutes—off scale
200	300
300	90
500	20
700	1.5
1000	0.045
2000	0.01

This fuse will adequately protect circuit wiring, since equipment is rated for these times to blow. However, this fuse will not protect a 100 A SCR. The SCR will blow first, at 100 A. A special class of fuse is needed to protect SCRs and other thyristors. Such fuses can be found listed in manufacturers' catalogs. A knowledge of various fuse classes is required in this and other cases. What are some of the considerations in choosing a fuse?

1. rated current for melting or blow
2. rated voltage of the application
3. interrupt capability (A 20 A fuse will not stop 20,000 A. The current will arc over.)
4. temperature of the environment in which the fuse will operate (Higher temperature means faster action and blowing below rated value.)
5. type of mounting
6. replaceable internal link or one-time operation
7. time-delay or regular type
8. other special requirements

Characteristics 1–4 are straightforward. The others will be explained further.

Various types of mounting are illustrated in Figure 16–2. The mountings are purposely different so that the replacement fuse must be the same class of fuse as the original.

Replaceable-link fuses have advantages and disadvantages. The advantage is that only a link needs replacement after the fuse blows. Replacing the link saves the cost of a whole fuse. Also, only links need to be stocked, not whole fuses. There is one major disadvantage: the link of a 150 A fuse can be replaced with the

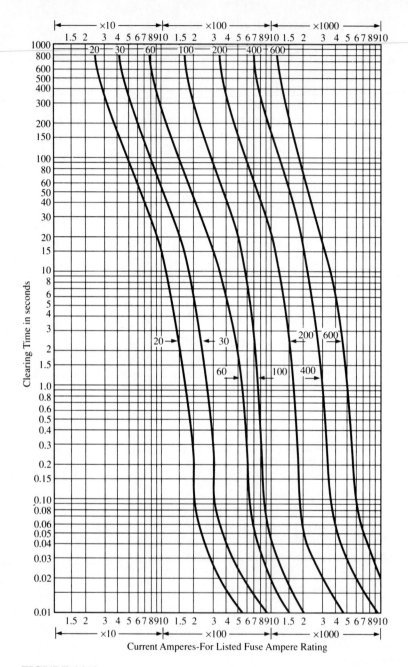

FIGURE 16–1
Fuse Curve for Current versus Blowing (Melting) Time (Courtesy of Bussmann Division, Cooper Industries)

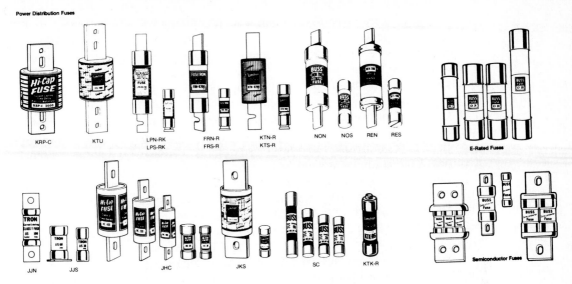

Power Distribution Fuses

KRP-C KTU LPN-RK / LPS-RK FRN-R / FRS-R KTN-R / KTS-R NON NOS REN RES E-Rated Fuses

JJN JJS JHC JKS SC KTK-R Semiconductor Fuses

FIGURE 16–2
Various Types of Fuses (Courtesy of Bussmann Division, Cooper Industries)

link of a 200 A fuse. The links are the same size. Without good control of maintenance procedures, this can happen. There have even been cases in which a link was replaced by a solid copper wire, resulting in injury to operating personnel. A picture of a replaceable-link fuse is shown in Figure 16–3.

What is a time-delay fuse? It is essentially one that lets short-duration, harmless, large, transient spikes through without blowing. A major use of time-delay fuses is for electric motors. When the motors start, they draw five to ten times rated current for a short time. If a regular fuse is used, a 20 A motor will

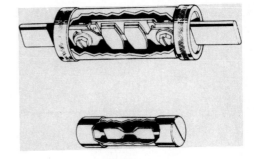

FIGURE 16–3
Replaceable-Link Fuse (Courtesy of Bussmann Division, Cooper Industries)

need a 100–200 A fuse. With a 100–200 A fuse, the whole distribution system must be constructed for the greater current. This is a rule for electrical feeder circuit construction. Needless to say, the feeder system would be larger and bulkier for the greater current—bigger wires, transformers, and disconnects. Therefore, the time-delay fuse makes good sense for motor circuits. Figure 16–4 shows the difference in operation of 25 A regular and time-delay fuses for a 20 A motor during starting.

One word of caution: The time-delay fuse is not a good idea for electronic circuits. Electronic devices do not like transient voltage spikes.

In all cases of fusing, the rules for size and type are covered in the National Electrical Code and in state and local codes.

There are many other types of fuses, three of which will be illustrated. Figure 16–5 shows the small cartridge fuses used to protect electronic devices. They have ratings from 1/10 A to 5 A. They may be of the time-delay type if need be. They usually have a glass case so that one can see whether they have been blown.

Some fuses of a milliamp rating may be installed on a PC board like resistors. These are shown in Figure 16–6. They are color-coded as shown in the figure. They may also have their values printed on them instead of having the color code. If one blows, a new fuse must be soldered into the PC board connections. This resoldering is a disadvantage. The fuses' advantage is that they can protect a small, low-current portion of a circuit. Furthermore, they are available in many ratings, permitting a close match to the blowing value needed.

The final type of fuse to be shown is the plug fuse. Plug fuses are of three types, as shown in Figure 16–7. The regular plug fuse, type W, comes in ratings from 1/10 A up to 30 A. Typical ratings are 5 A, $7\frac{1}{2}$ A, 10 A, 15 A, 20 A, 25 A, and 30 A. These fuses also come in the time-delay type, type T, as shown.

The third type of plug fuse is the type S. The type S fuse has two parts. There is an insert that is put in the fuse-box receptacle. Once in, the insert is in permanently. Then the plug fuse is, in turn, inserted into the insert. Why the insert? Each rating of insert is a different size. The insert prevents a plug fuse of a

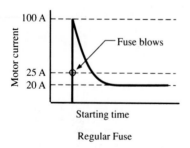
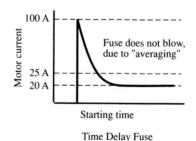

FIGURE 16–4
Time-Delay Fuse Operation

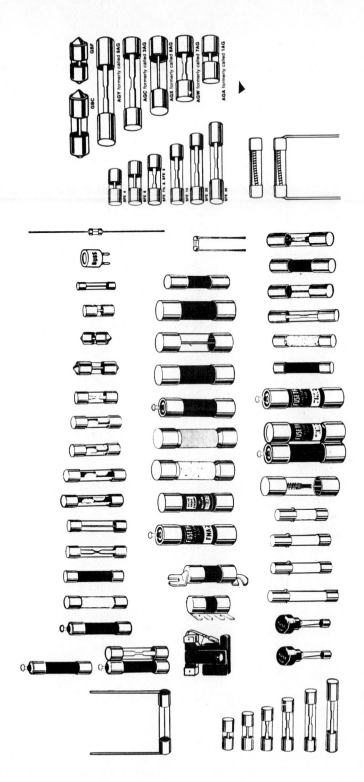

FIGURE 16–5
Small Cartridge Fuses (Courtesy of Bussmann Division, Cooper Industries)

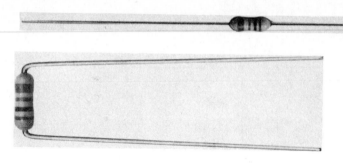

Rated Current MA	First Band	Second Band	Third Band	Fourth Band
62	Blue	Red	Black	Red
100	Brown	Black	Brown	Red
125	Brown	Red	Brown	Red
250	Red	Green	Brown	Red
375	Orange	Violet	Brown	Red
500	Green	Black	Brown	Red
750	Violet	Green	Brown	Red
1000	Brown	Black	Red	Red
1500	Brown	Green	Red	Red
2000	Red	Black	Red	Red
2500	Red	Green	Red	Red
3000	Orange	Black	Red	Red
3500	Orange	Green	Red	Red
4000	Yellow	Black	Red	Red
5000	Green	Black	Red	Red
7000	Violet	Black	Red	Red
10000	Brown	Black	Orange	Red
12000	Brown	Red	Orange	Red
15000	Brown	Green	Orange	Red

FIGURE 16–6
Small Pico Fuses and Color Code (Courtesy of Littlefuse® Tracor/Div. Westmark)

greater rating from being inserted. The inserts are sized so that a plug fuse of a smaller rating can be inserted. For example, suppose a regular plug fuse of 20 A is used in the fuse box for a circuit feeding an 18 A motor. If the 20 A fuse burns out, there is no way to prevent a 25 A or 30 A fuse from being used as a replacement. The motor would then not be properly protected. With the type S fuse, placing a larger-rating fuse in the insert is impossible. A fuse of 20 A or less must be used as a replacement.

Circuit breakers have one major obvious difference from fuses: they are resettable. When a circuit breaker trips, it can be reset by flipping its switch backward and then back ON. In some cases the circuit breaker is not as fast-acting as a fuse. However, it does not have to be replaced. Circuit breakers are available in five major types:

FIGURE 16–7
Plug Fuses (Courtesy of Bussmann Division, Cooper Industries)

1. *nonautomatic*—essentially an ON/OFF switch
2. *thermal*—trips when a certain level of heat is reached
3. *magnetic*—uses a magnetic field from the current for fast tripping
4. *thermal magnetic*—a combination of actions 2 and 3
5. *programmable solid-state*—adjustable for trip value and time rates

Figure 16–8 shows some typical circuit breakers. The first type, nonautomatic, is similar to a conventional wall switch. The thermal breaker is an overcurrent protective device. Its operation depends on heating from the current flowing through it. As current becomes excessive, its thermal element heats to the set level. It trips at the set level. The thermal breaker is a time-delay device suitable for motors. The time delay involved with thermal breakers is too long for electronic circuit protection.

The magnetic circuit breaker trips quickly because of the use of a magnetic field. The circuit current, when of a high short-circuit value, produces a large magnetic field. This large magnetic field is utilized to open the breaker contacts quickly.

The thermal magnetic breaker combines the constructions of the thermal and magnetic types. It combines the best characteristics of both types.

Commercially available circuit breakers come in ratings up to 2500 A and with interrupt capabilities of up to 100,000 A.

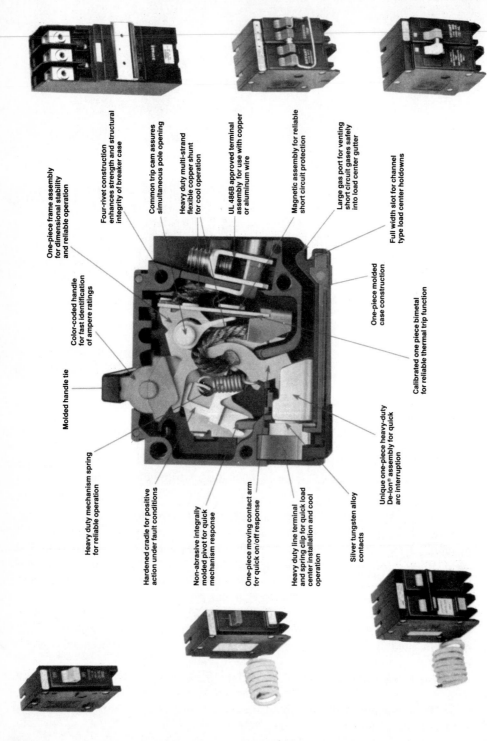

One-piece frame assembly for dimensional stability and reliable operation

Four-rivet construction enhances strength and structural integrity of breaker case

Common trip cam assures simultaneous pole opening

Heavy duty multi-strand flexible copper shunt for cool operation

UL 486B approved terminal assembly for use with copper or aluminum wire

Magnetic assembly for reliable short circuit protection

Large gas port for venting short circuit gases safely into load center gutter

Full width slot for channel type load center holdowns

Color-coded handle for fast identification of ampere ratings

One-piece molded case construction

Molded handle tie

Calibrated one piece bimetal for reliable thermal trip function

Heavy duty mechanism spring for reliable operation

Hardened cradle for positive action under fault conditions

Non-abrasive integrally molded pivot for quick mechanism response

One-piece moving contact arm for quick on/off response

Heavy duty line terminal and spring clip for quick load center installation and cool operation

Silver tungsten alloy contacts

Unique one-piece heavy-duty De-Ion® assembly for quick arc interruption

FIGURE 16-8
Circuit Breakers (Courtesy of Bryant Electric Div./Westinghouse Electric)

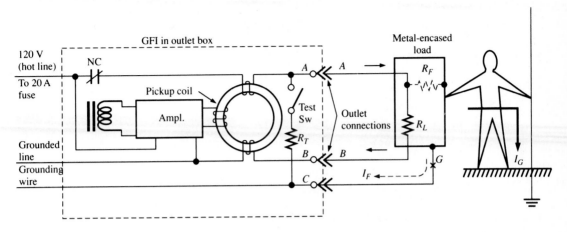

FIGURE 16–9
GFI Principle

The programmable breaker is very accurate in trip time. It is also very expensive. Various parameters may be set on a programmable breaker, such as trip setting, time delay, and value of allowable transient current.

A special type of breaker is the ground fault interrupter (GFI). It provides personnel protection from electrical shock. For example, an electrical circuit is protected by a 20 A fuse in the breaker box or fuse box. The metal case of a device becomes energized through internal insulation failure. If someone touches the case, that person may be injured or even killed before the 20 A fuse blows. If a GFI is installed in this situation, only 5 mA of current will flow through the person (for a very short time) before the GFI trips. This 5 mA is not injurious for the short time before the GFI trips. The operation of the GFI is illustrated by Figure 16–9. Suppose there is no short (ground) to case, a safe condition. Resistance R_f is infinite. There will be no ground current, I_g, through a person touching the case. The currents at A and B will be exactly equal and of opposite polarity. The resulting net voltage pickup current to the amplifier will be zero. The "doughnut" transformer will produce no voltage, since A and B cancel out. The GFI takes no action.

Now suppose that R_F is nearly 0 Ω because an internal bare wire is touching the case. If someone touches the metal case and stands on a moist floor, current will flow. It takes 20 A to trip the breaker (not shown). However, current B is now less than A. Current B is equal to current A minus the current flow through the person, I_G. The GFI transformer detects the difference between A and B. The differential voltage results in a voltage in the pickup coil. The pickup differential voltage is amplified. The amplifier sends a voltage to the GFI trip circuit. The circuit trips, opening the NC contact in the main line. When the current rises to

FIGURE 16–10
GFIs (Courtesy of Bryant Electric Div./Westinghouse Electric)

only 5 mA, the GFI trips. Therefore, the person is protected from any dangerous level of current.

Examples of GFIs are shown in Figure 16–10.

The final subject of this section is overload protection. Typical overload devices are illustrated in Figure 16–11. They can be thermal, current-sensitive, or a combination thermal-current type. Their mountings fall into two categories. They may be mounted in the device itself—an arrangement called *inherent protection*. (*Inherent* means the overload is installed on or inside the device being electrically protected.) Or, they may be mounted in the control panel for the device being protected.

Two types of on-device overload relays are illustrated in Figure 16–11. One has an automatic reset. Like any thermostat, when it cools off, the overload contact recloses. The advantage of the automatic type is that it recloses without being manually reset. An application would be in a remote location, such as a ceiling-mounted fan. A major disadvantage of this type can occur when the power is not turned OFF before the device is repaired. The overload can cool off while someone is working on the device. The device will then restart while the person is working on it. A motor could restart, resulting in possible injury.

The most common type of overload relay is the reset type (which is not shown). This type of overload remains open until the reset button (normally red) is manually depressed. While inconvenient and a little more costly, the reset type is a safer configuration.

Both internal overloads shown are series devices. The current flows through the coils and the contacts in series. In contrast, the panel-mount overload shown

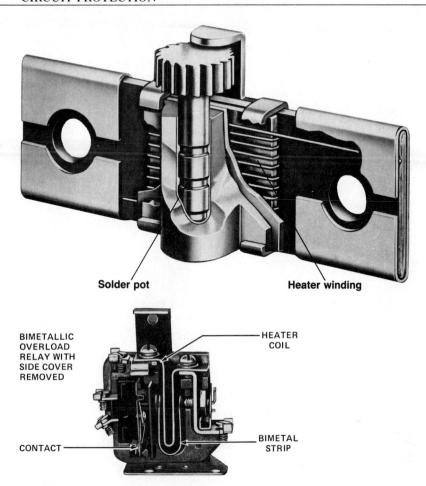

FIGURE 16–11
Overload Relays (Courtesy of Square D Company)

is a parallel device. The current of the device being protected flows through a coil as shown. The coil heats up. The overload opens a contact when the set trip rating current is reached. The overload's control contact is a separate circuit from the two terminals feeding line current through the overload device. The overload control contact terminals are then connected into the control circuit, as previously illustrated in Chapter 12.

One feature of all overload relays is that they are furnished in many ratings in small steps. These overloads ratings can be chosen to closely match the current limits of the device being protected. On motor-internal overload relays, there are small set screws. These set screws are adjusted by the manufacturer to the precise protection level required, and then they are sealed in place.

16–3　PROPER WIRING, CONDUIT, CONNECTORS, AND OTHER HARDWARE

This section will cover the basics of wiring and connecting of power systems. First, what size of wire is suitable for a given current-carrying requirement? Electrical wire is sized according to the current it is required to carry. Greater current requires a bigger cross-section of wire. The voltage to which the wire is to be subjected determines the type of insulation with which it is covered. Both current and voltage ratings must be adhered to for product and personnel safety.

Figure 16–12 is a standard table of current-carrying capabilities of various sizes and insulation types of copper and aluminum wire. At normal room temperature these capabilities are applicable. The temperatures given at the top of the table are the temperatures that the wire itself may reach. For example, #4 TW copper wire can reach a temperature of 60 °C. It can carry up to 70 A of current safely. Aluminum wire can reach the same temperature and carry up to 55 A safely.

As insulation type varies, the allowable wire temperature varies. As a result, the wire's current capability varies from different types of insulation. Note an important point: A given wire size does not always have the same current-carrying capability. The type of insulation must be considered also.

What happens if the room temperature rises in the summer, or for some other reason? The current capability goes down according to the correction factor portion of the table. For example, assume #3 type V copper wire in an environment of 58 °C. Current capability is 110 A, from the upper table, at normal temperature. From the lower table, the correction factor is 0.67 for 58 °C. The actual maximum allowable current at 58 °C is $0.67 \times 110 = 73.7$ A.

Other factors also reduce the allowable current in wires. Note that the table says it is for "not more than three single insulated conductors." If there are, for example, six wires in the conduit or raceway, a further reduction in current capability is needed. This is determined by other tables, not shown here.

A few words about aluminum conductors versus copper: Aluminum weighs less, so it is useful in high-voltage cross-country tower distribution lines. It is also less costly per pound. However, it has greater resistivity. Therefore, it can carry less current for a given wire size, as shown by the table. Another factor in using aluminum wire is the special precautions needed in connecting or terminating it to electrical terminals of receptacles or other devices. Improper termination and its resulting problems in mobile homes have been covered by the news media. The differences to look out for include the following:

- Aluminum and copper expand at different rates, causing problems in splicing. Special terminating connectors are required.
- Aluminum is softer and can "cold flow," causing open circuits.
- Aluminum is more subject to corrosion. The oxide film is an insulator, as with copper, and can cause a poor electrical connection to develop.
- Aluminum and copper, when abutting, form an electrolytic cell in moisture (with impurities). The aluminum is eaten away by electrolytic action.

Table 310-16.   Ampacities of Not More than Three Single Insulated Conductors, Rated 0 through 2000 Volts, in Raceway in Free Air and Ampacities of Cable Types AC, NM, NMC and SE

Based on Ambient Air Temperature of 30°C (86°F).

Size	Temperature Rating of Conductor. See Table 310-13.								Size
	60°C (140°F)	75°C (167°F)	85°C (185°F)	90°C (194°F)	60°C (140°F)	75°C (167°F)	85°C (185°F)	90°C (194°F)	
AWG MCM	TYPES †TW, †UF	TYPES †FEPW, †RH, †RHW, †THW, †THWN, †XHHW, †USE, †ZW	TYPE V	TYPES TA, TBS, SA, AVB, SIS, †FEP, †FEPB, †RHH, †THHN, †XHHW*	TYPES †TW, †UF	TYPES †RH, †RHW, †THW, †THWN, †XHHW †USE	TYPE V	TYPES TA, TBS, SA, AVB, SIS, †RHH, †THHN, †XHHW*	AWG MCM
	COPPER				ALUMINUM OR COPPER-CLAD ALUMINUM				
18	....	....	....	14	....	....	....	....	....
16	....	....	18	18	....	....	....	....	....
14	20†	20†	25	25†	....	....	....	....	....
12	25†	25†	30	30†	20†	20†	25	25†	12
10	30	35†	40	40†	25	30†	30	35†	10
8	40	50	55	55	30	40	40	45	8
6	55	65	70	75	40	50	55	60	6
4	70	85	95	95	55	65	75	75	4
3	85	100	110	110	65	75	85	85	3
2	95	115	125	130	75	90	100	100	2
1	110	130	145	150	85	100	110	115	1
1/0	125	150	165	170	100	120	130	135	1/0
2/0	145	175	190	195	115	135	145	150	2/0
3/0	165	200	215	225	130	155	170	175	3/0
4/0	195	230	250	260	150	180	195	205	4/0
250	215	255	275	290	170	205	220	230	250
300	240	285	310	320	190	230	250	255	300
350	260	310	340	350	210	250	270	280	350
400	280	335	365	380	225	270	295	305	400
500	320	380	415	430	260	310	335	350	500
600	355	420	460	475	285	340	370	385	600
700	385	460	500	520	310	375	405	420	700
750	400	475	515	535	320	385	420	435	750
800	410	490	535	555	330	395	430	450	800
900	435	520	565	585	355	425	465	480	900
1000	455	545	590	615	375	445	485	500	1000
1250	495	590	640	665	405	485	525	545	1250
1500	520	625	680	705	435	520	565	585	1500
1750	545	650	705	735	455	545	595	615	1750
2000	560	665	725	750	470	560	610	630	2000

AMPACITY CORRECTION FACTORS

Ambient Temp. °C	For ambient temperatures other than 30°C (86°F), multiply the ampacities shown above by the appropriate factor shown below.								Ambient Temp. °F
21-25	1.08	1.05	1.04	1.04	1.08	1.05	1.04	1.04	70-77
26-30	1.00	1.00	1.00	1.00	1.00	1.00	1.00	1.00	79-86
31-35	.91	.94	.95	.96	.91	.94	.95	.96	88-95
36-40	.82	.88	.90	.91	.82	.88	.90	.91	97-104
41-45	.71	.82	.85	.87	.71	.82	.85	.87	106-113
46-50	.58	.75	.80	.82	.58	.75	.80	.82	115-122
51-55	.41	.67	.74	.76	.41	.67	74	.76	124-131
56-60	....	.58	.67	.71	....	.58	.67	.71	133-140
61-70	....	.33	.52	.58	....	.33	.52	.58	142-158
71-80	....	....	.30	.41	....	....	.30	.41	160-176

† Unless otherwise specifically permitted elsewhere in this Code, the overcurrent protection for conductor types marked with an obelisk (†) shall not exceed 15 amperes for 14 AWG, 20 amperes for 12 AWG, and 30 amperes for 10 AWG copper; or 15 amperes for 12 AWG and 25 amperes for 10 AWG aluminum and copper-clad aluminum after any correction factors for ambient temperature and number of conductors have been applied.

* For dry and damp locations only. See 75°C column for wet locations.

FIGURE 16-12

Current-Carrying Capabilities for Various Wire Sizes and Insulations (Courtesy of the National Fire Protection Association. Reprinted with permission from NFPA 70, National Electric Code, Copyright © 1987, National Fire Protection Association, Quincy, MA 02269. This reprinted material is not the complete and official position of the NFPA on the referenced subject, which is represented only by the standard in its entirety. National Electric Code® and NEC® are Registered Trademarks of the National Fire Protection Association, Inc. Quincy, MA.)

Table 3B. Maximum Number of Conductors in Trade Sizes of Conduit or Tubing

Type Letters	Conductor Size AWG, MCM	Conduit Trade Size (Inches)											
		½	¾	1	1¼	1½	2	2½	3	3½	4	5	6
THWN,	14	13	24	39	69	94	154	164	160	106	136		
	12	10	18	29	51	70	114	104	79				
	10	6	11	18	32	44	73	51					
	8	3	5	9	16	22	36						
THHN, FEP (14 thru 2), FEPB (14 thru 8), PFA (14 thru 4/0), PFAH (14 thru 4/0), Z (14 thru 4/0)	6	1	4	6	11	15	26	37	57	76	98	154	137
	4	1	2	4	7	9	16	22	35	47	60	94	116
	3	1	1	3	6	8	13	19	29	39	51	80	97
	2	1	1	3	5	7	11	16	25	33	43	67	72
	1		1	1	3	5	8	12	18	25	32	50	
XHHW (4 thru 500MCM)	1/0	1	1	1	3	4	7	10	15	21	27	42	61
	2/0	1	1	1	2	3	6	8	13	17	22	35	51
	3/0	1	1	1	1	3	5	7	11	14	18	29	42
	4/0			1	1	2	4	6	9	12	15	24	35
	250			1	1	1	3	4	7	10	12	20	28
	300			1	1	1	3	4	6	8	11	17	24
	350				1	1	2	3	5	7	9	15	21
	400						1	3	5	6	8	13	19
	500				1	1	1	2	4	5	7	11	16
	600					1	1	1	3	4	5	9	13
	700					1	1	1	3	4	5	8	11
	750					1	1	1	2	3	4	7	11
XHHW	6	1	3	5	9	13	21	30	47	63	81	128	185
	600				1	1	1	1	3	4	5	9	13
	700					1	1	1	3	4	5	7	11
	750					1	1	1	2	3	4	7	10

Note: This table is for concentric stranded conductors only. For cables with compact conductors, the dimensions in Table 5A shall be used.

FIGURE 16-13

Conduit Size for Number of Conductors (Courtesy of the National Fire Protection Association. Reprinted with permission from NFPA 70, National Electric Code, Copyright © 1987, National Fire Protection Association, Quincy, MA 02269. This reprinted material is not the complete and official position of the NFPA on the referenced subject, which is represented only by the standard in its entirety. National Electric Code® and NEC® are Registered Trademarks of the National Fire Protection Association, Inc. Quincy, MA.)

Wiring is run through conduit, in raceways, and sometimes, incorrectly, directly through space. Figure 16–13 is a typical table for the allowable number of conductors in conduit or tubing. For example, suppose sixteen #6 TWH wires are to run through an electrical conduit. According to the table, a 3 in. conduit must be used, since 2 1/2 in. conduit allows only fifteen conductors. If wires of mixed sizes are run in a conduit, special tables are used. Such tables are included in books listed in the bibliography. If the requirements of the table in Figure 16–13 are not met, wiring insulating deterioration due to excessive heating will occur. One of the biggest field problems with conduit is the addition of more wires to an

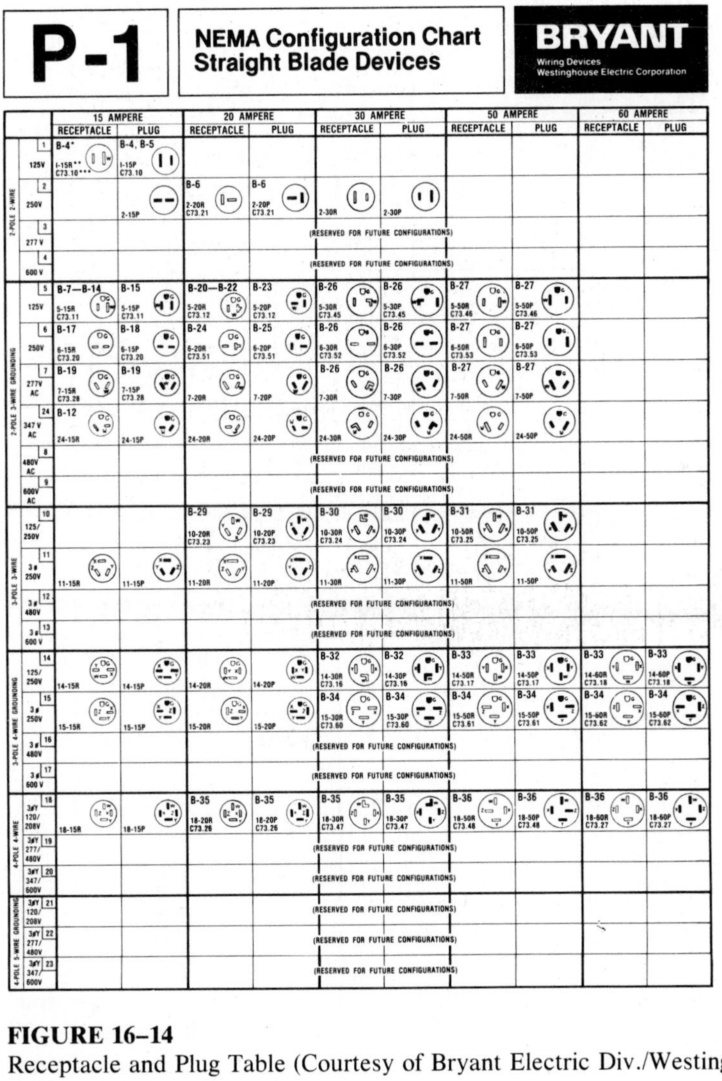

FIGURE 16–14
Receptacle and Plug Table (Courtesy of Bryant Electric Div./Westinghouse Electric)

NEMA Configuration Chart
Locking Devices

Category	Voltage	Code	15 AMPERE Receptacle	15 AMPERE Plug	20 AMPERE Receptacle	20 AMPERE Plug	30 AMPERE Receptacle	30 AMPERE Plug
2-POLE 2-WIRE	125V	ML1	G-4 ML-1R	G-4 ML-1P				
	125V	L1	G-5 L1-15R C73.31	G-5 L1-15P C73.31				
	250V	L2			G-6 L2-20R C73.32	G-6 L2-20P C73.32		
2-POLE 3-WIRE GROUNDING	125V	ML2	G-7 ML-2R	G-7 ML-2P				
	125V	L5	G-8, G-9 L5-15R C73.42	G-8, G-9 L5-15P C73.42	G-12 L5-20R C73.72	G-12 L5-20P C73.72	G-16 L5-30R C73.73	G-16 L5-30P C73.73
	250V	L6	G-10 L6-15R C73.74	G-10 L6-15P C73.74	G-13 L6-20R C73.75	G-13 L6-20P C73.75	G-16 L6-30R C73.76	G-16 L6-30P C73.76
	277V AC	L7	G-11 L7-15R C73.43	G-11 L7-15P C73.43	G-13 L7-20R C73.77	G-13 L7-20P C73.77	G-16 L7-30R C73.78	G-16 L7-30P C73.78
	480V AC	L8			G-14 L8-20R C73.79	G-14 L8-20P C73.79	G-17 L8-30R C73.80	G-17 L8-30P C73.80
	600V AC	L9			G-14 L9-20R C73.81	G-14 L9-20P C73.81	G-17 L9-30R C73.82	G-17 L9-30P C73.82
	28V DC	FSL1					G-15 FSL1	G-15 FSL1
	400Hz 120V	FSL2					G-15 FSL2	G-15 FSL2
3-POLE 3-WIRE	125/250V	ML3	G-18 ML-3R	G-18 ML-3P				
	125/250V	L10			G-22 L10-20R C73.96	G-22 L10-20P C73.96	G-26 L10-30R C73.97	G-26 L10-30P C73.97
	3Ø 250V	L11	L11-15R C73.98	L11-15P C73.98	G-23 L11-20R C73.99	G-23 L11-20P C73.99	G-27 L11-30R C73.100	G-27 L11-30P C73.100
	3Ø 480V	L12			G-23 L12-20R C73.101	G-23 L12-20P C73.101	G-28 L12-30R C73.102	G-28 L12-30P C73.102
	3Ø 600V	L13					G-28 L13-30R C73.103	G-28 L13-30P C73.103
3-POLE 4-WIRE GROUNDING	125/250V	L14			G-31 L14-20R C73.83	G-31 L14-20P C73.83	G-33 L14-30R C73.84	G-33 L14-30P C73.84
	3Ø 250V	L15			G-32 L15-20R C73.85	G-32 L15-20P C73.85	G-34 L15-30R C73.86	G-34 L15-30P C73.86
	3Ø 480V	L16			G-32 L16-20R C73.87	G-32 L16-20P C73.87	G-35 L16-30R C73.88	G-35 L16-30P C73.88
	3Ø 600V	L17					G-35 L17-30R C73.89	G-35 L17-30P C73.89
	400Hz 3ØΔ 120V	FSL3					G-30 FSL3	G-30 FSL3
4-POLE 4-WIRE	3ØY 120/208V	L18			G-42 L18-20R C73.104	G-42 L18-20P C73.104	G-45 L18-30R C73.105	G-45 L18-30P C73.105
	3ØY 277/480V	L19			G-43 L19-20R C73.106	G-43 L19-20P C73.106	G-46 L19-30R C73.107	G-46 L19-30P C73.107
	3ØY 347/600V	L20			G-43 L20-20R C73.108	G-43 L20-20P C73.108	G-46 L20-30R C73.109	G-46 L20-30P C73.109
4-POLE 5-WIRE GROUNDING	3ØY 120/208V	L21			G-48 L21-20R C73.90	G-48 L21-20P C73.90	G-53 L21-30R C73.91	G-53 L21-30P C73.91
	3ØY 277/480V	L22			G-49 L22-20R C73.92	G-49 L22-20P C73.92	G-53 L22-30R C73.93	G-53 L22-30P C73.93
	3ØY 347/600V	L23			G-49 L23-20R C73.94	G-49 L23-20P C73.94	G-53 L23-30R C73.95	G-53 L23-30P C73.95
	400Hz 3ØY 120/208V	FSL4					G-52 FSL4	G-52 FSL4

FIGURE 16–14 (*continued*)

already full conduit. This cramming can cause shorted wires and shorts due to overheated and deteriorated insulation.

There are other hardware considerations, which are covered in the textbooks listed in the bibliography. Only two of these considerations can affect the operation and installation of electrical and electronic control equipment, and therefore they will be discussed here. The first is the types of receptacles and plugs. Figure 16–14 illustrates the major types of receptacles and plugs in use. These are standardized by NEMA and listed in the National Electrical Code. One common field problem is an equipment plug that does not match the receptacle in the plant. There can be two to five wires to be connected. Voltages can differ, as well as current-carrying capacity. Plugs and receptacles can be for single-phase or three-phase power. Do not assume that because the plug is correct for the application, the outlet will match it.

The other consideration is use in hazardous locations. In any place where the atmosphere contains material that can explode, special equipment enclosures and installation are required. In Chapter 11, hazardous-location requirements for starters were illustrated in Figure 11–2. The hazardous-location requirements also apply to the complete power distribution system. For example, suppose an electronic control unit is going to be used near a paint booth. The enclosure must meet Class I, Division I, Group D requirements. It would be a heavy metal enclosure that would not let an internal explosion get out of the enclosure. A special-design enclosure to proper specifications is required. Furthermore, the openings where wiring leaves and enters would need special construction to meet code. The hazardous-location requirements also apply to other devices in the area, such as switch boxes and light fixtures.

16–4   CAUSES OF LOW VOLTAGE: LINE LOSS AND POOR POWER FACTOR

All electronic and most electrical devices have a low-voltage limit at which they will function properly. These devices also have upper limits of voltage to keep from damaging components. A typical specified equipment operational voltage is 120 V ± 10%. Low voltage is the more common problem. In this section two major causes of excessively low voltage will be discussed. The power into the factory or office main distribution center will be assumed to have a constant, or reasonably constant, voltage. Line voltage losses due to resistance voltage drop will be discussed. The effects of poor power factor, which add to the line voltage drop, will also be covered.

Wire resistance is often not considered significant, as it is small in relation to the load resistance. As electrical power feeder lines become longer and longer, the wire resistance factor must be considered. Figure 16–15 is a table for standard electrical wire by wire gauge (size). For reference, the wire diameter and area are given in two different dimensional systems. The resistance values at room temperature, 20 °C, will be used for illustrations in this section.

To illustrate the effects of line drop on voltage, the example shown in Figure 16–16 will be used. In Figure 16–16A the input voltage at the power panel is 115 V and remains constant as load current is delivered. The distribution center feeds power to a remote location 175 ft away. The remote location contains an electric motor and a computer, as shown. Other loads at the remote location, such as lighting, will not be considered for this example. Assume that the remote location is a barn with a fan motor and a computer to monitor livestock eating habits.

With only the computer running, the voltage at the remote location is very nearly 115 V. The computer draws only 1 A at 115 V. If the motor is running, its

AWG #	Area (CM)	Ω/1000 ft at 20°C	AWG #	Area (CM)	Ω/1000 ft at 20°C
0000	211,600	0.0490	19	1,288.1	8.051
000	167,810	0.0618	20	1,021.5	10.15
00	133,080	0.0780	21	810.10	12.80
0	105,530	0.0983	22	642.40	16.14
1	83,694	0.1240	23	509.45	20.36
2	66,373	0.1563	24	404.01	25.67
3	52,634	0.1970	25	320.40	32.37
4	41,742	0.2485	26	254.10	40.81
5	33,102	0.3133	27	201.50	51.47
6	26,250	0.3951	28	159.79	64.90
7	20,816	0.4982	29	126.72	81.83
8	16,509	0.6282	30	100.50	103.2
9	13,094	0.7921	31	79.70	130.1
10	10,381	0.9989	32	63.21	164.1
11	8,234.0	1.260	33	50.13	206.9
12	6,529.0	1.588	34	39.75	260.9
13	5,178.4	2.003	35	31.52	329.0
14	4,106.8	2.525	36	25.00	414.8
15	3,256.7	3.184	37	19.83	523.1
16	2,582.9	4.016	38	15.72	659.6
17	2,048.2	5.064	39	12.47	831.8
18	1,624.3	6.385	40	9.89	1049.0

FIGURE 16–15
American Wire Gage (AWG) Sizes for Solid Round Copper

current will reduce the remote voltage, as its current is 8.6 A. As was seen in Chapter 9, a motor will draw five to ten times rated current when it starts. A value of five times rated current will be assumed for this example. When the motor is first started, the voltage will dip even more significantly than when the motor is up to running speed. Actual voltage reduction can be calculated as shown in Figure 16–16A. It is carried out as follows:

The value of wire resistance is obtained from Figure 16–15. It is 2.525 Ω/1000 ft.

The wire resistance is calculated. Total wire length is 350 ft. Resistance value is calculated as 0.89 Ω.

Line current when both computer and water motor are running is the sum of computer current and rated motor current. This is 9.6 A.

Line current when the motor is started is calculated similarly. It totals 44 A.

Voltage drop due to the wiring is computed by Ohm's law, $V = I \times R$. Drops are 8.5 V and 39 V.

M = motor (1 1/2 HP, 1800 RPM, 4-pole): 115 V/230 V $^{8.6 A}/_{4.3 A}$ C = computer: 115 V/230 V $^{1 A}/_{0.5 A}$

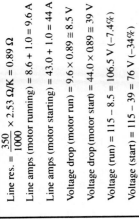

Line res. = $\frac{350}{1000}$ × 2.53 Ω/K = 0.89 Ω
Line amps (motor running) = 8.6 + 1.0 = 9.6 A
Line amps (motor starting) = 43.0 + 1.0 = 44 A
Voltage drop (motor run) = 9.6 × 0.89 ≅ 8.5 V
Voltage drop (motor start) = 44.0 × 0.89 ≅ 39 V
Voltage (run) = 115 – 8.5 = 106.5 V (–7.4%)
Voltage (start) = 115 – 39 = 76 V (–34%)

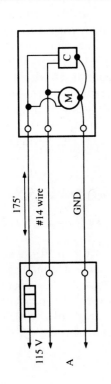

A
115 V
175'
#14 wire
GND

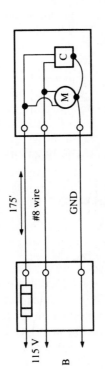

Line res. = $\frac{350}{1000}$ × 0.63 Ω/K = 0.22 Ω
Line amps (motor running) = 8.6 + 1.0 = 9.6 A
Line amps (motor starting) = 43.0 + 1.0 = 44 A
Voltage drop (motor run) = 9.6 × 0.22 ≅ 2 V
Voltage drop (motor start) = 44.0 × 0.22 ≅ 10 V
Voltage (run) = 115 – 2 = 113 V (–2%)
Voltage (start) = 115 – 10 = 105 V (–8.7%)

B
115 V
175'
#8 wire
GND

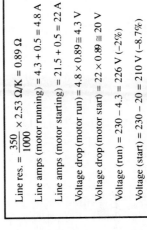

Line res. = $\frac{350}{1000}$ × 2.53 Ω/K = 0.89 Ω
Line amps (motor running) = 4.3 + 0.5 = 4.8 A
Line amps (motor starting) = 21.5 + 0.5 = 22 A
Voltage drop (motor run) = 4.8 × 0.89 ≅ 4.3 V
Voltage drop (motor start) = 22 × 0.89 ≅ 20 V
Voltage (run) = 230 – 4.3 = 226 V (–2%)
Voltage (start) = 230 – 20 = 210 V (–8.7%)

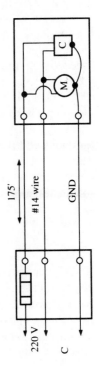

C
220 V
175'
#14 wire
GND

FIGURE 16–16
Remote-Location Voltage Drop

Actual terminal volts for each case are calculated both for voltage value and for % drop as shown.

Obviously, in the system of Figure 16–16A, the computer will malfunction when the motor is started, because of the big voltage dip. What are some solutions to prevent the big voltage drop? Moving the barn closer to the house, or vice versa? This, of course, is not feasible. The two best actual solutions are shown in Figure 16–16B and C. Figure 16–16B shows what happens if more money is invested in a larger wire size. The voltage drops are calculated in the same manner as for part A. Operating voltages are now within the normal 10% variation.

Another solution is shown in part C with the use of 220 V. When double the voltage is used, only half the current is needed to produce the same horsepower or watts. Of course, the motor and the computer must have been configured, when purchased, for 220 V operation. Even when #14 wire is used, the voltage remains within normal limits, as shown by the calculations.

A second cause of low operating voltage is poor power factor. For a purely resistive load, power factor is unity, and load current is a minimum value. In actual practice, plant loading is a combination of resistance and inductance. This combination causes the load current to be out of phase with respect to load voltage. To accomplish a given amount of real work in watts, extra current must be supplied. Figure 16–17 illustrates how much extra current is required. There is 30 kW electrical load. If the load is pure resistance (100% PF), only 30 kVA is needed, and load current is 68 A, a minimum. This 68 A requires #4 wire. There is a certain amount of line loss because of this current, as explained previously.

Now, as the load becomes partially inductive, it takes more kilovolt-amps and more current to do the same amount of work. As the current goes up, line loss rises, and bigger wiring must be installed. The alternative to bigger wiring is more line loss and the possibility of exceeding allowable code current for the wire. As the quantity of kilovolt-amps rises, a larger transformer, and larger switch disconnects, and so on are required. This increase will cause the cost, size, and bulk of the installation to go up considerably. As PF goes from the desired 100% down to a poor 30%, the required current goes from 68 A to 227 A. This is a 233% increase in amps.

What can be done to keep the current down by improving PF to 80% or better, the normally accepted economical limit? Correction is accomplished by adding capacitors at the inductive load locations. Capacitors of the correct value counteract the inductance. The correct capacitor values (which can be calculated) for PF correction for a given motor size are found in tables in textbooks listed in the bibliography. Will one large master capacitor for a whole plant work well? No. When only some of the motors are running, PF will go out of limits in the other, capacitive direction—with a corresponding increase in line current. It is best to install correction at each motor or other device. When the device is turned on, the right amount of corrective capacitance is also turned on. An illustration of capacitor correction of motor PF is shown in Figure 16–18. Cost payback for PF correction is usually one to three years.

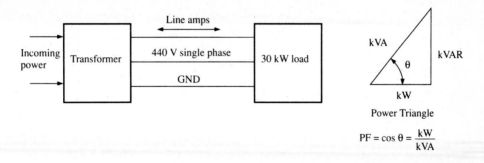

$$PF = \cos\theta = \frac{kW}{kVA}$$

kW	PF (%)	PF Angle (°)	$kVA = \frac{kW}{PF}$	$Current = \frac{VA}{A}$ (amps)	Required TW Wire Size
30	100	0	30	68	#4
30	80	36	38	86	#2
30	70	46	43	98	#1
30	60	53	50	114	#0
30	50	60	60	137	#00
30	40	66	75	171	#0000
30	30	72	100	227	300 MCM

FIGURE 16–17
Power Factor versus Required Line Current

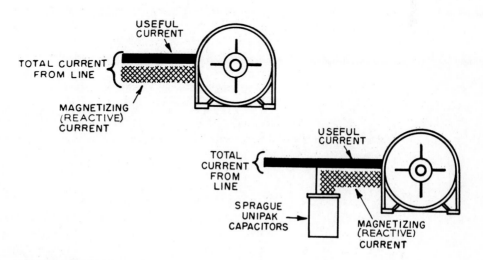

FIGURE 16–18
Motor Power-Factor Correction

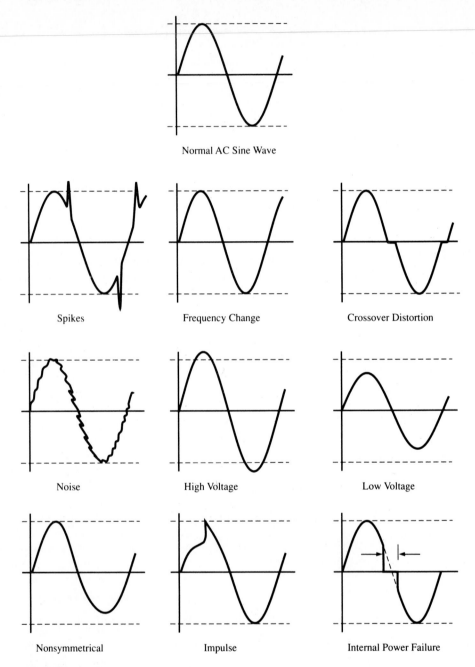

Normal AC Sine Wave

Spikes   Frequency Change   Crossover Distortion

Noise   High Voltage   Low Voltage

Nonsymmetrical   Impulse   Internal Power Failure

FIGURE 16–19
Possible AC Waveform Distortion

16–5   EFFECTS OF POWER SURGES AND TRANSIENTS

An often-neglected cause of electronic equipment failures is operational variations caused by variations of the power input waveform. Some of the possible variations of waveforms are shown in Figure 16–19. In one year, at 60 Hz, a plant receives more than 1.9 billion sine waves of voltage or power. If just one of these sine waves has a large variation, as shown, computer and control systems can malfunction.

There are many possible causes of waveform distortion, including the following:

- □ surges from lightning hitting the power lines
- □ switching of power-control disconnect switches by the power utility
- □ disturbances from adjacent plants
- □ internal factory power waveform disturbances

When a large motor is turned OFF and ON, its inductive effect can affect the waveform in a plant. When a computer and a motor are on the same distribution line, the computer can be affected by the motor-induced electrical surges.

To prevent such electrical problems from getting to equipment, *line purifiers,* or *conditioners,* may be added. The incoming electrical line goes to the purifier. The output of the purifier then goes to the equipment. Two types of purifiers are shown in Figure 16–20.

Line purifiers perform two functions. They act as voltage regulators, keeping the voltage to the proper, set level. They also act as filters that reduce wave variations to negligible variations. The conditioners come in various sizes, from a small, computer size to very large units. Some conditioners have recorders for recording the errant sine waves when and if they occur. The smaller three units shown in Figure 16–20 have recorders included on their front panels. The cost of these conditioners is proportional to their ratings and the level of performance required. For example, suppose only 5% regulation is needed. There is no need to go to the expense of obtaining a unit with 1% regulation.

Note that there is no need for a conditioner for electric heaters, motors, or lights. Such devices are unaffected by sine wave variations. The use of purifiers is in keeping electronic devices, which are susceptible to sine wave variations, operating properly.

16–6   THREE-PHASE VERSUS SINGLE-PHASE

Lower-power electronic devices and systems are powered by single-phase power, typically 115 V. In control systems of large ratings, the incoming power is very often three-phase, typically 208 V, 220 V, or 440 V. For example, a technician called a former instructor about installing a 220 V three-phase laser system. Could it be run from the available 220 V single-phase line and the 110 V center tap? The center tap was 110 V to either 220 V line. The answer was no. A three-phase line was needed. Alternatively, a costly single-phase-to-three-phase converter would have to be procured. Operating the three-phase laser device on single-phase power, with the tap, would cause the laser to malfunction. Furthermore, the laser system would probably be damaged. This section will show in general terms why three-phase devices will not run properly on single-phase power.

On the other hand, single-phase devices can be run on any one of the three phases of a three-phase input power system, provided the voltage is correct.

Note that, in rare instances, two-phase power systems are used. They are four-wire systems, two for each phase. The phases are 90° apart—in contrast to the 120° phase difference of three-phase systems. Two-phase is used for high-torque, two-phase motors, in special applications. A two-phase power generator is of course required.

Specifications

VA RATINGS: 7.5 kVA to 75 kVA.

INPUT/OUTPUT: all common three-phase AC voltages.

FREQUENCY: 60 Hz Standard; 50 Hz available.

VOLTAGE REGULATION:

Line-to-Line:

  A. ±5% over a regulating range of +10%, -20% with load unbalance up to ±30% of the average of the total load.

  B. ±7% over a regulating range of +10%, -20% with load unbalance up to ±75% of the average of the total load.

  Average of total load is calculated by dividing total line amps by 3. This figure should then be used as the reference to determine acceptable load unbalances.

Line-to-Neutral:

  ±3% over a regulating range of +10%, -20%.

ISOLATION: complete isolation from power line. .001pf capacitance between windings.

COMMON MODE NOISE REJECTION: better than 120 dB.

TRANSVERSE MODE NOISE REJECTION: better than 60 dB.

VOLTAGE SPIKE ATTENUATION: 250:1.

HARMONICS: typically 5%.

LINE INTERRUPTIONS: output will be maintained with line interruptions of up to 3 milliseconds.

RESPONSE TIME: instantaneous. Complete within 25 milliseconds (1.5 cycles).

EFFICIENCY: up to 92%.

OPERATING TEMPERATURE: -40°C to +40°C.

SHORT-CIRCUIT PROTECTION: output voltage drops to zero, output current is limited to 200% of rated value. No transformer damage will occur.

FIGURE 16–20
Power Purifier, or Conditioner, Units (Courtesy of Best Power Technology, Inc.)

There are a number of differences between single-phase and three-phase power. Some are advantages, and some are disadvantages. The major advantages of three-phase are as follows:

- ☐ The cost of utility power lines and distribution equipment per watt is about 25% less. There is less line loss per watt also.
- ☐ Three-phase motors are more efficient, as discussed in Chapter 9. They are easily reversible. Above $7\frac{1}{2}$ HP, three-phase is mandatory to run motors.
- ☐ Multiple voltages are directly available—for example, 120 V and 208 V.

Some disadvantages of three-phase are as follows:

- ☐ It initially costs more to furnish three-phase to a store, office building, or factory.
- ☐ The three phases can become unbalanced. For example, a 440 V line may become 415 V, 435 V, and 450 V across the three phases.
- ☐ The possible unbalance can affect connected single-phase devices with its high or low voltage.
- ☐ Electric motor efficiency drops rapidly with a small three-phase unbalance, increasing energy costs.
- ☐ Single-phase loads across one of the three phases can cause low voltage across that phase, caused by greater line drop in the one phase.
- ☐ Three-phase electrical grounding is more difficult than single-phase, especially in delta-connected loads.

Single-phase voltages are always 120/240 three-wire. The common three-phase voltages in use in the United States are:

Customer-use voltages:

120/208	For general use and special 208 motors
220	Obsolete but still found in older factories
277/480	Commonly used in factories mainly for 440/460 motors and other devices
2400/4160	Special voltage for high-horsepower motors

Distribution voltages—transmission line to customer substation:

7200/12,470
14,400/25,000

Transmission-line voltages—Delta:

46,000
69,000
115,000
345,000

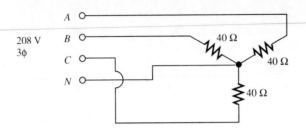

Given:

V_{supply} = 208 V, 3-phase = V_{LL}

Load = 40 Ω per phase

Find:

① V_{phase} ③ I_{line}

② I_{phase} ④ Total watts

⑤ Neutral current

Solution:

① $V_{phase} = \dfrac{V_{LL}}{\sqrt{3}}$ for Wye systems = 120 V

② $I_{phase} = \dfrac{V_{phase}}{R_{phase}} = \dfrac{120}{40} = 3$ A

③ $I_{line} = I_{phase}$ for Wye systems = 3 A

④ Phase watts $= \dfrac{V^2}{R} = I^2R = VI = 360$ W

Total watts = 3 × 360 = 1080 W

⑤ There is no neutral current, as the voltages and phase resistances are balanced (the same).

FIGURE 16–21
Balanced Three-Phase Wye Resistive Load

There are two ways to connect the three basic three-phase loads. These are wye (Y) and delta (Δ). Wye-connected loads have one end of each load connected together. The other ends of the loads are connected to the three incoming lines. For delta-connected loads, each load is connected directly across a different pair of the three incoming lines.

The relationship between three-phase line and load electrical values will be explained by going through six examples. For all six it is assumed that all three incoming lines have exactly the same matched value of voltage. It is also assumed that the three voltages are exactly 120° apart. In actual practice, incoming voltages are very closely balanced and *are* almost exactly 120° apart.

The first example is for three equal loads, 40 Ω, connected in wye. Line voltage is assumed to be 208 V three-phase. The calculations to determine line and phase values are shown in Figure 16–21. Three important characteristics for wye-connected loads are illustrated:

1. Line current equals phase current (Kirchhoff's law).
2. Line voltage is larger than phase voltage by the square root of 3.
3. For balanced loads there is no neutral current.

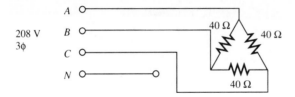

Given:

V_{supply} = 208 V, 3-phase = V_{LL}

Load = 40 Ω per phase

Find:

① V_{phase} ③ I_{line}

② I_{phase} ④ Total watts

⑤ Neutral current

Solution:

① $V_{phase} = V_{LL}$ for delta systems = 208 V

② $I_{phase} = \dfrac{V_{phase}}{R_{phase}} = \dfrac{208}{40} = 5.2$ A

③ $I_{line} = \sqrt{3} \times I_{phase}$ for delta systems = $\sqrt{3} \times 5.2 = 9.0$ A

④ Phase watts = $\dfrac{V^2}{R} = I^2R = VI = 1080$ W

Total watts = 3 × 1080 = 3240 W

⑤ There is no neutral current, as the neutral is not connected to a delta system.

FIGURE 16–22
Balanced Three-Phase Delta Resistive Load

Figure 16–22 illustrates the calculations and relationships for the same three resistive loads connected in delta. Three important characteristics for delta-connected loads are illustrated:

1. Line voltage and load voltage are the same (Kirchhoff's law).
2. Line current is larger than phase current by the square root of 3.
3. There is no place to connect the neutral.

Most three-phase loads are not purely resistive. They are resistance-inductance combinations. For electric motors, the three windings have identical *RL* values, so their load is a balanced one. Figure 16–23 shows the calculations for a balanced three-phase wye impedance load. The principle of calculation is the same as for the resistive load. The added characteristics of the impedance load calculations are as follows:

☐ Phase angle must be considered.
☐ There are reactive volt-amps as well as watts to consider in power calculations.

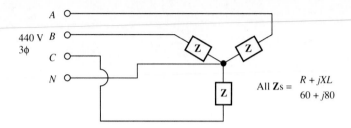

Given:

V_{supply} = 440 V, 3-phase = V_{LL}

Load = $60 + j80 = 100 \angle +53°$

Find:

① V_{phase} ④ Total watts

② I_{phase} ⑤ Total VAR

③ I_{line} ⑥ Power factor

⑦ Neutral current

Solution:

① $V_{phase} = \dfrac{V_{LL}}{\sqrt{3}}$ for Y systems = 254 V

② $I_{phase} = \dfrac{V_{phase}}{Z_{phase}} = \dfrac{254}{100} = 2.54$ A

③ $I_{line} = I_{phase}$ for Wye systems = 2.54 A

④ Phase watts = $(I_{phase})^2 \times R_{phase} = 2.54^2 \times 60 = 387$ W

Total watts = $3 \times 387 = 1161$ W

⑤ Phase VAR = $(I_{phase})^2 \times$ net impedance = $2.54^2 \times 80 = 516$ VAR

Total VAR = $3 \times 516 = 1548$ VAR

⑥ Power factor = cosine of the impedance angle

$$= \cos 53° = 0.60 \text{ (inductive)}$$

⑦ There is no neutral current in a balanced
load system.

FIGURE 16–23
Balanced Three-Phase Wye Impedance Load

□ There is no neutral current for balanced impedance loads, even though a
neutral is connected.

Figure 16–24 shows the calculations for the three impedances of Figure
16–23, now connected in delta. The calculation characteristics are similar to the
Figure 16–23 calculation characteristics.

What happens when the loads become unbalanced? Many of the characteristics of the balanced calculations do not hold. Figure 16–25 is for an unbalanced
three-phase resistive load. Some of the characteristics of this circuit are as follows:

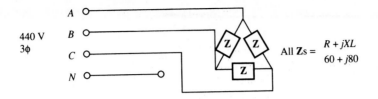

Given:

$V_{\text{supply}} = 440$ V, 3-phase $= V_{LL}$

Load $= 60 + j80 = 100 \angle +53°$

Find:

① V_{phase} ④ Total watts

② I_{phase} ⑤ Total VAR

③ I_{line} ⑥ Power factor

         ⑦ Neutral current

Solution:

① $V_{\text{phase}} = V_{LL}$ for delta systems = 440 V

② $I_{\text{phase}} = \dfrac{V_{\text{phase}}}{Z_{\text{phase}}} = \dfrac{440}{100} = 4.4$ A

③ $I_{\text{line}} = \sqrt{3} \times I_{\text{phase}}$ for delta systems

      $= \sqrt{3} \times 4.4$

      $= 7.62$ A

④ Phase watts $= (I_{\text{phase}})^2 \times R_{\text{phase}}$

            $= 4.4^2 \times 60$

            $= 1162$ W

   Total watts $= 3 \times 1162$

            $= 3484$ W

⑤ Phase VAR $= (I_{\text{phase}})^2 \times$ net impedance

            $= 4.4^2 \times 80$

            $= 1548$VAR

   Total VAR $= 3 \times 1548 = 4646$

⑥ Power factor $=$ cosine of the impedance angle

            $= \cos 53°$

            $= 0.60$ (inductive)

⑦ Neutral current $= 0$ (not connected)

FIGURE 16–24
Balanced Three-Phase Delta Impedance Load

□ Angles of the three input voltages must be considered.
□ Line currents are unbalanced. They have different magnitudes.
□ Without the neutral, phase voltages become unbalanced (circuit not shown).
□ With a neutral, the neutral line carries a current, and voltages are balanced.
□ Neutral current is smaller than line current. Consequently, in large systems the neutral wire may be smaller in gauge than the main lines.

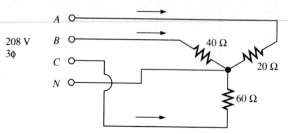

Note: Arrows (———→) denote assumed current direction.

Given:

$V_{LN} = V_{LL} / \sqrt{3}$

$V_{AN} = 120\angle 0°$ V

$V_{BN} = 120\angle 120°$ V

$V_{CN} = 120\angle 240°$ V

$R_{AN} = 20\ \Omega$

$R_{BN} = 40\ \Omega$

$R_{CN} = 60\ \Omega$

Find:

① V_{phase} for A, B, and C

② I_{phase} for AN, BN, and CN

③ I_{line} for A, B, and C

④ Total watts

⑤ Neutral current

Neutral is connected.
If not connected, $\sqrt{3}$ does not apply and voltages are not balanced
(example: 110 $\angle 5°$, 127 $\angle 136°$, 104 $\angle 231°$).
Solution is determined by simultaneous equations.

Solution:

① V_{phase} given

② $I_{AN} = \dfrac{V_{AN}}{R_{AN}} = \dfrac{120\ \angle 0°}{20} = 6\ \angle 0° = 6 + j0\ \Omega$

$I_{BN} = \dfrac{V_{BN}}{R_{BN}} = \dfrac{120\ \angle 120°}{40} = 3\ \angle 120° = -1.5 + j2.6\ \Omega$

$I_{CN} = \dfrac{V_{CN}}{R_{CN}} = \dfrac{120\ \angle 240°}{60} = 2\ \angle 240° = -1 - j1.7\ \Omega$

③ Same as ②

④ Watts = watts A + watts B + watts C = $\Sigma I^2 R$

= $(6^2 \times 20) + (3^2 \times 40) + (2^2 \times 60)$

= $720 + 360 + 240 = 1320$ W

⑤ Neutral current = $I_{AN} + I_{BN} + I_{CN}$

= $(6 + j0) + (-1.5 + j2.6) + (-1 - j1.7)$

= $3.5 + j0.8$

= $3.59\angle +12.8 = \overline{3.6}$ A

FIGURE 16–25
Unbalanced Three-Phase Wye Resistive Load

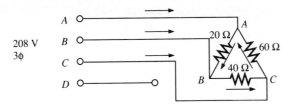

Note: Arrows (⟶) denote assumed current direction.

Given:

$V_{LN} = V_{LL}$

$V_{AB} = 208\angle 0°$

$V_{BC} = 208\angle 120°$

$V_{CA} = 208\angle 240°$

$R_{AB} = 20\ \Omega$

$R_{BC} = 40\ \Omega$

$R_{CA} = 60\ \Omega$

Find:

① I_{phase} for AB, BC, and CA

② I_{line} for A, B, and C

③ Total watts

Solution:

① $I_{AB} = \dfrac{V_{AB}}{R_{AB}} = \dfrac{208\ \angle 0°}{20} = 10.4\ \angle 0° = 10.4 + j0$ A

$I_{BC} = \dfrac{V_{BC}}{R_{BC}} = \dfrac{208\ \angle 120°}{40} = 5.2\ \angle 120° = -2.6 + j4.5$ A

$I_{CA} = \dfrac{V_{CA}}{R_{CA}} = \dfrac{208\ \angle 240°}{60} = 3.5\ \angle 240° = -1.75 + -j3.0$ A

② $I_A = I_{AB} - I_{CA} = (10.4 + j0 - (-1.75 + j3.0) = 12.15 - j3.0 = 12.5\angle -13.8°$ A

$I_B = I_{BC} - I_{AB} = (-2.6 + j4.5) - (10.4 + j0) = -13.0 + j4.5 = 13.7\ \angle +161°$ A

$I_C = I_{CA} - I_{BC} = (-1.75 - j3.0) - (-2.6 + j4.5) = 0.85 - j6.65 = 6.7\ \angle -82.7°$ A

③ Watts = watts A + watts B + watts C = $\Sigma I^2 R$

= $(10.4^2 \times 20) + (5.2^2 \times 40) + (3.5^2 \times 60)$

= 2163 + 1081 + 735 = 3979 W

FIGURE 16–26
Unbalanced Three-Phase Delta Resistive Load

  □ Line current and phase current are the same.
  □ Phase currents are unequal.

Figure 16–26 shows the impedances of Figure 16–25 connected in delta. Some of the characteristics of this circuit are as follows:

  □ Angles of the three input voltages must be considered.
  □ Phase currents are unequal.
  □ Line currents are also unequal.
  □ Current flow directions must be assumed before making calculations. If the assumed directions are wrong, the calculated values come out negative.
  □ There is no neutral current possible, as the neutral is not connected.

Sample calculations for unequal impedance loads will not be carried out. They are more complicated and are not normally encountered by the technician. Refer to the textbooks listed in the bibliography for this type of calculation.

16–7 ELECTRICAL GROUNDING SYSTEMS AND PROTECTION

There are two reasons that proper electrical grounding systems must be installed in electrical distribution systems. The first, and primary, reason is to protect operating personnel and other people around the equipment from receiving electrical shocks. Proper electrical grounding can minimize or eliminate these shock hazards. The second reason for electrical grounding involves the equipment. Proper grounding can protect equipment from failure. Also, some electronic equipment will not work properly without a good, almost perfect grounding system. Most grounding systems are connected to building grounds. A few others have a relative ground wire, which runs directly from device to device. Examples of equipment requiring good electrical grounds are computer systems and communications equipment.

  A conventional 115/230 V distribution system is shown, including its ground, in Figure 16–27. There are three wires from the distribution transformer secondary to the breaker or fuse panel. Two wires have 230 V across them. Half of the voltage, 115 V, appears between each 230 V wire and the center wire. The center wire is electrically grounded at the distribution transformer. This ground consists of a wire run from the transformer center tap to a spike in the ground soil. At the distribution panel, the voltage is fed to various 115 V and 230 V loads through circuit breakers or fuses. An electrical ground is run from the neutral bar in the distribution panel to earth ground via a water pipe. For good soil conditions, the resistance between grounds is effective near 0 Ω.

  Figure 16–28 illustrates what can happen if the frame or case of an electrical device is installed without a ground. Assume that there is only two-conductor wiring to the device. The electrical device can be a motor, computer, lamp, or some other such device, or it can be the conduit, receptacle, or similar element. Improper equipment grounding would occur if a three-wire system with a three-

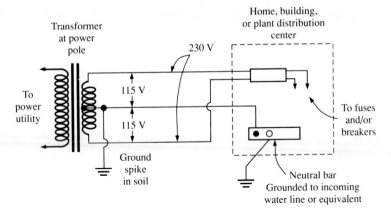

FIGURE 16–27
Basic 115/230 V Distribution System

prong, grounding receptacle plug were not used. Metal conduit and receptacles are to be grounded by a separate wire, which is missing in this instance. To compound the safety problem, the device case is not grounded with a separate ground wire. There is no GFI, either.

An example of this situation could be a fan motor in a basement. The fan motor is installed on a plastic, nonconducting frame. The basement floor has moisture on it. Suppose that the wire connected to terminal *B* has an insulation crack. The copper of the wire with the cracked insulation touches the frame. Now someone who wants to clean the motor touches the frame. The person is in luck. The frame is nearly at ground potential, and nothing happens. Now, suppose the "hot" wire, *A*, touches the frame. When the person touches the frame, he or she

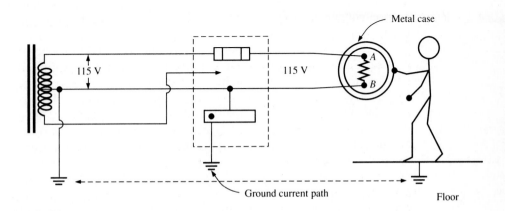

FIGURE 16–28
Load with No Grounding System

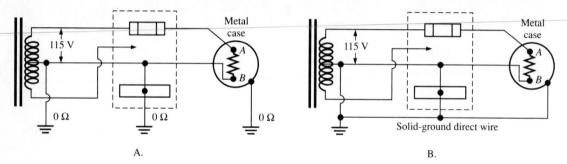

FIGURE 16–29
Low-Resistance Ground Circuits

becomes a parallel load to 115 V. A person has about 1500 Ω resistance. Then, the race is on as to which goes first —the fuse or the person. The person will probably be badly hurt or even killed. A few milliamps is enough to cause serious injury. A current of 120 mA is enough to kill a person.

In contrast, the proper grounding system shown in Figure 16–29 would prevent injury. As soon as B becomes shorted, the fuse blows. The circuit is therefore deenergized. A person touching the frame is protected. Figure 16–29 has an ideal grounding system of 0 Ω. Such an ideal cannot always be attained. One sure way to make sure the ground has very low resistance is to run a separate ground wire as shown.

One word of caution: You may assume that the larger, round grounding prong on the plug works infallibly. It does not. There have been cases where the cord ground wire has opened up. In such a case there is no ground shock protection. In another situation the receptacle ground is not connected to anything. This connection can be checked visually or with an ohmmeter. Ungrounded plug ground connections are especially possible in older building installations. An appropriate ohmmeter check is always good insurance that the proper ground circuit exists.

Grounds are not always perfect. A ground wire has resistance depending on its size. Connections have some resistance. The earth soil itself has resistance, which varies with soil type and moisture content. The distribution-panel ground-bar connection can also have some resistance value. Figure 16–30 shows some imperfect grounds, which can be typical of actual installations. The ground at the transformer, the ground at the panel, or the ground from the load device to building ground can have some resistance. Three possibilities are shown. Assume there is a short from wire A to the metal case, as before. It might be assumed that the fuse will blow, deenergizing the circuit. It may not. By Ohm's law, the current for A in Figure 15–30A is 115 V/16 Ω = 7.2 A. The fuse will not necessarily blow—only if the load current is an added 7.8 A or more. By voltage divider, the frame will have 15/16 × 115 V, or 108 V above ground potential. This is dangerous.

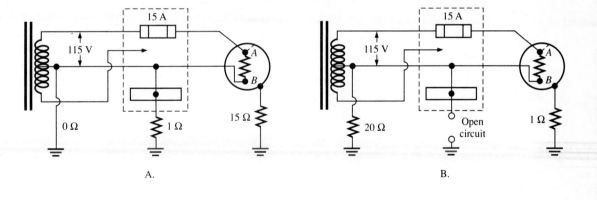

A.   B.

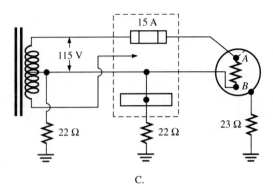

C.

FIGURE 16–30
Imperfect Grounds

The other examples, in Figure 16–30*B* and *C*, can have values calculated similarly. For part *B* there will be 5.5 A and only 5 V (a better situation) at the frame. For *C* there are two ground resistances in parallel to consider. The wire will have only 3.4 A, but there will be 78 V at the frame. The whole point is that a poor ground will not give good personnel hazard protection. Improved grounding systems are needed in these cases.

Sometimes, no ground is used in a transformer secondary circuit. The grounding of the supply voltage can cause instrumentation or operational problems. Ungrounded AC supplies are generally satisfactory at low voltages, such as 115 V or 24 V. There can be some hazard with no ground of the supply voltage as shown in Figure 16–31. There is a 2300 V primary of a transformer. Its feeder line is grounded back at the previous transformer. Suppose that the primary becomes shorted to the secondary as shown. Now lines *A* and *B* are at 2300 V above ground—a dangerous situation. If the transformer secondary were grounded at one terminal, the 2300 V protective fuse would blow. In any case, the frame of the load should be grounded as shown, even if the 115 V circuit is not.

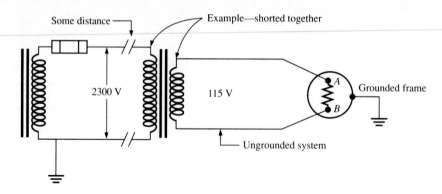

FIGURE 16–31
Winding-to-Winding Short with No Secondary Ground

As a further ground shock protection, newer plug receptacle combinations have one of the two load current prongs larger than the other. The plug and its proper wiring are shown in Figure 16–32. The high, or 115 V, side is required to be connected to the fuse. If the fuse blows, the device will then be at ground potential, near 0 V. If either the receptacle or the plug is wired incorrectly, there is a problem. If the fuse blows, the device's inner wiring will remain at 115 V above ground, as illustrated.

Single-phase circuit grounding has been discussed. Grounding of three-phase circuits is smaller, but more involved and complicated. For proper construction of these ground circuits, references to other books and codes is necessary. Figure 16–33 shows two basic methods of wye circuit grounding. In Figure 16–33A is a system in which the neutral doubles as the ground. This arrangement generally functions safely. What if the neutral were to open up? There would be no safety ground even though the main lines were still energized.

A more costly but surer wye grounding system is shown in part B. A separate grounding wire is run in addition to the neutral. If neutral opens up, there is still protection. This system is called the three-phase, five-wire distribution system.

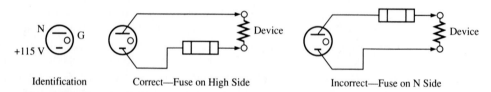

FIGURE 16–32
Polarized Plug Connection

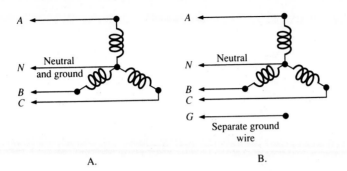

FIGURE 16–33
Wye Circuit Grounding

Delta systems are grounded in various ways. Three of the delta grounding methods are shown in Figure 16–34. These are no ground, ground at one corner, and ground in the middle of one phase. These grounds must match the grounding systems of parallel delta connections. Details may be found in the textbooks listed in the bibliography.

There are automatic systems available to monitor three-phase grounds. These systems sound an alarm if a ground fault should occur. Details of these alarm systems are given in textbooks listed in the bibliography.

Sometimes electrical grounding of a circuit is not desired. A typical example is in the use of an ungrounded 115 V for an electronic instrument. In this case a 1 : 1 transformer could be used without grounding the secondary. A typical example of this situation is shown in Figure 16–35. There is a three-phase incoming power line. The system is wye with a ground/neutral at the tie point, as shown. The waveform of each phase of the incoming line is to be observed. For waveform observation an oscilloscope is used. First, assume that the scope is plugged into a building 115 V receptacle. One side of the building 115 V receptacle is at ground potential, as previously discussed. As shown, the oscilloscope has one side of its

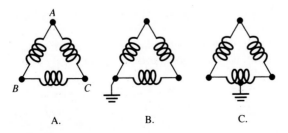

FIGURE 16–34
Delta Circuit Grounding

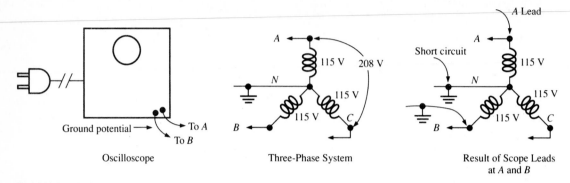

FIGURE 16–35
Unisolated Oscilloscope Causing a Short Circuit

measuring leads grounded to the incoming 115 V electrical ground. To observe the three-phase waveforms, the two scope lead probes are used. A direct short across one 208 V phase can be created between the scope ground and the three-phase ground, as illustrated. To prevent this shorting, the scope should be supplied through an ungrounded 1 : 1 isolation transformer.

16–8   POWER COST CONSIDERATIONS

Four major components make up the electrical power costs for a plant or building:

1. monthly flat service fee
2. electrical usage in kilowatt-hours
3. power factor charge or credit
4. peak-demand charge

The control engineer or technician has little or no control over costs 1 and 2. Control engineers and technicians can, however, help control costs 3 and 4. Cost 3, power factor, can be improved as discussed in Section 16–4. Cost 4 can be greatly reduced in some cases. The technician and engineer can take steps to reduce peak demand by proper overall control-system design, as will be explained. First, each of these four cost elements will be discussed briefly.

The monthly *flat service fee* is a fixed charge that covers the power company's overhead and equipment maintenance costs. It varies in value from utility to utility. It is normally a small portion of the overall bill.

The *usage,* or *energy, charge* varies between about $0.02/kWh and $0.12/kWh in the United States. The base charge rate varies with the location in the country and in each state. The charge for this usage may be lower than the base rate during off-peak hours. Peak hours are typically 8 A.M. to 9 P.M. in the winter and 8 A.M. to 6 P.M. in the summer. Power used at times other than these peak hours may cost less per kilowatt-hour, depending on the utility pricing policy. The utility has unused generating capacity available at off-peak times, so it can offer a discount. For the base rate, there is a stepped rate per kilowatt-hour. Essentially, this is a quantity discount. As usage reaches certain rate levels, the unit charge for each succeeding kilowatt-hour used is decreased.

Power factor has been previously discussed. Poor power factor affects the power utility's efficiency. The utility must supply more current to poor power-factor installations. Larger wires, transformers, and so on must be used. The electric meter only reads kilowatt-hours and does not pick up the extra current. The utility company determines average power factor by a separate monitoring meter on the power line. A surcharge is often made for the extra current supplies for PF below 80%. For PF below 80%, a credit is often given. A typical rate schedule for varying PF is as follows:

- □ For each percentage point above 80% power factor, a credit is given, typically 0.15% (of the total bill).
- □ For each percentage point below 80% power factor, a charge is made, typically 0.30% (of the total bill).

Peak-demand charge is based on peaks of usage. Demand peaks that occur during off-peak hours are often not billed. Peak demand occurring during peak hours is always billed. Monthly peak demand is determined by a separate monitoring meter in the plant incoming power line. The peak demand meter records the largest kilowatt demand of the plant during the month. An example of a large peak of power is in a sawmill with three 100 HP saws. If all three saws are started simultaneously, a very high peak of power is needed. Even though it occurs for only a very short time, this usage can be recorded as the peak-kilowatt-demand reading. This peak-power kilowatt value is obviously much greater than the average power used. The demand charge can exceed the energy charge for some plant operations. Demand peaks, if everyone has them, can require the utility to build more generating capacity, adding to their costs. These costs must ultimately be passed along to the users.

In a paper mill or processing plant, these preventable peak costs can run into thousands of dollars per month. Proper planning of machine starting and running can reduce the peak demand and the resulting costs. The newer building heating-and-ventilating-control systems use computers, which analyze demand. A good computer control can reduce peaks, thus lowering power costs considerably. The computer system sequences machines, heaters, and other devices that have large electrical demand on starting and running. In some operations, a peaking standby generator in the plant is used to reduce large peaks from the power company (at a cost). The standby generator comes ON to help cover the power demand when it is above a certain value. This reduces the peak demand from the power company.

For reference for this section, a typical power bill (somewhat simplified) is shown in Figure 16–36. The $55.00 is for the flat charge. The energy charge is $4391.95. The peak-demand charge is $1060.90. This peak-demand cost might be improved with some controls analysis, as previously explained. The power factor charge is $198.28.

Customer Charge $ 55.00

Energy Charge (from kWh meter)

    All at $ 0.063 per kWh
    Used 27 days, 2 shifts, 16 h/day
    Averaged 168.6 kW
            27 days   ×   16 h/day   =   432 h
               432   ×   168.6   =   72,835 kWh (meter reading)
        72,835 kWh   × $ 0.0603/kWh   = $4391.95 $4391.95

Peak Demand (from demand meter)

    All at $ 5.15 per kW

    206.0 kW (meter reading)

    206.0 × $ 5.15 = $1060.90 $1060.90

Power Factor (from PF meter)   Subtotal 5507.85

    Average 0.68 (meter reading)

    80 − 68 = 12

    12 × 0.003 × $5507.85 = $198.28 $198.28

    Grand Total $5706.13

FIGURE 16–36
Typical Power Bill

16–9   STANDBY POWER UNITS—GENERATOR AND ELECTRONIC

Should the engineer and the technician be concerned about standby power units? One example might illustrate why they should be. A large service organization spent hundreds of thousands of dollars on their new computer system. They did not choose to include a standby power backup for the system. One day the power went OFF for 3 min. When it came back ON, they found that they had lost a lot of data. It is estimated that the cost to redo the lost data ran into multiple tens of thousands of dollars. Dissatisfaction of some customers also ensued. A relatively inexpensive standby power module could have prevented this loss of data and customer confidence.

The engineer and the technician should consider the feasibility of standby power for each electrical or electronic installation. There are many possible choices:

1. no backup (not needed for personnel safety or system needs)
2. lighting backup only
3. second independent line from the power company
4. major standby generator backup for critical areas
5. electronic, instantaneous backup power

No backup, choice 1, would be for small, noncritical operations. Lighting backup only, choice 2, is for control rooms, hallways, public areas, and similar places. The familiar lighting standby packages are battery-backup low-power devices. They switch to standby power in a matter of 1 s or less.

A second line from the power company, choice 3, is an added expense for the utility and the customer. When one line fails, the plant switches over to the second line. It takes, typically, an hour or so to reconnect and switch over to the alternate incoming line, unless a very expensive transfer switch has been installed. Another problem with a separate line is that both incoming lines could be OFF during a major emergency.

The most feasible standby choices for process and computer control systems are types 4 and 5.

A standby generator, choice 4, is a high-power-rated machine. Ratings can go as high as 1500 kW, for a price. Such generators switch ON within a minute after main power goes off. The power from these units usually is not switched to an entire plant or building. The power is switched to only a portion of the total plant. This might include elevators, critical lighting, computers, ventilating fans, and the boss's office. Such equipment as air conditioning, annealing furnaces, and other heavy, noncritical loads would not be included. It would be too expensive to have a unit large enough to supply the whole plant. A typical standby generator system is shown in Figure 16–37. Note that this generator system would not have prevented the data loss in the original example.

The controls of the generator include an electronic voltage regulator, frequency monitoring, safety interlocks, and other controls. A basic standby system layout is given in Figure 16–38. When the main power goes OFF, a sensing relay contact starts the engine. There is some delay before starting. This keeps the engine from starting during very short outages. When the generator gets up to speed, the generator power contactor closes, furnishing the required standby power.

In most systems a time clock turns the standby system ON periodically. This operation is done during low plant-usage times. The reason for it is to make sure the unit is operational and ready to function when needed. During a recent complete power blackout in a large city, many standby generators did not function. The outage was found to be the first time the power had failed in six years. The generators had not been test-run even once in the six years. Functional problems that had developed because of idleness had not been corrected.

What would happen to a computer or electronic control device if it switched to the standby power of choice 2, 3, or 4? It could lose its data, some programs, and its position in the operational sequence. Therefore, the electronic type of

TYPICAL MAGNA ONE® GENERATOR
Cutaway

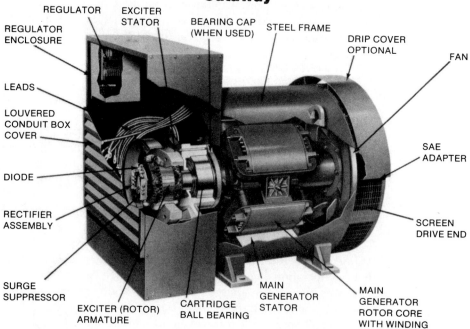

REGULATOR

EXCITER STATOR

REGULATOR ENCLOSURE

BEARING CAP (WHEN USED)

STEEL FRAME

DRIP COVER OPTIONAL

FAN

LEADS

LOUVERED CONDUIT BOX COVER

DIODE

RECTIFIER ASSEMBLY

SURGE SUPPRESSOR

EXCITER (ROTOR) ARMATURE

CARTRIDGE BALL BEARING

MAIN GENERATOR STATOR

SAE ADAPTER

SCREEN DRIVE END

MAIN GENERATOR ROTOR CORE WITH WINDING

FIGURE 16–37
Diesel-Driven Standby Generator Unit (Courtesy of Marathon Electric)

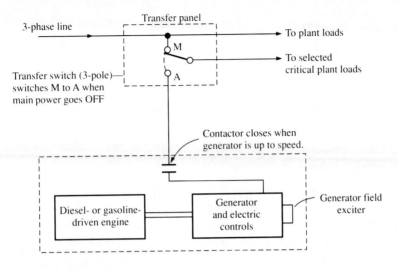

FIGURE 16–38
Standby Generator System Layout

Type of Equipment	Short-Duration Disturbances					Long-Duration Disturbances			
	Transients, Noise	Spikes	Momentary Interruption Flicker	Voltage Dips/ Sags	Surges	Undervoltage, Brownouts	Overvoltage, High Line	Outage, Blackouts	Frequency Variation
Voltage regulator	No	No	No	Yes	Yes	Yes	Yes	No	No
Isolation transformer	Yes	No	No	No	No	No	No	No	No
Power conditioner	Yes	Yes	No	Minimal	Yes	Yes	Yes	No	No
Motor generator	Yes	Yes	Yes	Yes	Yes	No	No	No	No
Standby power system	No	No	Minimal	No	No	No	No	Yes	No
Uninterruptible power system	Yes	Yes	Yes	Yes	Yes	Yes	Yes	Yes	Yes

FIGURE 16–39
Comparison of Standby Power Systems

Normal Mode

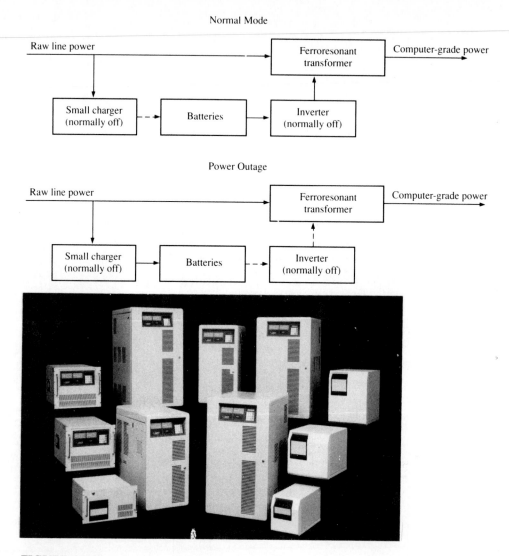

FIGURE 16–40
Electronic Standby Power Units (Courtesy of Best Power Technology, Inc.)

standby power, choice 5, is mandatory in many cases. The table in Figure 16–39 is a comparison of the uninterruptable electronic standby unit with other types of alternative standby power sources.

Figure 16–40 shows electronic standby units. These come in typical ratings from 250 VA up to 15 kVA and higher. Their cost per watt is many times that of the generator type. Their advantage, in spite of the cost, is that they respond instantly to a power loss. The output sine wave continues unabated when input

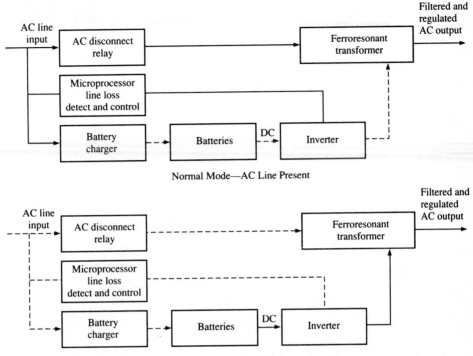

FIGURE 16–41
Electronic Standby Power Unit Power Flow

power is interrupted. Many of these units are also line purifiers (previously discussed) as well as power standby units.

A typical block-diagram layout is shown in Figure 16–41. The diagram shows the power flow for line ON and line OFF conditions.

SUMMARY

□ Fuse and circuit breaker actuation clearing values are not absolute. A given value of device has a given time-versus-value curve, which must be considered for application of fuses, circuit breakers, and overload relays.

□ Fuses come in a large variety of types to match equipment applications. However, fuses must be replaced, whereas a circuit breaker can be reset manually.

□ Circuit breakers may be magnetic or thermal types. They come in a large number of ratings and poles.

□ Overload relays may be thermal, current-sensitive, or both. They may be installed inside or outside the device being protected.

☐ Improper wiring and electrical distribution hardware can adversely affect equipment performance.

☐ Low voltage, constant or during short intervals, can affect electronic equipment operation. The two major causes of low voltage in a plant are line loss and poor power factor of other adjacent electrical devices.

☐ Electronic equipment can malfunction because of power utility surges, transients, and poor waveforms. Line purifiers, or conditioners, can eliminate these utility-line imperfections and, consequently, eliminate equipment malfunctions.

☐ Three-phase power is more effective in power transmission and also for three-phase motors operation. There are advantages and disadvantages to using three-phase electrical values.

☐ Electrical grounding of electrical and electronic equipment is mandatory for personnel safety. Proper grounding is often needed for proper operation of electronic equipment, and special grounding may be needed.

☐ Monthly power costs have a number of elements. Two of these elements, power factor charges and peak-demand charge, can be controlled by proper plant operational-control-system design.

☐ Standby power systems have many forms. Two are of major importance for equipment operation: generators and electronic units. Generators come on line in up to a minute. Electronic units act essentially instantly.

EXERCISES

1. List the major comparative advantages and disadvantages of fuses and circuit breakers as they apply to electronic control and process equipment.

2. Refer to Figure 16–1. Make up tables of current versus blow time for 30 A and 400 A fuses. Comment on the significance of these tables.

3. List some applications for time-delay fuses. List some devices and equipment that should not be protected by a time-delay fuse.

4. A pico fuse has a color code of orange, black, red, red. What is the current rating of the fuse?

5. What are some advantages and disadvantages of plug fuses?

6. Refer to Figure 16–9. The person illustrated touches the ends of a wire with a resistance of 7 Ω from the case to ground. Which will blow, the GFI or the 20 A fuse in the distribution panel?

7. What are some advantages of the overload built into a motor, or other device, over an overload installed in the starter panel?

8. What is the advantage of an overload relay over a fuse?

9. What is the current-carrying capacity of each of the following types of wire?
   a. #6   TW   copper
   b. #2   TA   aluminum
   c. 3/0   RH   copper
   d. 400MCM   RHW   aluminum

10. What conduit size is needed for each of the following groups of wires? (Copper and aluminum are the same.)
    a. 16   #3   TWH
    b. 15   350MCM   RHH
    c. 22   #6   TW

11. What model of locking plug and receptacle is needed for 20 A, three-phase, four-pole, five-wire grounding applications at 60 Hz, 480 volts? What straightblade plug and receptacle are needed for two-pole, three-wire grounding 30 A, 250 V?

12. Propane is in the area of an application. What class of hazardous-location equipment should be used? Refer to Figure 11–2.

13. Refer to Figure 16–16. The $1\frac{1}{2}$ HP motor is replaced by a 2 HP, 1800 RPM motor. The motor rating is 11.4 A/5.7 A for 115 V/230 V. Make an analysis of voltage drop similar to the figure. Use same wire sizes, distances, and voltages as shown.

14. Refer to Figure 16–17. Make an analysis for 20 kW at 220 V the same as that shown in the figure and described in the text.

15. For what types of application is a line conditioner desirable? For what types of application is it not necessary?

16. Refer to Figure 16–21. Repeat the problem for 120 V and 12 Ω.

17. Refer to Figure 16–22. Repeat the problem for 440 V and 3.4 kΩ.

18. Refer to Figure 16–23. Repeat the problem for 208 V and $Z = 73 - j24$.

19. Refer to Figure 16–24. Repeat the problem for 220 V and $Z = 73 + j24$.

20. Refer to Figure 16–25. Repeat the problem for 440 V and loads of 1 kΩ, 2 kΩ, and 3 kΩ.

21. Refer to Figure 16–26. Repeat the problem for 440 V and loads of 150 Ω, 275 Ω, and 425 Ω.

22. A person is, on the average, electrically equivalent to 1500 Ω plus a small capacitance. Based on the resistance only, what would be the initial current through a person, standing on a ground potential surface, when he or she touched the case in each of the following circuits? (Metal case is erroneously touching incoming line A.)
    a. Figure 16–29A
    b. Figure 16–29B
    c. Figure 16–30A
    d. Figure 16–30B
    e. Figure 16–30C

23. Determine the monthly power cost for a plant with the following characteristics:

    Monthly maintenance charge of $85
    Three shifts for 25 days—average usage of 142 kW
    Energy charge of $0.075 per kWh
    Peak demand of 287 kW
    Average power factor of 0.9 (Use the PF schedule in the text.)

24. In the example in Exercise 23, at what peak kW load would the peak-demand charge equal the energy charge?

25. There are two large motors in a plant. Only one should be in the start mode at once, to reduce peak demand. Spin-up time for start is 55 s. Develop a PLC control circuit to accomplish the required motor interlocking.

26. Outline a standby power system plan for each of the following:
   a. 200-bed full-service hospital
   b. 10-story office building
   c. 2000-student school

APPLICATION ASSIGNMENT ANSWER

1. For the relays that keep dropping out, check the voltage. The relay coils are probably experiencing periodic low voltage due to line loss or some other cause. A voltage-versus-time plot on a strip recorder may have to be made.
2. The computer malfunctions are possibly caused by line surges. Install a surge-proof, regulated power course. Its cost will be offset by improved machine performance if line surges are the cause.
3. The motor problem will not necessarily be prevented by a lower-value fuse. Use a motor with internally installed overload protection (at slightly greater cost). Also check the motor's feeder line for low or high voltage.
4. Check the power bill for the reason for the cost increase. Compare a number of months' relative charges by category. Which category or categories have gone up a lot? If the problem is peak power, machine sequencing would help. If possible, interlock machines so that two large ones cannot run at the same time. If the problem is power-factor charges, consider installing power-factor correction packs at the machines. If the problem is overall power usage, a more in-depth analysis must be made, possibly with the assistance of your electrical utility personnel.

VI

SAFETY AND AUTOMATION

17

Safety Considerations— Personnel and Equipment

CHAPTER OUTLINE

17–1 Introduction □ 17–2 General Rules and Procedures □ 17–3 Eye, Ear, and Skin Protection □ 17–4 Electrical and Electronic Device Safety □ 17–5 Power Feeder Handling Safety □ 17–6 Tools and Equipment □ 17–7 Special Safety Considerations for Robots and Work Cells □ 17–8 Postaccident Procedures

LEARNING OBJECTIVES

□ Outline the requirements for safety of the OSHA law and other regulations.
□ List the safety precautions for eye, ear, and skin protection.
□ Discuss some of the basic procedures for safety around electrical and electronic equipment.
□ Outline basic safety procedures in dealing with incoming power to a device you are near or are working on.
□ List major safety considerations when choosing tools for work on and around electrical and electronic equipment. Discuss the situations requiring safety-toe shoes and hard hats for personnel safety.
□ List the added safety precautions to observe around robots or work cells.
□ List the considerations in assisting an injured person or persons if an accident does occur.

APPLICATION ASSIGNMENT

You have been assigned to repair an electrical control panel for a robotic paint booth. The robot does not function consistently. The cause of the problem has been determined to be in the robot's control system. What equipment will you take to the site, and what safety procedures will you follow?

17-1   INTRODUCTION

This chapter outlines some general safety precautions to follow in working with electrical and electronic equipment. The listed precautions do not by any means cover all situations. Each individual situation is different. Therefore, additional safety measures will be needed for each and every local condition.

First, safety rules and regulations, such as those of the Occupational Safety and Health Administration (OSHA), will be covered. General procedures will be covered also, in the same section. Protection for ears through ear coverings, eye protection through proper glasses or shields, and skin protection by gloves will next be discussed. Special eye-safety protection around lasers is included.

Specific operational safety procedures for working on or around electrical and electronic equipment will be outlined. General safety procedures for working with incoming power to the equipment will also be covered. Fundamentals for choosing and using proper tools and equipment will be discussed next. A section on special safety procedures around robots and work cells follows. Finally, such things as first aid kits and CPR will be discussed in general terms.

Note that this chapter does not deal with product safety in any way. That is to say, it does not tell how to design a safe product. Neither does it tell how to test a device that will be sold, to predict its safe operation.

17-2   GENERAL RULES AND PROCEDURES

Accidents can be caused by ignorance, worker fatigue, pressure to hurry up, faulty or improper tools, improper procedures or instructions, and just plain carelessness. To prevent accidents for any of these causes, there must be a company policy to address each of these factors, and other factors as needed. In addition to company rules, you may develop some of your own practices for safe conduct on the job.

Some safety hazards are obvious, such as a spinning belt or wheel. Others are not, such as an inactive machine that goes on only periodically. New workers must be given safety indoctrinations in these specific areas and others as required.

Persons under mental or physical stress and pressure are more likely to have an accident. Distractions or horseplay can also contribute to the likelihood of accidents.

It is essential that any potentially unsafe situations be reported and corrected. If accidents do occur, a record of each accident, large or small, should be kept. A periodic review of the accident log will highlight any unsafe trends on a specific machine or in a specific area that need correction. Another important safety rule is not to work alone if the situation is possibly dangerous. If one person is injured, there is another person to take action or summon help. In one situation it was necessary to keep testers overtime to test motors in a remote area. It was essential that two or more testers be assigned, even though the company had to pay two people overtime rates.

A company must have a well-defined safety policy, which addresses the listed situations and other factors. A safety policy includes such concerns as safety training, safety updating, and first aid. Safe operation can be at odds with the fast and efficient operation needed for higher profit. Safety should come first. Many companies have a regularly scheduled safety meeting to keep safety in the minds of those who work there.

In the United States there is a safety oversight agency, the Occupational Safety and Health Administration (OSHA). The agency inspects, usually at random, workplaces of all kinds. Its job is to enforce safety and health standards that research and experience have proven to be conducive to personal safety and health. The health effects considered are both short-range and long-range. An example of a hazard with a long-range effect on health is asbestos contamination. In addition, many trade associations have their own safety standards for their member organizations. Examples of special trade organizations are Underwriters Laboratory and the Fire Underwriters organization, which establishes the National Electrical Code. Both organizations establish equipment standards for safe operation of products and wiring systems.

In summary, there is no substitute for training, knowledge, alertness, and common sense in providing for àccident-free, safe operations.

17–3   EYE, EAR, AND SKIN PROTECTION

The safety procedures for eye protection fall into three general classifications: protection from foreign objects, protection from excessive light or radiation levels, and protection from chemicals or corrosive atmospheres. Note that eye protection is basic in electrical and electronic work. You may not have to protect your ears, skin, or lungs in most cases of electrical or electronic operation. You will, however, want to have eye protection in all cases.

There are many types of safety glasses available for different types of operating environments. Some of the major types are shown in Figure 17–1. It is important to select the proper type of safety glasses for the work in which you are involved. Basic safety glasses are soft lenses with plastic side shields. These glasses will protect you from objects flying at you from in front or from the side. Soldering, wire clipping, drilling, sawing, and other activities can produce splashing material or flying pieces of metal or plastic. Remember, the eye cavity presents a fast passage to the brain. Flying material might not only enter the eye; it might also penetrate into the brain.

When doing work on grinders or heavy machinery, a different type of safety glasses is needed. This type has metal mesh side shields. Other types of glass and side shields are available. Check the application and safety glasses catalogs for the types needed. Safety glasses may be obtained in prescription form or in plain glass.

Eyes must also be protected from excessive light. For welding, more complex eye protection is needed than that previously discussed. In addition to safety

FIGURE 17–1
Various Types of Safety Glasses (Courtesy of Wardco Safety, Inc., and Glendale Protective Technologies, Inc.)

glasses, a face shield with a small, very dark glass opening must be utilized. The welding arc is of high intensity. Looking at the arc can permanently damage the retina because of excessive light in one spot. Welding should not be looked at directly, even from a distance. Canvas or dark material is normally placed around welding areas in factories. This prevents possible eye damage to passersby who might otherwise look directly at the welding arc.

A relatively new field of unsafe high levels of radiation or light is lasers. Lasers run at different levels, rated in watts or kilowatts. Using very low-level lasers requires no special eye protection. As the laser radiation levels increase, the need for eye protection increases accordingly. Different eye protection is required for different types of lasers. Figure 17–2 outlines the major wavelength of each of nine major types of lasers. The figure also lists the different types of glasses for use in working with these different lasers. The figures on the right of the table indicate the percentage of light transmitted by the glass types. Note in the table that there are some types of laser safety glasses that may be used with more than one type of laser. Such versatile safety glasses are called multiuse eye protection. A major rule in laser eye protection is never to look directly back at the laser source under any circumstances.

Laser safety glasses come in different configurations. One configuration is physically the same as conventional safety glasses, except for the glass. Another type is for situations where no light should leak in from any angle. This type is completely sealed all the way around.

The third type of eye protection is the one for work in areas where corrosive chemicals or liquids are used. It has a fitted, vinyl, molded body. Chemical-protection glasses will not normally stop flying material as well as basic safety glasses, since they have softer lenses.

Note that there is danger when wearing contact lenses in a welding or laser area. A large arc or burst of light is concentrated by the contact lenses. The contact lenses could be "welded" to the eyeball.

Ear protection is needed generally for two situations. The first is exposure to a very high noise level for a short time. The second is exposure to a highly noisy environment on a continuing basis. The need for ear protection depends on both the noise level and the time of exposure to the high noise level. The ear-damaging noises fall into two categories: industrial or business situations and those which are self-inflicted. Industrial situations such as paper mills impose varying needs for ear protection, depending on the section of the mill you work in. Self-inflicted noises include loud audio levels, noise from snowmobiles or motorcycles (where you wear a noise-trapping helmet), and the sounds of trapshooting. Appropriate ear protection is required in these situations as well as on the job.

For ear protection at medium noise levels, the simple wearing of ear plugs can be effective. At higher levels you should wear the "earmuff"—over-the-ears—type. Higher noise levels include those from high-speed drills, jackhammers, and jet engines. Many manufacturing plants, such as punch press operations, have the ear protection built around the machinery. The machines are

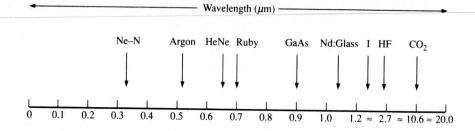

Wavelength, mμ				
Range	Principal	Recommended Eyewear	Optical Density	Approximate Luminous Transmission, %
0.20–0.36	uv	Employee: UV 400* Visitor: UV 400*	>3	90
0.22–0.42	uv	Employee: GG-9** Visitor: GG-9**Plano or UVC-303***	>5 4	85 95
0.20–0.55	uv and argon (o.51)	Employee: OG Series Visitor: LGS-A†	>5 >6	60-50 45
0.20–0.61	dyes	Employee: RG Series** Visitor: RG Series**	>5	40-5
0.58–0.67	HeNe (0.63)	Employee: BG-1** Visitor: LGS-HN†	>5 5	20 20
0.67–1.30	Krypton (0.67) ruby (0.60)	Employee: BG-18** Visitor: LGS-R†	5 6	45 20
0.90 20.0	Nd:glass (1.06)	Employee: KG-3** Visitor: KG-3**	4.5	85
	HF (2.70)	Employee: KG-3** Visitor: KG-3**	3	85
	CO_2 (10.60) (any material absorbs CO_2)	Employee: KG-3** Visitor: GL 1112A†	>10 >10	85 95

Doubled Frequency (simultaneously)

0.53 and 1.06	Nd:glass (1.06)	Employee: KG-3** with Clipsite A† Visitor: AO-698††	5 6	60 5
Other wavelength combinations (Flip-down spectacle)*			>5	(varies)

FIGURE 17–2

Laser Eye Protection for Different Laser Types (Courtesy of Reed Optical)

enclosed in sound-deadening booths. The insurance carrier or state agencies can assist in assessing noise levels and the type of ear protection to be worn.

Sometimes there are situations in which electronic equipment operation involves toxic liquids. For example, batteries contain toxic acids that can harm the skin. Appropriate gloves should be worn in these cases. In very rare situations, fumes may be involved, as in electroplating or painting. In such cases, the fumes must be exhausted, or else collected so as not to pollute the general atmosphere. In extreme cases of possible toxic fumes, masks, often with oxygen supplies, must be worn for protection.

17–4 ELECTRICAL AND ELECTRONIC DEVICE SAFETY

This section will discuss safety practices to follow when working on electrical or electronic devices. The next section will cover safety practices for handling the incoming power to these various devices. The topics covered in this section are only a sampling of major considerations. To determine additional, specific safety precautions, consult the manufacturers' equipment manuals, and review other applicable procedures. Of course, the first rule is to wear safety glasses at all times. Follow eye, ear, skin, and lung safety measures as previously outlined. Also, the equipment you are to work on may be in an area requiring the wearing of hard hats or safety shoes. These two items will be discussed in Section 17–6.

If you are installing, removing, or troubleshooting equipment, make sure that the power to the equipment is OFF. It only takes a few milliamps to injure or kill a person. If the power disconnection is not controlled by removing a plug from a receptacle, a lock-out system should be used. The lock-out system involves the use of a keyed lock assigned to each individual worker. The electrical disconnect is locked in the OFF position, as shown in Figure 17–3. Only the worker and his supervisor have keys to the lock. Tags may also be used to indicate the power should be left OFF. When the work is completed, the lock and tags are removed.

Do not feel free to work on equipment just because someone else's lock is already on the disconnect switch. Add yours. The other lock could be removed without your knowledge and the equipment energized while you are working on it.

Should you assume there is no power to the equipment when the power is locked out? No. There may be alternate power from other voltage sources. Also, the plant could be wired differently from the way the prints show. Assume the equipment is still live until you test that it is OFF, with a meter. If a test of the incoming line shows no voltage, you may proceed. Some companies require testing with two meters in case the first meter is not working properly.

Make sure all capacitors in the equipment are discharged before working with it. Some capacitors can hold a charge after the equipment has been OFF for a long time, even days in some cases. Most capacitors that can remain charged are in the DC power-supply sections. When working with cathode-ray tubes (CRTs), remember that they can implode. They are under vacuum, so if they are broken, the pieces accelerate inward and then outward at high speeds. For safety, cover

FIGURE 17–3
Electrical Equipment Lock-Out System (Courtesy of Marathon Electric)

the CRT in equipment you are working on with a heavy canvas, or canvaslike, blanket.

If it is necessary to work on live equipment, added precautions may include the following:

1. Work with proper tools (see Section 17–6).
2. Work with one hand only, if at all possible.
3. Do not work alone.
4. When measuring an unknown voltage, turn power OFF, install meter, step back, and then turn power back ON to observe the voltage reading.
5. Do not disable interlocks on doors and other devices.
6. Install warning signs as appropriate. For example, you have removed the cover of a control panel. Install signs and barriers as appropriate to keep others from getting too near and possibly getting an electrical shock.
7. Check for radiation in the area you are working in. For example, if RF radiation is present, it may be harmful over an extended exposure period.
8. Do not wear neckties or scarves when working around rotating equipment. The tie or scarf, if caught in a rotating device, can pull you into the device. If you must wear a tie, use the clip-on type.
9. Do not wear loose clothing—sleeves, for example—around rotating devices. It can become tangled and pull you into the machinery. In some situations, it is a good idea not to wear watches or rings for the same reason, as well as to prevent electrical shocks.

This list contains general precautions. Your specific situation will probably require additional precautions.

A person who ignored precautions 4 and 5 in the preceding list was very badly injured. The person stepped inside a deenergized RF induction-heating control station. He held a volt ohm meter (VOM) across two points to check for the presence of 120 V AC. He then had someone push the door interlock limit switch so that he could obtain a voltage reading. The 120 V AC was superimposed on 10,000 V DC, a fact he did not know. The 10,000 V DC arced across the meter and then through his two hands.

17–5   POWER FEEDER HANDLING SAFETY

In the previous section, safety procedures when working on and around electrical and electronic equipment were discussed. This section will cover a few of the safety procedures concerning turning OFF incoming power to the equipment being worked on.

When the electrical breaker lever is turned to OFF, make sure it works. There are rare occasions when the breaker mechanism breaks. Even though the handle is on OFF, the electrical contacts are then still closed. An observation window is provided on large breakers to observe their proper operation. Check to see that the main circuit breaker contacts are indeed open.

Many electrical operations require that the power be turned OFF at two points in the line—for example, by flipping a breaker and by disconnecting the main wires. Observe both operations visually in these cases. If the breaker cannot be locked out (see the previous section), add appropriate signs and tags to prevent its being turned ON. Testing for zero voltage with a meter is a good added check after disconnecting.

In some cases, fuses must be removed to replace a defective fuse. A crescent wrench, screwdrivers for prying, and hand removal are not desirable methods. A properly sized fuse puller (see next section) should be used. Fuses should be replaced with fuses of the same type and rating. Putting in a fuse of the next higher rating to prevent another blown fuse is an unacceptable practice. Occasionally you may have to replace fuses in a live voltage situation. In this case, remove the "hot" side of the fuse first. When reinserting, reverse the procedure.

Where higher voltages feed the equipment, safety grounding is desirable even though the power is disconnected. Large clipped cables are run from each phase of power at the equipment to a solid building ground. Why do this? If there is a line voltage surge or electrical storm, a voltage surge can arc across the open main breaker contacts. When the surge gets to the device, it has three possible paths. You, the device, or the grounding leads. Without the grounding leads, you are the primary path if you are working on the machine at the time.

One final note. If you are not sure how to shut the power OFF properly, consult the person or persons familiar with the power distribution system you are working on.

17–6   TOOLS AND EQUIPMENT

A good voltmeter or voltage indicator device will tell you if a circuit or device is electrically live. Meters should be periodically checked for proper operation and voltage indication. If a meter failed to indicate the presence of voltage, you might assume incorrectly that the line was dead. Meter leads should have plastic sleeves over the end tips or clips. Otherwise, when you hold the ends of uninsulated tips and touch a live circuit, you will receive an electrical shock.

Fuse pullers have already been mentioned. The proper size should be used to match the fuse diameter. A typical fuse puller is shown in Figure 17–4. Also shown in the figure are other electrical tools.

Electrical tools differ from conventional tools in that the handles are covered with electrically insulating plastic. These tools include screwdrivers, pliers, wire cutters, and even saws. Insulated handles are guaranteed for comfort of use, not protection from electrical shock. Therefore, you should not depend on tools with insulated handles to protect you from electrical shock. You could probably remove a terminal nut on unenergized equipment with conventional, noninsulated pliers. However, why take a chance? It is good practice to treat unenergized circuits as though they were energized—as an extra precaution. Insulated handles can help. Figure 17–4 illustrates a number of hand tools with insulated handles.

The workbench on which you work on electrical or electronic equipment should be of wood or another nonconducting material. If the workbench is metal, a nonconductive material should be permanently applied to the working surfaces. Metal workbench surfaces conduct electricity and can cause shock to you, line-to-line or line-to-ground. Having a GFI receptacle on the workbench electrical supply is also a good practice (see Section 16–2).

Appropriate gloves should be worn as warranted. Gloves are of two types. One type is rubber to keep you separated from electrical voltages. The higher the voltage, the thicker the gloves must be. The second type protects the hands and fingers from mechanical danger. Examples of mechanical danger include chips from drilling and the hazards of working on grinding devices. Some leather mechanical-protection gloves fit over the rubber gloves. This arrangement gives you double protection. Gloves are often clumsy but are well worth using when needed.

If it is necessary to use ladders, a fiberglass ladder is electrically safer than a metal ladder. If you somehow come in contact with electricity on a fiberglass ladder, there is no path to ground, so the shock hazard is reduced. Place the ladder at the proper angle, 75.5°. Make sure the bottom of the ladder is on solid ground, not on an oil slick. If the base friction is questionable, tie the bottom of the ladder to the wall in some manner. Do not use ladders outside on an extremely windy day, especially long ones. Finally, look up before you position the ladder. Look for electrical lines that could make contact with the ladder.

In construction or large process areas, hard hats must be worn when specified. There are various types of hard hats. A hat of the proper classification must be worn.

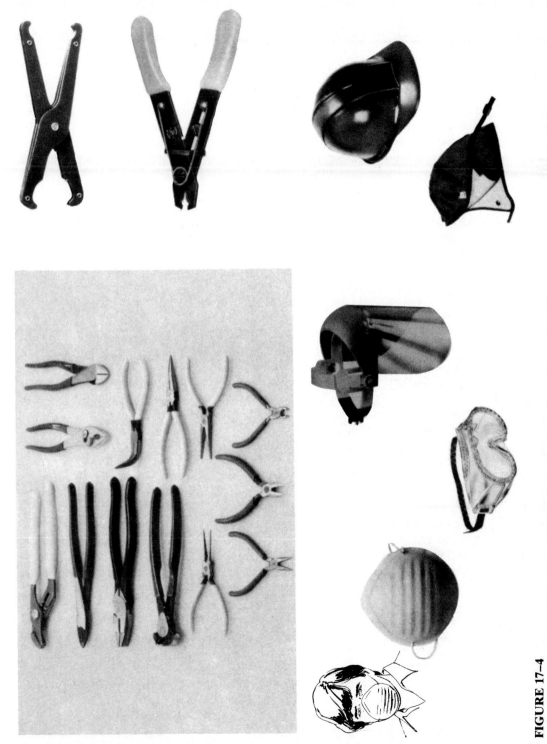

FIGURE 17-4
Fuse Puller, Electrical Tools with Insulated Handles, and Other Safety Equipment (Courtesy of Klein Tools, Inc.)

Safety equipment in some types of electrical work should include steel-toed shoes. Steel-toed safety shoes are good to wear if an electronic or electrical module or some other object could drop on your foot. Conventional shoes or athletic shoes are comfortable to wear, but are no help in protecting your feet.

An actual situation shows the need for following the rules of this section. The situation involved plastic-handled tools. It also shows the need for safety glasses. A house was being torn down. The electrical fixtures, boxes, and receptacles were being salvaged. The person doing the salvage wanted to remove the stove 220 V AC receptacle. He removed all fuses from the service box, including the two main fuses as well as the individual fuses. When he cut into the feeder wires leading to the receptacle, there was a big flash and a bang. The wire cutters melted, and there were sparks all over. Fortunately, the person doing the salvage had used wire cutters with plastic handles and had worn safety glasses with side shields. He was not injured in any way. He had looked away when cutting, by force of habit, so the flash did not affect his eyes.

The homeowner had installed the electric stove a few years after the home was originally built and wired. He had found no place to connect the three stove wires in the electrical box. Therefore, he had connected the stove wires ahead of the fuse box, directly to the incoming line. The only fusing of the stove circuit was back at the power company's primary fuse, a few hundred amps. Fortunately, no one was hurt as a result of that arrangement.

Finally, one good way to point out hazardous areas or equipment is to post appropriate signs in the area. Such signs are especially effective in warning persons entering an unfamiliar area. A collection of various warning signs is shown in Figure 17–5.

17–7 SPECIAL SAFETY CONSIDERATIONS FOR ROBOTS AND WORK CELLS

The new technology of robotic devices and work cells has created additional safety hazards. *Robots* are special machine tools with sophisticated programming. *Work cells* are combinations of robots, conveyors, and machines working together to make products. These devices will be discussed in detail in Chapter 18.

This section will emphasize robotic safety. The robotic safety procedures can be expanded to cover other similar devices and work cells. The first robotic safety procedure is to define the work envelope. The envelope then defines the horizontal operational floor area. Typical robot work envelopes will be illustrated in Chapter 18. If you are in the work area while the robot is operational, it could hit and injure you. Appropriate warning signs can be posted in the area to caution personnel in the area to stay clear.

Once the work area is defined, safety devices should be added to make sure no one is in the operating area when the robot is running. These devices may be barriers, safety mats, *light curtains,* or *volumetric ultrasonic curtains.*

Barriers can include walls, fences, and markings on the floor. The fences or walls are not exactly at the perimeter of the horizontal work envelope. Allowance

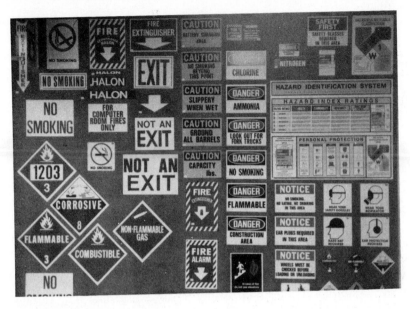

FIGURE 17–5
Typical Hazard Warning Signs

must be made for the protrusion of any working tools or fixtures attached to the end of the robot arm. Then the barrier must be placed an additional 3 ft or so farther out. The extra 3 ft is included to eliminate a "pinch point." A person must have space to back away against the fence or wall and not be touched by an operating robot end tool.

Safety mats may be placed on the floor to cover the work area and possibly an extra adjacent area. If anyone steps on the mat to enter the area, the robot is stopped. Light curtains are vertical sheets of light about the operational area. The light is emitted in the direction of thin vertical receivers. If something or someone breaks the light path, an alarm sounds or machinery is turned OFF. Ultrasonic volume systems function in a similar manner. The volume level is set for the volume of the robot, its attachments, and the work pieces. If someone enters the area, the envelope volume changes. The ultrasonic detector senses the volume change, and the robot stops.

There are some cases when you have to be in the operating envelope while the robot or machine is operational—for example, during setup, troubleshooting, or training of the robot. When you do enter the operational area and operate the robot, take a control pendant with you. The control pendant should be the "dead-man" type. In dead-man control, you keep one finger on an Operate button. If you remove your finger from the button for any reason, the robot or machine stops where it is. In some cases there are pressure sensors associated with the robot arm. These sensors stop motion when an obstacle is encountered. The force settings on these sensors are fairly high. You cannot depend on sensor action to completely protect you from injury.

Other robot safety considerations include the following:

☐ Do not turn your back on an operational robot under any circumstances when you are in the work envelope.

☐ The robot can move and react faster than you can.

☐ Parts should be held inward or downward by the robot grippers, not outward. From outward-facing grippers, the part can be flung outward if the grippers fail or do not hold the part tightly.

☐ Set the robot arm on blocks when you stop it for long periods of time. The arm can creep downward, and you might hit your head when you stand up under it.

☐ An immobile robot is not an OFF robot. It may be in a pause mode.

☐ Just because a robot has performed one pattern of movement many consecutive times does not mean that it will necessarily do the same thing next time it operates.

☐ Do not assume the robot will follow the program you have put in. It could skip to another program by mistake. The first time through a new program, keep your hand near the stop button.

☐ Pushing the halt button or the stop button may not stop the robot. It may return the robot arm to the home position. Make sure you know which control button performs each function.

☐ Some robots have breakaway tooling. These robots are designed so that end-of-arm tooling breaks free when an obstacle is encountered. This action saves the expensive tooling from great damage. For safe operation, the tooling should have a substantial safety chain firmly attached to it from the arm. If the tooling should break loose, the chain keeps the tool from being flung outward into someone.

A final general consideration. Who is likely to be injured by a robot or automatic machinery? Someone who knows nothing about the robot will normally stay away. Someone who is well trained on a machine will normally proceed deliberately and safely. Someone with a little knowledge who goes beyond his or her capabilities is a good candidate for injury. Make sure you have complete knowledge of a robot or machine before operating it, especially if you must be in, or even near, the operating envelope.

17–8   POSTACCIDENT PROCEDURES

In spite of training in all the safety precautions possible, there still may be accidents. These come from failure to follow procedures and from unforeseen occurrences. What should you do if an accident happens? This section will briefly touch on a number of actions.

If there is an electrical fire, do not get a hose and spray building water on the fire. The water is a conductor from the object through your body to electrical ground. You could be electrocuted. Use appropriate fire extinguishers. If the fire is of any magnitude, get out of the area and call the fire brigade or fire department.

For small cuts, clean the cut or scrape, and add antiseptic and an adhesive bandage from the first aid kit. Then see the company nurse or doctor immediately. Any cut can become infected if not treated effectively with suitable ointment. The possibility of tetanus is always present and must be considered. For larger cuts, abrasions, or bruises, follow the company procedures, and see the nurse right away. Note that many companies require an employee to have professional attention for a cut or abrasion, no matter how small the wound is.

For severe injury, such as broken bones, do not move the injured person until professional help arrives. If the person must be moved away from a fire, do so as carefully as possible, preferably on a blanket to reduce movement of bones. Moving the person may make the injury worse. The injured person may be in shock (not electrical). Keep the person in shock warm, horizontal, and comfortable until help arrives.

What if someone you work with becomes entangled in an electrical circuit? The person may be thrown free. If the person cannot let go because muscles have tightened up, he or she may still be touching the electrical circuit. Your first impulse will be to rush over and pull the person away from the electricity. If you do, there will be two victims, the second of whom will be you. The best procedure is to turn the power OFF as quickly as possible. If the power disconnect is not nearby, be careful. You must pull the person away by using an insulated pole, a rubber hose, or some insulating material.

Once the injured person is away from the power, see if he or she is breathing and conscious. If the person is not breathing, cardiopulmonary resuscitation (CPR) may be applied by someone qualified. Cardiopulmonary resuscitation, a combination of mouth-to-mouth resuscitation and closed-chest heart massage, can be used until help arrives. Brain damage will occur in 4–6 min if the person is not revived by CPR or other means. It is essential that all people working around electricity be trained in CPR.

SUMMARY

- ☐ General rules for operational safety are provided by each company, by professional associations, by local and state governments, and by the federal government through OSHA.
- ☐ Eye protection has many forms, to protect from high light levels and flashes and from flying or splashing foreign material.
- ☐ Ear protection is needed where there is exposure to high noise levels for an extended time or to a very high noise level for a short time.
- ☐ Skin is protected from electrical shock and from irritating or corrosive material by use of proper gloves and protective clothing.
- ☐ There are many safety procedures to be followed when working on or around un-energized electrical or electronic equipment.

☐ When working on electrical or electronic equipment that is energized, there are many other safety rules to follow.

☐ Working safely with feeder electric power requires special procedures, such as a lock-out system.

☐ Proper tools, with plastic handles, and special equipment are required for safety around electrical modules and electrical power. Safety shoes and hard hats are sometimes required also.

☐ Working with robots and work cells requires special safety procedures because of the complexity of their operation.

☐ The work envelope for a robot or work cell must be defined. Then appropriate safety interlocking systems must be installed to prevent the robot or work cell from operating if someone enters the area by mistake.

☐ Accidents can occur, from the minor through the critical. Advanced training in the procedures to follow after an accident is necessary before working in an area.

☐ Knowledge of appropriate responses to injury, such as CPR and fundamental first aid, is a good safety practice.

☐ Training in safety, attention to what you are doing, and common sense are basic to a safe environment on the job.

EXERCISES

1. List ten major safety rules that you would follow when performing electrical/electronic maintenance or installation in an industrial plant. Compare the ten rules you have listed with those of others in your class or group.

2. You have been put in charge of running a departmental monthly safety meeting. What might you use as an agenda?

3. The text discussed the problem of scarves, neckties, and clothing. What are your comments on loose clothing?

4. Why do safety glasses have side shields?

5. Is it possible to purchase a general laser type of safety glasses?

6. What major wavelengths need to be filtered out for the argon and ruby lasers?

7. Under what situations should you wear form-fit safety glasses?

8. What are the two major types of ear protection? Where would they typically be used?

9. What are some typical liquids or pastes for which you would need glove and tight clothing protection?

10. Describe how you would make sure that equipment power of a machine you are going to work on is turned OFF and stays OFF.

11. Describe the electrical lock-out system. When might it not be foolproof?

12. List, in your own words, the major precautions to follow when it is necessary that you must work on electrically "live" equipment.

13. You wish to shut the power OFF twice in a fixed electrical installation that has no plug. How could you do so?

14. List tools and equipment that should have plastic-covered handles for electrical and electronic work. Are insulated handles guaranteed to protect against electric shock?

15. Why should you electrically ground electrical equipment when it is OFF and you are working on it?

16. You have some rubber gloves that you know are safe for handling 120 V AC. What do you see if you must handle 440 V AC?

17. You are driving an auger-equipped tractor in an open area near a farm or industrial building. What safety precautions should you observe when moving from place to place?

18. List, in your own words, some of the special safety precautions needed when working around robots or work cells.

19. You have observed a robot working on a part continuously all day. It has stopped for 5 minutes. Is it safe to enter the work area? Why or why not?

20. You have set up a work area for a robot, based on manufacturer-furnished drawings. When the robot arrives, is it safe to use the designated area based on the drawings? Why or why not?

21. What postaccident knowledge and training should you have when working in a departmental area in a manufacturing plant?

22. Contact a company, insurance company, or the local electrical public utility for their safety procedures or pamphlets. List their major points.

APPLICATION ASSIGNMENT ANSWER

First you will assemble a set of appropriate tools to take with you. You will then refer to manuals and drawings for any safety procedures specified by the equipment manufacturers, which must be followed. Wear safety glasses at all times. Ear and skin protection would not be needed except in special cases. For a robot that handles heavy materials, you should wear safety shoes and possibly a hard hat.

You will observe the faulty operation, if possible, from a safe distance. To check or remove electrical or electronic parts, you will make sure the power to them is OFF. Follow a lock-out procedure. Know where safety equipment, such as first aid kits, is located. Do not work on the equipment by yourself. Finally, work slowly and deliberately, making sure that you are alert.

18

Control of Robots and Work Cells

CHAPTER OUTLINE

LEARNING OBJECTIVES

- ☐ List and describe the five major types of robots and their work envelopes.
- ☐ Tell what type of power system is used with each type of robot, and why.
- ☐ Describe the possible angular variations in orienting the base of a robot.
- ☐ List specific applications for each type of robot.
- ☐ List four types of end effectors. Give one or two applications for each type.
- ☐ Explain how vision systems are used with robots and how they improve the robot's effectiveness.
- ☐ Describe the type of computer control needed for each type of robot.
- ☐ Write a control program for a pick-and-place robot operational sequence.
- ☐ Define *work cell*. Give examples of work cells.
- ☐ Outline the general control system for operating a work cell.
- ☐ Write an ON/OFF control program for a four-unit work cell.
- ☐ Discuss CIM and the controllers that it uses.
- ☐ Discuss network communications principles, including OSI.

APPLICATION ASSIGNMENT

Your manufacturing company has decided to procure two robots. An economic study has determined that the robots will pay for themselves. You are working with a group to determine the type and configuration of robot to buy. The first robot is to be installed in an assembly area. It will pick a chassis (6 × 8 × 3 in.) off a conveyor. There are five different variations of the chassis. It will put the chassis in a machine that performs an assembly operation. Next, the part will be turned over and placed in another machine, which performs a second assembly operation. The finished chassis will then be put back on the same conveyor.

The second robot's function is to put a compound (not straight) bead weld on the assembly at a point later in the operation. What should you consider in choosing the robots?

18–1   INTRODUCTION

This chapter covers the principles of controlling robots and work cells. Some examples of programs for each are included. Before the control systems are illustrated, configurations and work areas for the five major types of robots will be described. The types of operations and control systems for which each of the robots is best suited will be identified.

Working robots require *end effectors,* otherwise known as *end-of-arm tooling,* in order to function. End effectors include grippers, welding torches, and assembly devices such as automatic screwdrivers. These are attached to the end of the robot's working arm. Robot auxiliary systems include tactile feeler and vision guidance systems. End effectors and auxiliary systems and their control will be described. Next, the methods of controlling robots with computers will be discussed in general. A program for controlling a pick-and-place robot will be detailed.

The makeup of a work cell will be covered in the next section of the chapter. Work cell control will then be discussed in general terms. Then, a specific program for a small work cell will be illustrated.

In the last sections, CIM, CIM controllers, and network communications will be discussed.

18–2   ROBOT TYPES AND THEIR USES

A robot can be defined as a mechanical device programmed to perform a manipulative task under automatic control. If a mechanical device carries out manipulation that is never changed, it is not strictly a robot. A true robot can be quickly and easily reprogrammed to do different tasks. Robots are normally used in conjunction with auxiliary devices, such as machines and fixtures. Auxiliary devices will be discussed in Section 18–3.

Robots are used effectively in applications where a complicated process is going to be repeated, not necessarily continuously. They are also used effectively in hazardous areas, where a person might be harmed by fumes, high temperature, or other harmful factors. Monotonous human endeavors or extremely complicated human procedures are other areas where robots are effective.

Robots do not need heat, light, coffee breaks, overtime pay, workers' compensation insurance, or kind words. They do need competent service people to fix them and plenty to do to pay for their cost.

One misconception about robots is that they are good only for long product runs of the same part number. On the contrary, the advantage of robots is that mixed products can simultaneously flow through production efficiently, thanks to the robot's fast reprogramming capabilities. A production run of 1 is as easy to set up for as a run of 5000.

Robots come in many shapes and forms. The forms fall into five major categories:

1. cylindrical
2. rectilinear
3. spherical
4. jointed spherical
5. selective-compliance-assembly robot arm (SCARA)

The first two are the simplest, least expensive types of robots. Their configurations are shown in Figure 18–1. The *cylindrical* robot has two axes of motion, up-down and rotational. Additionally, the horizontal arm can move in and out, giving a limited third axis of movement. The *rectilinear* robot has three axes of motion, *x, y,* and *z.* For this reason, the rectilinear robot is sometimes called a Cartesian robot. Both of these basic robots are almost always operated by pneumatic air cylinders. The air cylinders are electrically controlled by appropriate programming to be either ON or OFF. The end of travel of each axis in both types is controlled by mechanical stops. For this reason, these two robots are often called "bang-bang" robots (from the noise they make when operating). There are sometimes intermediate stops included in one axis of the rectilinear robot's operation. The intermediate stops are controlled by limit switches.

Other axes of operation can be added to either robot in Figure 18–1 by using an end-of-arm manipulator that has various rotational capabilities. These motions are *roll, pitch,* and *yaw,* which simulate the capabilities of the human wrist. Section 18–3 of this chapter will discuss these manipulators in some detail. The two basic robot types, in their simplest form, have their travel distances programmed and reprogrammed by the resetting of mechanical stops. The sequencing

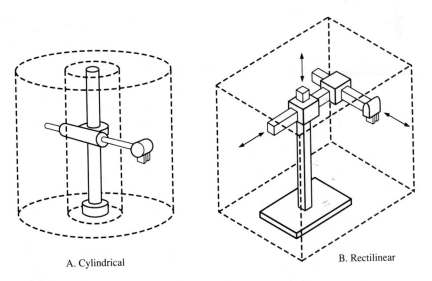

A. Cylindrical

B. Rectilinear

FIGURE 18–1
Basic Robots

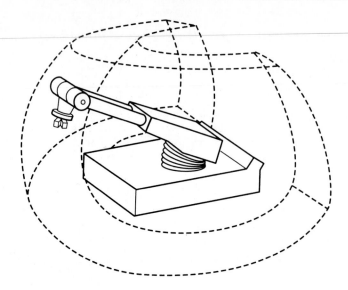

FIGURE 18–2
Spherical Robot

of motions is controlled by basic controllers, including relay systems, drum controllers, and small-scale PLCs (see Chapter 15).

The third type of robot is the *spherical* robot, shown in Figure 18–2. The spherical robot was common in the 1970s. The spherical robot was replaced in the 1980s for many applications by the jointed spherical robot. The spherical robot is usually large, with an arm that telescopes in and out. It is almost always hydraulically powered. Programming is accomplished at a medium level with a PLC or equivalent control. Spherical robots are used often for part manipulation for loading and unloading of lathes or milling machines. They are also used for general part handling, such as process dipping and die casting.

The two basic motions of the spherical robot are rotational (at the base) and angularly up and down (at the arm). As stated, the arm is usually telescopic in nature. For example, the arm may retract to 4 ft and extend to 7 ft. The manipulator at the end of the arm can add additional motion, as with the basic robots of Figure 18–1.

Spherical robots and the types that follow in this section are very often programmed in the *Train mode*. A control pendant Train button is depressed. The robot is then led through its motion and operational sequence to process a part. The robot controller "remembers" the motions, as long as Train is depressed. The motions are stored under a program name or number. The program can then be recalled and run whenever the given part is to be processed. This train system takes the place of complicated desktop programming. It also saves time because it is not necessary to retrain the robot each time a given part is to be processed, days or even weeks later.

FIGURE 18-3
Jointed Spherical Robot (Courtesy of Greenheck
Fan and ABB Robotic, Inc.)

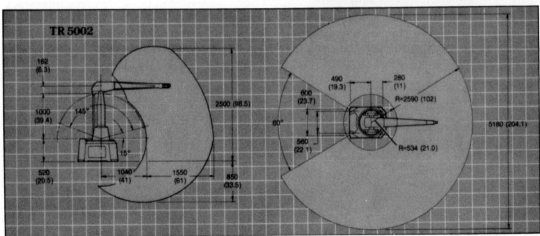

The fourth type of robot, the *jointed spherical* robot, is shown in Figure
18–3. It is a very common type of medium or large robot. The robot and a typical
work envelope are shown in the figure. This class of robots is alternatively called
jointed spherical resolute or *anthropomorphic articulated. Anthropomorphic*
means humanlike. The robot arm simulates the human arm by including an elbow.
Some special forms of this type of robot have more than one elbow, even as many

as five. The special forms are used for hard-to-get-at places—for example, behind an automobile dashboard. These special multi-elbow robots are very expensive, for both the robot and, especially, the control package.

The jointed spherical robot is usually powered by AC or DC servomotor systems. The end of the arm usually contains manipulators with two, three, or more additional motion capabilities. The control of jointed spherical robots is fairly complicated. The motion of one joint or axis affects the position and motion of the other axes. Consequently, special computer controls with special programs are needed to operate these robots. This type of robot has good accuracy and very good repeatability under load conditions. Jointed spherical robots are used in applications for spot welding, seam welding, spray painting, inspection, and sealant application, as well as in many material-handling applications.

A final type of robot to be discussed is the *selective-compliance-assembly robot arm* (SCARA), shown in Figure 18–4. It is usually small or medium-sized. It is a cylindrical robot with a very versatile arm arrangement. The SCARA has limited up-down motion, usually only a few inches. The movement in the horizontal plane is very fast and covers a large area. The end-of-arm tool can be straight up-and-down or can be manipulated sideways. The SCARA control requirement is in the medium-capability area. A SCARA is used mainly in assembly applications, such as part insertion, drilling, tapping, and wire wrapping.

How can robots be made more versatile? One way is to put the robot base on a track. The robot work envelope can be elongated by the effective length of the track, for larger operational capabilities. The base can be moved along the track manually, for switching between two work stations. The robot can also be moved end to end by pneumatic cylinder action. Alternatively, the track movement can be in precise segments to various intermediate positions. This precise control is accomplished by servo control, at considerable extra cost.

The track can be mounted on the floor horizontally. Also, in some applications the robot may be mounted on a sloping track or on a 90° vertical track. In still other applications the robot may be mounted upside down on an overhead track. The overhead track presents an added and expensive dimension to holding tolerances. The robot can be mounted upside down with no track. An upside-down robot is called a Gantry robot, whether it is on a track or mounted in a stationary position. Robots can also be mounted in fixed positions with the base at an angle with the horizontal anywhere from 1° to 180°.

What about the robots in *Star Wars* and similar entertainment media? Such robots are not used industrially and make up less than 1% of the robots in use. It will be a number of years before these robots are developed to act as personal servants, maids, or dish washers. Therefore, their configurations and uses are not covered in this book.

18–3   ROBOT AUXILIARY CONTROL DEVICES AND SYSTEMS

Effectively utilizing a robot in an industrial application requires more than just the robot itself and the robot controller. In many applications, the cost of the robot is

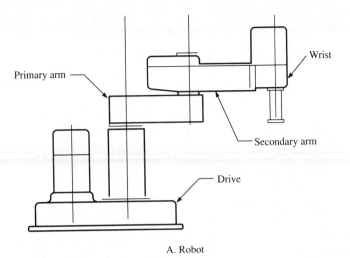

A. Robot

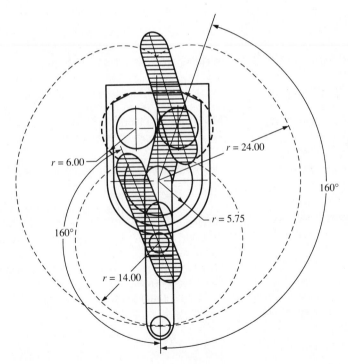

B. Work Envelope

FIGURE 18–4
SCARA

only a fraction of the total cost of the installation. The added costs include those for end effectors, part-holding fixtures, conveyors, sensors, and, in some cases, vision systems. The robot's associated safety systems, which add to cost also, were covered in Chapter 17.

In this section various end effectors will be described, and the basic principles of machine vision will be covered in Section 18–4. Sensors have been covered, and conveyors are straightforward ON/OFF devices. Therefore, neither will be discussed in this section.

A listing of the major applications of robots and the type or types of end effectors required for each is shown in Figure 18–5. The applications are listed generally in order of frequency in industry, starting with the most frequent.

Robots are widely used to spot-weld body assemblies in automotive plants. Some spot welding is the fusion type. An electrode is moved to the proper place by the robot arm, and then power is applied. Other spot-welding processes and all arc-welding systems use a wire-feed electrode. The wire is fed to the point where the electrode and the material being welded meet. The wire melts and forms the

Application	End Effector(s) Used
Spot welding	Welding electrodes, can be multiple
Arc welding	Welding torch in a breakaway holder
Material handling	Mechanical grippers, inside or outside
	Vacuum cup, often multiple
Machine loading	Mechanical grippers
Spray painting	Paint nozzle, can be multicolor
Manufacturing processes	Grinding wheel
	Deburring tools
	Polishers
	Metal-cutting torches
	Laser—cutting or heat-treating
	Water-jet cutter
Assembly	Screwdrivers
	Inserters
	Wire wrappers
	Soldering devices
	Sealant or glue application
Packaging and palletizing	Grippers
Inspection	Calibrated grippers
	Color discriminators
	Retroreflective laser (to check presence of spot welds)
	Alignment checking

FIGURE 18–5
Robot Applications and Some Typical End Effectors Used for Them

weld. The control of the spot or arc welder is coordinated precisely with the control of the robot arm movement for these applications. Figure 18–6 shows the end fixtures of the robot for welding applications. The welding torch is moved by robot movement. The torch is moved to appropriate positions and angles by the roll, pitch, and yaw of the end effector. In some arc-welding applications, the part being welded is also rotated by a movable holding fixture. This dual-motion system is used for complex welds in hard-to-reach areas.

When a robot is used for material handling or machine loading, appropriate grippers are used for holding parts or boxes. The grippers can be inside-diameter or outside-diameter, as shown in Figure 18–7. The grippers have two or more fingers to hold the part tight enough to keep it in place without damaging the part. Holding end effectors can also be in the form of vacuum grippers or vacuum tables, also shown in the figure.

The robot is also used effectively in spray painting. Money is saved by the elimination of operator time and by the saving of paint normally lost by hand spraying. The operator carries out an initial paint-spraying pattern with the robot in the Train mode. The robot controller remembers the pattern. The robot system then repeats the pattern each time the same part is painted. A painting robot is shown in Figure 18–8. For each paint color a different spray head is used. The spray heads are picked up by a spray-head-holder end effector. Parts that vary in configuration and color can be sent by conveyor to the robot spray system in any order. A code on the part, picked up by a sensor, designates the color and spray pattern stored in the control computer memory. The robot then chooses the correct color and follows the prescribed pattern.

There are many applications in manufacturing processes for robots. For material finishing, the end effector can take the form of a grinding wheel, a deburring tool, or a polishing wheel. The part being processed is held stationary in a fixture. The wheel or tool is then moved appropriately around the part's surface by the robot. In other manufacturing processes, the robot moves a cutting torch to cut shapes out of flat metal plates. Materials cut include metal, plastic, wood, and even multiple layers of cloth for suits. The cutting head can be a plasma arc, a torch, a laser, a water jet, or another cutting device.

Robots can be used for the various types of assembly operations listed in Figure 18–5. Appropriate end effectors are used singularly or in multiples for the operation being performed. In some cases the robot performs more than one operation. In those cases, a tool holder is attached to the end of the robot arm. For each operation, the robot moves to a tool rack. The tool holder deposits the tool it has been using in the proper holder spot. Then the tool holder picks up the next tool in the sequence. The next operation is then carried out. Two or more operations can thereby be performed by only one robot. A tool-changing system for end effectors is shown in Figure 18–9.

Robots can be used for packaging or palletizing. A palletizing layout is shown in Figure 18–10. As parts come down a conveyor, they are picked up and placed at appropriate sequential positions on the pallet. When the pallet is filled, it moves down a conveyor. Then a new pallet is moved into position and started.

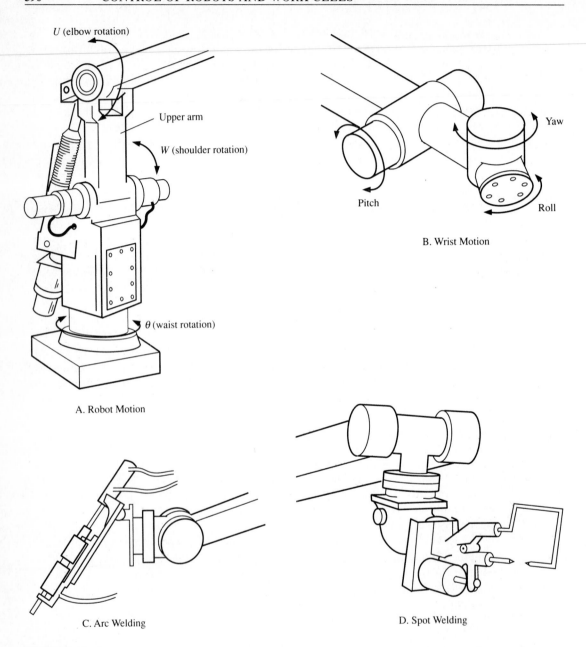

U (elbow rotation)

Upper arm

W (shoulder rotation)

θ (waist rotation)

A. Robot Motion

Yaw

Pitch

Roll

B. Wrist Motion

C. Arc Welding

D. Spot Welding

FIGURE 18–6
Motions and End Effectors of Welding Robot

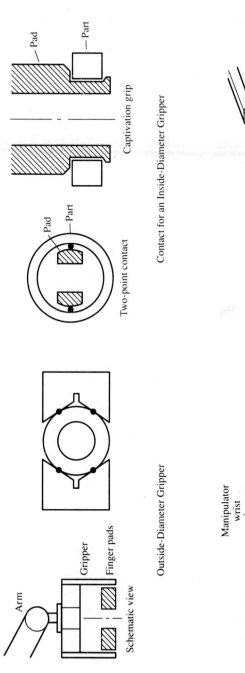

Arm

Gripper

Finger pads

Schematic view

Two-point contact

Outside-Diameter Gripper

Pad

Part

Pad

Part

Captivation grip

Contact for an Inside-Diameter Gripper

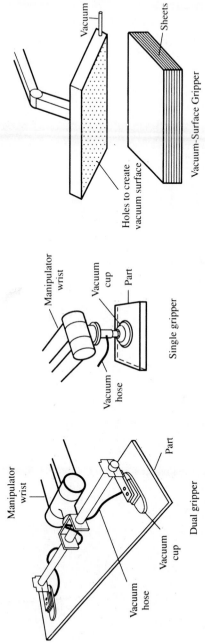

Manipulator wrist

Vacuum cup

Part

Vacuum hose

Single gripper

Manipulator wrist

Part

Vacuum cup

Vacuum hose

Dual gripper

Vacuum Grippers

Vacuum

Sheets

Holes to create vacuum surface

Vacuum-Surface Gripper

FIGURE 18–7
Material-Handling Grippers

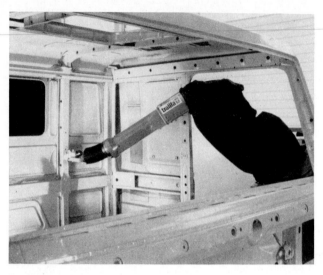

FIGURE 18–8
Painting (Coating) Robot (Courtesy of ABB Robotics, Inc.)

FIGURE 18–9
Multiple-Tool Robot System (Courtesy of ABB Robotics, Inc.)

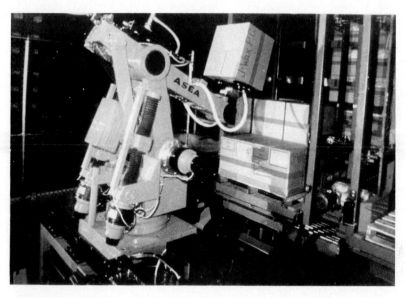

FIGURE 18–10
Robot Palletizing System (Courtesy of ABB Robotics, Inc.)

Advanced palletizing systems can mix and match parts coming down a conveyor. A dual system can work on two different pallets for two different parts. The system recognizes the parts and puts them sequentially on the correct pallet lines.

Robots can be used for inspection, with the proper fixturing, end effectors, and control programs. Figure 18–5 listed some of the inspection possibilities. The numerical results of the inspections can be fed into computers and analyzed by part number for characteristic-dimension conformance and variation.

18–4 MACHINE VISION

Machine vision systems are an increasingly used robot aid. Vision systems are very expensive to procure and program, so they are used only in special applications. Note that use of vision systems is not confined to robotic applications. The majority of vision systems in use are not on robots. Vision systems are used in such applications as sorting oranges and lemons for size, color, and quality. However, vision systems are very valuable when properly used with some robots and work cells.

A vision system consists of the parts illustrated in Figure 18–11. A camera or cameras look at the part or parts involved. The camera may be stationary, and parts may be brought into its range. In other cases, a camera control moves the camera appropriately. Visual information is sent by the camera to the vision computer. The computer sends display information to a visual CRT output. The

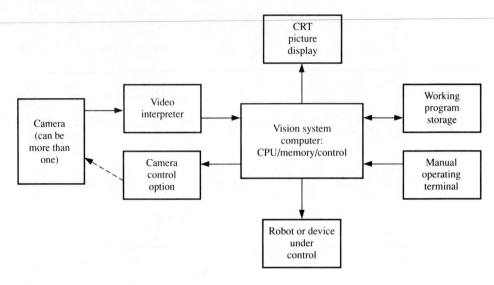

FIGURE 18–11
Fundamental Machine Vision System

computer develops appropriate control signals for the robot or machine being visually controlled. The machine or robot responds according to the program used. The vision computer is also connected to a manual control terminal and to a memory bank containing the various programs.

Some of the major machine vision tasks are as follows:

Part identification. The camera identifies the part seen by comparing it with part outlines programmed in the computer memory.

Part location and orientation. If a robot is to pick up a piece, it needs to know the part location and the angular movement required. The robot reacts to the signal by rotating its gripper and sending it to the proper location to pick up the part.

Part inspection. A vision system can check parts for conformance to dimensions and tolerances. It can also check for proper part outline. A major part inspection application is to check for the presence or absence of parts, holes, and other elements.

Alignment. A vision system can check for proper alignment between two or more parts or assemblies. In robotic welding it can check for the proper alignment of the weld bead with the edges of parts being welded.

The applications just listed are mostly two-dimensional. A few vision systems are set up to "see" three-dimensionally. Such vision can be accomplished by using two cameras and analyzing what they see stereoscopically. The complexity and cost of three-dimensional vision systems are much greater than those for two dimensions. Some advanced machine vision systems are also designed to discriminate colors.

How does the vision system "see"? Machine vision is accomplished in a manner similar to human vision. A lens focuses the incoming light onto a surface. Then photosensitive devices feed information about the light to computers.

There are two types of machine vision analysis. In the first, shown in Figure 18–12*A*, the camera develops a picture analysis in the same manner as a conventional television system. This system has some inherent problems. Precise time analysis of points in the sweep is needed to make a computer analysis of light and dark. This requirement creates a possibility of error. The more commonly used system is the one shown in Figure 18–12*B* and *C*. The camera "retina" is divided into sections. Part *B* shows the arrangement for linear vision, and part *C* the arrangement for two-dimensional vision. In simple vision systems each section is ON or OFF, depending on the light level falling on that section from the camera lens. In most vision systems, each square is sensitive to gradations of light—not just ON or OFF. The gradations, called the gray scale, can have 8, 16, or up to 256 steps. The more steps, the more complex the vision programming, and the more costly the system becomes. The vision system analyzes the light patterns and takes action according to the application involved. The individual squares are normally called *pixels*.

As an example, consider an analysis of a one-dimensional vision system. In Figure 18–12*A* the array is 12 pixels wide. In this example the array is assumed to be 8 pixels wide. The individual pixels of the vision system in this example can recognize nine light levels. The levels are 0 through 8, with 0 representing no light

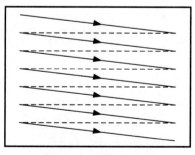

A. Conventional TV (Sweep)

B. Linear, One-Dimensional Charge-Coupled
Display (CCD)

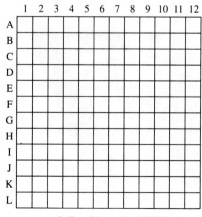

C. Two-Dimensional CCD

FIGURE 18–12
Vision Display Layouts

and 8 maximum light. A strip of unknown width is placed perpendicular to the array. The resulting levels from the strip's shadow as reported by the vision system computer are 8, 7, 8, 5, 1, 0, 4, 8. Note that the readings are not absolutely accurate, because of system noise and errors. An analysis is to be made of the vision system levels to determine how wide the strip is.

Assume that each pixel represents 2.5 cm of object width. On the left the shadow is almost certainly in the fourth pixel, and on the right the shadow is in the seventh pixel. Using interpolation, the left side of the strip is at $3 + 3/9$ and on the right $6 + 5/9$. The shadow spans the difference, $6.55 - 3.33$, or 3.22 pixels. At 2.5 cm/pixel, 3.22 pixels represents 8.05 cm.

In Figure 18–13, the same principles are applied to a figure seven (7) in a 16-step, 12 × 12 array. The numbers (1 through 16) of the light levels are written in the array. Note that the light levels are, again, not absolute.

The actual pixels are created by a charge-coupled display (CCD). The CCD principle is illustrated in Figure 18–14. The individual ''squares'' are coated spots on a semiconductor base. The more light on a CCD, the more it is charged. The charge signals are not connected individually as outputs. The level signals are removed by being shifted electronically to a buffer row by row. When all rows have been scanned, the total information in the buffer is sent to the vision computer. The number of pictures per second depends on the speed of the electronic system used.

Tactile systems determine part configuration by feel with an array of sensors. Sensors are ON or OFF depending on whether something is touching them. Multistep, variable-pressure systems can be used. The harder the pressure, the

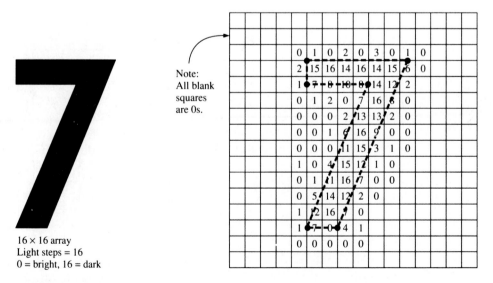

16 × 16 array
Light steps = 16
0 = bright, 16 = dark

Note:
All blank
squares
are 0s.

FIGURE 18–13
Pixel Array for the Number Seven

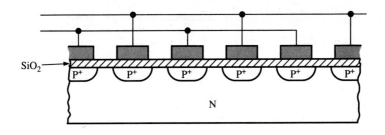

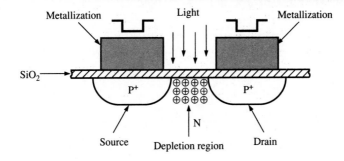

FIGURE 18–14
CCD Principle

higher the recognition number. The principles involved in tactile part recognition are the same as in the vision pixel systems.

18–5   CONTROLLING AND PROGRAMMING A PICK-AND-PLACE ROBOT

In this section the principles of programming a relatively simple *pick-and-place robot* will be explained. Programming of the more complex, *continuous-path robots* will not be covered. The continuous-path robot can be programmed to move in complex motion patterns. Continuous-path robots move on one to eight axes in a simultaneous, coordinated manner. Their programming is quite complex. The pick-and-place robot, on the other hand, moves on each member axis in fixed (usually straight) paths to fixed positions, one or two at a time.

A pick-and-place robot may be operated automatically by a drum control with pegs, by a microprocessor-based program, or by a PLC. Figure 18–15 is a diagram of a basic cylindrical robot that can move parts from one position to another.

This robot starts its operation from the initial position, which is the lower-left, at-rest position. Suppose the robot is to be used to move a part from position *A* to position *B*. The operational sequence to do so is as follows:

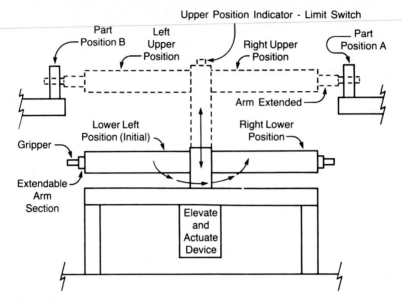

FIGURE 18–15
Basic Pick-and-Place Robot

1. Initial position is as shown at left with arm not extended and gripper open.
2. Arm moves to upper position.
3. Arm rotates right.
4. Arm (hand) extends to position *A*.
5. Gripper closes, gripping part.
6. Arm swings to left to position *B*.
7. Gripper opens, releasing part.
8. Arm (hand) retracts.
9. Arm lowers to initial position.

Assume that the robot's moving parts are controlled by the action of four pneumatic cylinders, which are electrically operated. Energizing each of the four solenoids causes the robot to take one of the following actions:

1. Rotate (Arm rotates full right.)
2. Raise (Arm moves to upper position.)
3. Extend (Arm extends hand.)
4. Grip (The gripper closes.)

For this robot, if the solenoid is *not* energized, the robot member is in the *opposite* position. Note that more than one solenoid may be actuated at a time to achieve a given position. An operation matrix for solenoid ON/OFF is illustrated in Figure 18–16. This ON/OFF pattern will accomplish the task, previously described, of moving a part from position *A* to position *B*.

Step	Up	Rotated Right	Hand Out	Grip Closed
Initialized	O	O	O	O
1	X	O	O	O
2	X	X	O	O
3	X	X	X	O
4	X	X	X	X
5	X	O	X	X
6	X	O	X	O
7	X	O	O	O
8	O	O	O	O

An O indicates the opposite position: down, left, in, or open.

FIGURE 18–16
Robot Operation Matrix

How is the robot motion controlled through the sequence of moving the part from *A* to *B*? Four switches could be used, one for each function, as shown in Figure 18–17*A*. Someone could manually turn switches ON (X) and OFF (O) per the Figure 18–16 operational matrix. Note that an X in the matrix means that the robot is in the listed position. The operator would have to pay attention to timing, making sure each step was complete before starting the next. Utilizing this manual operation would defeat the purpose of installing a robot. Therefore, a more automatic control system is needed. Figure 18–17*C* and *D* shows the program for control of this process by a typical PLC. The program goes from one step to the next each time a timing device energizes the sequencer. The PLC sequencer (DR) bit patterns in Figure 18–17*D* are identical in layout to the pattern of Figure 18–16. A mechanical control drum with the peg pattern of Figure 18–17*D* could also be used. Any electrical or electronic device with the same stepping pattern could step the robot through the proper sequence.

The program shown in Figure 18–17 is a simplified program. The simplified program does not take into account the time for an operation to be completed. For example, you could not successfully turn step 3 ON until step 4 is complete. The basic program also does not take into account the possibility of jamming, cylinder failure, or a slipping part in the gripper. Sensors would have to be added for more positive control. One example of a sensor used for positive control is shown in Figure 18–17*B*. A limit switch is added in the third line. The sensor ensures that the arm is indeed in the upper position before the process continues. Other similar programming additions can be included for safer and more positive operation. These can include including emergency stop, other interlocks, and additional position sensors.

A typical, small, more advanced industrial pick-and-place robot is shown in Figure 18–18. This robot has the same four functions as the previous robot. It also has two other functional capabilities: gripper rotation and left-right movement with four intermediate stops. The total robot capabilities are as follows:

1. arm movement up or down (Elevate)
2. arm rotation 180°
3. gripper hand extension and retraction
4. gripper opening and closing
5. gripper rotation 180°
6. slide to left or right with five stations—two ends and three intermediate

Input Switches

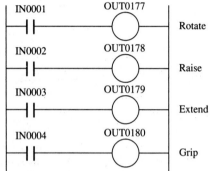

A. **Simple Coil-Contact Robot Control Program**

Input Switches

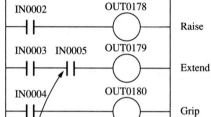

└ Upper Position Limit Switch

B. **Coil-Contact Robot Control Program with Interlock**

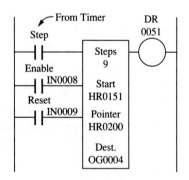

C. **DR/SQ Robot Control Function**

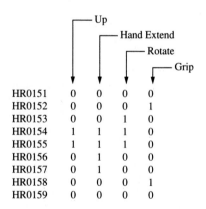

Register Pattern - 4 Bits of 8 or 16

D. **DR/SQ Robot Control Register Patterns**

FIGURE 18–17
Robot Sequence Control

FIGURE 18–18
Industrial Pick-and-Place Robot (Courtesy of TII Robotics)

This robot is operated by pneumatic air cylinders, electrically actuated. In some cases the robot does not return to the original position when its pneumatic cylinder is deenergized. Another, complementary cylinder must be actuated for the corresponding opposite motion. The control of this robot may be by manual operation, mechanical drum, EPROM-programmed IC, or PLC.

A wiring scheme must be used so that the robot's controller terminals match the functional inputs and outputs of the robot. An input and output scheme for this robot is shown in Figure 18–19. Assume that a PLC is used for control. The first column is for the numbers chosen for the PLC's 16 inputs and 16 outputs. The chosen interconnect cable wire letter number is given in the next column. The robot pin number and the corresponding robot function are shown in the two right-hand columns.

For an example of input control, suppose the gripper is to be rotated. Output 24 of the PLC is energized by suitable PLC programming and control. The signal voltage from output 24 goes through cable wire A to robot pin 9. Robot pin 9 voltage is fed to the robot gripper-rotate-control pneumatic solenoid.

Next, look at a robot function station-status indication. Suppose the robot is at station 4. There is thus continuity at robot pin 5. Therefore, there is continuity

Below is a listing of the I/O numbers and the letter that corresponds to the I/O on the cable between the I/O rack and control panel designed for the PLC. Also listed are the robot pin # and the corresponding function.

PLC Input #	Cable Letter	Robot Pin #	Robot Function
1	A	—	—
2	B	—	—
3	C	—	—
4	D	—	—
5	E	—	—
6	F	—	—
7	G	—	—
8	H	22	Aux. Input
9	J	23	Aux. Input
10	K	24	Aux. Input
11	L	25	Aux. Input
12	M	2	Station 1
13	N	3	Station 2
14	P	4	Station 3
15	Q	5	Station 4
16	R	6	Station 5
	S	7	Common Input

PLC Output #	Cable Letter	Robot Pin #	Robot Function
17	T	8	Grip
18	U	10	Elevate
19	V	15	Extend
20	W	11	Rotate CW
21	X	12	Rotate CCW
22	Y	13	Slide Right
23	Z	14	Slide Left
24	a	9	Rotate Grip
25	b	19	Aux. Output
26	c	20	Aux. Output
27	d	21	Aux. Output
28	e	—	—
29	f	—	—
30	g	—	—
31	h	—	—
32	j	—	—
	k	16	Common Output

FIGURE 18–19
Robot-Control Cable-Connection Scheme

from cable wire Q to common. This is recognized as a closed status at PLC input pin 15. This information is used by the PLC logic.

Now that the robot is connected to the controller, an appropriate computer program can be created. It is initially created on a work sheet, as shown in Figure 18–20. For each step, an X is placed in each box representing a robot function to be energized. The functions are identified by the robot positions required. For example, in step 1 the robot is to slide right and rotate the arm. Therefore, Xs are put in the squares under the appropriate functions: rotate CCW and slide right. In this case two operations are to be combined. Sometimes operations can be com-

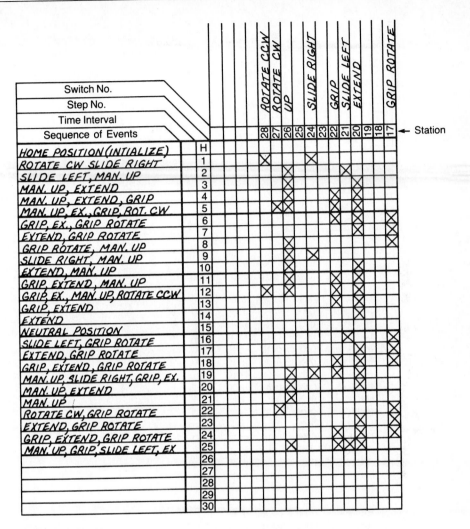

FIGURE 18–20
Robot Program Code Sheet (Courtesy of TII Robotics)

bined, and sometimes they cannot, depending on the particular robot. Note that in step 3 the robot can go up and also extend at the same time. However, it must do so before performing step 4. Combining steps 3 and 4 could cause the arm to run into a holding fixture. Steps 3 and 4 must therefore be separate steps.

As with the previous robot, the control matrix is placed in a control program. The corresponding program for the matrix in Figure 18–20 is illustrated in Figure 18–21. A PLC sequencer-system program (see Chapter 15) is also shown in the figure. The same matrix could be used with any other control scheme chosen to perform the sequential operation of Figure 18–20.

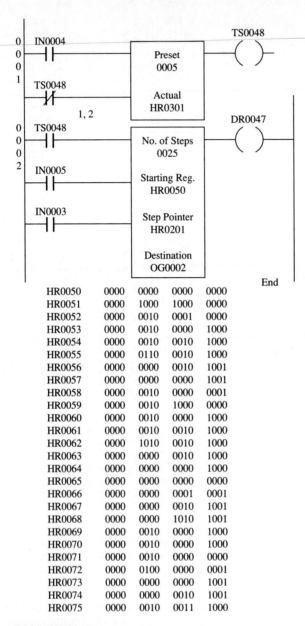

HR0050	0000	0000	0000	0000
HR0051	0000	1000	1000	0000
HR0052	0000	0010	0001	0000
HR0053	0000	0010	0000	1000
HR0054	0000	0010	0010	1000
HR0055	0000	0110	0010	1000
HR0056	0000	0000	0010	1001
HR0057	0000	0000	0000	1001
HR0058	0000	0010	0000	0001
HR0059	0000	0010	1000	0000
HR0060	0000	0010	0000	1000
HR0061	0000	0010	0010	1000
HR0062	0000	1010	0010	1000
HR0063	0000	0000	0010	1000
HR0064	0000	0000	0000	1000
HR0065	0000	0000	0000	0000
HR0066	0000	0000	0001	0001
HR0067	0000	0000	0010	1001
HR0068	0000	0000	1010	1001
HR0069	0000	0010	0000	1000
HR0070	0000	0010	0000	1000
HR0071	0000	0010	0000	0000
HR0072	0000	0100	0000	0001
HR0073	0000	0000	0000	1001
HR0074	0000	0000	0010	1001
HR0075	0000	0010	0011	1000

FIGURE 18–21
PLC Robot-Sequence Control Scheme

18–6 WORK CELLS

A robot is often utilized in an industrial process that also involves two or more other machines. The resulting system is called a work cell. A work cell may include more than one robot or more than two other machines. In addition, most work cells also include one or more conveyors. The system is called a *work station* if only one part is continuously manufactured by the combination of machines and robots. The combination is called a *work cell* if various similar parts can be produced by the work station. A major requirement of a work cell is that it be quickly and easily programmed to make another part. Complex work cells can mix and match parts easily. In such work cells more than one part may be in process at one time.

The implementation of work cells in a manufacturing flow is best accomplished by the "bottom up" approach, according to Square D and other automation specialists. Start small, with only two or three machines and one robot. Group them into a cell. Once the cell is functioning properly, set up another small group in a cell format. Once the groups all work properly, you may combine small groups into larger groups. Eventually you may completely automate a production line. Many work cell projects fail because the designers start with the whole production line at once, not with small groups of machines and robots.

Some of the benefits of work cells are as follows:

1. Work cells coordinate various activities, such as processing, robots, vision, and measurement.
2. Work cell control, often by a PLC, is adaptable to new programs for different tasks with rapid reprogramming.
3. Consistency of product and productivity can be improved.
4. Machine setup time is reduced, speeding up overall production.
5. Production control (PC) and quality control (QC) data can be obtained directly and quickly by tapping into the operating program. Commercial software packages are available for PC and QC data collection and analysis.

Figure 18–22 illustrates a typical work cell layout. Parts (rods) come to the cell on conveyor A or B and go out on either A or B. Incoming parts have two possible diameters. Each rod that comes to the cell has a hole drilled in its middle. The hole is either $\frac{1}{2}$ in. or $\frac{3}{4}$ in. in diameter, depending on the part being produced. The rod's holes, after drilling, may or may not have a countersink (counterbore) of 1 in. diameter for a short depth, as shown. Another optional operation is the chamfer, as shown. The robot is programmed to move the part being processed from one station to another. If there is enough transfer work to be done simultaneously, a second optional robot (not included here) might be practical.

A program layout scheme for the Figure 18–22 work cell is shown in Figure 18–23. There are 32 possible combinations of operations for the rod. These programs are stored in memory. The required program is called up to operate the four

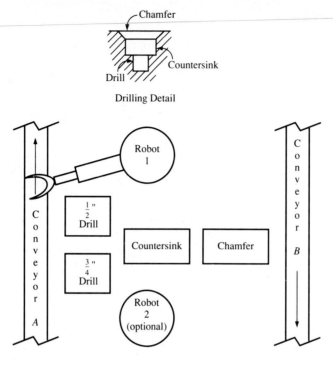

FIGURE 18–22
Drilling Work Cell Layout

machines and the robot to make the proper part. The program is chosen either by a code on the part being processed or by a master production-control program.

Why are there 32 possible program combinations? Figure 18–24 shows the possible combinations. Parts can come in or go out on either conveyor. The rod may be drilled for either a $\frac{1}{2}$ in. or a $\frac{3}{4}$ in. diameter hole. These choices give 8 possibilities. Countersink (counterbore) and chamfer give another two possible combinations each. Note that the total number of possibilities is not the sum of the possibilities for all the operations. It is the product, as shown in the figure.

A training and industrial work cell is shown in Figure 18–25. The cell contains two CNC-controlled machines, a lathe and a milling machine. The abbreviation *CNC* stands for *computer numerical control*. It applies to machine tools of all types, such as lathes and milling machines. The CNC designation indicates that a machine is controlled by computer, not manually or electrically. The parts of this work cell are as follows:

1. a computer-controlled conveyor
2. a CNC lathe, loaded and unloaded by a robot with PLC control
3. a CNC milling machine, loaded and unloaded by another robot with PLC control

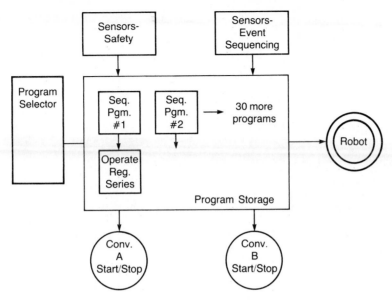

FIGURE 18–23
Work-Cell Programming Scheme

4. a SCARA for drilling or small assembly operations on the conveyor
5. a master PLC to coordinate the conveyor and the other two PLCs
6. a computer-controlled storage rack
7. a computer-program to keep track of the location of each part in the storage rack
8. a master computer to oversee the whole operation
9. safety devices and interlocks as required

An industrial work cell is shown in Figure 18–26. This large work cell is part of a manufacturing operation for the fabrication of large computer cabinets. The

In on Conveyor	Drill Size	Counter-Sink	Counter-Drill	Out on Conveyor	
A	¼"	Yes	Yes	A	
or	or	or	or	or	
B	½"	No	No	B	
	× 2 =	× 2 =	× 2 =	× 2 =	} Combinational
2	4	8	16	32	} Possibilities

FIGURE 18–24
Work Cell Programs—Possible Combinations

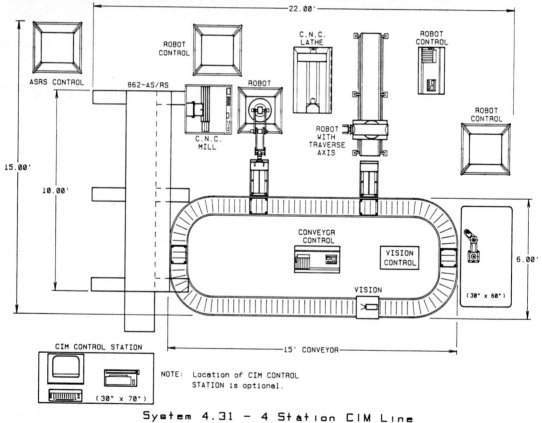

FIGURE 18–25

Industrial and Training Work Cell (Courtesy of Amatrol, Inc.)

FIGURE 18–26
Industrial Work Cell (Courtesy of ABB Robotics, Inc.)

equipment shown is a multistation robotic-fabrication work cell. In this work cell, two robots are utilized for final weld assembly of the computer cabinets. The robots have individual controllers. Their individual controllers are under the over-all control of the cell master computer. Other cell equipment, such as the convey-ors, is also under the control of the cell's master computer. The robot on the left performs spot welding. The other robot, on the right, subsequently performs arc (bead) welding on the cabinets. The cell is reprogrammable for different models of computer cabinets as they come into the cell area.

18–7   CONTROLLING AND PROGRAMMING A WORK CELL

To illustrate basic work-cell control and programming, the basic work cell of Figure 18–27 will be used. It consists of one pick-and-place robot moving bar-stock rods from an incoming conveyor. For this program, assume that parts come in only on conveyor A. The rods are moved, one at a time, from conveyor A to the lathe. At the lathe the bar stock is turned to the specified finish diameter. Next the robot moves the part to the mill. At the mill, one end of the turned rod has a flat

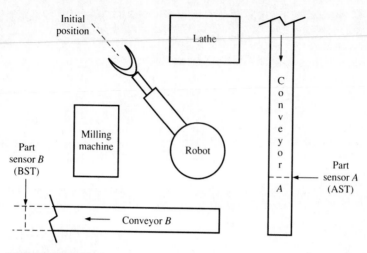

FIGURE 18–27
Sample Work Cell

surface milled on it. When the milling is completed, the robot places the finished shaft on conveyor B, which then moves the part to the next manufacturing station.

A specific program for the work cell's operation will not be written here. Instead, flowcharts will be created, using the standard flowchart symbols illustrated in Figure 18–28. From a flowchart a specific control program can be written that uses the specific programming formats of the robot, lathe, and milling machine used.

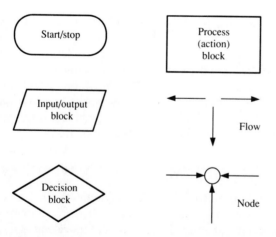

FIGURE 18–28
Standard Programming Flowchart Symbols

First, for illustration, a flowchart will be made for a straight-through part flow through the two machining operations. The program for this sequence of operations is presented in flowchart form in Figure 18–29. Conveyor *A* runs until a rod reaches the part sensor. Conveyor *A* stops, and the robot rotates and picks up the rod. The rod is then placed in the lathe chuck. The robot withdraws, and the rod is turned to diameter by the lathe. When turning is completed, the robot picks the part out of the lathe and places it in the milling machine fixturing. The robot withdraws, and a flat surface is milled on one end of the turned rod by the milling machine. The rod is now a finished turned-and-milled shaft. Finally, the robot moves the finished shaft from the milling machine to conveyor *B*. Conveyor *B* starts and moves the shaft to the next operation.

This straight-through sequence is obviously not the most efficient use of the work cell. The lathe and the milling machine should somehow operate simultaneously for faster processing of parts.

Figure 18–30 shows one possible scheme for accomplishing simultaneous operation. Once the mill is loaded, a parallel programming branch allows the robot to load the lathe from conveyor *A*. Later, once the mill is unloaded, a part can be moved from the lathe to the mill. The next programming operation after the mill is loaded from the lathe would depend on the relative elapsed times of turning and milling. With this more complex programming scheme, both the lathe and mill can run parts at the same time.

Parallel operation is just one basic program modification needed to make this a completely flexible work cell. Some other possible product- and processing-flexibility variations that could be included are as follows:

☐ different part diameters, lengths, and mill depths and lengths
☐ different turning times and milling times
☐ parts with no milled flat (a skipped operation)
☐ parts with milled flats on both ends (same or different dimensions)

The programming for these added variations can become very involved.

18–8   NETWORK COMMUNICATIONS

Computer-Integrated Manufacturing

A relatively recent concept in the field of industrial control is *computer-integrated manufacturing (CIM)*. There are no explicit definitions of CIM, such as one might find with a communications protocol. Rather, CIM is a philosophy for integrating hardware and software in such a way as to achieve total automation. Although each company has its own idea of what CIM really means, most follow a pattern similar to that in Figure 18–31. In this diagram dedicated processing tasks are shown distributed around a factory. As computers are further removed from the actual manufacturing area, their function shifts from real-time control toward supervision.

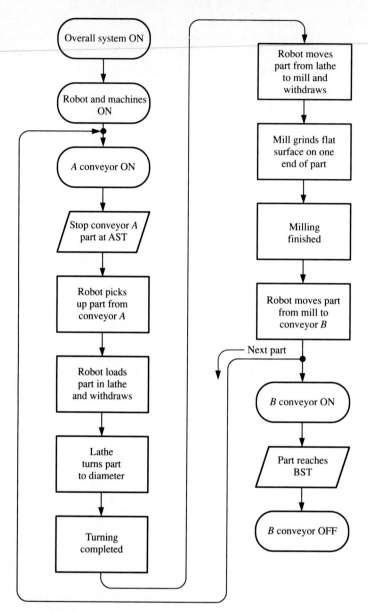

FIGURE 18–29
Straight-Through Work-Cell Operation

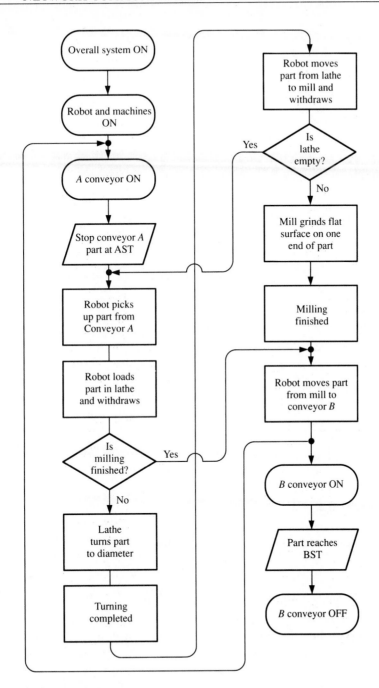

FIGURE 18–30
Work Cell Parallel-Operation Program

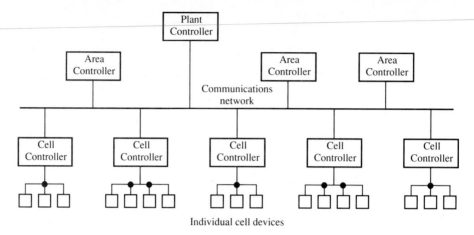

FIGURE 18–31
Distribution of Processing Tasks in a Plant According to Function

It is generally agreed that at least three levels of computer integration are required for CIM to work: the *cell level,* the *area level,* and the *plant level.* Each level has certain tasks within its range of responsibility. Cell controllers, for example, are generally responsible for data acquisition and direct machine control. Area controllers are assigned the tasks of machine and tool management, maintenance tracking, material handling and tracking, and computer-assisted simulation and design facilities. The plant-level computers are responsible for such things as purchasing, accounting, materials management, resource planning, and report generation.

Plant Controllers

Technically speaking, any computer is capable of performing both control and supervisory tasks. The trend, however, is toward customizing computers for specific applications. It can be seen from the hierarchical structure of CIM that specialized computers would be advantageous. In fact, many companies are already producing computers to satisfy the specific requirements of every CIM level. Invariably, mainframe computers are implemented at the plant level in large facilities. In some cases, those computers were already in place to handle such tasks as inventory and payroll. To promote them to the top of the CIM pyramid, one must simply select the proper software. These computers are usually in a location remote from the actual plant floor. All the information they require is available from the host of subordinate area controllers throughout the plant.

Area Controllers

Area controllers are usually on the plant floor and are therefore subjected to a harsh environment. A special breed of computers, called *industrial computers,* has been developed specifically for this application. The Allen-Bradley 6120 and 6121, manufactured by IBM, are examples of such machines. Like most industrial computers, these are very similar to medium-sized personal computers (PCs). In fact, the 6120 is designed to be compatible with the IBM PC/AT (and all of its "clones"). The main difference between a PC and an industrial computer lies in the physical construction. Very simply, industrial computers are built to withstand higher temperatures, greater vibration, electromagnetically noisy environments, and rough handling. Since their task is still supervisory in nature, their operating hardware does not differ significantly from that of a PC. Again, it is the software that turns a generic computer into an industrial computer. The PC should not be ruled out as a strong contender for the position of area controller. Personal computers are inexpensive, and with the myriad clones available, they are plentiful. An added advantage of using PCs for area control is that there is a large software base already established for such computers.

Cell Controllers

The most specialized computers are those used to control cells. As already indicated in this chapter, a cell is defined as a group of machine tools or equipment integrated to perform a unit of the manufacturing process. Therefore, the computer that coordinates action within the cell has special hardware requirements as well as special software requirements. Specifically, such a computer must have multiple data paths (I/O ports) through which it may communicate with the various cell devices.

Traditionally, cell control has been accomplished with PLCs. Programmable logic controllers were in fact designed for this very purpose. A CIM structure, however, requires that the cell coordinators communicate with other cell controllers, as well as with the area controller. The language of the PLC is somewhat restricting in this respect. In addition, most PLC languages do not lend themselves well to analysis and record-keeping tasks.

With these problems in mind, most PLC manufacturers are developing PLCs with greater software capability. A recent outgrowth of this effort is a computer called a *cell controller.* Cell controllers combine the software sophistication of a PC with the I/O handling capability of a PLC. Systems such as the Allen-Bradley Vista 2000 or the Eclipse MV/4000, produced by Data General, are marketed specifically as cell-control computers.

Again, the PC should not be overlooked. Many third-party manufacturers produce I/O cards designed to plug directly into the expansion slots of most PCs. These cards perform data-acquisition functions as well as data conversion and power control. A PC equipped with special software packages and I/O cards can emulate many of the popular PLCs in use today, and it can also perform the computational and communication tasks required of a cell controller.

Microcontrollers

Computers are even being implemented at the device level. That is, single-chip *microcontrollers* are being embedded in "intelligent" machines. Basically, a microcontroller is the marriage of a CPU to program and data memory and various I/O components, all on a single piece of silicon. The Intel 8797, for example, contains a 16-bit CPU, along with 8192 bytes of EPROM (erasable-programmable read-only memory), 232 bytes of random-access memory (RAM), a 10-bit analog-to-digital converter, a full-duplex serial I/O port, a pulse-width modulated output, and four 8-bit I/O ports, all in a single 68-pin package. Devices of this type are, in the truest sense, complete computer systems, having all the power of their larger counterparts (PCs). Their capability is, of course, limited by the number of elements integrated onto the chip.

The concept of a microcontroller is nothing new. In fact, microcontrollers have been used in such devices as keyboards and printers for years. Their application in industry, however, is quite new. They may be found in applications that range from intelligent temperature transmitters to multi-axis robotic end effectors. With high-speed 32-bit microcontrollers now being developed, the future should prove very interesting.

OSI Network Architecture

The key to successful implementation of CIM is communication compatibility between all the computers involved in the process. This, in fact, poses quite a problem. Although communication standards do exist, each computer system may use a different standard. In addition, as the level of sophistication of communication increases, the need for more sophisticated standards arises. In anticipation of this problem, the International Standards Organization (ISO) has developed a model for what is called the Open Systems Interconnection (OSI). The OSI is a seven-layer model for communication network architecture (see Figure 18–32). Each layer represents a different level of communication sophistication and provides the necessary support for the layers above it. Implementation of such an architecture would allow the user to connect virtually any data communication device to the network and be assured of compatibility.

Buried within each OSI layer are communication protocols. A protocol is nothing more than an agreed-upon set of rules by which communication will take place. Protocols exist at all communication levels and are assigned to the appropriate OSI layer according to level of sophistication.

To illustrate the need for protocols at different levels, consider the simple action of making a telephone call. There must be some convention for the way the wires are connected, the allowable bandwidth of the channel, and the signal levels used. Next, the sequence of signal exchange must be established. That is, the

FIGURE 18–32

Open System Interconnection (OSI) Model for
Network Architecture

Layer	
7	Application
6	Presentation
5	Session
4	Transport
3	Network
2	Data-link
1	Physical

sequence of dial-tone transmission, dial pulse or touch-tone detection, and ringing or busy signal transmission must be clearly defined. Finally, when connection has been made, the rules that govern human dialogue take over. There are three levels of protocols in this example: physical, transmission, and user. Note that the physical protocol provides the necessary support for the transmission protocol, which in turn supports the user protocol. Additionally, if the physical protocol is changed, the transmission and user protocols should be unaffected. That is, changes within layers should be transparent to the other layers. This principle is the essence of OSI.

To date, only layers 1, 2, and 3 of the OSI model have been clearly defined and successfully implemented. Layers 4 through 7 will, in the future, contain the more complex application protocols that are sure to be developed. Layer 1 contains physical-link protocols—those that define such things as signal level and connection conventions. Protocols RS-232C and RS-422 are examples that would fit into this layer. Layer 2, the data-link layer, contains protocols that address such problems as circuit establishment, transmission sequence, and error control. Protocols such as SDLC, HDLC, and BISYNC belong in this layer. The network layer (layer 3) defines procedures for data routing, packet switching, and error recovery. The X.25 protocol is the most pervasive in this layer.

Several protocols can exist in each OSI layer. This ability, in fact, is what allows the network to support such a variety of different computers. A particular industrial computer may require X.25, SDLC, and RS-232C to accomplish its communication link; another may require X.25, BISYNC, and RS-422. Although the communication paths through the OSI layers would differ, both computers would use three layers, and—more important—both could use the network.

Proprietary Network Architectures

The OSI network-architecture model is by no means the only attempt at network standardization. Several proprietary network architectures are currently being used. International Business Machines (IBM) promotes its Systems Network Architecture (SNA), Digital Equipment Corporation (DEC) markets a system called Digital Network Architecture (DNA), and Burroughs Corporation has introduced Burroughs Network Architecture (BNA). These network architectures resemble the OSI model, differing only in the number of layers involved and, of course, the nomenclature. One disadvantage of these proprietary architectures is that they tend to lock the user into products from a single vendor. For example, SNA is designed around IBM mainframes, and DNA is designed to support DEC mini-computers. The OSI architecture, on the other hand, is designed to accommodate products from any vendor.

The Physical Network

Generally, networks can be classified as *wide-area networks* (*WANs*) or *local-area networks* (*LANs*). The difference lies in the transmission distance and, therefore, in the proximity of the stations to each other. Virtually all plants and factories fall into the LAN category, although different factories within the same company may communicate over a WAN.

A LAN may be characterized by its method of station access. Two access methods are in common use today: *random access* and *token access*. Random access is a method by which all stations "listen" to the transmission line and wait until the line is free before attempting a transmission. If a collision occurs, all transmitting stations shut down and wait for the next opportunity to transmit. This method is called *carrier-sense multiple access with collision detection* (*CSMA/CD*). Protocols such as ETHERNET (introduced by XEROX) use this type of access at the transport level of the OSI model.

Token access is a method by which possession of a token (a special code sent along the transmission line) allows a station to transmit. The token is passed from station to station until it reaches one with a message to send. That station then removes the token and replaces it with a frame of data to be transmitted. After successful transmission, the station places the token back on the line, and the token proceeds to the next station. This method allows each station an equal chance to send data. General Motors Corporation's Manufacturing Automation Protocol (MAP) is a transport protocol that uses token access.

SUMMARY

☐ Robots are automatic manipulative devices that are quickly and easily reprogrammable.
☐ There are five major types of robots. The robot type chosen for a particular task's motion requirement is based on the work envelope, the accuracy required, and the size and complexity of the part being produced.
☐ Robot auxiliary devices include the robot end effector, fixtures to hold the part being processed, and conveyors to transport the parts to and from the work area.
☐ End effectors are of many different configurations. These include grippers, tool holders, tools, welding torches and spot welders, paint guns, and other special types.

□ A major auxiliary system for a robot is the machine vision system, which is expensive but can be very effective when applied properly.

□ The machine vision pixel analysis system is the type most often used.

□ Pick-and-place robots can be controlled from drum control or PLC sequencer control. The control system is an ON/OFF system.

□ Pick-and-place robot programs are created on a work sheet and then transferred to a controller program matrix.

□ Continuous-path robot programming is quite complicated. However, many continuous-path robots are easily programmed in the Train mode.

□ A work cell is a group of machines that includes at least one robot.

□ Work cells have a master computer to control the individual unit programs in the proper sequence.

□ An effective work cell can be quickly and easily reprogrammed to produce another part. The work cell program should also be flexible enough to enable more than one part to be processed at once. The parts in process may be of different configurations in more advanced work cells.

□ Computer-integrated manufacturing (CIM) is an industry-wide philosophy for implementing computers in industrial processes. Although there is no concrete model for this concept, it is generally agreed that computers should have dedicated functions.

Microcontrollers control individual devices within work cells.

Cell controllers coordinate overall operation of work cells.

Area controllers manage data from many cell controllers.

Plant controllers coordinate the exchange of data among all the subordinate computers.

□ For CIM to be successful, communication among all the computers in a facility must be structured. The International Standards Organization (ISO) has established a seven-layer Open Systems Interconnection (OSI) model for computer network protocol.

EXERCISES

1. List the five basic types of robots. Describe the work envelope of each.

2. Describe the size range of each of the five types of robots.

3. Describe the program control usually found with each type of robot.

4. What types of power systems are used to actuate the robot—electrical, electronic, pneumatic, or hydraulic? Where, in general terms, is each type of actuator power used?

5. List two possible applications for each type of robot.

6. List two applications in which a robot would not be appropriate.

7. What is the difference in programming between "bang-bang" operation and continuous-path operation of a robot?

8. Describe the axes of movement for a typical robot and its end effector. How many axes are there, typically? Can there be more axes?

9. Sketch the gripper to use for each of the following objects:
   a. a triangular bar
   b. a cylinder
   c. a metal ring
   d. a sheet of plywood

10. Figure 18–5 lists end effectors and their applications. List a few other possible applications for some of the end effectors.

11. What are the advantages and disadvantages of installing and using a painting robot?

12. Describe the advantage of using the multiple-tool approach with a robot.

13. Describe the general programming needed for a palletizing robot.

14. Is a pixel array absolutely accurate? Explain.

15. Refer to Figure 18–13. Make up a typical pixel number chart for the number 18.

16. Refer to Figure 18–15. Make up an operational matrix like Figure 18–16 to move a part from the lower right (directly below position *A*) to position *B*.

17. For the matrix in the preceding exercise, what would the PLC registers in Figure 18–17*D* have as a pattern?

18. Refer to the work cell shown in Figure 18–22. Could the second robot speed up the production rate? If so, describe its use in coordination with the other robot.

19. Refer to Figures 18–27 and 18–29. Assume that the manufacturing cycle is: (1) turn in lathe, (2) mill a flat, and (3) return to lathe for finish turning. Assume that all three operations have approximately the same time interval. Draw a program flowchart for this process.

    Extra credit: Repeat the above assuming that the milling time is three times the turn and the finish times. Turn and finish times are about equal. Include a part-staging station. Develop a different operational flowchart to speed up the overall operation, as compared to the previous flowchart.

20. Refer to Figure 18–27. Assume that the parts are milled and then turned. Draw a flowchart for a parallel-operation program, like that in Figure 18–30, for this process.

21. As a project, obtain the layout of a work cell from an industry. Describe its makeup and operation. Draw a program flowchart for its operation.

22. What differentiates a cell controller from an area controller?

23. What is one advantage of distributing processing tasks around the plant?

24. What does the acronym *CIM* stand for? What does it mean?

25. Why have computer network communications become so important in industry?

26. What is a network protocol?

APPLICATION ASSIGNMENT ANSWER

You must first choose the types of robots. The first one should be a pick-and-place, pneumatic robot with appropriate grippers. It could be programmed very easily with a PLC. The second would have to be a continuous-path robot. It could be relatively small and be operated by electrical servomotors. Its program would have to be fairly complex. Hydraulic drive would not be required for such a small robot.

Other considerations are the weight of the parts handled and the speed of operation required. Heavier weights or faster speed requires a larger and more expensive robot. A consideration for the welding robot is the precision required. More precision requires more programming capability, also at greater cost. In either case, do not buy more capability than you need.

Be sure to consider possible future products. What additional part or parts will be run on these two robots in the future? A greater variety of parts results in a need for more computer control capability. If the new parts are going to be bigger or heavier, buy a larger robot now. Another consideration is whether you might want to add vision or another peripheral auxiliary system in the future.

Glossary

AC   See *alternating current*.

across-the-line starter   A switch or contactor that connects a motor directly to the incoming power line for both starting and running. Sometimes called the full-voltage starter.

address   A specific location in a computer, usually in the memory. Represented by a label, number, or name.

AIEEE   American Institute of Electrical and Electronic Engineers. A trade organization that sets standards and disseminates technical information.

air gap   In electric motors and some other magnetic devices, the distance between the outside diameter of the rotor and the inside diameter of the stator. Otherwise, the distance between a moving magnetic part and a stationary magnetic part.

alternating current   Current that periodically changes direction. For electrical power systems, a sine wave, centered at zero, alternating at 60 Hz (cycles per second).

analog   Variable continuously between two limits. Can apply to voltage, angle, current, and other electrical or mechanical parameters.

ANSI   American National Standards Institute. An organization that issues standards in various industrial categories.

antiplugging   Of a control system, preventing the application of power to change motor direction until the motor has come to a complete stop from either direction.

ASI   American Standards Institute. An organization that represents trade organizations and issues standards.

automatic control   A device or process control system that is controlled without human involvement. Can be electrical, electronic, or mechanical.

BCD   Binary coded decimal. A number system in which each decimal digit, 0–9, is represented by a pattern of four binary bits.

bipolar   Of stepper motors, having a staggered pole arrangement for stepper operation. Current may flow through stator in both directions.

bit   From *bi*nary dig*it*. A single digit. Can have a value of 1 or 0.

bit pattern   The arrangement of 1s and 0s in an address or register.

brake   A friction device, coupled to a shaft, that stops and holds a load in a stationary position. Releases when energized.

breakover voltage   The voltage below which a device will not conduct electricity. The device has infinite resistance below this voltage. When the voltage applied to the device reaches the breakover voltage, the device abruptly begins to conduct. The device then has pratically zero resistance.

brushes   Solid, stationary pieces of carbon that ride on a rotating commutator or slip ring in a motor or generator. The facing surface is appropriately curved. Embedded in each brush is an electrical wire (pigtail), which is connected to the electrical circuitry.

capacitor-start motor   A single-phase motor that is started from a standstill by means of a capacitor in series with start windings and a centrifugal switch. The centrifugal switch opens for normal, full-speed running, disconnecting the capacitor and the start winding from the circuit.

cardiopulmonary resuscitation   A procedure for attempting to restore breathing in a person in shock. Requires training in the special procedures involved.

CCD   Charge-coupled device. A very small, light-sensitive semiconductor device. Used for pixel light-level indication in machine vision.

central processing unit (CPU)   The central control unit of a programmable logic controller.

centrifugal switch   Most commonly, an internal motor switch that opens the start winding circuit at about 75% of running speed. Also, an external motor switch that indicates zero speed to the control circuit.

circuit breaker   A protective electrical device for interrupting current at the breaker's rated value. Is resettable and is normally in a molded plastic case.

closed loop   In a control circuit, an arrangement involving one or more feedback circuits.

cold-junction compensator   A electronic circuit used to simulate the presence of a reference thermocouple junction held at 0°C.

commutator   Groups of copper bars around a motor shaft. Bars are insulated from each other and the shaft. May be axial or radial. Each bar is connected by individual wires to coils in the rotor. Brushes riding on the commutator are used for electrically connecting the rotor coils to the motor circuitry.

Compare function   In PLC control, a function that compares two numbers to see if they meet a specified criterion.

conduit   Metal pipe through which wires are run from location to location. Protects the wires. May be conventional thickness, thin-wall, or flexible braided.

constant current limiting   A method of overcurrent protection whereby the output current of a power supply ceases to increase as the load increases beyond the maximum specified limit.

contactor   A device for opening and closing a heavy-duty power or current circuit. A specially constructed high-current relay.

continuous-path robot   A robot that moves its working end effector through a complex continuous path that includes the angle of the end effector.

control circuit   An electrical circuit that carries the signals for controlling a device and its logic system. Does not carry operational power. Also called control circuit system.

counter electromotive force (CEMF)   The back induced voltage in an AC or DC motor or generator. Opposes the applied voltage, thus reducing the motor or generator current at higher speeds. Can also be called back EMF.

CPR   See *cardiopulmonary resuscitation.*

CPU   See *central processing unit.*

CUJT   Complementary unijunction transistor. A UJT with directly opposite polarity.

current transformer   A transformer used primarily to step down current.

cylindrical robot   A robot whose work envelope has the form of a vertical cylinder or partial cylinder. Arm moves up and down, in and out, and angularly in a horizontal plane.

data highway   A computer data transmission path. Carries data from any device connected to it to one or more other connected devices. Sometimes called a bus. However, *bus* can also denote a power-connecting network.

DC   See *direct current.*

deadband   The process variable range for which an ON/OFF controller remains in one of its stable states.

delta   In a three-phase power system, an arrangement of generator or secondary transformer windings that results in identical phase and line voltages.

derating   A decimal or percentage that indicates the reduced capability of a device when operating under abnormal conditions. For example, an electrical device is rated at 15 HP. It is operating at a higher temperature than the rated temperature. At the higher temperature, it can operate continuously at only 85% capacity—at 12.75 HP.

diac   A small, two-lead bilateral semiconductor that starts to conduct in a given direction when the rated voltage for that direction is applied.

diffuse proximity optical sensing   A retroreflective optical sensing arrangement in which the sensed object acts as the reflector.

digital   Having discrete states—ON or OFF, high or low, or 1 or 0.

digital gate   A device that analyzes the digital states of its inputs to determine its appropriate output state, high or low. Also called digital logic gate.

DIN   Deutsches Institut für Normung. A German agency that establishes industrial standards.

diode   An electronic device that conducts well in one direction and very little or not at all in the other. Forms include the tube diode (almost obsolete), control and power semiconductor diodes, and the reverse-region-utilizing zener diode.

direct current (DC)   Current that flows in one direction only. Can be irregular or nonperiodic, but is normally a steady-state, constant-value current. Steady-state, constant-value DC current normally has a slight periodic variation called ripple.

discrete device   A device that can be ON or OFF only.

drum controller   A device, normally mechanical in operation, that by cam action operates through a sequence of multiple-ON/OFF electrical contact states.

dual operation   The capability of a device, usually a motor, to be used in place of either of two or more similar devices. Dual operation includes voltage, torque, speed, and other characteristics. The actual mode of operation depends on the device's internal and external connections.

duty cycle   The ratio of device ON time to cycle time, which is ON time plus OFF time. The greater the duty cycle, the larger a motor must be for the application. For example, higher duty cycle means less OFF time. That gives less cool-down time and higher-temperature motor operation.

dynamic braking   A method of rapidly decelerating a DC motor. Involves switching the armature from across the line to across a resistor, with the field voltage still applied. This armature switching causes the motor to go from motor action to generator action, causing rapid stopping.

electromechanical relay   A device that uses electrical energy to move a plunger mechanically. The plunger, in turn, opens and closes electrical contacts.

EMR   Electromagnetic radiation.

enclosure   A metal box used to enclose electrical devices and wiring. There are several types. The type chosen depends on the operating environment.

energy-saving motor starter/controller   An electronics-based motor controller that saves electrical energy by varying the voltage applied to the motor as the load varies.

error   The difference between the system set-point and the system process variable.

fail-safe   Able to prevent unsafe operating conditions if a control component fails or if the control power goes OFF.

feedback   In a control system, a signal indicating output status. The status signal is fed to a controller, which appropriately modifies the input signal's action for proper system operation.

FET   Field-effect transistor. A transistor that operates by varying device conduction by the action of varying voltage to a terminal. The voltage, in turn, varies the conduction by controlling a device internal field.

flowchart   A block diagram with interconnecting directional lines to show the flow of a computer program, step by step.

foldback current limiting   A method of overcurrent protection whereby the output current of a power supply decreases as the load increases beyond a specified limit.

frame size   A standardized number (from NEMA) indicating the dimensions of a motor. Motors of the same frame size that are made by different manufacturers are mechanically interchangeable.

fuse   A device that rapidly interrupts current. The interruption is achieved by the melting of a thin strip of metal inside the fuse.

gauge factor   The ratio of resistance change to length change for a strain gauge element.

ground fault interrupter (GFI)   A device that, unlike a fuse, opens a circuit when it detects a small, set amount of ground current (often 5 mA). Must be manually reset to restore circuit voltage. Used primarily for personnel electrical shock protection.

grounding   A wire or system of wires that connects a device metal case or one side of the power line to the earth ground potential (0 V). Used for personnel safety or for proper operation of electronic equipment.

GTO   Gate turn-on switch. A thyristor that can be turned ON or OFF from a single gate. It is inefficient and is available only in low power ratings.

Hall effect   A phenomenon observed in semiconductors, whereby current flowing perpendicular to a magnetic field produces a potential difference perpendicular to both the field and the current.

hazardous location   An environment containing flammable gas or flammable particles in the air. Hazardous areas require electrical equipment with specially constructed rugged enclosures.

holding current   In thyristors, the minimum current required to continue conduction when the firing voltage is removed from the thyristor gate.

hot junction   The junction of a thermocouple that is exposed to the material whose temperature is being measured.

hydraulic   Operated mechanically by fluid pressure. The fluid used is usually oil-based.

infrared pyrometer   A device used to determine temperature from a distance by measuring the intensity of infrared radiation emitted from the object.

input group register   A single register in which a PLC CPU records the statuses of 4, 8, or 16 discrete input registers.

input module   An electrical unit or circuit used to connect input devices electrically to a PLC CPU. The module sends coded signals to the CPU indicating the status of each of the inputs.

instrumentation amplifier   A special differential amplifier used for high-precision measurements.

integrated circuit (IC)   A small, rectangular or round electronic device with multiple leads. It contains a chip consisting of layered semiconductor materials. The chip is connected to the case leads and contains subminiature transistors, resistors, and capacitors.

interface   An electronic system that connects one or more devices or systems to one or more other devices or systems. Often signal conversion takes place in the interface.

interlock   An electrical or mechanical system that prevents one device from operating while another device is operational. May be applied multiply.

interpole   In large motors, a smaller, intermediate pole midway between main poles. Used to improve the alignment of the main-pole magnetic field.

intrinsic stand-off ratio (ISTOR)   In UJTs, the ratio of triggering voltage to total applied voltage (approximately).

inverter   An electronic circuit used to convert direct current to alternating current.

IRED   Infrared-emitting diode.

isolation   For an electrical device or circuit, lack of connection to ground or to another circuit. Often accomplished by added electrical insulation to ground or between parts.

JFET   A special, high-gain FET.

JIC   Joint Industry Council. An organization that establishes standards of various types for its member manufacturing firms and organizations.

jog   A state of being momentarily ON or in motion. In motor controls momentary ON is created by depressing a control push button. When the spring-return button is released, the device goes OFF. The device is not sealed in the ON state, as it would be with conventional three-wire control.

jointed spherical robot   A robot that has an operating arm, wrist, and gripper that simulate the human arm by including an elbow.

lamination   A thin, metal precision punching. Laminations are stacked and jointed to form magnetic field-producing rotors and stators for motors and other devices.

LASCR   An SCR that is triggered by pulses of light or radiant energy, not by voltage to a gate.

latch   An electrical or mechanical system that causes a relay to remain ON after the ON power is removed. Another circuit must have power applied to turn the relay OFF.

lathe   A machine tool that turns metal or other material from cylindrical raw stock to a specified diameter.

limit switch   An electromechanical switch specifically designed to operate when an object comes in physical contact with the actuator.

line loss   A reduction in voltage at the termination of a power line, due to the resistance of the line's wires. Volts lost is the product of I and R.

linear acceleration or deceleration   The speeding up or slowing down of a device at a constant rate throughout the speed range.

linear motor   A motor whose motion is in a straight line, not rotary.

linear variable differential transformer (LVDT)   A transformer whose output is proportional to the position of its core.

line regulation   A figure of merit that defines how well a voltage regulator maintains its output with respect to variations in input voltage.

line unbalance   The differences in the three voltages of a three-phase feeder system. Ideally would be zero.

line voltage or current   The voltage or current that is available between any two legs of a three-phase system.

load regulation   A figure of merit that defines how well a voltage regulator maintains its output with respect to variations in load.

lock-out   A procedure in which the disconnecting device of an electrical system is locked in the OFF position by a personal padlock.

LVDT   See *linear variable differential transformer*.

machine vision An artificial intelligence system that simulates eye-brain functions for process control. The eye is simulated by photosensitive devices, and the brain is simulated by computer analysis.

magnetic flow sensor A flow sensor that operates by the principle that a conductive fluid flowing in the presence of a magnetic field will generate a potential difference perpendicular to both the field and the flow.

master control relay function (MCR) In a programming system, a function that irrevocably turns OFF a specified group of other functions.

microprocessor A complete computer contained on a single IC. It has all the capabilities of the digital computer.

microstepping For stepper motors, the process of dividing steps into substeps by using appropriately varying adjacent-pole voltage ratios.

milling machine A machine tool that machines flat or formed surfaces on metal and other materials.

motor field An electrically produced magnetic field in a motor. One motor field reacts with another motor field to produce motion.

Move function In a PLC, a function that moves information from a specified register or registers to another specified register or registers.

multiple-station control A process control system that has controls, such as ON/OFF switches, at more than one location.

NEC National Electric Code. A construction code for electrical equipment and distribution system installation. Formulated by the National Fire Protection Association.

NEMA National Electric Manufacturers Association. A trade organization made up of various member manufacturers and organizations. Formulates standards for electrical equipment, such as motors.

nonshorting switch A switch that has no continuity from the incoming line to either outgoing line when the switch is in the intermediate position. Sometimes called break-before-make (BBM).

normally closed (NC) Of an electrical contact, closed when the associated relay coil is unenergized.

normally open (NO) Of an electrical contact, open when the associated relay coil is unenergized.

ON/OFF control A method of control that attempts to maintain system equilibrium by turning the output actuator alternately ON and OFF.

open loop A process in which the output is not monitored and there is no feedback to modify the input to correct output inaccuracies.

operational amplifier (op amp) An integrated amplifier whose characteristics can be controlled primarily by external components.

opposed optical sensing An optical sensing arrangement in which the transmitter and the receiver are set in opposition.

optointerrupter An IRED and a phototransistor set in opposition with a narrow gap between them.

optoisolator An enclosed dual-semiconductor device in which the input is coupled to the output by light variations. The input semiconductor emits varying light depending on the input signal. The output semiconductor picks up the varying light levels and puts out a corresponding signal. The input and the output are electrically isolated from each other.

OSHA Occupational Safety and Health Administration. A federal agency responsible for enhancing worker health in the workplace.

output group register A PLC register that can control multiple (4, 8, or 16) outputs—ON or OFF.

output module An electrical unit or circuit used to connect the CPU of a PLC to outside devices that are to be varied or turned ON or OFF.

overload relay A small device, similar to a circuit breaker, that protects a motor or other electrical device from electrical or thermal overload. Can be inside the device or in the device starter. Available in more rating steps than circuit breakers.

PC A personal computer.

peak demand The highest-kilowatt use of electricity in a month by a given power user. The utility company assesses a charge if the peak demand is considerably higher than the average use by that customer.

phase voltage or current The voltage or current that is available from each output winding in a three-phase system.

photoconductive cell A device whose conductivity changes with variations in light intensity.

photodiode A diode used to detect the presence of (or to measure the intensity of) electromagnetic radiation in the infrared (IR) range.

phototransistor A transistor whose base current can be altered by the application of electromagnetic radiation in the infrared (IR) range.

photovoltaic cell A diode that develops a potential difference in response to the application of electromagnetic radiation in the infrared (IR) range.

pick-and-place robot A robot configured to move parts from one point to another. Not designed to do other, more complicated tasks.

PID control See *proportional-integral-differential (PID) control.*

piezoelectricity A property of certain materials, whereby mechanical strain results in the production of a potential difference across the material.

piezoresistivity A property of certain materials, whereby mechanical strain results in a change in the conductivity of the material.

pilot light A light that indicates process status—for example, ON, OFF, or over limit.

pilot run A preliminary, small-lot production run to work ''bugs'' out of the process and the product design before full-scale operation begins.

pixel In machine vision or tactile systems, one of many small indicating spots in a large array. Pixel arrays are usually square—for example, 16 by 16.

PLC See *programmable logic controller.*

plugging Slowing or stopping a motor by applying reverse-rotation power. Used primarily for three-phase AC motors.

pneumatic   Operated by high-pressure air.

point-to-point robot   A robot that moves from one position to another, remote position. Cannot be programmed to follow a complicated path between positions.

pole   In a motor, a single coil of wire wrapped on a metal frame protrusion. Poles normally work in pairs to produce working magnetic fields.

potentiometer (pot)   A variable three-terminal resistor. Two connections are to the ends of the resistor. One connection (the wiper) goes to a rotating contact that rides on the resistor to vary resistance from wiper to either end.

power factor   The cosine of the angle between voltage and current in AC power systems. Ideally near 1.00.

process variable   The variable being controlled in an industrial process.

programmable logic controller (PLC)   A user-friendly computer that can easily be programmed to control many functions of various types and complexity.

programmer-monitor   In a PLC, a device containing the keyboard and some type of visual display.

proportional band   The process variable range for which the output of a proportional controller remains proportional to its input.

proportional control   A method of control that attempts to maintain system equilibrium by producing output corrections that are proportional to the system error.

proportional-integral-derivative (PID) control   A sophisticated analog control designed to move from one point or process variable value to another as accurately and as quickly as possible.

pulse-width modulation (PWM)   A method whereby the effective power delivered to a load is controlled by varying the duty cycle of the constant voltage or current applied to that load.

PUT   Programmable unijunction transistor. A thyristor that performs as a UJT does but is programmable for variations in operating points. Programming is accomplished by varying the ratios of external resistor values.

PWM   See *pulse-width modulation*.

rectifier   An electrical or electronic device or system that converts incoming AC voltage to an output DC voltage. May include a filter to produce relatively constant-value DC. May also include a regulator to keep voltage constant as load current varies.

rectilinear robot   A robot with a work envelope that is square horizontally and rectangular vertically (box-shaped).

reduced-voltage motor starter   A motor starter that applies reduced voltage to a motor at starting. Can have one step of reduced voltage, or more. Accomplished by resistors, by transformers, by a part-winding, by a wye-delta arrangement, or electronically.

register   A location in a computer for storing information, permanently or temporarily, usually in bit form. Also called an address.

regulation   A measure of the relative constancy of a parameter such as voltage, current, or speed. Percentage regulation = [full-load value/(no-load value − full-load value)] × 100. Ideally zero.

resistance temperature detector (RTD) Metallic wire or film transducer, used to measure temperature. Has nearly linear responses and has positive temperature coefficients.

resolver A position indicator that uses quadrature voltages for reference.

response time The time a device takes to reach a new position or value after the signal for that change has been applied.

retroreflective optical sensing An optical sensing arrangement in which the sensing beam is reflected from transmitter to receiver by a remote reflector.

rheostat (1) A two-terminal variable resistor. The variable tap is internally connected to one end. (2) A large, high-wattage wire-wound variable resistor.

ripple The small variation often present in the steady-state value of DC voltage or current. Ideally zero.

robot A programmable electromechanical manipulative device. Can be rapidly reprogrammed.

rotor The rotating member of a motor. Consists of stacked laminations pressed onto a shaft. The laminations may contain aluminum or copper bars or may have coils of wire wound in insulated slots in the lamination stack.

RTD See *resistance temperature detector.*

safety glasses Eyeglasses, plain or prescription, designed to protect the eyes. Injury prevention includes protection against flying particles, chemicals, and excess radiation from such things as welding arcs and lasers.

SAS Silicon asymmetrical switch. A bilateral triggering thyristor that fires at a different voltage in each direction.

SBS Silicon bilateral switch. A bilateral triggering thyristor that fires at the same voltage in both directions.

SCARA Selective-compliance assembly robot arm. A robot whose work envelope is approximately circular in the horizontal plane and only a few inches high. Used for small assembly tasks.

Schottky diode A four-layer diode with a small forward resistance.

SCR Silicon-controlled rectifier. A four-layer semiconductor that conducts in only one direction. It starts conduction only when appropriately triggered.

SCS Silicon-controlled switch. A one-way thyristor triggering device.

seal In motor control, an auxiliary contact that seals a control circuit ON after the start button is released. Sometimes incorrectly called a latch.

Seebeck effect A phenomenon in which a current is produced in a loop of two dissimilar metals when the two junctions are at different temperatures.

self-heating A condition of thermal transducers, in which current through the transducer increases the device temperature.

sensor A transducer and all the signal-conditioning circuitry needed to make it compatible with other electronic circuits.

series voltage regulator A circuit that attempts to maintain voltage across a load by varying the conductivity of an element in series with the load.

sequence listing   A listing, in order, of occurrence in a control system operation. May have branching and repetitive sections.

sequencer   A control system that controls multiple outputs and devices. Each step has an ON/OFF pattern for each device under control. May be accomplished by a drum switch, by mechanical pegs (as in a music box), or by electronic means (as with a PLC).

sequencing   In control systems, the circuitry that ensures a certain order of operation or nonoperation of the devices under control. Normally accomplished by interlocks.

service factor   A decimal multiplier greater than one (for example, 1.15) that indicates how much a motor can be loaded above its nameplate rating. Note that a motor can be run above rating only under ideal operating and environmental conditions.

servomotor   A motor, AC or DC, that has precise control of both speed and position. Used to control a process. Requires precise electronic control.

servovalve   A precise, analog position valve with two-way control. Its two openings (ports), which control fluid flow rate, are positioned in proportion to control voltage.

set-point   The operating point at which a system is to function. Indicated to the controller by a potentiometer, a computer value, or another appropriate method.

Shockley diode   A diode that is fired in the forward direction by the application of breakover voltage.

shorting switch   A switch whose switching wiper makes contact with the next position before it leaves its starting position. In the intermediate position both outputs are in contact with the input. Sometimes called make-before-break (MBB).

single-phase   Of an incoming power system, having only two wires, with one voltage.

Skip function   In PLCs and other computer systems, a function that skips a specified number of functions that follow it in the programming sequence. The skipped functions are left in their previous states, ON or OFF, as long as Skip is activated.

slew rate   Speed of voltage or current change, which indicates the speed with which a device turns ON or OFF. Ideally, infinity, indicating immediate change.

slip   The percentage of synchronous speed that actual motor speed lags synchronous speed ([difference/synchronous speed] × 100). Normally 3% to 5% in AC induction motors for normal full-load running speed.

slip ring   A solid ring on a motor shaft, connected to windings in the rotor. A brush rides on the ring to connect the rotor voltages electrically to the motor circuitry.

soft-start motor control   An electronic motor-starting system that increases voltage at an adjustable constant rate until full running voltage is achieved.

solar cell   A photovoltaic cell used specifically for direct energy conversion.

solenoid   A device that, using magnetic action from applied current in a coil wound on its frame, creates linear mechanical motion of a plunger from one extreme position to the other. Usually spring-return. Can be single- or double-acting.

solid-state   Made of solid-state material, such as silicon or germanium. Implies small size in comparison with other electronic configurations. Applies to semiconductors, thyristors, integrated circuits, and so on.

solid-state relay   A semiconductor device that performs the functions of an electrome-chanical relay. Relatively small and low in rating. No moving parts.

solid-state switch   An electromechanical switch in which the contacts have been re-placed by Hall-effect devices.

speed-torque curve   A curve showing the torque output of a motor versus its speed. Used to choose a suitable motor to match an application.

spherical robot   A robot whose work envelope is a part of an oblong sphere.

split-phase motor   A single-phase AC motor that starts without the use of a capacitor. The start-winding phase shift required is accomplished by having different induc-tance values in the main winding and the start winding.

standby power unit   A power unit that comes ON when line power fails. May be an engine-driven generator or an electronic unit, depending on the amount of standby power needed and the urgency of the need.

starter   An assembly of electrical components used for motor starting. The basic starter devices are proper-sized contactors and correct motor-overload relays. May also include low-voltage starting devices and mechanical or electrical interlocks.

stator   The stationary part of a motor. Has insulated slots in its lamination stack. Coils of wire are inserted in the slots to produce operating magnetic fields.

stepper motor   A solenoid-action motor whose rotor moves in discrete steps. A step occurs as a voltage pulse is appropriately applied to the stator of the motor.

strain gauge   A transducer that changes its electrical conductivity with changes in me-chanical strain.

SUS   Silicon unilateral switch. One type of one-way thyristor triggering device.

switch   Any device—electrical, mechanical, or electronic—that makes and breaks a current path.

synchro   A class of positioning devices that utilize three-phase voltages for position reference.

synchronous speed   The speed of the rotating field of an AC motor. Depends on the line frequency and the number of motor main pole pairs (pole pairs per phase in three-phase motors).

tandem electrical contactors   Two or more electrical contactors that alternately operate the same device or motor. Electrically and often mechanically interlocked so that only one contactor at a time can be ON.

temperature transmitter   An electronic circuit that accepts RTDs or thermocouples as inputs and produces as an output a voltage or current proportional to the tempera-ture of the input device.

thermal flow sensor   A sensor used to measure fluid flow by measuring the difference in temperature between two elements in the flow path.

thermistor   A sintered semiconductor or metal oxide transducer used to measure tem-perature. Has a very nonlinear response and (usually) a negative temperature coefficient.

thermocouple   Two dissimilar metal wires joined at one end. The voltage produced at the other end of the wires is proportional to the temperature of the junction.

three-phase   Having three supply voltages 120° apart. May have a three-wire or four-wire configuration. Normally the three voltages are exactly equal in magnitude.

three-wire motor control   A control system that has three wires for control between the control panel and the device—two for power and a third to complete the seal circuit.

thyristor   A class of semiconductors that have four layers of alternating n and p materials.

timer   A device that controls the time sequence of an operation. May be mechanical, electrical, pneumatic, or electronic. Most timers have adjustable time intervals. Some have multiple patterns available, and some have multiple functions.

torque   The magnitude of a rotational force. The turning effect. The product of the force applied and the radius on which the force acts. The force acts perpendicular to the radius.

torque motor   A small motor that does not rotate. A metal bar pivots one way or the other in proportion to applied voltage.

transducer   A device that converts one form of energy to another.

transfer curve   A graph that describes the response of a control system to variations in system error.

transformer   A static device with two or more windings. Transforms voltages and currents to different values by winding interaction. Transformed values depend on the coil turns ratios. Power transformers convert AC power. Other types can convert pulses and other electronic signals. Many different transformer configurations are available, such as multiple windings or multiple voltages.

transient voltage   A short interval of higher-than-normal voltage. Usually undesirable and hazardous to connected devices. Also sometimes called a surge.

transistor   A three-layer, three-terminal solid-state semiconductor.

triac   A three-terminal thyristor that conducts in both directions when appropriately triggered. Essentially the same as two SCRs back to back. Has lower power and frequency response than an SCR.

triggering   An application of voltage pulses of very short duration. Triggering pulses are used to activate gates or other terminals of semiconductors and especially thyristors.

troubleshooting chart   An organized guide to steps for analyzing equipment malfunctions. Usually a block diagram with lines and arrows indicating the next step to take if further analysis is needed.

turbine flow sensor   A sensor used to measure fluid flow. Fluid passing the sensor causes a turbine inside to spin, producing an electrical signal proportional in magnitude to the flow.

two-phase   Having power and voltage furnished by four wires, two for each phase. Voltages are 90° apart. Normally used only for high-torque electric motors.

ultrasonic flow sensor   A sensor used to measure fluid flow through pipe walls. It emits ultrasonic energy and measures the Doppler frequency shift of the energy.

ultrasonic level sensor   A sensor used to measure distance by emitting ultrasonic energy and calculating the echo time.

uninterruptable power system (UPS)   A special type of inverter that is automatically switched into service when the AC line fails.

UJT   Unijunction transistor. A thyristor triggering device that transforms a small-value input pulse into a high-current output pulse.

Underwriters Laboratory (UL)   A trade organization that tests consumer and commercial electrical equipment designs and products for operational safety. Approved products are appropriately labeled. The organization is supported by member manufacturers and organizations.

user-friendly   Of a computer program operation, capable of being carried out by a relatively untrained person. A user-friendly program includes a sequence of instructions and menus on the screen, which can easily be followed to carry out the operation.

vane sensor   A magnetic sensor used to detect slits in thin metal cards (called vanes) as the cards pass through the sensor opening.

variable-frequency control   Control of three-phase AC motor speed by varying the frequency of the motor's input power. A conversion unit varies the input frequency between about 2 Hz and about 120 Hz. The AC motor speed is directly proportional to the converted frequency.

vortex flow sensor   A sensor that measures fluid flow by monitoring the production of vortices as the fluid passes the flow element.

work cell   A group of machines that work together to manufacture a product. Normally includes one or more robots. The machines are programmed to work together in appropriate sequences. Master and individual control computers are quickly reprogrammable to manufacture different parts.

wye   In three-phase power systems, an arrangement of generator or secondary transformer windings that results in identical phase and line currents.

zener   A special type of semiconductor diode used as a voltage stabilizer and regulator. The reverse-breakdown zener characteristic is used for voltage reference. Not normally interchangeable with a conventional diode.

Bibliography

General Electronic Control Textbooks

Chute, George, and Robert Chute. *Electronics in Industry*. New York: McGraw-Hill, 1979.

Harrington, John. *Automated Control Electronics*. Albany, N.Y.: Delmar, 1989.

Humphries, James T., and Leslie P. Sheets. *Industrial Electronics*. 2d ed. Boston: Breton, 1986.

Hunter, Ronald. *Automated Process Control Systems*. Englewood Cliffs, N.J.: Prentice-Hall, 1987.

Jacob, Michael. *Industrial Control Electronics*. Englewood Cliffs, N.J.: Prentice-Hall, 1988.

Maloney, Timothy J. *Industrial Solid State Electronics*. Englewood Cliffs, N.J.: Prentice-Hall, 1986.

Patric, Dale, and Stephen Fardo. *Industrial Electronics, Devices and Systems*. Reston, Va.: Reston, 1984.

Chapter 2: Electrical Control Devices

Driscoll, Edward. *Industrial Electronics*. Chicago: American Technical Society, 1980.

Chapter 3: Control Diagrams

Kryshor, Cyrus, and Kurt Stone. *Electronics Drafting Workbook*. New York: McGraw-Hill, 1985.

Lamit, G., and S. Lloyd. *Drafting for Electronics*. Columbus, Ohio: Merrill, 1985.

Chapter 4: Solid-State Devices

Boylestad, Robert, and Louis Nashelski. *Electronic Devices and Circuit Theory*. Englewood Cliffs, N.J.: Prentice-Hall, 1987.

Floyd, Thomas. *Electronic Devices*. 3d ed. New York: Merrill/Macmillan, 1992.

Chapter 5: Transducers and Sensors

Flow and Level Measurement Handbook and Encyclopedia. Vol. 26. Stamford, Conn.: Omega Engineering Co., 1988.

Honeycutt, Richard D. *Electromechanical Devices*. Englewood Cliffs, N.J.: Prentice-Hall, 1986.

Pressure, Strain and Force Measurement Handbook and Encyclopedia. Vol. 26. Stamford, Conn.: Omega Engineering Co., 1988.

Seippel, Robert G. *Transducers, Sensors and Detectors*. Reston, Va.: Reston, 1983.

Temperature Measurement Handbook and Encyclopedia. Vol. 26. Stamford, Conn.: Omega Engineering Co., 1988.

Chapter 6: Signal Conditioning

Honeycutt, Richard. *Op Amps and Linear Integrated Circuits*. Albany, N.Y.: Delmar, 1988.

Stanley, William. *Operational Amplifiers with Linear Integrated Circuits*. 2nd ed. Columbus, Ohio: Merrill, 1989

Chapter 7: SCRs, Triacs, and Other Thyristors

Boylestad, Robert, and Louis Nashelski. *Electronic Devices and Circuit Theory*. Englewood Cliffs, N.J.: Prentice-Hall, 1987.

Floyd, Thomas. *Electronic Devices*. 3d ed. New York: Merrill/Macmillan, 1992.

Chapter 9: Industrial-Use Motors

Humphries, James T. *Motors and Controls*. Columbus, Ohio: Merrill, 1988.

Patric, Dale, and Stephen Fardo. *Rotating Electrical Machines and Power Systems*. Englewood Cliffs, N.J.: Prentice-Hall, 1985.

Rosenberg, Robert, and August Hand. *Electric Motor Repair*. New York: CBS College Publishing, 1987.

Chapter 10: Special-Purpose Motors

Humphries, James T. *Motors and Controls*. Columbus, Ohio: Merrill, 1988.

Patric, Dale, and Stephen Fardo. *Rotating Electrical Machines and Power Systems*. Englewood Cliffs, N.J.: Prentice-Hall, 1985.

Rosenberg, Robert, and August Hand. *Electric Motor Repair*. New York: CBS College Publishing, 1987.

Chapter 11: Electrical Power-Control Devices

Herman, Stephen, and Walter N. Alerich. *Industrial Motor Control*. Albany, N.Y.: Delmar, 1985.

Rexford, Kenneth B. *Electrical Control for Machines*. Albany, N.Y.: Delmar, 1987.

Rockis, Gary, and Glen Mazur. *Electric Motor Controls*. Alsip, Ill.: American Technical Publishers, 1982.

Wiring Diagrams Bulletin. SM 304. Milwaukee: Square D Co.

Chapter 12: Control of Motors

Herman, Stephen, and Walter N. Alerich. *Industrial Motor Control*. Albany, N.Y.: Delmar, 1985.

Rexford, Kenneth B. *Electrical Control for Machines*. Albany, N.Y.: Delmar, 1987.

Rockis, Gary, and Glen Mazur. *Electric Motor Controls*. Alsip, Ill.: American Technical Publishers, 1982.

Wiring Diagrams Bulletin. SM 304. Milwaukee: Square D Co.

Chapter 13: Analog Controllers

Bateson, Robert. *Introduction to Control System Technology*. 3d ed. New York: Merrill/Macmillan, 1989.

Jacob, Michael. *Industrial Control Electronics—Applications and Design*. Englewood Cliffs, N.J.: Prentice-Hall, 1988.

Chapter 14: Closed-Loop Systems

Bateson, Robert. *Introduction to Control System Technology*. 3d ed. New York: Merrill/Macmillan, 1989.

Jacob, Michael. *Industrial Control Electronics—Applications and Design*. Englewood Cliffs, N.J.: Prentice-Hall, 1988.

Chapter 15: Programmable Logic Controllers

Kissell, Thomas E. *Understanding and Using Programmable Controllers*. Englewood Cliffs, N.J.: Prentice-Hall, 1986.

Petruzella, Frank D. *Programmable Logic Controllers*. New York: McGraw-Hill, 1989.

Webb, John W. *Programmable Logic Controllers: Principles and Applications*. 2d ed. New York: Merrill/Macmillan, 1992.

Chapter 16: Power Distribution Effects on Industrial Electronic Controls

Kaiser, Joe. *Electrical Power—Motors, Controls, Generators, Transformers*. South Holland, Ill.: Goodheart-Willcox, 1982.

Naden, John, Bert Gelmine, and Edward McLaughlin. *Industrial Electricity*. Albany, N.Y.: Delmar, 1984.

National Electric Code, 1987. Portland, Ore.: Building Tech Bookstore, 1987.

Wildi, Theodore. *Electrical Power Technology*. New York: Wiley, 1981.

Chapter 17: Safety Considerations—Personnel and Equipment

Lacey, Edward. *Handbook of Electronic Safety*. Englewood Cliffs, N.J.: Prentice-Hall, 1976.

Chapter 18: Control of Robots and Work Cells

Hoekstra, Robert. *Robotics and Automated Systems*. Cincinnati: South-Western, 1986.

Hunt, Daniel. *Industrial Robotics Handbook*. New York: Industrial Press, 1983.

Malcolm, Douglas R. *Robotics: An Introduction*. Boston: PWS-Kent, 1988.

Staugaard, Andrew. *Robotics and AI: An Introduction to Applied Machine Technology*. Englewood Cliffs, N.J.: Prentice-Hall, 1987.

Appendix

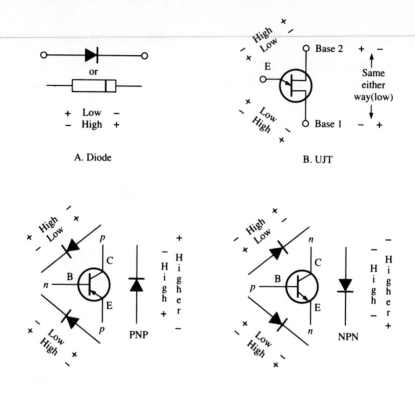

A. Diode

B. UJT

C. Transistors

Ohmmeter Checks for Solid State Devices

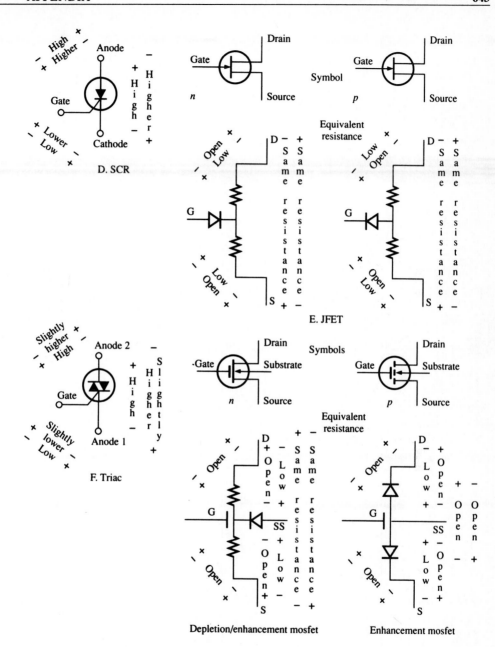

D. SCR

E. JFET

F. Triac

Depletion/enhancement mosfet

Enhancement mosfet

G. MOSFETs

TRIAC Typical Specification Sheet

I_T RMS Maximum Forward Current (Amps)

V_{RRM} DC or Peak Volts	0.8 A	2.5 A	2.5 A	4 A	4 A	8 A	8 A	10 A	10 A	10 A
50				ECG5601						ECG5631
100		ECG5640	ECG5650	ECG5602				ECG5612	ECG5622	ECG5632
200	ECG5655	ECG5641	ECG5651	ECG5603				ECG5613	ECG5623	ECG5633
400	ECG5656	ECG5642	ECG5652	ECG5605	ECG5629	ECG5608		ECG5614	ECG5624	ECG5635
600	ECG5657	ECG5643	ECG5653	ECG5607		ECG5609	ECG5638	ECG5616	ECG5626	ECG5637
800						ECG5610		ECG5618	ECG5628	ECG5645
I_{GT} Min. (mA) quadrants I & III	5.0	25	3.0	30	3.0	10	10	50	50	50
I_{GT} Min. (mA) quadrants II & IV	5.0	40	3.0	---	3.0	10	10	75	50*	75
V_{GT} Max. (V)	2.0	2.2	2.2	2.5	2.0	2.5	2.0	2.5	2.0	2.5
I_{surge} Max. (A)	8.0	25	25	30	40	80	80	100	100	120
I_{hold} Min. (mA)	20	35	25	30	5.0	15	10	50	50	500
V_{ON} Max. (V)	1.5	1.8	1.8	2.0	1.6	1.5	1.6	1.8	1.65	1.6
V_{GM} (V)	±5.0	±5.0	±5.0	±5.0	±5.0	±10	±5.0	±5.0	±10	±5.0
PG Av (W)	.01	.05	.05	.5	.3	.5	.4	.5	.5	.5
Operating temperature T_J (°C)	-40 to +110	-65 to +100	-40 to +90	-40 to +110	-40 to +110	-40 to +110	-40 to +110	-65 to +100	-40 to +100	-40 to +110
Operating quadrants	I,II,III,IV	I,II,III,IV	I,II,III,IV	I, III	I,II,III,IV	I,II,III,IV	I,II,III,IV	I, III.	I, II, III	I,II,III,IV
Fig. no.	Z36	Z61	Z61	Z38	Z40	Z41	Z41	Z39	Z41	Z41
Package	TO-92	TO-5	TO-5	TO-126	TO-202	TO-220	TO-220 Isolated tab	TO-127	TO-220	TO-220 Isolated tab

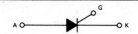

REVERSE BLOCKING TRIODE THYRISTORS

These devices are glassivated planar construction designed for gating operation in mA/μA signal or detection circuits.

- Low-Level Gate Characteristics —
  I_{GT} = 10 mA (Max) @ 25°C

- Low Holding Current —
  I_H = 5.0 mA (Typ) @ 25°C

- Glass-to-Metal Bond for Maximum Hermetic Seal

SILICON CONTROLLED RECTIFIERS

1.6 AMPERE RMS
50 thru 400 VOLTS

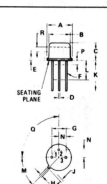

*MAXIMUM RATINGS (T$_J$ = 125°C unless otherwise noted).

Rating	Symbol	Value	Unit
Repetitive Peak Reverse Blocking Voltage	V_{RRM}		Volts
2N1595		50	
2N1596		100	
2N1597		200	
2N1598		300	
2N1599		400	
Repetitive Peak Forward Blocking Voltage	V_{DRM}		Volts
2N1595		50	
2N1596		100	
2N1597		200	
2N1598		300	
2N1599		400	
RMS On-State Current (All Conduction Angles)	$I_{T(RMS)}$	1.6	Amps
Peak Non-Repetitive Surge Current (One Cycle, 60 Hz, T$_J$ = -65 to +125°C)	I_{TSM}	15	Amps
Peak Gate Power	P_{GM}	0.1	Watt
Average Gate Power	$P_{G(AV)}$	0.01	Watt
Peak Gate Current	I_{GM}	0.1	Amp
Peak Gate Voltage — Forward	V_{GFM}	10	Volts
Reverse	V_{GRM}	10	
Operating Junction Temperature Range	T_J	-65 to +125	°C
Storage Temperature Range	T_{stg}	-65 to +150	°C

*Indicates JEDEC Registered Data.

STYLE 3:
PIN 1. CATHODE
2. GATE
3. ANODE (CONNECTED TO CASE)

DIM	MILLIMETERS		INCHES	
	MIN	MAX	MIN	MAX
A	8.89	9.40	0.350	0.370
B	8.00	8.51	0.315	0.335
C	6.10	6.60	0.240	0.260
D	0.406	0.533	0.016	0.021
E	0.229	3.18	0.009	0.125
F	0.406	0.483	0.016	0.019
G	4.83	5.33	0.190	0.210
H	0.711	0.864	0.028	0.034
J	0.737	1.02	0.029	0.040
K	12.70	–	0.500	–
L	6.35	–	0.250	–
M	45° NOM		45° NOM	
P	–	1.27	–	0.050
Q	90° NOM		90° NOM	
R	2.54	–	0.100	–

All JEDEC dimensions and notes apply.

CASE 79-02
TO-39

©MOTOROLA INC., 1977

ELECTRICAL CHARACTERISTICS (T_C = 25°C unless otherwise noted).

Characteristic	Symbol	Min	Typ	Max	Unit
*Peak Reverse Blocking Current (Rated V_{RRM}, T_J = 125°C)	I_{RRM}	--	--	1000	μA
*Peak Forward Blocking Current (Rated V_{DRM}, T_J = 125°C)	I_{DRM}	--	--	1000	μA
*Peak On-State Voltage (I_F = 1.0 Adc, Pulsed, 1.0 ms (Max), Duty Cycle ≈ 1%)	V_{TM}	--	1.1	2.0	Volts
*Gate Trigger Current (V_{AK} = 6.0 V, R_L = 12 Ohms)	I_{GT}	--	2.0	10	mA
*Gate Trigger Voltage (V_{AK} = 6.0 V, R_L = 12 Ohms) (V_{AK} = 6.0 V, R_L = 12 Ohms, T_J = 125°C)	V_{GT}	-- 0.2	0.7 --	3.0 --	Volts
Reverse Gate Current (V_{GK} = 10 V)	I_{GR}	--	17	--	mA
Holding Current (V_{AK} = 12 V)	I_H	--	5.0	--	mA
Turn-On Time (I_{GT} = 10 mA, I_F = 1.0 A) (I_{GT} = 20 mA, I_F = 1.0 A)	t_{gt}	-- --	0.8 0.6	-- --	μs
Turn-Off Time (I_F = 1.0 A, I_R = 1.0 A, dv/dt = 20 V/μs, T_J = 125°C)	t_q	--	10	--	μs

*Indicates JEDEC Registered Data.

CURRENT DERATING

FIGURE 1 – CASE TEMPERATURE REFERENCE

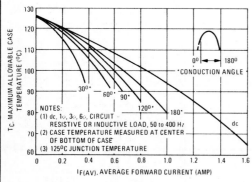

NOTES:
(1) dc, 1ϕ, 3ϕ, 6ϕ, CIRCUIT — RESISTIVE OR INDUCTIVE LOAD, 50 to 400 Hz
(2) CASE TEMPERATURE MEASURED AT CENTER OF BOTTOM OF CASE
(3) 125°C JUNCTION TEMPERATURE

FIGURE 2 – AMBIENT TEMPERATURE REFERENCE

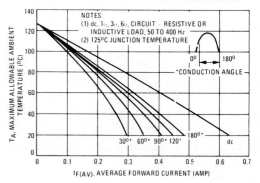

NOTES:
(1) dc, 1ϕ, 3ϕ, 6ϕ, CIRCUIT — RESISTIVE OR INDUCTIVE LOAD, 50 TO 400 Hz
(2) 125°C JUNCTION TEMPERATURE

MOTOROLA

2N4877

MEDIUM-POWER NPN SILICON TRANSISTOR

. . . designed for switching and wide band amplifier applications.

- Low Collector-Emitter Saturation Voltage — $V_{CE(sat)}$ = 1.0 Vdc (Max) @ I_C = 4.0 Amp

- DC Current Gain Specified to 4 Amperes

- Excellent Safe Operating Area

- Packaged in the Compact TO-39 Case for Critical Space-Limited Applications.

4 AMPERE POWER TRANSISTOR

NPN SILICON
60 VOLTS
10 WATTS

JANUARY 1971 – DS 3189

* MAXIMUM RATINGS

Rating	Symbol	Value	Unit
Collector-Emitter Voltage	V_{CEO}	60	Vdc
Collector-Base Voltage	V_{CB}	70	Vdc
Emitter-Base Voltage	V_{EB}	5.0	Vdc
Collector Current — Continuous	I_C	4.0	Adc
Base Current	I_B	1.0	Adc
Total Device Dissipation @ T_C = 25°C Derate above 25°C	P_D	10 57.2	Watts mW/°C
Operating and Storage Junction Temperature Range	T_J, T_{stg}	–65 to +200	°C

*Indicates JEDEC Registered Data

THERMAL CHARACTERISTICS

Characteristic	Symbol	Max	Unit
Thermal Resistance, Junction to Case	θ_{JC}	17.5	°C/W

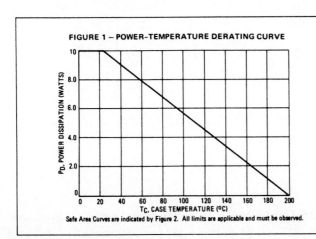

FIGURE 1 — POWER-TEMPERATURE DERATING CURVE

P_D, POWER DISSIPATION (WATTS)

T_C, CASE TEMPERATURE (°C)

Safe Area Curves are indicated by Figure 2. All limits are applicable and must be observed.

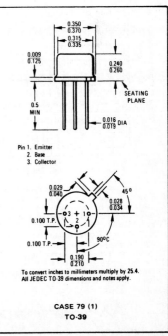

Pin 1. Emitter
2. Base
3. Collector

To convert inches to millimeters multiply by 25.4.
All JEDEC TO-39 dimensions and notes apply.

CASE 79 (1)
TO-39

© MOTOROLA INC., 1971

*ELECTRICAL CHARACTERISTICS (T_C = 25°C unless otherwise noted)

Characteristic	Symbol	Min	Max	Unit
OFF CHARACTERISTICS				
Collector-Emitter Sustaining Voltage (1) (I_C = 200 mAdc, I_B = 0)	$V_{CEO(sus)}$	60	–	Vdc
Collector Cutoff Current (V_{CE} = 70 Vdc, $V_{EB(off)}$ = 1.5 Vdc) (V_{CE} = 70 Vdc, $V_{EB(off)}$ = 1.5 Vdc, T_C = 100°C)	I_{CEX}	– –	100 1.0	µAdc mAdc
Collector Cutoff Current (V_{CB} = 70 Vdc, I_E = 0)	I_{CBO}	–	100	µAdc
Emitter Cutoff Current (V_{BE} = 5.0 Vdc, I_C = 0)	I_{EBO}	–	100	µAdc
ON CHARACTERISTICS(1)				
DC Current Gain (I_C = 1.0 Adc, V_{CE} = 2.0 Vdc) (I_C = 4.0 Adc, V_{CE} = 2.0 Vdc)	h_{FE}	30 20	– 100	–
Collector-Emitter Saturation Voltage (I_C = 4.0 Adc, I_B = 0.4 Adc)	$V_{CE(sat)}$	–	1.0	Vdc
Base-Emitter Saturation Voltage (I_C = 4.0 Adc, I_B = 0.4 Adc)	$V_{BE(sat)}$	–	1.8	Vdc
DYNAMIC CHARACTERISTICS				
Current-Gain-Bandwidth Product (I_C = 0.25 Adc, V_{CE} = 10 Vdc, f = 1.0 MHz) (I_C = 0.25 Adc, V_{CE} = 10 Vdc, f = 10 MHz)**	f_T	4.0 30	– –	MHz
SWITCHING CHARACTERISTICS				
Rise Time (V_{CC} = 25 Vdc, I_C = 4.0 Adc, I_{B1} = 0.4 Adc)	t_r	–	100	ns
Storage Time (V_{CC} = 25 Vdc, I_C = 4.0 Adc,	t_s	–	1.5	µs
Fall Time I_{B1} = I_{B2} = 0.4 Adc)	t_f	–	500	ns

*Indicates JEDEC Registered Data.
**Motorola guarantees this value in addition to JEDEC Registered Data.
Note 1: Pulse Test: Pulse Width ≤ 300 µs, Duty Cycle ≤ 2.0%.

FIGURE 2 – ACTIVE-REGION SAFE OPERATING AREA

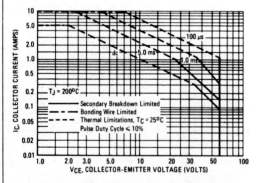

FIGURE 3 – SWITCHING TIME TEST CIRCUIT

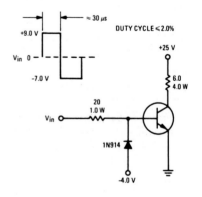

DUTY CYCLE ≤ 2.0%

There are two limitations on the power handling ability of a transistor: average junction temperature and second breakdown. Safe operating area curves indicate I_C–V_{CE} limits of the transistor that must be observed for reliable operation; i.e., the transistor must not be subjected to greater dissipation than the curves indicate.

The data of Figure 2 is based on $T_{J(pk)}$ = 200°C; T_C is variable depending on conditions. Second breakdown pulse limits are valid for duty cycles to 10% provided $T_{J(pk)}$ ≤ 200°C. At high case temperatures, thermal limitations will reduce the power that can be handled to values less than the limitations imposed by second breakdown. (See AN-415)

MOTOROLA *Semiconductor Products Inc.*

BOX 20912 • PHOENIX, ARIZONA 85036 • A SUBSIDIARY OF MOTOROLA INC.

MOTOROLA

Designers Data Sheet

"SURMETIC" RECTIFIERS

. . . subminiature size, axial lead mounted rectifiers for general-purpose low-power applications.

Designers Data for "Worst Case" Conditions

The Designers Data Sheets permit the design of most circuits entirely from the information presented. Limit curves — representing boundaries on device characteristics — are given to facilitate "worst case" design.

**LEAD MOUNTED
SILICON RECTIFIERS**

**50-1000 VOLTS
DIFFUSED JUNCTION**

*MAXIMUM RATINGS

Rating	Symbol	1N4001	1N4002	1N4003	1N4004	1N4005	1N4006	1N4007	Unit
Peak Repetitive Reverse Voltage Working Peak Reverse Voltage DC Blocking Voltage	V_{RRM} V_{RWM} V_R	50	100	200	400	600	800	1000	Volts
Non-Repetitive Peak Reverse Voltage (halfwave, single phase, 60 Hz)	V_{RSM}	60	120	240	480	720	1000	1200	Volts
RMS Reverse Voltage	$V_{R(RMS)}$	35	70	140	280	420	560	700	Volts
Average Rectified Forward Current (single phase, resistive load, 60 Hz, see Figure 8, $T_A = 75^oC$)	I_O	1.0							Amp
Non-Repetitive Peak Surge Current (surge applied at rated load conditions, see Figure 2)	I_{FSM}	30 (for 1 cycle)							Amp
Operating and Storage Junction Temperature Range	T_J, T_{stg}	–65 to +175							^oC

*ELECTRICAL CHARACTERISTICS

Characteristic and Conditions	Symbol	Typ	Max	Unit
Maximum Instantaneous Forward Voltage Drop ($i_F = 1.0$ Amp, $T_J = 25^oC$) Figure 1	v_F	0.93	1.1	Volts
Maximum Full-Cycle Average Forward Voltage Drop ($I_O = 1.0$ Amp, $T_L = 75^oC$, 1 inch leads)	$V_{F(AV)}$	—	0.8	Volts
Maximum Reverse Current (rated dc voltage) $T_J = 25^oC$ $T_J = 100^oC$	I_R	0.05 1.0	10 50	μA
Maximum Full-Cycle Average Reverse Current ($I_O = 1.0$ Amp, $T_L = 75^oC$, 1 inch leads	$I_{R(AV)}$	—	30	μA

*Indicates JEDEC Registered Data.

MECHANICAL CHARACTERISTICS

CASE: Transfer Molded Plastic
MAXIMUM LEAD TEMPERATURE FOR SOLDERING PURPOSES: 350^oC, 3/8" from case for 10 seconds at 5 lbs. tension
FINISH: All external surfaces are corrosion-resistant, leads are readily solderable
POLARITY: Cathode indicated by color band
WEIGHT: 0.40 Grams (approximately)

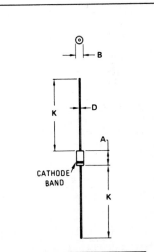

DIM	MILLIMETERS		INCHES	
	MIN	MAX	MIN	MAX
A	5.97	6.60	0.235	0.260
B	2.79	3.05	0.110	0.120
D	0.76	0.86	0.030	0.034
K	27.94	—	1.100	—

CASE 59-04
Does Not Conform to DO-41 Outline.

Trademark of Motorola Inc.

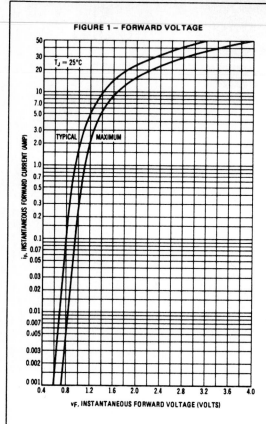

FIGURE 1 – FORWARD VOLTAGE

i_F, INSTANTANEOUS FORWARD CURRENT (AMP)

v_F, INSTANTANEOUS FORWARD VOLTAGE (VOLTS)

$T_J = 25°C$

TYPICAL   MAXIMUM

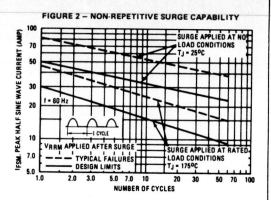

FIGURE 2 – NON-REPETITIVE SURGE CAPABILITY

I_{FSM}, PEAK HALF SINE WAVE CURRENT (AMP)

NUMBER OF CYCLES

SURGE APPLIED AT NO LOAD CONDITIONS
$T_J = 25°C$

$f = 60$ Hz

1 CYCLE

V_{RRM} APPLIED AFTER SURGE
TYPICAL FAILURES
DESIGN LIMITS

SURGE APPLIED AT RATED LOAD CONDITIONS
$T_J = 175°C$

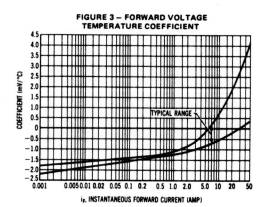

FIGURE 3 – FORWARD VOLTAGE
TEMPERATURE COEFFICIENT

COEFFICIENT (mV/°C)

i_F, INSTANTANEOUS FORWARD CURRENT (AMP)

TYPICAL RANGE

FIGURE 4 – TYPICAL TRANSIENT THERMAL RESISTANCE

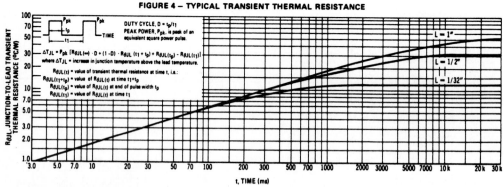

$R_{\theta JL}$, JUNCTION-TO-LEAD TRANSIENT THERMAL RESISTANCE (°C/W)

t, TIME (ms)

DUTY CYCLE, D = t_p/t_1
PEAK POWER, P_{pk}, is peak of an equivalent square power pulse.

$\Delta T_{JL} = P_{pk} [R_{\theta JL}(\infty) \cdot D + (1-D) \cdot R_{\theta JL}(t_1 + t_p) + R_{\theta JL}(t_p) - R_{\theta JL}(t_1)]$
where ΔT_{JL} = increase in junction temperature above the lead temperature.
$R_{\theta JL}(t)$ = value of transient thermal resistance at time t, i.e.:
$R_{\theta JL}(t_1 + t_p)$ = value of $R_{\theta JL}(t)$ at time $t_1 + t_p$
$R_{\theta JL}(t_p)$ = value of $R_{\theta JL}(t)$ at end of pulse width t_p
$R_{\theta JL}(t_1)$ = value of $R_{\theta JL}(t)$ at time t_1

$L = 1"$
$L = 1/2"$
$L = 1/32"$

The temperature of the lead should be measured using a thermocouple placed on the lead as close as possible to the tie point. The thermal mass connected to the tie point is normally large enough so that it will not significantly respond to heat surges generated in the diode as a result of pulsed operation once steady-state conditions are achieved. Using the measured value of T_L, the junction temperature may be determined by:

$$T_J = T_L + \Delta T_{JL}.$$

 MOTOROLA *Semiconductor Products Inc.*

MOTOROLA

Designers▲ Data Sheet

ONE WATT HERMETICALLY SEALED GLASS SILICON ZENER DIODES

- Complete Voltage Range — 2.4 to 100 Volts
- DO-41 Package — Smaller than Conventional DO-7 Package
- Double Slug Type Construction
- Metallurgically Bonded Construction
- Nitride Passivated Die

Designer's Data for "Worst Case" Conditions

The Designers▲ Data sheets permit the design of most circuits entirely from the information presented. Limit curves — representing boundaries on device characteristics — are given to facilitate "worst case" design.

1.0 WATT
ZENER REGULATOR DIODES
3.3–100 VOLTS

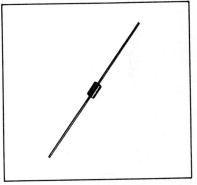

*MAXIMUM RATINGS

Rating	Symbol	Value	Unit
DC Power Dissipation @ T_A = 50°C	P_D	1.0	Watt
Derate above 50°C		6.67	mW/°C
Operating and Storage Junction Temperature Range	T_J, T_{stg}	–65 to +200	°C

MECHANICAL CHARACTERISTICS

CASE: Double slug type, hermetically sealed glass

MAXIMUM LEAD TEMPERATURE FOR SOLDERING PURPOSES: 230°C, 1/16" from case for 10 seconds

FINISH: All external surfaces are corrosion resistant with readily solderable leads.

POLARITY: Cathode indicated by color band. When operated in zener mode, cathode will be positive with respect to anode.

MOUNTING POSITION: Any

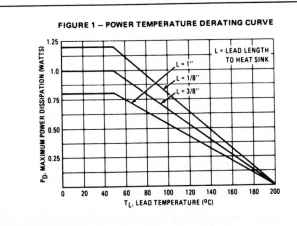

FIGURE 1 — POWER TEMPERATURE DERATING CURVE

L = LEAD LENGTH TO HEAT SINK

L = 1"
L = 1/8"
L = 3/8"

P_D, MAXIMUM POWER DISSIPATION (WATTS)

T_L, LEAD TEMPERATURE (°C)

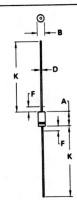

NOTE:
1. POLARITY DENOTED BY CATHODE BAND
2. LEAD DIAMETER NOT CONTROLLED WITHIN "F" DIMENSION.

DIM	MILLIMETERS		INCHES	
	MIN	MAX	MIN	MAX
A	4.07	5.20	0.160	0.205
B	2.04	2.71	0.080	0.107
D	0.71	0.86	0.028	0.034
F	–	1.27	–	0.050
K	27.94	–	1.100	–

All JEDEC dimensions and notes apply.

CASE 59-03
(DO-41)

©MOTOROLA INC., 1978

*Indicates JEDEC Registered Data
▲Trademark of Motorola Inc.

*ELECTRICAL CHARACTERISTICS (T_A = 25°C unless otherwise noted) V_F = 1.2 V max, I_F = 200 mA for all types.

JEDEC Type No. (Note 1)	Nominal Zener Voltage V_Z @ I_{ZT} Volts (Notes 2 and 3)	Test Current I_{ZT} mA	Maximum Zener Impedance (Note 4)			Leakage Current		Surge Current @ T_A = 25°C
			Z_{ZT} @ I_{ZT} Ohms	Z_{ZK} @ I_{ZK} Ohms	I_{ZK} mA	I_R μA Max	V_R Volts	i_r – mA (Note 5)
1N4728	3.3	76	10	400	1.0	100	1.0	1380
1N4729	3.6	69	10	400	1.0	100	1.0	1260
1N4730	3.9	64	9.0	400	1.0	50	1.0	1190
1N4731	4.3	58	9.0	400	1.0	10	1.0	1070
1N4732	4.7	53	8.0	500	1.0	10	1.0	970
1N4733	5.1	49	7.0	550	1.0	10	1.0	890
1N4734	5.6	45	5.0	600	1.0	10	2.0	810
1N4735	6.2	41	2.0	700	1.0	10	3.0	730
1N4736	6.8	37	3.5	700	1.0	10	4.0	660
1N4737	7.5	34	4.0	700	0.5	10	5.0	605
1N4738	8.2	31	4.5	700	0.5	10	6.0	550
1N4739	9.1	28	5.0	700	0.5	10	7.0	500
1N4740	10	25	7.0	700	0.25	10	7.6	454
1N4741	11	23	8.0	700	0.25	5.0	8.4	414
1N4742	12	21	9.0	700	0.25	5.0	9.1	380
1N4743	13	19	10	700	0.25	5.0	9.9	344
1N4744	15	17	14	700	0.25	5.0	11.4	304
1N4745	16	15.5	16	700	0.25	5.0	12.2	285
1N4746	18	14	20	750	0.25	5.0	13.7	250
1N4747	20	12.5	22	750	0.25	5.0	15.2	225
1N4748	22	11.5	23	750	0.25	5.0	16.7	205
1N4749	24	10.5	25	750	0.25	5.0	18.2	190
1N4750	27	9.5	35	750	0.25	5.0	20.6	170
1N4751	30	8.5	40	1000	0.25	5.0	22.8	150
1N4752	33	7.5	45	1000	0.25	5.0	25.1	135
1N4753	36	7.0	50	1000	0.25	5.0	27.4	125
1N4754	39	6.5	60	1000	0.25	5.0	29.7	115
1N4755	43	6.0	70	1500	0.25	5.0	32.7	110
1N4756	47	5.5	80	1500	0.25	5.0	35.8	95
1N4757	51	5.0	95	1500	0.25	5.0	38.8	90
1N4758	56	4.5	110	2000	0.25	5.0	42.6	80
1N4759	62	4.0	125	2000	0.25	5.0	47.1	70
1N4760	68	3.7	150	2000	0.25	5.0	51.7	65
1N4761	75	3.3	175	2000	0.25	5.0	56.0	60
1N4762	82	3.0	200	3000	0.25	5.0	62.2	55
1N4763	91	2.8	250	3000	0.25	5.0	69.2	50
1N4764	100	2.5	350	3000	0.25	5.0	76.0	45

* Indicates JEDEC Registered Data.

NOTE 1 — Tolerance and Type Number Designation. The JEDEC type numbers listed have a standard tolerance on the nominal zener voltage of ±10%. A standard tolerance of ±5% on individual units is also available and is indicated by suffixing "A" to the standard type number.

NOTE 2 — Specials Available Include:

A. Nominal zener voltages between the voltages shown and tighter voltage tolerances,

B. Matched sets.

For detailed information on price, availability, and delivery, contact your nearest Motorola representative.

NOTE 3 — Zener Voltage (V_Z) Measurement. Motorola guarantees the zener voltage when measured at 90 seconds while maintaining the lead temperature (T_L) at 30°C ± 1°C, 3/8" from the diode body.

NOTE 4 — Zener Impedance (Z_Z) Derivation. The zener impedance is derived from the 60 cycle ac voltage, which results when an ac current having an rms value equal to 10% of the dc zener current (I_{ZT} or I_{ZK}) is superimposed on I_{ZT} or I_{ZK}.

NOTE 5 — Surge Current (i_r) Non-Repetitive. The rating listed in the electrical characteristics table is maximum peak, non-repetitive, reverse surge current of 1/2 square wave or equivalent sine wave pulse of 1/120 second duration superimposed on the test current, I_{ZT}, per JEDEC registration; however, actual device capability is as described in Figures 4 and 5.

APPLICATION NOTE

Since the actual voltage available from a given zener diode is temperature dependent, it is necessary to determine junction temperature under any set of operating conditions in order to calculate its value. The following procedure is recommended:

Lead Temperature, T_L, should be determined from

$$T_L = \theta_{LA} P_D + T_A$$

θ_{LA} is the lead-to-ambient thermal resistance (°C/W) and P_D is the power dissipation. The value for θ_{LA} will vary and depends on the device mounting method. θ_{LA} is generally 30 to 40°C/W for the various clips and tie points in common use and for printed circuit board wiring.

The temperature of the lead can also be measured using a thermocouple placed on the lead as close as possible to the tie point. The thermal mass connected to the tie point is normally large enough so that it will not significantly respond to heat surges generated in the diode as a result of pulsed operation once steady-state conditions are achieved. Using the measured value of T_L, the junction temperature may be determined by:

$$T_J = T_L + \Delta T_{JL}$$

ΔT_{JL} is the increase in junction temperature above the lead temperature and may be found as follows:

$$\Delta T_{JL} = \theta_{JL} P_D$$

θ_{JL} may be determined from Figure 3 for dc power conditions. For worst-case design, using expected limits of I_Z, limits of P_D and the extremes of $T_J(\Delta T_J)$ may be estimated. Changes in voltage, V_Z, can then be found from:

$$\Delta V = \theta_{VZ} \Delta T_J$$

θ_{VZ}, the zener voltage temperature coefficient, is found from Figure 2.

Under high power-pulse operation, the zener voltage will vary with time and may also be affected significantly by the zener resistance. For best regulation, keep current excursions as low as possible.

Surge limitations are given in Figure 5. They are lower than would be expected by considering only junction temperature, as current crowding effects cause temperatures to be extremely high in small spots resulting in device degradation should the limits of Figure 5 be exceeded.

 MOTOROLA *Semiconductor Products Inc.*

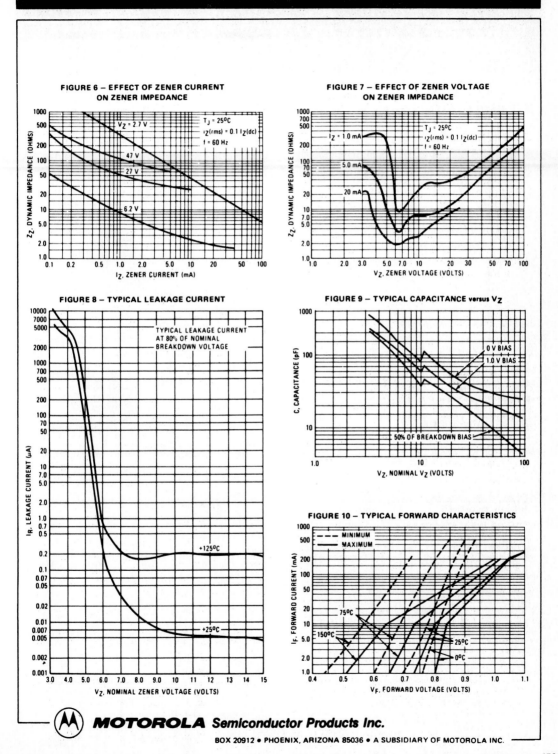

FIGURE 6 — EFFECT OF ZENER CURRENT
ON ZENER IMPEDANCE

FIGURE 7 — EFFECT OF ZENER VOLTAGE
ON ZENER IMPEDANCE

FIGURE 8 — TYPICAL LEAKAGE CURRENT

FIGURE 9 — TYPICAL CAPACITANCE versus V_Z

FIGURE 10 — TYPICAL FORWARD CHARACTERISTICS

MOTOROLA *Semiconductor Products Inc.*

BOX 20912 • PHOENIX, ARIZONA 85036 • A SUBSIDIARY OF MOTOROLA INC.

MOTOROLA

SEMICONDUCTORS

PO BOX 20912 • PHOENIX, ARIZONA 85036

MOC7811 MOC7821
MOC7812 MOC7822
MOC7813 MOC7823

OPTO SLOTTED COUPLER/INTERRUPTER MODULES

These devices consist of a gallium arsenide infrared emitting diode facing a silicon NPN phototransistor in a molded plastic housing. A slot in the housing between the emitter and the detector provides a means of interrupting the signal. They are widely used in position and motion indicators, end of tape indicators, paper feed controls and arcless switches.

- 1.0 mm Aperture
- Easy PCB Mounting
- Cost Effective
- Industry Standard Configuration
- Uses Long-Lived LPE IRED

OPTO SLOTTED COUPLER

TRANSISTOR OUTPUT

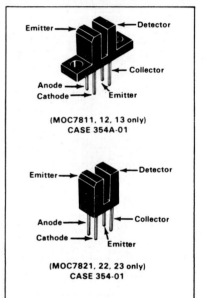

(MOC7811, 12, 13 only)
CASE 354A-01

(MOC7821, 22, 23 only)
CASE 354-01

ABSOLUTE MAXIMUM RATINGS: (25°C)

Rating	Symbol	Value	Unit
TOTAL DEVICE			
Storage Temperature	T_{stg}	–40 to +100	°C
Operating Temperature	T_J	–40 to +100	°C
Lead Soldering Temperature (5 seconds maximum)	T_L	260	°C
INFRARED EMITTING DIODE			
Power Dissipation	P_D	150*	mW
Forward Current (Continuous)	I_F	50	mA
Reverse Voltage	V_R	6	V
PHOTOTRANSISTOR			
Power Dissipation	P_D	150**	mW
Collector-Emitter Voltage	V_{CEO}	30	V

*Derate 2.0 mW/°C above 25°C ambient.
**Derate 2.0 mW/°C above 25° ambient.

INDIVIDUAL ELECTRICAL CHARACTERISTICS: (25°C) (See Note 1)

Characteristic	Symbol	Min	Typ	Max	Unit
EMITTER					
Reverse Breakdown Voltage ($I_R = 100\ \mu A$)	$V_{(BR)R}$	6	—	—	V
Forward Voltage ($I_F = 50$ mA)	V_F	—	1.3	1.8	V
Reverse Current ($V_R = 6.0$ V, $R_L = 1.0$ MΩ)	I_R	—	50	—	nA
Capacitance ($V = 0$, $f = 1$ MHz)	C_i	—	25	—	pF
DETECTOR					
Breakdown Voltage ($I_C = 10$ mA, $H \approx 0$)	$V_{(BR)CEO}$	30	—	—	V
Collector Dark Current ($V_{CE} = 10$ V, $H \approx 0$)	I_{CEO}	—	—	100	nA

Note 1 Stray irradiation can alter values of characteristics. Adequate shielding should be provided. ⓒ MOTOROLA INC. 1982

COUPLED ELECTRICAL CHARACTERISTICS: (25°C) (See Note 1)

Characteristics	Symbol	MOC7811/7821			MOC7812/7822			MOC7813/7823			Unit
		Min	Typ	Max	Min	Typ	Max	Min	Typ	Max	
I_F = 5.0 mA, V_{CE} = 5.0 V	$I_{CE(on)}$	0.15	—	—	0.30	—	—	0.60	—	—	mA
I_F = 20 mA, V_{CE} = 5.0 V	$I_{CE(on)}$	1.0	—	—	2.0	—	—	4.0	—	—	mA
I_F = 30 mA, V_{CE} = 5.0 V	$I_{CE(on)}$	1.9	—	—	3.0	—	—	5.5	—	—	mA
I_F = 20 mA, I_C = 1.8 mA	$V_{CE(sat)}$	—	—	—	—	—	0.40	—	—	0.40	V
I_F = 30 mA, I_C = 1.8 mA	$V_{CE(sat)}$	—	—	0.40	—	—	—	—	—	—	V
V_{CC} = 5.0 V, I_F = 30 mA, R_L = 2.5 kΩ	t_{on}	—	12	—	—	12	—	—	12	—	μs
V_{CC} = 5.0 V, I_F = 30 mA, R_L = 2.5 kΩ	t_{off}	—	60	—	—	60	—	—	60	—	μs

Note 1: Stray irradiation can alter values of characteristics. Adequate shielding should be provided.

FIGURE 1 — OUTPUT CURRENT versus INPUT CURRENT

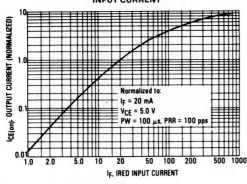

Normalized to:
I_F = 20 mA
V_{CE} = 5.0 V
PW = 100 μs, PRR = 100 pps

FIGURE 2 — t_{on}, t_{off} versus LOAD RESISTANCE

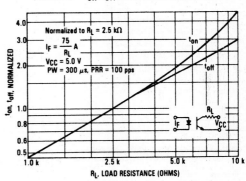

Normalized to R_L = 2.5 kΩ
$$I_F = \frac{75}{R_L} A$$
V_{CC} = 5.0 V
PW = 300 μs, PRR = 100 pps

FIGURE 3 — OUTPUT CURRENT versus POSITION OF SHIELD COVERING APERTURE

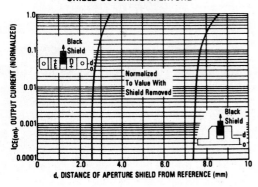

Normalized To Value With Shield Removed

Black Shield

Black Shield

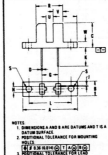

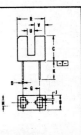

NOTES:
1. DIMENSIONS A AND B ARE DATUMS AND T IS A DATUM SURFACE.
2. POSITIONAL TOLERANCE FOR MOUNTING HOLES:
3. POSITIONAL TOLERANCE FOR LEAD DIMENSION J:
4. POSITIONAL TOLERANCE FOR LEAD DIMENSION D:
5. DIMENSIONING AND TOLERANCING PER Y14.5, 1973.

CASE 354A-01

NOTES:
1. DIMENSIONS R AND B ARE DATUMS AND T IS A DATUM SURFACE.
2. POSITIONAL TOLERANCE FOR LEAD DIMENSION J:
3. POSITIONAL TOLERANCE FOR LEAD DIMENSION D:
4. DIMENSIONING AND TOLERANCING ARE PER Y14.5, 1973.

CASE 354-01

 MOTOROLA *Semiconductor Products Inc.*

BOX 20912 • PHOENIX, ARIZONA 85036 • A SUBSIDIARY OF MOTOROLA INC.

MC1741, MC1741C
MC1741N, MC1741NC

INTERNALLY COMPENSATED, HIGH PERFORMANCE OPERATIONAL AMPLIFIERS

. . . designed for use as a summing amplifier, integrator, or amplifier with operating characteristics as a function of the external feedback components.

- No Frequency Compensation Required
- Short-Circuit Protection
- Offset Voltage Null Capability
- Wide Common-Mode and Differential Voltage Ranges
- Low-Power Consumption
- No Latch Up
- Low Noise Selections Offered — N Suffix

OPERATIONAL AMPLIFIER
SILICON MONOLITHIC
INTEGRATED CIRCUIT

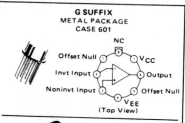

G SUFFIX
METAL PACKAGE
CASE 601

NC
Offset Null — VCC
Invt Input — Output
Noninvt Input — Offset Null
VEE
(Top View)

P1 SUFFIX
PLASTIC PACKAGE
CASE 626
(MC1741C, MC1741NC)

U SUFFIX
CERAMIC PACKAGE
CASE 693

Offset Null — NC
Invt Input — VCC
Noninvt Input — Output
VEE — Offset Null
(Top View)

MAXIMUM RATINGS (T_A = +25°C unless otherwise noted)

Rating	Symbol	MC1741C	MC1741	Unit
Power Supply Voltage	V_{CC}	+18	+22	Vdc
	V_{EE}	-18	-22	Vdc
Input Differential Voltage	V_{ID}	±30		Volts
Input Common Mode Voltage (Note 1)	V_{ICM}	±15		Volts
Output Short Circuit Duration (Note 2)	t_S	Continuous		
Operating Ambient Temperature Range	T_A	0 to +70	-55 to +125	°C
Storage Temperature Range Metal, Flat and Ceramic Packages Plastic Packages	T_{stg}	-65 to +150 -55 to +125		°C

Note 1. For supply voltages less than ± 15 V, the absolute maximum input voltage is equal to the supply voltage.

Note 2. Supply voltage equal to or less than 15 V.

 L SUFFIX
CERAMIC PACKAGE
CASE 632
TO-116

P2 SUFFIX
PLASTIC PACKAGE
CASE 646
(MC1741C, MC1741NC)

NC — NC
NC — NC
Offset Null — NC
Inputs — VCC
 — Output
VEE — Offset Null
NC — NC
(Top View)

EQUIVALENT CIRCUIT SCHEMATIC

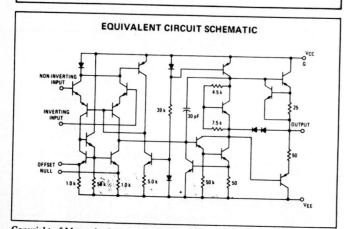

NON-INVERTING INPUT

INVERTING INPUT

OFFSET NULL

4.5 k
39 k
30 pF
7.5 k
25
OUTPUT
50
1.0 k   50 k   1.0 k   5.0 k   50 k   50
VCC G
VEE

F SUFFIX
CERAMIC PACKAGE
CASE 606-04
TO-91

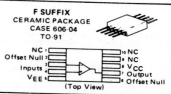

NC — NC
Offset Null — NC
Inputs — VCC
VEE — Output
 — Offset Null
(Top View)

ELECTRICAL CHARACTERISTICS (V_{CC} = 15 V, V_{EE} = 15 V, T_A = 25°C unless otherwise noted).

Characteristic	Symbol	MC1741 Min	MC1741 Typ	MC1741 Max	MC1741C Min	MC1741C Typ	MC1741C Max	Unit
Input Offset Voltage ($R_S \leqslant$ 10 k)	V_{IO}	—	1.0	5.0	—	2.0	6.0	mV
Input Offset Current	I_{IO}	—	20	200	—	20	200	nA
Input Bias Current	I_{IB}	—	80	500	—	80	500	nA
Input Resistance	r_i	0.3	2.0	—	0.3	2.0	—	MΩ
Input Capacitance	C_i	—	1.4	—	—	1.4	—	pF
Offset Voltage Adjustment Range	V_{IOR}	—	±15	—	—	±15	—	mV
Common Mode Input Voltage Range	V_{ICR}	±12	±13	—	±12	±13	—	V
Large Signal Voltage Gain (V_O = ±10 V, $R_L \geqslant$ 2.0 k)	A_v	50	200	—	20	200	—	V/mV
Output Resistance	r_o	—	75	—	—	75	—	Ω
Common Mode Rejection Ratio ($R_S \leqslant$ 10 k)	CMRR	70	90	—	70	90	—	dB
Supply Voltage Rejection Ratio ($R_S \leqslant$ 10 k)	PSRR	—	30	150	—	30	150	μV/V
Output Voltage Swing	V_O							V
($R_L \geqslant$ 10 k)		±12	±14	—	±12	±14	—	
($R_L \geqslant$ 2 k)		±10	±13	—	±10	±13	—	
Output Short-Circuit Current	I_{os}	—	20	—	—	20	—	mA
Supply Current	I_D	—	1.7	2.8	—	1.7	2.8	mA
Power Consumption	P_C	—	50	85	—	50	85	mW
Transient Response (Unity Gain — Non-Inverting)								
(V_I = 20 mV, $R_L \geqslant$ 2 k, $C_L \leqslant$ 100 pF) Rise Time	t_{TLH}	—	0.3	—	—	0.3	—	μs
(V_I = 20 mV, $R_L \geqslant$ 2 k, $C_L \leqslant$ 100 pF) Overshoot	os	—	15	—	—	15	—	%
(V_I = 10 V, $R_L \geqslant$ 2 k, $C_L \leqslant$ 100 pF) Slew Rate	SR	—	0.5	—	—	0.5	—	V/μs

ELECTRICAL CHARACTERISTICS (V_{CC} = 15 V, V_{EE} = 15 V, T_A = *T_{high} to T_{low} unless otherwise noted.)

Characteristic	Symbol	MC1741 Min	MC1741 Typ	MC1741 Max	MC1741C Min	MC1741C Typ	MC1741C Max	Unit
Input Offset Voltage ($R_S \leqslant$ 10 kΩ)	V_{IO}	—	1.0	6.0	—	—	7.5	mV
Input Offset Current	I_{IO}							nA
(T_A = 125°C)		—	7.0	200	—	—	—	
(T_A = -55°C)		—	85	500	—	—	—	
(T_A = 0°C to +70°C)		—	—	—	—	—	300	
Input Bias Current	I_{IB}							nA
(T_A = 125°C)		—	30	500	—	—	—	
(T_A = -55°C)		—	300	1500	—	—	—	
(T_A = 0°C to +70°C)		—	—	—	—	—	800	
Common Mode Input Voltage Range	V_{ICR}	±12	±13	—	—	—	—	V
Common Mode Rejection Ratio ($R_S \leqslant$ 10 k)	CMRR	70	90	—	—	—	—	dB
Supply Voltage Rejection Ratio ($R_S \leqslant$ 10 k)	PSRR	—	30	150	—	—	—	μV/V
Output Voltage Swing	V_O							V
($R_L \geqslant$ 10 k)		±12	±14	—	—	—	—	
($R_L \geqslant$ 2 k)		±10	±13	—	±10	±13	—	
Large Signal Voltage Gain ($R_L \geqslant$ 2 k, V_{out} = ±10 V)	A_v	25	—	—	15	—	—	V/mV
Supply Currents	I_D							mA
(T_A = 125°C)		—	1.5	2.5	—	—	—	
(T_A = -55°C)		—	2.0	3.3	—	—	—	
Power Consumption (T_A = +125°C)	P_C	—	45	75	—	—	—	mW
(T_A = -55°C)		—	60	100	—	—	—	

*T_{high} = 125°C for MC1741 and 70°C for MC1741C

T_{low} = -55°C for MC1741 and 0°C for MC1741C

MOTOROLA SEMICONDUCTORS

P.O. BOX 20912 • PHOENIX, ARIZONA 85036

MPC100

Advance Information

SMARTpower SERIES
10 AMPERES POSITIVE VOLTAGE REGULATOR

This fixed voltage regulator is a series pass monolithic integrated circuit capable of supplying current up to 10 amperes. SMARTpower technology, utilizing a combination of a high-power bipolar output transistor in conjunction with small-signal CMOS control circuitry offers a unique monolithic chip with the following features:

- Internal Thermal Protection
- Internal Short Circuit Protection
- Low Differential Voltage — Typ 1.5 V @ 10 A

POSITIVE 5.0 VOLT
FIXED VOLTAGE REGULATOR

10 AMPERES
80 WATTS

MAXIMUM RATINGS

Rating	Symbol	Value	Unit
Input Voltage	V_{in}	25	Vdc
Output Current	I_o	10	Adc
Total Power Dissipation @ T_C = 25°C Derate above T_C = 25°C	P_D	80 0.8	Watts W/°C
Storage Temperature Range	T_{stg}	0 to 150	°C
Operating Junction Temperature Range	T_J	0 to 125	°C

THERMAL CHARACTERISTICS

Thermal Resistance, Junction-to-Case	$R_{\theta JC}$	1.25	°C/W
Maximum Lead Temperature for Soldering Purposes 1/8" from Case for 5.0 sec	T_L	275	°C

STANDARD APPLICATION

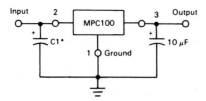

*C1 is required if the regulator is located an appreciable distance from the power supply main filter capacitor.

STYLE 4:
PIN 1. GROUND
2. INPUT
CASE OUTPUT

NOTES:
1. DIAMETER V AND SURFACE W ARE DATUMS.
2. POSITIONAL TOLERANCE FOR HOLE Q:
   ⊕ ∅ 0.25 (0.010) Ⓜ W V Ⓜ
3. POSITIONAL TOLERANCE FOR LEADS:
   ⊕ ∅ 0.30 (0.012) Ⓜ W V Ⓜ Q Ⓜ

DIM	MILLIMETERS MIN	MILLIMETERS MAX	INCHES MIN	INCHES MAX
A	–	39.37	–	1.550
B	–	21.08	–	0.830
C	6.35	7.62	0.250	0.300
D	0.97	1.09	0.038	0.043
E	1.40	1.78	0.055	0.070
F	30.15 BSC		1.187 BSC	
G	10.92 BSC		0.430 BSC	
H	5.46 BSC		0.215 BSC	
J	16.89 BSC		0.665 BSC	
K	11.18	12.19	0.440	0.480
Q	3.81	4.19	0.150	0.165
R	–	26.67	–	1.050
U	2.54	3.05	0.100	0.120
V	3.81	4.19	0.150	0.165

CASE 1-04
TO-204AA (TYPE)

ELECTRICAL CHARACTERISTICS (1)

Characteristic	Symbol	Min	Typ	Max	Unit
Output Voltage $(10\ \text{mA} \leqslant I_o \leqslant 10\ \text{A})$ $(10\ \text{mA} \leqslant I_o \leqslant 5.0\ \text{A},\ 6.5\ \text{V} \leqslant V_{in} \leqslant 20\ \text{V})$	V_O	4.75	—	5.25	Vdc
Line Regulation (2) $(6.5\ \text{V} \leqslant V_{in} \leqslant 20\ \text{V})$ $(6.5\ \text{V} \leqslant V_{in} \leqslant 10\ \text{V})$	Reg_{line}	 — —	 — —	 100 50	mV
Load Regulation (2) $(10\ \text{mA} \leqslant I_o \leqslant 10\ \text{A},\ T_C = 0\ \text{to}\ 125°\text{C})$	Reg_{load}	—	—	50	mV
Quiescent Current $(10\ \text{mA} \leqslant I_o \leqslant 10\ \text{A})$	I_B	—	—	25	mA
Ripple Rejection $(V_{in} = 8.0\ \text{V},\ f = 120\ \text{Hz})$	RR	—	45	—	db
Dropout Voltage $(I_o = 10\ \text{A},\ T_C = 0\ \text{to}\ 125°\text{C})$	$V_{in} - V_O$	—	1.5	2.0	Vdc
Short Circuit Current (3) $(V_{in} = 10\ \text{Vdc})$	I_{SC}	—	—	20	Adc
Averge Temperature Coefficient of Output Voltage $(I_o = 10\ \text{A})$	TCV_O	—	1.6	—	mV/°C
Output Noise Voltage $(10\ \text{Hz} \leqslant f \leqslant 100\ \text{kHz})$	V_N	—	5.0	—	mV
Output Resistance $(f = 120\ \text{Hz})$	R_O	—	2.0	—	mΩ

(1) Unless otherwise specified, test conditions are: $T_J = 25°\text{C}$, $V_{in} = 7.0\ \text{Vdc}$, $I_o = 5.0\ \text{Adc}$, $P_o \leqslant P_{max}$ and $C_O = 10\ \mu\text{F}$.

(2) Load and line regulation are specified at constant junction temperature. Changes in V_O due to heating effects must be taken into account separately. Pulse testing is used with pulse width $\leqslant 3.0$ ms and duty cycle $\leqslant 1.0\%$. Kelvin contacts must be used for these tests.

(3) Depending on heat sinking and power dissipation, thermal shutdown may occur.

SMARTpower VERTICAL PROFILE

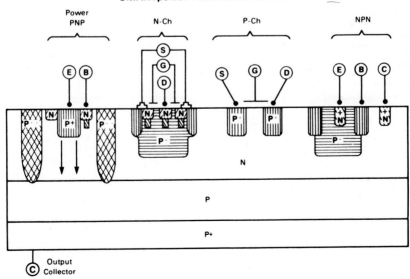

The PNP power transistor uses standard fabrication technology which assures a low saturation voltage and results in a low input-output differential voltage. The die bond is the output collector contact which results in superior load regulation.

 MOTOROLA *Semiconductor Products Inc.*

Index